Advances in Intelligent Systems and Computing

Volume 201

For further volumes:
http://www.springer.com/series/11156

Jagdish Chand Bansal ·
Pramod Kumar Singh · Kusum Deep
Millie Pant · Atulya K. Nagar
Editors

Proceedings of Seventh International Conference on Bio-Inspired Computing: Theories and Applications (BIC-TA 2012)

Volume 1

 Springer

Editors
Jagdish Chand Bansal
South Asian University
Chankya Puri, New Delhi
India

Pramod Kumar Singh
ABV-IIITM, Gwalior
Gwalior, Madhya Pradesh
India

Kusum Deep
Department of Mathematics
Indian Institute of Technology Roorkee
Roorkee
India

Millie Pant
Department of Applied Science
and Engineering
Indian Institute of Technology Roorkee
Roorkee
India

Atulya K. Nagar
Department of Mathematics
and Computer Science
Liverpool Hope University
Liverpool
UK

ISSN 2194-5357   ISSN 2194-5365 (electronic)
ISBN 978-81-322-1037-5   ISBN 978-81-322-1038-2 (eBook)
DOI 10.1007/978-81-322-1038-2
Springer New Delhi Heidelberg New York Dordrecht London

Library of Congress Control Number: 2012954374

Printed on acid-free paper

Springer is part of Springer Science+Business Media (www.springer.com)

Preface

Human beings have always been fascinated by nature and especially by biological diversity and their evolutionary process. This has resulted into inspirations drawn from natural or biological systems, and phenomenon, for problem solving and has seen an emergence of a new paradigm of computation known as Natural Computing with Bio-inspired Computing as its subset. The widely popular methods, e.g., evolutionary computation, swarm intelligence, artificial neural networks, artificial immune systems, are just some examples in the area. Such approaches are of much use when we need an imprecise, inaccurate but feasible solution in a reasonable time as many real-world problems are too complex to be dealt using traditional methods of finding exact solutions in a reasonable time. Therefore, bio-inspired approaches are gaining popularity as the size and complexity of the real-world problems require the development of methods which can give the solution within a reasonable amount of time rather than an ability to guarantee the exact solution. Bio-inspired Computing can provide such a rich tool-chest of approaches as it tends to be, just like its natural system counterpart, decentralized, adaptive and environmentally aware, and as a result have survivability, scalability and flexibility features necessary to deal with complex and intractable situations.

Bio-Inspired Computing: Theories and Applications (BIC-TA) is one of the flagship conferences on Bio-Computing bringing together the world's leading scientists from different branches of Natural Computing. Since 2006 the conferences have taken place at Wuhan (2006), Zhengzhou (2007), Adelaide (2008), Beijing (2009), Liverpool and Changsha (2010), Penang (2011). BIC-TA has attracted wide ranging interest amongst researchers with different backgrounds resulting in a seventh edition in 2012 at Gwalior. It is our privilege to have been part of this seventh edition of the BIC-TA series which is being hosted for the first time in India.

This volume in the AISC series contains papers presented at the Seventh International Conference on Bio-Inspired Computing: Theories and Applications (BIC-TA 2012) held during December 14–16, 2012 at ABV-Indian Institute of Information Technology and Management Gwalior (ABV-IIITM Gwalior), Madhya Pradesh, India. The BIC-TA 2012 provides a unique forum to researchers and practitioners working in the ever growing area of bio-inspired computing methods and their applications to solve various real-world problems.

BIC-TA 2012 attracted attention of researchers from all over the globe and we received 188 papers related to various aspects of bio-inspired computing with umpteen applications, theories, and techniques. After a thorough peer-review process a total of 91 thought-provoking research papers are selected for publication in the Proceedings, which is in two volumes (Volume 1 and 2). This thus corresponds to an acceptance rate of 48% and is intended to maintain a high standard in the conference proceedings. We hope that the papers contained in this proceeding will serve the purpose of inspiring more and more researchers to work in the area of bio-inspired computing and its application.

The editors would like to express their sincere gratitude to the authors, plenary speakers, invited speakers, reviewers, and members of international advisory committee, programme committee and local organizing committee. It would not have been possible to come out with the high quality and standard of the conference as well as this edited Proceeding without their active participation and whole hearted support. It would not be fair on our part if we forget to mention special thanks to the ABV – Indian Institute of Information Technology and Management Gwalior (ABV-IIITM Gwalior) and its Director Prof. S. G. Deshmukh for providing us all the possible help and support including excellent infrastructure of the Institute to make this conference a big success. We express our gratitude to the Department of Mathematics and Computer Science, Liverpool Hope University, Liverpool, UK headed by Prof. Atulya K. Nagar for providing us much valued and needed support and guidance. Finally, we would like to thank all the volunteers; their untiring efforts in meeting the deadlines and managerial skills in managing the resources effectively and efficiently which has ensured a smooth running of the conference.

It is envisaged that the BIC-TA conference series will continue to grow and include relevant future research and development challenges in this exciting field of Computing.

Jagdish Chand Bansal, South Asian University, New Delhi, India
Pramod Kumar Singh, ABV-IIITM, Gwalior, India
Kusum Deep, Indian Institute of Technology, Roorkee, India
Millie Pant, Indian Institute of Technology, Roorkee, India
Atulya K. Nagar, Liverpool Hope University, Liverpool, UK

Editors

Jagdish Chand Bansal
Pramod Kumar Singh
Kusum Deep
Millie Pant
Atulya K. Nagar

About Editors

Dr. Jagdish Chand Bansal is an Assistant Professor with the South Asian University New Delhi, India. Holding an excellent academic record, he is a budding researcher in the field of Swarm Intelligence at the International Level.

Dr. Pramod Kumar Singh is an Associate Professor with the ABV-Indian Institute of Information Technology and Management, Gwalior, India. He is an active researcher and has earned a reputation in the areas of Nature-Inspired Computing, Multi-/Many-Objective Optimization, and Data Mining.

Dr. Kusum Deep is a Professor with the Department of Mathematics, Indian Institute of Technology Roorkee, Roorkee, India. Over the last 25 years, her research is increasingly well-cited making her a central International figure in the area of Bio-Inspired Optimization Techniques, Genetic Algorithms and Particle Swarm Optimization.

Dr. Millie Pant is an Associate Professor with the Department of Applied Science and Engineering, Indian Institute of Technology, Roorkee, Roorkee, India. At this age, she has earned a remarkable International reputation in the area of Genetic Algorithms, Differential Algorithms and Swarm Intelligence.

Prof. Atulya K. Nagar is the Professor and Head of Department of Mathematics and Computer Science at Liverpool Hope University, Liverpool, UK. Prof. Nagar is an internationally recognized scholar working at the cutting edge of theoretical computer science, natural computing, applied mathematical analysis, operations research, and systems engineering and his work is underpinned by strong complexity-theoretic foundations.

Organizing Committees

BIC-TA 2012 was held at ABV- Indian Institute of Information Technology and Management Gwalior, India. Details of the various organizing committees are as follows:

Patron: S. G. Deshmukh, ABV-IIITM Gwalior, India

General Chairs: Atulya Nagar, Liverpool Hope Uuniversity Liverpool, UK
Kusum Deep, IIT Roorkee, India

Conference Chairs: Jagdish Chand Bansal, South Asian University New Delhi, India
Pramod Kumar Singh, ABV-IIITM Gwalior, India

Program Committee Chairs: Millie Pant, IIT Roorkee, India
T. Robinson, MCC Chennai, India

Special Session Chair: Millie Pant, IIT Roorkee, India

Publicity Chairs: Manoj Thakur, IIT Mandi, India
Kedar Nath Das, NIT Silchar, India

Best Paper Chair: Kalyanmoy Deb, IIT Kanpur, India
(Technically Sponsored by KanGAL,
IIT Kanpur, India)

Conference Secretaries: Harish Sharma, ABV-IIITM Gwalior, India
Jay Prakash, ABV-IIITM Gwalior, India
Shimpi Singh Jadon, ABV-IIITM Gwalior, India
Kusum Kumari Bharti, ABV-IIITM Gwalior, India

Local Arrangement Committee: Jai Prakash Sharma
(ABV-IIITM Gwalior, India) Narendra Singh Tomar
Alok Singh Jadon
Rampal Singh Kushwaha
Mahesh Dhakad
Balkishan Gupta

International Advisory Committee:　　Atulya K. Nagar, UK
　　　　　　　　　　　　　　　　　　　　Gheorghe Paun, Romania
　　　　　　　　　　　　　　　　　　　　Giancarlo Mauri, Italy
　　　　　　　　　　　　　　　　　　　　Guangzhao Cui, China
　　　　　　　　　　　　　　　　　　　　Hao Yan, USA
　　　　　　　　　　　　　　　　　　　　Jin Xu, China
　　　　　　　　　　　　　　　　　　　　Jiuyong Li, Australia
　　　　　　　　　　　　　　　　　　　　Joshua Knowles, UK
　　　　　　　　　　　　　　　　　　　　K G Subramanian, Malaysia
　　　　　　　　　　　　　　　　　　　　Kalyanmoy Deb, India
　　　　　　　　　　　　　　　　　　　　Kenli Li, China
　　　　　　　　　　　　　　　　　　　　Linqiang Pan, China
　　　　　　　　　　　　　　　　　　　　Mario J. Perez-Jimenez, Spain
　　　　　　　　　　　　　　　　　　　　Miki Hirabayashi, Japan
　　　　　　　　　　　　　　　　　　　　PierLuigi Frisco, UK
　　　　　　　　　　　　　　　　　　　　Robinson Thamburaj, India
　　　　　　　　　　　　　　　　　　　　Thom LaBean, USA
　　　　　　　　　　　　　　　　　　　　Yongli Mi, Hong Kong

Special Sessions:

Session 1: Computational Intelligence in Power and Energy Systems, Amit Jain, IIIT Hyderabad, India

Session 2: Bio-Inspired VLSI and Embedded System, Balwinder Raj, NIT Jalandhar, India

Session 3: Recommender System: Design Using Evolutionary & Natural Algorithms, Soumya Banerjee Birla Institute of Technology Mesra, India & Shengbo Guo, Xerox Research Centre Europe, France

Session 4: Image Analysis and Pattern Recognition, K. V. Arya, ABV-IIITM Gwalior, India

Session 5: Applications of Bio-inspired Techniques to Social Computing, Vaskar Raychoudhury, IIT Roorkee, India

Keynote Speakers:

Title: Spiking Neural P Systems
Speaker: Pan Linqiang

Title: Advancements in Memetic Computation
Speaker: Yew-Soon Ong

Title: Machine Intelligence, Generalized Rough Sets and Granular Mining: Concepts, Features ans Applications
Speaker: Sankar Kumar Pal

Title: Of Glowworms and Robots: A New Paradigm in Swarm Intelligence
Speaker: Debasish Ghose

Title: Advances in Immunological Computation
Speaker: Dipankar Dasgupta

Title: Selection of Machinery Health Monitoring Strategies using Soft Computing
Speaker: Ajit Kumar Verma

Title: Can Fuzzy logic Formalism via Computing with Words Bring Complex Environmental Issues into Focus?
Speaker: Ashok Deshpande

Technical Program Committee:

Abdulqader Mohsen, Malaysia
Abhishek Choubey, India
Adel Al-Jumaily, Australia
Aitor Rodriguez-Alsina, Spain
Akila Muthuramalingam, India
Alessandro Campi, Italy
Amit Dutta, India
Amit Jain, India
Amit Pandit, India
Amreek Singh, India
Anand Sharma, India
Andre Aquino, Brazil
Andre Carvalho, Brazil
Andrei Paun, USA
Andres Muñoz, Spain
Anil K Saini, India
Anil Parihar, India
Anjana Jain, India

Antonio J. Jara, Spain
Anupam Singh, India
Anuradha Fukane, India
Anurag Dixit, India
Apurva Shah, India
Aradhana Saxena, India
Arnab Nandi, India
Arshin Rezazadeh, Iran
Arun Khosla, India
Ashish Siwach, India
Ashraf Darwish, Egypt
Ashwani Kush, India
Atulya K. Nagar, UK
B.S. Bhattacharya, UK
Bahareh Asadi, Iran
Bala Krishna Maddali, India
Balaji Venkatraman, India
Balasubramanian Raman, India

Banani Basu, India
Bharanidharan Shanmugam, Malaysia
Carlos Coello Coello, Mexico
Carlos Fernandez-Llatas, Spain
Chang Wook Ahn, Korea
Chi Kin Chow, Hong Kong
Chu-Hsing Lin, Taiwan
Chun-Wei Lin, Taiwan
Ciprian Dobre, Romania
D.G. Thomas, India
Dakshina Ranjan Kisku, India
Dana Petcu, Romania
Dante Tapia, Spain
Deb Kalyanmoy, India
Debnath Bhattacharyya, India
Desmond Lobo, Thailand
Devshri Roy, India
Dipti Singh, India
Djerou Leila, Algeria
Asoke Nath, India
K K Shukla, India
Kavita Burse, India
Mrutyunjaya Panda, India
Shirshu Varma, India
Raveendranathan K.C., India
Shailendra Singh, India
Eduard Babulak, Canada
Eric Gregoire, France
Erkan Bostanci, UK
F N Arshad, UK
Farhad Nematy, Iran
Francesco Marcelloni, Italy
G.R.S. Murthy, India
Gauri S. Mittal, Canada
Ghanshyamsingh Thakur, India
Gheorghe Paun, Romania
Guoli Ji, China
Gurvinder Singh-Baicher, UK
Hasimah Hj. Mohamed, Malaysia
Hemant Mehta, India
Holger Morgenstern, Germany
Hongwei Mo, China
Hugo Proença, Portugal
Ivica Boticki, Croatia
Jaikaran Singh, India
Javier Bajo, Spain
Jer Lang Hong, Malaysia
Jitendra Kumar Rai, India
Joanna Kolodziej, Poland
Jose Pazos-Arias, Spain
Juan Mauricio, Brazil
K K Shukla, India
K V Arya, India
K.G. Subramanian, Malaysia
Kadian Davis, Jamaica
Kamal Kant, India

Kannammal Sampathkumar, India
Katheej Parveen, India
Kazumi Nakamatsu, Japan
Kedar Nath Das, India
Khaled Abdullah, India
Khelil Naceur, Algeria
Khushboo Hemnani, India
Kittipong Tripetch, Thailand
Kunal Patel, USA
Kusum Deep, India
Lalit Awasthi, India
Lam Thu Bui, Australia
Li-Pei Wong, Malaysia
Lin Gao, China
Linqiang Pan, China
M.Ayoub Khan, India
Madhusudan Singh, Korea
Manjaree Pandit, India
Manoj Saxena, India
Manoj Shukla, India
Marian Gheorghe, UK
Mario Koeppen, Japan
Martin Middendorf, Germany
Mehdi Bahrami, Iran
Mehul Raval, India
Michael Chen, China
Ming Chen, China
Mohammad A. Hoque, United States
Mohammad Reza Nouri Rad, Iran
Mohammed Abdulqadeer, India
Mohammed Rokibul Alam Kotwal, Bangladesh
Mohd Abdul Hameed, India
Monica Mehrotra, India
Monowar T, India
Mourad Abbas, Algeria
Mps Chawla, India
Muhammad Abulaish, Saudi Arabia
N.Ch.Sriman Narayana Iyengar, India
Nand Kishor, India
Narendra Chaudhari, India
Natarajamani S, India
Navneet Agrawal, India
Neha Deshpande, India
Nikolaos Thomaidis, Greece
Ninan Sajeeth Philip, India
O. P. Verma, India
P. G. Sapna, India
P. N. Suganthan, Singapore
Philip Moore, U.K
Pierluigi Frisco, UK
Ponnuthurai Suganthan, Singapore
Pramod Kumar Singh, India
Vidya Dhamdhere, India
Kishan Rao Kalitkar, India
Punam Bedi, India
Qiang Zhang, China

R. K. Singh, India
R. N. Yadav, India
R. K. Pateriya, India
Rahmat Budiarto, Malaysia
Rajeev Srivastava, India
Rajesh Sanghvi, India
Ram Ratan, India
Ramesh Babu, India
Ravi Sankar Vadali, India
Rawya Rizk, Egypt
Razib Hayat Khan, Norway
Reda Alhajj, Canada
Ronaldo Menezes, USA
S. M. Sameer, India
S. R. Thangiah, USA
Sami Habib, Kuwait
Samrat Sabat, India
Sanjeev Singh, India
Satvir Singh, India
Shan He, UK
Shanti Swarup, India
Shaojing Fu, China
Shashi Bhushan Kotwal, India
Shyam Lal, India
Siby Abraham, India
Smn Arosha Senanayake, Brunei
Darussalam Sonia Schulenburg, UK
Sotirios Ziavras, United States
Soumya Banerjee, India
Steven Gustafson, USA
Sudhir Warier, India
Sumithra Devi K A, India
Sung-Bae Cho, Korea

Sunil Kumar Jha, India
Suresh Jain, India
Surya Prakash, India
Susan George, Australia
Sushil Kulkarni, India
Swagatam Das, India
Thambi Durai, India
Thamburaj Robinson, India
Thang N. Bui, USA
Tom Hendtlass, Australia
Trilochan Panigrahi, India
Tsung-Che Chiang, Taiwan
Tzung-Pei Hong, Taiwan
Umesh Chandra Pati, India
Uzay Kaymak, Netherlands
V. Rajkumar Dare, India
Vassiliki Andronikou, Greece
Vinay Kumar Srivastava, India
Vinay Rishiwal, India
Vittorio Maniezzo, Italy
Vivek Tiwari, India
Wahidah Husain, Malaysia
Wei-Chiang Samuelson Hong, China
Weisen Guo, Japan
Wenjian Luo, China
Yigang He, China
Yogesh Trivedi, India
Yoseba Penya, Spain
Yoshihiko Ichikawa, Japan
Yuan Haibin, China
Yunong Zhang, China
Yuzhe Liu, US

Contents

Stochastic Algorithms for 3D Node Localization in Anisotropic Wireless Sensor Networks

Anil Kumar, Arun Khosla, Jasbir Singh Saini, and Satvir Singh

Abstract This paper proposes two range based 3D node localization algorithms using application of Hybrid Particle Swarm Optimization (HPSO) and Biogeography Based Optimization (BBO) for anisotropic Wireless Sensor Networks (WSNs). Target nodes and anchor nodes are randomly deployed with constraints over three layer boundaries. The anchor nodes are randomly distributed over top layer only and target nodes over middle and bottom layers. Radio irregularity factor, i.e., an anisotropic property of propagation media and an heterogenous property (different battery backup statuses) of devices are considered. PSO models provide fast but less mature convergence whereas the proposed HPSO algorithm provides fast and mature convergence. Biogeography is based upon the collective learning of geographical allotment of biological organisms. BBO has a new comprehensive energy based on the science of biogeography and apply migration operator to share selective information between different habitats, i.e., problem solutions. Due to size and complexity of WSN, localization problem is articulated as an NP-hard optimization problem . In this work, an error model in a highly noisy environment is depicted for estimation of optimal node location to minimize the location error using HPSO and BBO algorithms. The simulation results establish the strength of the proposed algorithms by equating the performance in terms of the number of target nodes localized with accuracy, and computation time. It has been observed that existing sensor networks localization algorithms are not significant to support the rescue operations

Anil Kumar
Panipat Institute of Engg. and Technoly, Panipat, Haryana, India e-mail: anil.rose@rediffmail.com

Arun Khosla
National Institute of Technology, Jalandhar, Punjab, India e-mail: khosla@nitj.ac.in

Jasbir Singh Saini
DCR Univ. of Sc. and Tech., Murthal, Sonepat, Haryana, India, e-mail: jssain@rediffmail.com

Satvir Singh
SBS State Technical Campus, Ferozpur, Punjab, India,e-mail: satvir15@gmail.com

J. C. Bansal et al. (eds.), *Proceedings of Seventh International Conference on Bio-Inspired Computing: Theories and Applications (BIC-TA 2012)*, Advances in Intelligent Systems and Computing 201, DOI: 10.1007/978-81-322-1038-2_1, © Springer India 2013

involving human lives. Proposed algorithms are beneficial for rescue operations too to find out the accurate location of target nodes in highly noisy environment.

Key words: Wireless Sensor Networks, Biogeography Based Optimization, Hybrid Particle Swarm Optimization, Anisotropic Network

1 Introduction

The WSNs play an important role in our society, as they have become the archetype of pervasive technology. WSNs consist of an array of sensors, either of same or diverse types, interconnected by communication network. Central aims of the sensor networks admit ease of deployment, reliability, accuracy, flexibility, cost and effectiveness . Sensors perform routing function to create single or multi-hop wireless networking to convey data from one to other sensor nodes. The rapid deployment, self-organization and fault-tolerance characteristics of WSN make them promising for a number of military and civilian applications [1, 2, 3]. In most of the applications, main role of a WSN is to detect and report events which can be meaningfully ingested and reacted to only if the exact location of the event is known. The locations of sensor nodes are often needed when identifying where the collected information comes from. The determination of coordinates of the sensors is one of challenging problems and is referred to as the localization problem, in WSNs.

Localization techniques are employed to estimate the location of the sensor nodes where coordinates are not known in a network (termed as target nodes) using available a priori knowledge of positions of typically a few specific sensor nodes called anchors, based on inter-sensor parameters/measurements such as connectivity distance, Time of Arrival (TOA), Time Difference of Arrival (TDOA), Angle of Arrival (AOA), etc. [4, 5].

WSN localization is a two-phase process, i.e., ranging and position estimation process. 2D localization assumptions are violated in underwater, atmospheric and space applications where height of the network can be significant and nodes are distributed over a three-dimensional (3D) space [6]. For example, underwater ad hoc sensor networks, which are 3 dimensional, have attracted a lot of attention recently [6, 7]. In underwater sensor networks, nodes may be deployed at different depths of an ocean and thus the network becomes three-dimensional. Better weather forecasting and climate monitoring can be done by deploying three-dimensional networks in the atmosphere.

This paper proposes the application of HPSO and BBO algorithms for range based 3D node localization in anisotropic WSNs. Both algorithms performed better in terms of number of nodes localized, localization accuracy and computation time. Nodes are randomly deployed with constraints over three layer boundaries. However, the anchor nodes are randomly distributed over top layer only and target nodes over lower layers beneath. Radio irregularity factor i.e. anistropic properties of propagation media (background noise and environmental factors) and an heterogenous properties (different battery backups) of devices are considered.

The rest of the paper is organized as follows: Literature Survey on WSN Localization is presented in Section II. Section III ushers the readership into a gentle overview of PSO and BBO algorithms used for localization in this work. This is followed with implementation of above said algorithms in section IV. Section V presents simulation results and comparative study. Finally, section VI presents conclusions and makes a acoustic projection on potential future research paths.

2 Literature Survey

A detailed survey of the relevant literature is available in [8, 9, 10, 11]. An efficient localization system with Accurate Positioning System (APS) extends the GPS capabilities to non-GPS nodes in ad hoc networks as anchors flood their location information to all nodes in the networks proposed in [12]. Then each target node performs a triangulation to three or more anchors to find its position. Node localization accuracy is improved by measuring anchor distances from their neighbors by introducing a refinement phase [13]. The issue of error accumulation is addressed in [14] through Kalman filter based least square estimation in [15, 16] to simultaneously locate the position of all sensor nodes. Node localization problem is addressed using convex optimization based on semidefinite programming. The semidefinite programming approach is further extended to nonconvex inequality constraints [17]. In [18], the gradient search technique demonstrates the use of a data analysis technique called *multidimensional scaling* (MDS) in estimating the position of unknown nodes. The algorithm localizes an individual patch by first computing all pair wise shortest paths between sensors in the patch. Then it applies MDS to these distances to get an initial layout. Finally, an absolute map is obtained by using the known node positions. These techniques work well with few anchors and reasonably high connectivity.

Soft computing plays a crucial role in optimization problems. WSN is treated as multi-modal and multidimensional optimization problem and addressed through population based stochastic techniques. A few GA-based node localization algorithms are presented in [19, 20, 21], that estimate optimal node locations of all one-hop neighbors. Simulated Annealing Algorithm (SAA) and GA based two phase centralized localization scheme is presented in [22]. PSO-based algorithm is proposed in [23, 24] to minimize the localization error. In [25], two intelligent localization schemes for WSNs are introduced for range-free localization, which utilize received signal strength (RSS) from the anchor nodes. In the first scheme, the edge weight of each anchor node is separately calculated and combined to calculate the location of sensor nodes. Fuzzy Logic System (FLS) is used to model edge weights and further optimized by the GA. In the second scheme, the localization is approximated as a single problem where the entire sensors' locations from the anchor node signals are mapped by a Neural Network (NN) [26]. In [27] a two-objective evolutionary algorithm which takes at the same time into account, both the localization accuracy and certain topological constraints induced by connectivity are considered during the evolutionary process, using metaheuristic approach, namely Simulated Annealing (SA), is proposed. An empirical study of the performance of several variants of the guiding functions and several metaheuristic are used to solve real Localization Distance (LD) problem presented in [23]. Each target node is localized under imprecise measurement of distances from three or more neighboring anchors/settled nodes. The methods proposed in this paper have following advantages:

1. There is better trade off between localization accuracy and fast convergence in highly noisy environments.
2. Energy efficiency of networks increased, due to minimum use of hardware (minimum number of anchor nodes)
3. Scalability for large scale deployment is possible.

3 Stochastic Algorithms (HPSO and BBO) for WSN Localization

Widespread acceptance of bio-inspired algorithms is credited to their correctness, and their fair computational load [23, 24, 26, 28, 29, 30, 31, 32, 33, 34, 35]. To get better and fast solution, an improved variant of the PSO, i.e., HPSO, and a recent optimization algorithm, i.e., BBO is applied for range-based distributive node localization in this paper. Ease of implementation and fast convergence are the qualities of *global best* PSO; however, it is likely to get trapped in local optima that leads to pre-matured convergence. The proposed HPSO and BBO algorithms provide matured convergence and better accuracy as compared to the PSO method proposed in [23, 24]. The following sections present an overview of HPSO and BBO.

3.1 Particle Swarm Optimization

The PSO method employs a number of practicable solutions within the search space, called a *Swarm of Particles* with random initial locations. The value of the objective function (which reflects error) corresponding to each particle location is evaluated. These particles move in the search space obeying rules inspired by bird flocking behavior [36, 37] to find new locations with better fitness. Each particle is induced to move towards the best position, the particle has come across so far (*pbest*) and the best position encountered by the entire swarm (*gbest*). To get an accurate solution, the whole swarm is subdivided into sub-swarms and the particle with the best fitness within the local swarm is termed as *lbest*. The *lbest* PSO model provides matured but slow convergence, whereas, in our proposed PSO variant named *HPSO*, the *ith* particle belonging to a sub-swarm feels collective attraction towards its past *pbest* location, P_i, the locally best location within the sub-swarm, P_l, and the overall best location P_g as explained below.

Consider a search space is d-dimensional and ith particle in the swarm can be represented as $X_i = [x_{i1}, x_{i2}, \ldots, x_{id}]$ and its velocity can be represented by another d-dimensional vector $V_i = [v_{i1}, v_{i2}, \ldots, v_{id}]$. Let the best position ever visited in the past by the ith particle be denoted by $P_i = [p_{i1}, p_{i2}, \ldots, p_{id}]$. Many a times, the whole swarm is subdivided into smaller groups and each group/sub-swarm has its own local best particle denoted $P_l = [p_{l1}, p_{l2}, \ldots, p_{ld}]$, and an overall best particle, denoted as $P_g = [p_{g1}, p_{g2}, \ldots, p_{gd}]$, where subscripts l and g are particle indices. The particle iterates in every unit time according to (1) and (2):

$$v_{id} = wv_{id} + c_1 r_1 (p_{id} - x_{id}) + c_2 r_2 (p_{gd} - x_{id}) + c_3 r_3 (p_{ld} - x_{id}) \tag{1}$$

$$x_{id} = x_{id} + v_{id} \tag{2}$$

(The eq. (2) is dimensionally valid in unit time) The parameters w, c_1, c_2 and c_3 termed as inertia weight, cognitive, social and neighborhood learning parameters, respectively, and have a critical role in the convergence characteristics of HPSO. The particles randomize the attraction with uniform random numbers r_1, r_2, and r_3 in the range [0, 1]. The weight factor w should be neither too large, (which results in an early convergence), nor too small, (which, on the contrary, slows down the convergence process). A value of $w = 0.7$ and $c_1 = c_2 = c_3 = 1.494$ were recommended for fast convergence by Eberhart and Shi after experimental tests in [34].

3.2 Biogeography Based Optimization

Biogeography is the study of migration, speciation, and extinction of species, that has often been considered as a process which applies equilibrium in the number of species in habitats [35, 38, 39, 40]. A habitat is an ecological space that is inhabited by plant or animal species and which is geographically isolated from other habitats. Each habitat is classified by Habitat Suitability Index (HSI) that is termed as fitness in other EAs. The features that characterize the habitat are called Suitability Index Variables (SIVs). Habitats with high HSI have a large population, high emigration rate, μ, simply by virtue of the large number of species that migrate to other habitats. The immigration rate, λ, is low for these habitats as these are already saturated with species. Habitats with low HSI have high immigration, λ, and low emigration, μ, because of sparse population. The suitability index of habitats with low HSI is likely to improve with the influx of species from other habitats as it is a function of its biological diversity. However, if HSI does not increase and remains low, species in that habitat go extinct, and this leads to additional immigration. For the purpose of simplicity, it is safe to accept a linear relationship between a habitat HSI and its immigration and emigration rates and, further the rates are same for all habitats. The immigration and emigration rates depend upon the number of species in the habitats. The values of immigration and emigration rates are respectively given as:

$$\lambda = I\left(1 - \frac{k}{n}\right) \tag{3}$$

and

$$\mu = \frac{E}{n} \tag{4}$$

where I is the maximum possible immigration rate; E is the maximum possible emigration rate (I is not necessarily equal to E); k is the number of species of the kth individual and n is the number of species and S_{max} is maximum number of species in a habitat. For a pseudo-code of the algorithm, one may refer to [35].

4 HPSO and BBO Based Node Localization

The main objective in WSN localization is to find out the coordinates of maximum number of target nodes by using M anchor nodes with range-based distributed technique. To estimate coordinates of N target nodes, the process followed is as below:

1. N Target nodes are randomly deployed over middle layer and bottom layer and M anchor nodes are randomly deployed at the top layer. Each target node and anchor node has their transmission range R. Anchor nodes compute their location awareness and transmit their coordinates. The nodes, which get settled at the end of iteration, serve as pseudo anchors or as reference nodes during the next iteration and behave like anchors.
2. The node that falls within transmission range of four or more anchors is considered as localizable node.
3. Each localizable node measures its distance from each of its neighboring anchors. The distance measurements are corrupted with gaussian noise, n_i, due to environment consideration and due to DOI. A node estimates its distance from ith anchor as $\hat{d}_i = [d_i + n_i]$ where d_i is actual distance given by (5)

$$d_i = \sqrt{(x_i - x_{ai})^2 + (y_i - y_{ai})^2 + (z_i - z_{ai})^2} \tag{5}$$

whereas (x_i, y_i, z_i) is the location of the target node and (x_{ai}, y_{ai}, z_{ai}) is the location of the ith anchor node in the neighborhood. The Gaussian assumption for range measurement is valid on practical experimental result of [41] and, therefore, localization results depend on the noise variance, σ_d^2, too.
4. HPSO and BBO-based two separate case studies are conducted, where each localizable target node runs HPSO and BBO algorithms seperately to localize itself. Both HPSO and BBO find the coordinate (x, y, z) that minimize the objective function that represents the error defined in (6).

$$f(x, y, z) = \frac{1}{M} \sum_{i=1}^{M} (\sqrt{(x - x_i)^2 + (y - y_i)^2 + (z - z_i)^2} - \hat{d}_i) \tag{6}$$

where $M \geq 4$ (3D location of a node needs minimum 4 anchors) is the number of anchors within transmission range, R, of the target node.

5. HPSO and BBO evolve the optimal location of target nodes, i.e., (x_i, y_i, z_i) by minimizing the error function (6).

6. After coordinates of all localizable nodes (say, N_L) are determined, the total localization error is computed as the mean of square of distances of computed node coordinates (x_i, y_i, z_i) and the actual node coordinates (X_i, Y_i, Z_i), for $i = 1, 2, \ldots, N_L$ determined for both cases of HPSO and BBO, as in (7):

$$E_l = \frac{1}{N_L} \sum_{i=1}^{L} (\sqrt{(x_i - X_i)^2 + (y_i - Y_i)^2 + (z_i - Z_i)^2}) \qquad (7)$$

7. Steps 2 to 6 are repeated until all target nodes get localized or no more nodes can be localized. The performance of the localization algorithm is based on E_l and N_{NL}, where $N_{NL} = [N - N_L]$ is the number of nodes that could not be localized. Lesser the value of N_{NL} and E_l, the better the performance is.

As the iterations progress, the number of localized nodes increases. This increases the number of references available for already localized nodes. A node that localizes using just four references in an iteration k may have more references in iteration $k + 1$. This decreases the probability of the flip ambiguity. On the other hand, if a node has more references in iteration $k + 1$ than in iteration k, the time required for localization increases, to reduce the time and energy consumption we considered the nearest four anchor nodes/ pseudo nodes to localize each of the unlocalized nodes. It has been observed from implementations of the above proposed algorithms that the maximum number of anchor nodes can be safely restricted to eight, with a view to minimize the hardware cost, increase the network scalability and increase energy efficiency of the network.

5 Simulation Results and Discussion

The WSN localization simulations are conducted using HPSO and BBO in MATLAB environment. 20 target nodes are deployed over middle layer and 20 target nodes are deployed over bottom layer (thus 40 target nodes) and 10 anchor nodes are randomly deployed at top layer of sensor field of $10 \times 10 \; l$ units. Each anchor has a transmission range of $R = 4$ units. Other strategic settings are specific to HPSO and BBO algorithms as discussed below:

5.1 HPSO based node localization

In the proposed framework, each target node that can be localized runs HPSO algorithm to localize itself. HPSO parameters for node localization are fixed as:

1. Population size = 20
2. Max iteration = 100
3. Noise variance (σ_d^2) = 0.02, 0.06 and 0.08 (for three sets of simulation experiments).
4. DOI = 0.01
5. Distance between middle and each of top and bottom layers = 2, 2.5 and 3m (for three sets of simulation experiments).

To localize each node, HPSO runs thirty tials (each trials consisting of 100 iterations) with Gaussian noise. Average of total localization errors defined in (6) for all 30 trials are computed.

5.2 BBO based node localization

For each target node that can be localized, a BBO algorithm is run. BBO strategy parameters for node localization are taken as:

1. Population size = 20
2. Max iteration = 100
3. Probability of mutation = 0.05
4. Noise variance (σ_d^2) = 0.02, 0.06 and 0.08 (for three sets of simulation experiments).
5. DOI = 0.01
6. Distance between middle and each of top and bottom layers = 2, 2.5 and 3m (for three sets of simulation experiments).

Thirty trials (each trials consisting of 100 iterations) experiment of BBO based localization is conducted with Gaussian noise. Average of total localization error defined as fitness function in (6) for all 30 trials is computed and minimized using BBO algorithm.

Both the proposed algorithms are stochastic; so, one can't expect the same solution in all trials even with identical deployment. This is the reason why the results of 30 trial runs are averaged. The initial deployment is random, so, the number of localizable nodes in each iteration is not expected to be the same, which makes the total computing time variable.

The actual nodes, anchors locations and coordinates estimated by PSO (implementation parameters are same as HPSO, except the division of swarm into sub swarm, i.e., *lbest* parameter), HPSO and BBO in a trial run are shown in Fig. 1.3. The distance between actual nodes and estimated nodes is shown in Fig. 2. It has been observed that performance of both the algorithms depends upon the gaussian noise level; lower the E_L more the localized nodes. It can be observed that PSO requires less memory and gives fast convergence but yields less accuracy. Proposed HPSO algorithm gives better accuracy and fast convergence. BBO gives better accuracy than HPSO, however, convergence is significantly slower than for HPSO. A choice between HPSO and BBO is dependent upon the trade off between accuracy

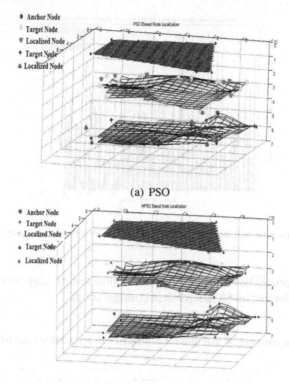

(a) PSO

(b) HPSO

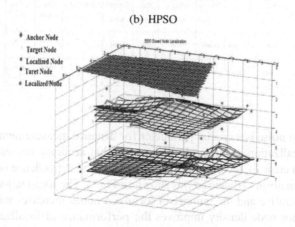

(c) BBO

Fig. 1 Node localizations with different stochastic algorithms

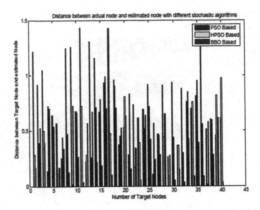

Fig. 2 Distance between actual Node and estimated Node with different stochastic algorithms

and fast convergence. Each point in the simulation results after 30 repetitions of experiments performed on 30 independent configurations.

Table 1 Simulation results of 30 trial runs for comparison of PSO, HPSO and BBO based WSN node location

SAs	$\sigma_d^2 = 0.08$			$\sigma_d^2 = 0.06$			$\sigma_d^2 = 0.02$		
	Mean of No. of un-localized node (N_{SL})	Mean Error (E_L)	Total Computing Time (sec)	Mean of No. of un-localized node (N_{SL})	Mean Error (E_L)	Total Computing Time(sec)	Mean of No. of un-localized node (N_{SL})	Mean Error (E_L)	Total Computing Time(sec)
PSO	1.0924	0.03078	75.006	1.027	0.02018	56.721	0.862	0.00951	32.70
HPSO	1.0493	0.01594	61.3870	0.749	0.00911	39.692	0.437	0.00219	27.06
BBO	1.0379	0.01483	80.8421	0.482	0.00353	73.154	0.371	0.00109	49.38

The Gaussian noise is a crucial parameter for distance measurements, which influences the localization accuracy. It is observed that accuracy decreases as noise increases (mean error E_L increases as σ_d increases). The dependence of mean error (E_L) on Gaussian noise variance (σ_d) is shown in Fig. 3. The localization algorithms discussed are iterative and the number of localized nodes increases with each iteration. The anchor node density improves the performance of localizability of the target node that can be observed in Fig. 4. It can be seen in Fig. 4 that with 4 anchor nodes (minimum 4 anchor nodes are required to get 3D coordinates of the target node), the percentage of localized target node increases sharply. It can also be noticed that as the number of anchor nodes increases, the percentage of localized nodes increases as shown in Fig. 4. The implemented algorithms provide clear insight into the cost trade-off between a WSN with all target nodes and anchor nodes equipped with GPS devices. As shown in Fig. 4, with 10 anchor nodes almost 100 percent

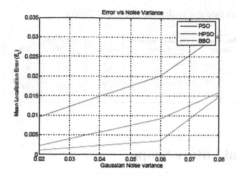

Fig. 3 Error Vs Noise variance (standard deviation)

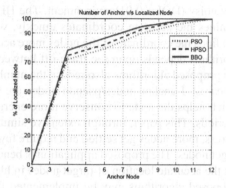

Fig. 4 Number of anchors Vs localized nodes

target nodes are localized. Simulation result (refer to Table 1) of Target Nodes = 40, transmission range = 4 unit, Gaussian noise = 0.02, field size = 10x10 units show that 10 GPS enabled anchor nodes can localize all the 40 target nodes. This results in about 75-80 percent saving in cost of GPS hardware. It is noticed that beyond a particular threshold, the effect of anchors becomes negligible as shown in Fig. 4.

5.3 Effect of Distance between Layers

The distance between the layers is varied keeping the radio range constant. The inter-layer spacing affects the success rate (in percentage) as shown in Table 2. It has been noticed from the Table 2 that HPSO and BBO based localization algorithms have better success rate as compared to PSO based localization algorithm.

Table 2 Effect on success rate due to distance between layers

Max. Radio Range	Distance Between Layers	Success Rate		
		PSO	HPSO	BBO
4	2.0	90.1	92.4	94.6
4	2.5	86.2	88.7	91.4
4	3.0	84.7	87.1	90.2
4	3.5	82.9	85.8	88.1
4	4.5	81.3	83.4	86.2

6 Conclusion and Future Scope

Stochastic range-based distributed node localization algorithms namely HPSO and BBO have been presented. The proposed algorithms have better accuracy and fast convergence in highly noisy (DOI = 0.01)environment. The HPSO-based localization algorithm determines the accurate coordinates quickly, whereas BBO-based localization algorithm finds the coordinate of the nodes more accurately. The choice between the two algorithms depends upon the trade-off between accuracy and the fast convergence. The proposed algorithms also reduce the number of transmissions to the base station (nearest 4 anchor nodes/pseudo nodes are selected to localize the target nodes), which helps the node to conserve more energy, so, the node can perform for longer periods. This paper, through extensive simulations, emphasizes that as iterations progress, more nodes get settled and require few anchors to find the coordinates of the target nodes. The proposed application is beneficial for the rescue operations to find out the accurate location of target nodes in highly noisy environment. Further, the proposed algorithms may be implemented for range-free localization and a comparison can be made for energy awareness. A hybrid stochastic algorithm may be proposed to achieve both more accuracy and faster convergence.

References

1. D. Estrin, D. Culler, K. Pister, G. Sukhatme, "connecting the physical world with pervasive networks", Pervasive Computing, IEEE 1 (2002) 59–69.
2. G. J. Pottie, W. J. Kaiser, "wireless integrated network sensors", Communications of the ACM 43 (5) (2000) 51–58.
3. I. Akyildiz, W. Su, Y. Sankarasubramaniam, E. Cayirci, A survey on sensor networks, IEEE Communications magazine 40 (8) (2002) 102–114.
4. L. Doherty, et al., Convex position estimation in wireless sensor networks, in: INFOCOM 2001. Twentieth Annual Joint Conference of the IEEE Computer and Communications Societies. Proceedings., Vol. 3, 2001, pp. 1655–1663.
5. A. Pal, Localization algorithms in wireless sensor networks: Current approaches and future challenges, Network Protocols and Algorithms 2 (1) (2010) 45–73.
6. S. Alam, Z. Haas, Topology control and network lifetime in three-dimensional wireless sensor networks, Arxiv preprint cs/0609047.
7. I. Akyildiz, D. Pompili, T. Melodia, Underwater acoustic sensor networks: research challenges, Ad hoc networks 3 (3) (2005) 257–279.

8. I. Akyildiz, W. Su, Y. Sankarasubramaniam, E. Cayirci, Wireless sensor networks: a survey, Computer networks 38 (4) (2002) 393–422.
9. J. Wang, R. K. Ghosh, S. K. Das, A survey on sensor localization, Journal of Control Theory and Applications 8 (1) (2010) 2–11.
10. A. Boukerche, H. Oliveira, E. Nakamura, A. Loureiro, Localization systems for wireless sensor networks, wireless Communications, IEEE 14 (6) (2007) 6–12.
11. J. Hightower, G. Borriello, Location systems for ubiquitous computing, Computer 34 (8) (2001) 57–66.
12. D. Niculescu, B. Nath, Ad hoc positioning system (aps), in: Global Telecommunications Conference, 2001. GLOBECOM'01. IEEE, Vol. 5, IEEE, 2001, pp. 2926–2931.
13. N. Bulusu, D. Estrin, L. Girod, J. Heidemann, Scalable coordination for wireless sensor networks: self-configuring localization systems, in: International Symposium on Communication Theory and Applications (ISCTA 2001), Ambleside, UK, 2001.
14. A. Savvides, H. Park, M. Srivastava, The bits and flops of the n-hop multilateration primitive for node localization problems, in: Proceedings of the 1st ACM international workshop on Wireless sensor networks and applications, ACM, 2002, pp. 112–121.
15. M. Di Rocco, F. Pascucci, Sensor network localisation using distributed extended kalman filter, in: IEEE/ASME international conference on Advanced intelligent mechatronics, 2007, IEEE, 2007, pp. 1–6.
16. R. Kalman, A new approach to linear filtering and prediction problems, Journal of basic Engineering 82 (Series D) (1960) 35–45.
17. P. Biswas, T. Lian, T. Wang, Y. Ye, Semidefinite programming based algorithms for sensor network localization, ACM Transactions on Sensor Networks (TOSN) 2 (2) (2006) 188–220.
18. Y. Shang, W. Ruml, Improved mds-based localization, in: Twenty-third Annual Joint Conference of the IEEE Computer and Communications Societies INFOCOM 2004, Vol. 4, IEEE, 2004, pp. 2640–2651.
19. S. Yun, J. Lee, W. Chung, E. Kim, S. Kim, A soft computing approach to localization in wireless sensor networks, Expert Systems with Applications 36 (4) (2009) 7552–7561.
20. Q. Zhang, J. Wang, C. Jin, Q. Zeng, Localization algorithm for wireless sensor network based on genetic simulated annealing algorithm, in: 4th International Conference on Wireless Communications, Networking and Mobile Computing, 2008. WiCOM'08., IEEE, 2008, pp. 1–5.
21. Q. Zhang, J. Huang, J. Wang, C. Jin, J. Ye, W. Zhang, A new centralized localization algorithm for wireless sensor network, in: Third International Conference on Communications and Networking in China, 2008. ChinaCom 2008., IEEE, 2008, pp. 625–629.
22. Y. Li, J. Xing, Q. Yang, H. Shi, Localization research based on improved simulated annealing algorithm in wsn, in: 5th International Conference on Wireless Communications, Networking and Mobile Computing, 2009. WiCom'09, IEEE, 2009, pp. 1–4.
23. R. Kulkarni, G. Venayagamoorthy, M. Cheng, Bio-inspired node localization in wireless sensor networks, in: IEEE International Conference on Systems, Man and Cybernetics, 2009. SMC 2009., pp. 205–210.
24. A. Gopakumar, L. Jacob, Localization in wireless sensor networks using particle swarm optimization, in: IET International Conference on Wireless, Mobile and Multimedia Networks, 2008., IET, 2008, pp. 227–230.
25. R. Stoleru, J. A. Stankovic, Probability grid: A location estimation scheme for wireless sensor networks, in: First Annual IEEE Communications Society Conference on Sensor and Ad Hoc Communications and Networks, 2004. IEEE SECON 2004., IEEE, 2004, pp. 430–438.
26. P. Chuang, C. Wu, An effective pso-based node localization scheme for wireless sensor networks, in: Ninth International Conference on Parallel and Distributed Computing, Applications and Technologies, 2008. PDCAT 2008, IEEE, 2008, pp. 187–194.
27. G. Mao, B. Fidan, B. Anderson, Wireless sensor network localization techniques, Computer Networks 51 (10) (2007) 2529–2553.
28. Y. del Valle, G. Venayagamoorthy, S. Mohagheghi, J. Hernandez, R. Harley, Particle swarm optimization: basic concepts, variants and applications in power systems, IEEE Transactions on Evolutionary Computation 12 (2) (2008) 171–195.

29. R. Schaefer, H. Telega, Foundations of global genetic optimization, Springer Verlag, 2007.
30. K. Price, R. Storn, J. Lampinen, Differential evolution: a practical approach to global optimization, Springer-Verlag New York Inc, 2005.
31. Y. Chen, W. Peng, M. Jian, Particle swarm optimization with recombination and dynamic linkage discovery, IEEE Transactions on Systems, Man, and Cybernetics, Part B: Cybernetics 37 (6) (2007) 1460–1470.
32. D. Kim, A. Abraham, J. Cho, A hybrid genetic algorithm and bacterial foraging approach for global optimization, Information Sciences 177 (18) (2007) 3918–3937.
33. J. Kennedy, R. Eberhart, Particle swarm optimization, in: Proceedings., IEEE International Conference on Neural Networks, 1995, Vol. 4, 1995, pp. 1942–1948.
34. Y. Shi, et al., Particle swarm optimization: developments, applications and resources, in: Proceedings of the 2001 Congress on Evolutionary Computation, 2001., Vol. 1, IEEE, 2001, pp. 81–86.
35. D. Simon, Biogeography-based optimization, Evolutionary Computation, IEEE Transactions on 12 (6) (2008) 702–713.
36. M. Noel, P. Joshi, T. Jannett, Improved maximum likelihood estimation of target position in wireless sensor networks using particle swarm optimization, in: Third IEEE International Conference on Information Technology: New Generations, 2006. ITNG 2006., 2006, pp. 274–279.
37. Y. Chen, V. Dubey, Ultrawideband source localization using a particle-swarm-optimized capon estimator, in: IEEE International Conference on Communications, 2005., Vol. 4, 2005, pp. 2825–2829.
38. A. Wallace, The Geographical Distribution of Animals, MA: Adamant Media Corporation, 2005.
39. C. Darwin, The Origin of Species, New York: Gramercy, 1859.
40. R. MacArthur, E. Wilson, The theory of island biogeography, Princeton Univ Press, 1967.
41. N. Patwari, J. Ash, S. Kyperountas, A. Hero III, R. Moses, N. Correal, Locating the nodes: cooperative localization in wireless sensor networks, Signal Processing Magazine, IEEE 22 (4) (2005) 54–69.

An Evaluation of Classification Algorithms Using Mc Nemar's Test

Betul Bostanci and Erkan Bostanci

Abstract Five classification algorithms namely J48, Naive Bayes, Multilayer Perceptron, IBK and Bayes Net are evaluated using Mc Nemar's test over datasets including both nominal and numeric attributes. It was found that Multilayer Perceptron performed better than the two other classification methods for both nominal and numerical datasets. Furthermore, it was observed that the results of our evaluation concur with Kappa statistic and Root Mean Squared Error, two well-known metrics used for evaluating machine learning algorithms.

Key words: Classifier Evaluation, Classification algorithms, Mc Nemar's test

1 INTRODUCTION

Evaluating the performance of machine learning methods is as crucial as the algorithm itself since this identifies the strengths and weaknesses of each learning algorithm. This paper investigates the usage of Mc Nemar's test as an evaluation method for machine learning methods.

Mc Nemar's test has been used in different studies in previous research. Dietterich [1] examined 5 different statistical tests including Mc Nemar's test to identify how these tests differ in assessing the performances of classification algorithms. A similar evaluation was performed on a large database by Bouckaert [2]. Demsar [3] has evaluated decision tree, naive bayes and k-nearest neighbours methods

Betul Bostanci
School of Computer Science and Electronic Engineering, University of Essex, Colchester, UK
e-mail: bbosta@essex.ac.uk

Erkan Bostanci
School of Computer Science and Electronic Engineering, University of Essex, Colchester, UK
e-mail: gebost@essex.ac.uk

J. C. Bansal et al. (eds.), *Proceedings of Seventh International Conference on Bio-Inspired Computing: Theories and Applications (BIC-TA 2012)*, Advances in Intelligent Systems and Computing 201, DOI: 10.1007/978-81-322-1038-2_2, © Springer India 2013

using other non-parametric tests including ANOVA (ANalysis Of VAriance) [4] and Friedman test [5, 6].

Other studies have evaluated classifiers using this test over a large set but our method differs in that we use a different criterion that compares how the individual instances are classified and how this is reflected in the whole dataset.

Five different machine learning methods namely J48 (Decision Tree), Naive Bayes [7], Multilayer Perceptron [7] IBK [8] and Bayes Net [9] were used in the experiments. WEKA [10] was used to obtain the classification results of these algorithms. These classification methods are used to classify samples from different datasets. Later, the classification results are analyzed using a non-parametric test in order to identify how a pair of learning methods differ from each other and which of the two performs better.

The rest of the paper is structured as follows: Section 2 presents the nominal and numeric datasets used in the experiments. Section 3 introduces Mc Nemar's test which is the main evaluation method proposed in this study followed by Section 4 where the experimental design is presented. Section 5 presents Mc Nemar's test results and compares them with two conventional evaluation criteria. Finally, the paper is drawn to a conclusion in Section 6.

2 DATASETS

In order to perform a fair evaluation, a relatively large number of datasets obtained from UCI Machine Learning Repository [11] are used. The datasets are selected from the ones including nominal (Table 1) and numeric data (Table 2).

Table 1 Nominal Datasets

Dataset	Number of Instances	Number of Attributes	Number of Classes
Car	1728	7	4
Nursery	12960	9	5
Tic-Tac-Toe	958	10	2
Zoo	101	18	7

Table 2 Numeric Datasets

Dataset	Number of Instances	Number of Attributes	Number of Classes
Diabetes	768	9	2
Glass	214	10	7
Ionosphere	351	35	2
Iris	150	5	3
Segment-Challenge	1500	20	7
Waveform-5000	5000	41	3

3 Mc NEMAR'S TEST

Mc Nemar's test [12, 13] is a variant of χ^2 test and is a non-parametric test used to analyse matched pairs of data. According to Mc Nemar's test, two algorithms can have 4 possible outcomes arranged in a 2×2 contingency table [14] as shown in Table 3.

Table 3 Possible results of two algorithms [13]

	Algorithm A failed	Algorithm A succeeded
Algorithm B failed	N_{ff}	N_{sf}
Algorithm B succeeded	N_{fs}	N_{ss}

N_{ff} denotes the number of times (instances) when both algorithms failed and N_{ss} denotes success for both algorithms. These two cases do not give much information about the algorithms' performances as they do not indicate how their performances differ. However, the other two parameters (N_{fs} and N_{sf}) show cases where one of the algorithms failed and the other succeeded indicating the performance discrepancies.

In order to quantify these differences Mc Nemar's test employs z score (Equation 1).

$$z = \frac{(|N_{sf} - N_{fs}| - 1)}{\sqrt{N_{sf} + N_{fs}}} \tag{1}$$

z scores are interpreted as follows: When $z = 0$, the two algorithms are said to show similar performance. As this value diverges from 0 in positive direction, this indicates that their performance differs significantly. Furthermore, z scores can also be translated into confidence levels as shown in Table 4.

Table 4 Confidence levels corresponding to z scores for one-tailed and two-tailed predictions [13]

z score	One-tailed Prediction	Two-tailed Prediction
1.645	95%	90%
1.960	97.5%	95%
2.326	99%	98%
2.576	99.5%	99%

Following the table, it is worth mentioning that *One-tailed Prediction* is used to determine when one algorithm is better than the other where *Two-tailed Prediction* shows how much the two algorithms differ.

Mc Nemar's test is known to have a low *Type-I* error which occurs when an evaluation method detects a difference between two learning algorithms when there is no difference [1].

4 EVALUATION CRITERION

By adopting the Mc Nemar's test to evaluate classification algorithms, the following criterion is defined: An algorithm is regarded as "successful" if it can identify the class of an instance correctly. Conversely, it is regarded as "failed" when it performs an incorrect classification for an instance.

Using this criterion, the z scores are calculated using Mc Nemar's test for the five classification algorithms. All the algorithms were used with their default parameters as parameter tuning may favor one algorithm to produce better results.

The null hypothesis (H_0) for this experimental design suggests that different classifiers perform similarly whereas the alternative hypothesis (H_1) claims otherwise suggesting that at least one of the classifiers performs differently as shown in Equation 2.

$$H_0 : C_1 = C_2 = C_3 = C_4 = C_5$$
$$H_1 : \exists C_i : C_i \neq C_j, (i, j) \in (1, 2, 3, 4, 5), i \neq j \tag{2}$$

At the end of the experiment, the z scores will indicate whether we should accept H_0 and reject H_1 or vice versa. In order to calculate the z scores, the classification results of the three classifiers must be identified for each individual instance.

This operation is performed for all instances in the given datasets. In WEKA, there are two options to see whether an instance is correctly classified or not. The first option is the graphical one (shown in Figure 1 with the squares while crosses denote correct classifications). The second option to show the incorrect classifications is via the "Output predictions" option of the classifier which displays a "+" in the output next to the instance which has been incorrectly classified.

10-fold cross-validation is used in the evaluation which works as folllows: First the data is separated into 10 sets each having $n/10$ instances. Then, the training is performed using 9 of these sets and testing is performed on the remaining 1 set. This process is repeated 10 times to consider all of the subsets created and the final result for the accuracy is obtained by taking the average of these iterations.

The first option is quite useful to see the result graphically, however in order to calculate the number of correct and incorrect classifications by the classifiers, one needs to export these results into a spreadsheet (e.g. Excel). For this reason, the second method was used to calculate number of instances where the classifiers succeeded and failed. Using these figures, the z scores were calculated using Equation 1.

In order to decide which classifier performed better, N_{sf} and N_{fs} values for two classifiers are examined. For example, classifier A is said to perform better than classifier B if N_{sf} is larger than N_{fs} according to Table 3.

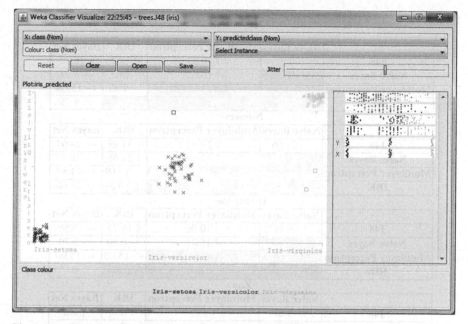

Fig. 1 Visualization of Classification Errors in WEKA

5 RESULTS

This section presents the results of the experiment. Results for the Mc Nemar's test
will be given first and then these results will be compared with two other evaluation
criteria namely Kappa statistic and Root Mean Squared Error (RMSE).

5.1 McNemar's Test Results

In Tables 5 and 6, the arrowheads ($\leftarrow, \uparrow$) denote which classifier performed better
in the given datasets. z scores are given next to the arrowheads as a measure of how
statistically significant the results are.

By looking at the Mc Nemar's test results for the nominal datasets (Table 5),
one can deduce that Multilayer Perceptron has produced significantly better results
than J48 and Naive Bayes classifiers (H_1 is accepted with a confidence level of
more than 99.5%). J48 classifier performed better than the Naive Bayes for *Nursery*
and *Tic-Tac-Toe* datasets. For the *Zoo* dataset, Naive Bayes performed better than
J48 and equally to the Multilayer perceptron (H_0 is not rejected.). The performance
differences between IBK and all other classifiers were not found to be statistically
significant for the *Zoo* dataset but for the rest of the nominal datasets, there were significant differences. Bayes Net shows a poor performance overall except for the *Zoo*

Table 5 Mc Nemar's Test Results for Nominal Datasets

Car

	Naive Bayes	Multilayer Perceptron	IBK	Bayes Net
J48	0	↑ 10.63	↑ 1.62	← 6.93
Naive Bayes		↑ 10.63	↑ 1.62	← 6.93
Multilayer Perceptron			← 9.82	← 15.08
IBK				← 9.75

Nursery

	Naive Bayes	Multilayer Perceptron	IBK	Bayes Net
J48	← 24.66	↑ 17.32	↑ 34.89	← 24.64
Naive Bayes		↑ 34.89	↑ 31.68	0
Multilayer Perceptron			← 12.09	← 34.87
IBK				← 31.66

Tic-tac-toe

	Naive Bayes	Multilayer Perceptron	IBK	Bayes Net
J48	← 8.44	↑ 10.06	↑ 15.73	← 8.56
Naive Bayes		↑ 15.73	← 0.70	← 0.70
Multilayer Perceptron			← 15.90	← 15.90
IBK				0

Zoo

	Naive Bayes	Multilayer Perceptron	IBK	Bayes Net
J48	← 0.67	↑ 1.23	0	↑ 0.5
Naive Bayes		0	0	0
Multilayer Perceptron			0	0
IBK				0

dataset where it performed better than J48 although the result was not statistically significant.

Many differences in the classification performance are noticeable in the numeric dataset results (Table 6). For the *Glass* and *Segment-Challenge* datasets J48 has given better classification performance than Naive Bayes. For the former dataset, the Multilayer Perceptron performed equally with J48 and Naive Bayes produced a poorer classification result than these two. IBK and Bayes Net shows better performance over J48, Naive Bayes and Multilayer Perceptron, however there was no statistically significant performance difference between these two classification methods.

It is interesting to see that the first three (J48, Naive Bayes and Multilayer Perceptron) classifiers performed similarly on the *Ionosphere* dataset (H_0 is not rejected for all pairs.). Some differences can noticeable between these classifiers and Bayes Net however the results are not significant ($z = 0.75$ for Naive Bayes and Multilayer Perceptron) A similar result is also visible when the *Iris* dataset is consided since the values are quite close to zero. For the *Diabetes* dataset, Naive Bayes showed better performance over J48 yet the difference was not very significant for the latter (with a confidence level less than 95%) whereas Naive Bayes performs significantly better than J48 for the *Waveform-5000* dataset.

We can also see that the Multilayer Perceptron did not produce good results for the *Ionosphere* dataset where a relatively large number of attributes are present. This

Table 6 McNemar's Test Results for Numeric Datasets

Diabetes

	Naive Bayes	Multilayer Perceptron	IBK	Bayes Net
J48	↑ 1.61	↑ 0.96	← 0.56	↑ 0.26
Naive Bayes		← 0.56	← 3.40	← 1.29
Multilayer Perceptron			← 2.97	← 0.59
IBK				↑ 2.16

Glass

	Naive Bayes	Multilayer Perceptron	IBK	Bayes Net
J48	← 4.07	0	↑ 4.07	↑ 0.97
Naive Bayes		↑ 4.07	↑ 5.05	↑ 5.24
Multilayer Perceptron			↑ 0.95	↑ 0.97
IBK				0

Ionosphere

	Naive Bayes	Multilayer Perceptron	IBK	Bayes Net
J48	0	0	0	← 1.05
Naive Bayes		0	← 2.71	← 0.75
Multilayer Perceptron			← 2.71	← 0.75
IBK				↑ 1.37

Iris

	Naive Bayes	Multilayer Perceptron	IBK	Bayes Net
J48	0	↑ 0.41	↑ 0.5	← 1.51
Naive Bayes		↑ 0.5	0	← 1.51
Multilayer Perceptron			← 1.16	← 2.00
IBK				← 1.23

Segment

	Naive Bayes	Multilayer Perceptron	IBK	Bayes Net
J48	← 12.95	↑ 1.69	↑ 14.22	← 6.58
Naive Bayes		↑ 14.22	↑ 13.48	↑ 8.53
Multilayer Perceptron			← 0.86	← 7.92
IBK				← 7.05

Waveform-5000

	Naive Bayes	Multilayer Perceptron	IBK	Bayes Net
J48	↑ 6.90	↑ 12.40	← 1.84	↑ 6.71
Naive Bayes		↑ 5.78	← 8.77	← 0.54
Multilayer Perceptron			← 13.89	← 6.05
IBK				↑ 8.59

lower performance can be due to an underfitting problem as the default parameters were used without any parameter tuning.

5.2 Comparison with Other Evaluation Criteria

Mc Nemar's test result showed that there are significant discrepancies in the performances of the classifiers. Additional experiments were carried out to see how the

results for Mc Nemar's test conform with other evaluation criteria namely Kappa
Statistic and Root Mean Squared Error.

5.2.1 Kappa Statistic

Kappa Statistic is a measure of the agreement between the predicted and the actual
classifications in a dataset [15]. For this reason, we expect a higher value for a
classifier which has more overlapping predictions and observations.

By looking at the nominal datasets in Figure 2, we see that Multilayer Perceptron
has the highest value in 3 out of 4 datasets. J48 is better than Naive Bayes except for
the *Zoo* dataset (Figure 2(d)). IBK shows good performance in all nominal datasets,
although the poorest performance can be seen in the *Car* dataset.

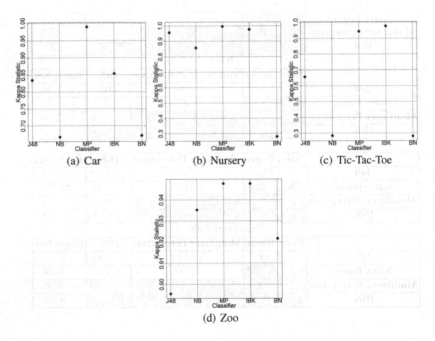

Fig. 2 Kappa Statistics for Nominal Datasets. NB: Naive Bayes, MP: Multilayer Perceptron, BN:
Bayes Net

The ranking between J48 and Multilayer Perceptron changes significantly for the
Glass and *Segment-Challenge* datasets for the numeric datasets in Figure 3. IBK
has a good performance in these two datasets ($Kappa = 0.60$ and $Kappa = 0.95$ respectively). Naive Bayes produced good results only for the *Diabetes* dataset in this
group. We can also say the Bayes Net achieves higher classification performance
for the numeric datasets than the nominal datasets.

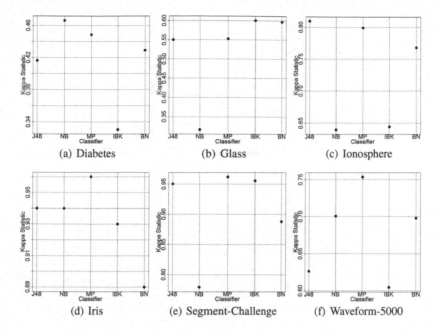

Fig. 3 Kappa Statistics for Numeric Datasets. NB: Naive Bayes, MP: Multilayer Perceptron, BN: Bayes Net

5.2.2 Root Mean Squared Error

Root Mean Squared Error (RMSE) [15] shows the error in the predicted and actual classes which the instances in a dataset belong to. RMSE should have lower values for more accurate classification results.

In nominal dataset results (Figure 4), Multilayer Perceptron had the lowest RMSE values for *Car*, *Nursery* and *Tic-Tac-Toe* datasets. J48 performed better than the Naive Bayes for the these datasets as well, while the ranking changed between them in the *Zoo* dataset shown in Figure 4(d). IBK shows the worst performance on the *Diabetes* and the best performance on the *Zoo* dataset. Bayes Net has poor performance in *Car* and *Tic-Tac-Toe* datasets.

A first look on the results in Figure 5 reveals that the Multilayer Perceptron results in lowest RMSE values for 4 out of 6 numeric datasets. Naive Bayes has a poor performance in *Glass*, *Ionosphere* and *Segment-Challenge* datasets. Naive Bayes showed the lowest performance in all datasets of the numeric dataset results except for the *Diabetes* dataset.

Table 7 show the mean results for all classifiers for the nominal and numeric datasets. Multilayer Perceptron has the highest values for Kappa statistic and lowest values for RMSE showing that the classification results using this classifier are accurate. IBK also shows a good classification performance for nominal and numeric data. Poor results are visible for Naive Bayes and Bayes Net.

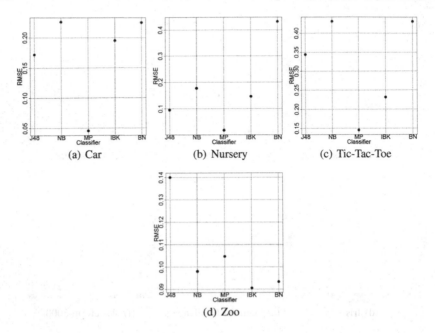

Fig. 4 RMSE for Nominal Datasets. NB: Naive Bayes, MP: Multilayer Perceptron, BN: Bayes Net

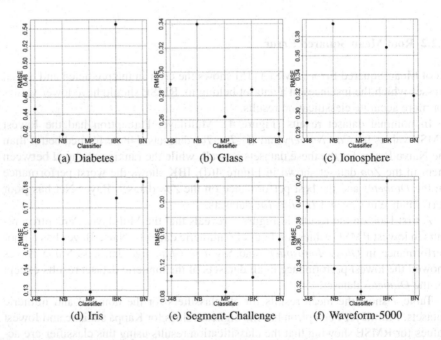

Fig. 5 RMSE for Numeric Datasets. NB: Naive Bayes, MP: Multilayer Perceptron, BN: Bayes Net

Table 7 Mean Kappa statistic and RMSE values for nominal and numeric datasets

	Nominal		Numeric	
	Kappa	RMSE	Kappa	RMSE
J48	0.82	0.19	0.72	0.36
Naive Bayes	0.69	0.23	0.64	0.31
Multilayer Perceptron	0.97	0.08	0.75	0.25
IBK	0.94	0.17	0.68	0.32
Bayes Net	0.54	0.30	0.71	0.28

From the three evaluation criteria (Mc Nemar's test, Kappa statistic and RMSE), Table 8 can be used to summarize the performance difference over the nominal and numeric datasets where + indicates performance grade. By looking at this summary table, it is evident that the Mc Nemar's test agrees with other evaluation criteria as an important result of the experiments. An exception can be seen for the comparison of Naive Bayes and J48 which is due to insignificance of the differences in Mc Nemar's test.

Table 8 Summary of the performances for nominal and numeric datasets

Mc Nemar's test					
	J48	Naive Bayes	Multilayer Perceptron	IBK	Bayes Net
Nominal	+++	++	+++++	++++	+
Numeric	+	++++	+++++	++	+++
Kappa statistic					
Nominal	+++	++	+++++	++++	+
Numeric	++++	+	+++++	++	+++
RMSE					
Nominal	+++	++	+++++	++++	+
Numeric	+	+++	+++++	++	++++

6 CONCLUSION

This study employed Mc Nemar's test in order to evaluate machine learning algorithms namely J48, Naive Bayes and Multilayer Perceptron, IBK and Bayes Net. By defining the success and failure criteria of Mc Nemar's test as correctly or incorrectly identifying the class of an instance in a dataset, the experiments presented the usage of a non-parametric test as a new method to evaluate classification algorithms.

The results showed that Multilayer Perceptron produced better results than the other methods for both nominal and numerical data. Bayes Net was placed in the lowest ranks for both types of data. Another interesting finding of the experiment is that the results of the Mc Nemar's test mostly conformed with Kappa statistic and RMSE as a justification of method's integrity.

The effect of parameter tuning is considered as future research. In this case, the classifiers will be tuned to achieve the optimal results and then the same tests can be applied to see whether there will be any changes in the rankings.

Acknowledgements The authors would like to thank Nadia Kanwal for discussions on Mc Nemar's test.

References

1. T. G. Dietterich, "Approximate statistical tests for comparing supervised classification learning algorithms," *Neural Computation*, vol. 10, pp. 1895–1923, 1998.
2. R. R. Bouckaert and E. Frank, "Evaluating the replicability of significance tests for comparing learning algorithms," in *Proceedings 8th Pacific-Asia Conference*, pp. 3–12, 2004.
3. J. Demsar, "Statistical comparisons of classifiers over multiple data sets," *Journal of Machine Learning Research*, pp. 1–6, 2006.
4. D. A. Berry, "Logarithmic transformations in ANOVA," *Biometrics*, vol. 43, no. 2, pp. 439–456, 1987.
5. M. Friedman, "The use of ranks to avoid the assumption of normality implicit in the analysis of variance," *Journal of the American Statistical Association*, vol. 32, no. 200, pp. 675–701, 1937.
6. M. Friedman, "A comparison of alternative tests of significance for the problem of m rankings," *The Annals of Mathematical Statistics*, vol. 1, no. 1, pp. 86–92, 1940.
7. P. Tan, M. Steinbach, and V. Kumar, *Introduction to Data Mining*. Pearson, 2006.
8. D. Aha and D. Kibler, "Instance-based learning algorithms," *Machine Learning*, vol. 6, pp. 37–66, 1991.
9. N. Friedman, D. Geiger, and Goldszmidt, "Bayesian network classifiers," *Machine Learning*, vol. 29, pp. 131–163, 1997.
10. M. Hall, E. Frank, G. Holmes, B. Pfahringer, P. Reutemann, and I. H. Witten, "The WEKA data mining software: An update," *SIGKDD Explorations*, vol. 11, 2009.
11. "UCI Machine Learning Repository." http://archive.ics.uci.edu/ml/, 2012.
12. Q. McNemar, "Note on the sampling error of the difference between correlated proportions or percentages," *Psychometrika*, no. 12, pp. 153–157, 1947.
13. A. F. Clark and C. Clark, "Performance Characterization in Computer Vision: A Tutorial."
14. D. Liddell, "Practical tests of 2 2 contingency tables," *Journal of the Royal Statistical Society*, vol. 25, no. 4, pp. 295–304, 1976.
15. I. H. Witten, E. Frank, and M. A. Hall, *Data Mining: Practical Machine Learning Tools and Techniques*. Morgan Kaufmann, 2011.

Permitting features in P systems generating picture arrays

K.G. Subramanian, Ibrahim Venkat, Linqiang Pan and Atulya K. Nagar

Abstract In the area of membrane computing, the biologically inspired model known as P system has proved to be a rich framework for studying several types of problems. Picture array generation is one such problem for which different P systems have been constructed in the literature. Incorporating the feature of permitting symbols in the rules, array P systems are constructed here for generating picture languages consisting of picture arrays. The advantage of this approach is that there is a reduction in the number of membranes used in the construction, in comparison to the existing array P system model.

Key words: Membrane Computing, P System, Picture Arrays, Permitting Features

1 Introduction

The computability model known as P system, introduced by Păun [10] inspired by the structure and functioning of living cells, has turned out to be a versatile frame

K.G. Subramanian
School of Computer Sciences, Universiti Sains Malaysia, 11800 Penang, Malaysia
e-mail: kgsmani1948@gmail.com

Ibrahim Venkat
School of Computer Sciences, Universiti Sains Malaysia, 11800 Penang, Malaysia
e-mail: ibrahim@cs.usm.my

Linqiang Pan
Department of Control Science and Engineering,Huazhong University of Science and Technology, Wuhan 430074, Hubei, China
e-mail: lqpan@mail.hust.edu.cn

Atulya K. Nagar
Department of Mathematics and Computer Science,
Liverpool Hope University, Liverpool L16 9JD UK

J. C. Bansal et al. (eds.), *Proceedings of Seventh International Conference on Bio-Inspired Computing: Theories and Applications (BIC-TA 2012)*, Advances in Intelligent Systems and Computing 201, DOI: 10.1007/978-81-322-1038-2_3, © Springer India 2013

work for studying computational problems in many different fields [11]. Picture grammar is one such area where different kinds of P systems for generating picture languages consisting of picture arrays, have been introduced and investigated. In [3], array P systems are introduced extending the string-objects P systems to array-objects P systems, thereby giving a link between picture grammars and P systems. Motivated by the study in [3], several variants of array P systems have been introduced (See for example, [1, 2, 16]).

On the other hand, regulating rewriting [6] in a grammar by permitting or forbidding the application of a rule based on the presence or absence of a set of symbols is known in formal language theory. Picture grammars that use this feature of permitting or forbidding symbols have also been introduced in [7, 8, 9].

Here we associate permitting symbols with rules in the regions of an array P system [3]. We call the resulting array P system as a permitting array P system and construct such a P system for generating picture languages consisting of picture arrays. The advantage of this approach is that the number of membranes used in the construction is reduced when compared to array P system [3]. The problem of generation of geometric figures such as squares, rectangles are of interest in the study of picture grammars (see for example [21, 20]). We consider in permitting array P system, the feature of $t-$communication in array P systems considered in [17] and this enables us to generate picture arrays representing solid squares with a reduced number of membranes in comparison to the $t-$communicating array P system given in [17] to generate such solid squares.

2 Preliminaries

For notions related to array grammars and array languages, we refer to [12, 13, 20], for notions on array P systems, we refer to [3, 16] and for notions of formal language theory to [14, 15].

Given an alphabet V, the set of all words over V, including the empty word λ, is denoted by V^* and $V^+ = V^* - \lambda$.

A picture array or simply an array in the two-dimensional plane consists of a finite number of labelled unit squares or pixels, with the labels belonging to an alphabet V and the unit squares not labelled with elements of V are considered to have the *blank symbol* $\# \notin V$. An array can be formally specified by listing the coordinates and the corresponding labels of the pixels. For example, for the T shaped array in Figure 1, this kind of specification is given as follows:

$$\{((0,5),a),\ ((1,5),a),\ ((2,5),a),\ ((3,5),a),\ ((4,5),a),\ ((5,5),a),$$

$$((6,5),a),\ ((7,5),a),\ ((8,5),a),\ ((9,5),a),\ ((10,5),a),$$

$$((5,0),a),\ ((5,1),a),\ ((5,2),a),\ ((5,3),a),\ ((5,4),a)\}$$

We note that only the relative positions of non-blank pixels in the array matter for us. The non-blank labels of the T shaped array are pictorially indicated in Figure 1.

$$
\begin{array}{c}
a\ a\ a\ a\ a\ a\ a\ a\ a\ a\ a \\
a \\
a \\
a \\
a \\
a
\end{array}
$$

Figure 1: T-shaped array with equal arms

We denote by V^{+2} the set of all two-dimensional non-empty finite arrays over V. The empty array is denoted by λ, and then the set of all arrays over V is $V^{*2} = V^{+2} \cup \{\lambda\}$. Any subset of V^{*2} is called an *array language*.

The array grammars [12, 13, 20] that involve array rewriting rules are extensions of string grammars [14, 15] to two dimensional picture arrays. We recall here the context-free and regular types of array rewriting grammars of the isometric variety which means that the rules preserve the geometric shape of the rewritten subarray.

An array grammar $G = (N, T, S, P, \#)$ where N, T are alphabets, $N \cap T = \phi$ and $S \in N$ is the start symbol. The elements of the finite set N are called nonterminals and those of T, terminals. P is a finite set of array rewriting rules of the form $r : \alpha \to \beta$ where α and β are arrays over $V \cup \#$ satisfying the following conditions:

1. the arrays α and β have identical shapes;
2. there is at least one element of N in α;
3. the symbols of T that occur in α are retained in their respective positions in β;
4. the application of the rule $r : \alpha \to \beta$ preserves the connectivity of the rewritten array.

For two arrays γ, δ over V and a rule r as above, we write $\gamma \Rightarrow_p \delta$ if δ can be obtained by replacing with β, a subarray of γ identical to α. The reflexive and transitive closure of the relation $\Rightarrow$ is denoted by $\Rightarrow^*$.

An array grammar is called:

1. *context-free*, if for all the rules $r : \alpha \to \beta$, the non$-\#$ symbols in α are not replaced by symbol $\#$ in β and for each rule $\alpha \to \beta$, α contains exactly one nonterminal with the remaining squares containing $\#$ and β contains no blank symbol $\#$;
2. *regular*, if the rules are of the following forms:

$$A\,\# \to a\,B,\ \#A \to B\,a,\ \begin{array}{c}\#\\A\end{array} \to \begin{array}{c}B\\a\end{array},\ \begin{array}{c}A\\\#\end{array} \to \begin{array}{c}a\\B\end{array},\ A \to B,\ A \to a,$$

where A, B are nonterminals and a is a terminal.

The array language generated by G is

$$L(G) = \{p \mid S \Rightarrow^* p \in T^{+2}\}.$$

Note that the start array is indeed $\{((0,0), S)\}$ and it is understood that this square labelled S is surrounded by #, denoting empty squares with no labels.

We denote by $AREG$ and ACF respectively the families of array languages generated by array grammars with regular and context-free array rewriting rules.

We now recall the basic model of a rewriting array-objects P system introduced in [3].

An array P system (of degree $m \geq 1$) [3] is a construct

$$\Pi = (V, T, \#, \mu, F_1, \ldots, F_m, R_1, \ldots, R_m, i_o),$$

where: V is the alphabet of nonterminals and terminals, $T \subseteq V$ is the terminal alphabet, $\# \notin V$ is the blank symbol, μ is a membrane structure with m membranes labelled in a one-to-one way with $1, 2, \ldots, m$, $F_1, \ldots, F_m$ are finite sets of arrays over V associated with the m regions of μ, $R_1, \ldots, R_m$ are finite sets of array rewriting rules over V associated with the m regions of μ; the array-rewriting rules (context-free or regular) of the form $\mathscr{A} \to \mathscr{B}(tar)$ have attached targets *here, out, in* (in general, we omit mentioning *here*); finally, i_o is the label of an elementary membrane of μ which is the output membrane.

A computation in an array P system is defined in the same way as in a string rewriting P system [10] with the successful computations being the halting ones. Every array, from each region of the system, which can be rewritten by a rule associated with that region (membrane), should be rewritten; the rewriting is sequential at the level of arrays which means that one rule is applied ; the array obtained by rewriting is placed in the region indicated by the target associated with the rule used (*here* means that the array remains in the same region, *out* means that the array exits the current membrane and thus, if the rewriting was done in the skin membrane, then it exits the system; (arrays leaving the system are "lost" in the environment), and *in* means that the array is immediately sent to one of the directly lower membranes, nondeterministically chosen if several exist; if no internal membrane exists, then a rule with the target indication *in* cannot be used).

A computation is successful only if it stops and a configuration is reached where no rule can be applied to the existing arrays. The result of an halting computation consists of the arrays composed only of symbols from T placed in the output membrane with label i_o in the halting configuration. The set of all such arrays computed or generated by a system Π is denoted by $AL(\Pi)$. The families of all array languages $AL(\Pi)$ generated by systems Π as above, with at most m membranes, with CF and regular array-rewriting rules are respectively denoted by $EAP_m(CF)$ and $EAP_m(REG)$.

We illustrate with an example the computation in an array P system.

Example 1. An array P system generating T shaped arrays (Figure 1) over $\{a\}$ is as follows:

$$\Pi_1 = (\{A,B,C,B',C',a\},\{a\},\#,[_1[_2[_3\ [_4\ ]_4]_3]_2]_1,$$

$$\left\{\begin{array}{c}A\,X\,B\\C\end{array}\right\},\emptyset,\emptyset,\emptyset,R_1,R_2,R_3,R_4,4),$$

with

$$R_1 = \{\#\,A \to A\ a(in),\},$$
$$R_2 = \{B\,\# \to a\,B'(in),\ B' \to B\ (out)\},$$
$$R_3 = \{\begin{array}{c}C\\\#\end{array} \to \begin{array}{c}a\\C\end{array}(out),\ \begin{array}{c}C\\\#\end{array} \to \begin{array}{c}a\\C'\end{array}(in)\},$$
$$R_4 = \{A \to a, B' \to a, C' \to a\}.$$

A computation in Π_1 starts with the initial array $\begin{array}{c}A\,X\,B\\C\end{array}$ in region 1, with other regions having no initial array. An application of the rule $\#\,A \to A\ a(in)$ grows the horizontal arm one step on the left, after which the array is sent to region 2, due to the target indication *in* in the rule. In region 2, the rule $B\,\# \to a\,B'(in)$ alone can be applied which grows the horizontal arm one step on the right, after which the array is sent to region 3, due to the target indication *in* in the rule. If the rule $\begin{array}{c}C\\\#\end{array} \to \begin{array}{c}a\\C\end{array}(out)$ is applied in region 3, then the vertical arm grows one step down and the array is sent back to region 2 due to the target indication *out*. In region 2, the primed version of the nonterminal B is changed into B and the array is brought back to region 1 and the process can repeat. If in region 3, the rule applied is $\begin{array}{c}C\\\#\end{array} \to \begin{array}{c}a\\C'\end{array}(in)$ then the array is sent to the output region 4 wherein all the nonterminals are changed into the terminal a and the computation halts yielding a T shaped array over $\{a\}$ with equal arms which is collected in the language generated.

3 Permitting array P systems

We consider permitting CF (respy. regular)array rewriting rule, which is a context-free array rewriting rules with permitting symbols. We then define a permitting array P system that makes use of such permitting CF array rewriting rules in its regions.

If $\mathscr{B}$ is a subarray of $\mathscr{A}$, then we denote by $\mathscr{A} \setminus \mathscr{B}$, the array formed by the labelled squares of $\mathscr{A}$ that are not labelled squares of $\mathscr{B}$. We denote by $l(\mathscr{A})$, the set of all symbols in the labelled squares of the array $\mathscr{A}$. Note that in a CF array rewriting rule $\mathscr{A} \to \mathscr{B}$, $\mathscr{A}$ contains exactly one labelled square with a nonterminal symbol as label.

A permitting CF (respy. regular) array rewriting rule is of the form $(\mathscr{A} \to \mathscr{B}, per)$ where $\mathscr{A} \to \mathscr{B}$ is a context-free array rewriting rule and $per \subseteq N$ with N being the set of nonterminals of the array grammar. If $per = \phi$, then we omit mentioning it in the rule. For any two arrays $\mathscr{C}, \mathscr{D}$, and a permitting CF array rule $(\mathscr{A} \to \mathscr{B}, per)$, the array $\mathscr{D}$ is derived from $\mathscr{C}$ by replacing $\mathscr{A}$ in $\mathscr{C}$ by $\mathscr{B}$, provided $per \subseteq l(\mathscr{C} \setminus \mathscr{A})$.

We now introduce the notion of an array P system with permitting symbols associated with the rules in the regions.

A permitting array P system (of degree $m \geq 1$ ($pEAPS_m(CF)$), is a construct

$$\Pi = (V, T, \#, \mu, F_1, \ldots, F_m, R_1, \ldots, R_m, i_o),$$

where the components $V, T, \#, \mu, F_1, \ldots, F_m, i_o$ are as in an array P system and the rules in the sets $R_1, \ldots, R_m$ are permitting CF array rewriting rules of the form $(\mathscr{A} \to \mathscr{B}, per)$ where $\mathscr{A} \to \mathscr{B}$ is a context-free array rewriting rule and $per \subseteq V - T$ with $V - T$ being the set of nonterminals.

A computation in $pEAPS_m(CF)$) is also as in an array P system except that the application of a permitting CF array rewriting rule in any $R_i, 1 \leq i \leq m$, is regulated by the associated permitting symbols as described earlier in deriving an array from a given array. The successful computations are the halting ones. The result of a computation is the set of arrays collected in the output elementary membrane i_0 in the halting configuration.

The family of all array languages generated by systems Π as above, with at most m membranes, with permitting array rewriting rules of type regular or CF is respectively denoted by $pEAP_m(REG)$ or $pEAP_m(CF)$.

We illustrate computation in a permitting array P system with an example.

Example 2. A permitting array P system generating T shaped arrays (Figure 1) over $\{a\}$ is as follows:

$$\Pi_2 = (\{A, B, C, A', B', C', D, a\}, \{a\}, \#, [_1[_2]_2]_1,$$

$$\left\{ \begin{matrix} A\,X\,B \\ C \end{matrix} \right\}, \emptyset, R_1, R_2, 2),$$

with

$$R_1 = \{ (\# A \to A'\, a, \{B, C\}), (B\, \# \to a\, B', \{A', C\}), \left(\begin{matrix} C \\ \# \end{matrix} \to \begin{matrix} a \\ C' \end{matrix}, \{A', B'\} \right),$$

$$(A' \rightarrow A, \{B', C'\}), (B' \rightarrow B, \{A, C'\}), (C' \rightarrow C, \{A, B\}),$$
$$(C' \rightarrow D(\{in\}), \{A, B\})\}$$
$$R_2 = \{A \rightarrow a, B \rightarrow a, D \rightarrow a\}$$

A computation in Π_2 starts with the initial array $\begin{matrix} AXB \\ C \end{matrix}$ in region 1, with region 2 having no initial array. The rule $\# A \rightarrow A' a$ alone is applicable as the permitting symbols B, C are present. The application of this rule grows one step on the left, the horizontal arm. The rule $B \# \rightarrow a B'$ can now be applied as the permitting symbols A', C are present in the array. The application of this rule grows one step on the right, the horizontal arm. Likewise the rule $\begin{matrix} C \\ \# \end{matrix} \rightarrow \begin{matrix} a \\ C' \end{matrix}$ can then be applied growing the vertical arm one step down. The primed versions of the nonterminals A, B, C are changed into their original versions A, B, C due to the application of the rules $A' \rightarrow A, B' \rightarrow B, C' \rightarrow C$ with the corresponding permitting symbols being present and the process can repeat. If the rule $C' \rightarrow D$ is applied instead of $C' \rightarrow C$, then the array is sent to the inner region 2, due to the target indication *in* In region 2, the nonterminals are changed into the terminal a with the computation coming to a halt yielding a T shaped array over $\{a\}$ with equal arms which is collected in the language generated.

Note that the array P systems in both the examples 1 and 2, generate the same picture language consisting of picture arrays representing T shaped figure. But the number of membranes used is only two in example 2 where permitting symbols are used in the rules whereas the number of membranes used in example 1 is four, where the feature of permitting symbols is absent. Although not entirely unexpected, this shows the power of permitting symbols in the rules in reducing the number of membranes.

Theorem 1.

1. $pEAP_m(\alpha) \subseteq pEAP_{m+1}(\alpha), \alpha \in \{REG, CF\}$
2. $pEAP_2(REG) - EAP_2(REG) \neq \emptyset$
3. $pEAP_2(REG) - AREG \neq \emptyset$

Proof. The statement 1 is immediate from the definition of the family $pEAP_m(\alpha), \alpha \in \{REG, CF\}$.

The statement 2 can be seen as follows: The picture language L consisting of T shaped arrays with equal arms is in the family $pEAP_2(REG)$, as seen in example 2 where a permitting array P system with two membranes generates L. But application of regular array rewriting rules just alternating between two membranes can not keep generating all three arms of equal length, namely the left horizontal arm, the right horizontal arm and the vertical arm of the T shaped array, together once the derivation reaches the 'junction' in the T shaped array. Hence without the feature of permitting symbols in the rules, any basic model array P system [3] with regular array rewriting rules will require at least three membranes to generate T.

The statement 3 is due to the fact that no regular array grammar by the nature of its rules can ensure that the arms are of equal length. In fact the regular array grammar rules cannot generate two arms together.   □

The maximal mode or $t-$mode of derivation has been studied in a cooperating distributed grammar system [4] which was developed as a language-theoretic model of distributed complex systems. In [5] the $t-$ communication mode is brought into string rewriting P systems [10] thereby linking cooperating distributed string grammar systems and string-objects P systems. As a natural extension of the study in [5], Subramanian et al [17] incorporated this $t-$mode of communication into array P systems [3].

We now briefly recall a $t-$communicating array P system of type tin introduced in [17].

A $t-$communicating array P system of degree $m \geq 1$ and of type tin, $(tEAPS_m(tin,CF))$, is a construct

$$\Pi = (V,T,\#,\mu,F_1,\ldots,F_m,R_1,\ldots,R_m,i_o),$$

where the components $V,T,\#,\mu,F_1,\ldots,F_m,i_o$ are as in an array P system and the rules in the sets $R_1,\ldots,R_m$ are CF array rewriting rules of the form $\mathscr{A} \to \mathscr{B}$.

The computation is done in the usual way starting with the initial arrays (if any) in the regions. The arrays are communicated among the regions in the following manner: If an array-rewriting rule with target indication out, is applied to an array, then the resulting array is sent to its immediately direct upper region. If an array-rewriting rule has no target indication, then the array to which it is applied remains in the same region if it can be further rewritten there but if no rule can be applied to it in that region, then it is sent to the immediately direct inner region if one such region exists. In other words the $t-$mode or maximal derivation performed enforces the in target command. If the membrane is elementary, the rewritten array remains there. Note that the system does not have rules with target indication in. The result of a computation is the set of arrays over T collected in the output elementary membrane in the halting configuration.

The family of all array languages generated by a $t-$communicating array P system of type tin Π as above, with at most m membranes, with rules of type $\alpha \in \{REG,CF\}$ is denoted by $tEAP_m(tin,\alpha)$.

The problem of generation of picture arrays representing geometric figures such as solid squares over $\{a\}$ is a problem of interest in the area of picture grammars. It is known that the set S_s of all $n \times n$ $(n \geq 2)$ solid squares over a, can be generated [21] by a regular array grammar but the number of rules required is very large. In [17], a $t-$communicating array P system of type tin is given to generate it. Here we endow the rules of a $t-$communicating array P system of type tin with permitting symbols and construct such a system to generate the set S_s of solid squares of $a's$. Such a solid square of $a's$ is shown in Figure 2.

$$a\ a\ a\ a\ a\ a$$
$$a\ a\ a\ a\ a\ a$$
$$a\ a\ a\ a\ a\ a$$
$$a\ a\ a\ a\ a\ a$$
$$a\ a\ a\ a\ a\ a$$
$$a\ a\ a\ a\ a\ a$$

Figure 2: A Solid square of $a's$

A permitting $t-$communicating array P system of type tin and of degree $m \geq 1$, is a $t-$communicating array P system of degree $m \geq 1$ and of type tin [17]except that the array-rewriting rules in the regions are of the form $(\mathscr{A} \rightarrow \mathscr{B}, per)$ where $\mathscr{A} \rightarrow \mathscr{B}$ is a context-free array rewriting rule and $per \subseteq V - T$ with $V - T$ being the set of nonterminals. In a computation in the system, application of the rules to arrays in the regions is done as in a permitting array P system and communication of arrays from one region to another is done as in the $t-$communicating array P system of type tin. As usual, a successful computation is a halting computation with the arrays collected in the output membrane constituting the language generated. The family of picture array languages generated by permitting $t-$communicating array P systems of type tin is denoted by $ptEAP_m(tin,CF)$ or $ptEAP_m(tin,REG)$ depending on the array rewriting rules in the system being context-free or regular.

In [17], a $t-$communicating array P system of type tin is given to generate the set S_s of all $n \times n$ ($n \geq 2$) solid squares over a and it is known that $S_s \in tEAP_4(tin,CF)$ [17] so that the number of membranes used is four and the number of rules in all the four membranes together is 12. Here we construct a $t-$communicating array P system of type tin with permitting symbols and regular array-rewriting rules to generate the set S_s of solid squares of $a's$, which requires only two membranes.

Theorem 2. $S_s \in ptEAP_2(tin,REG)$.

Proof. To prove the theorem, we construct a permitting $t-$communicating array P system $ptEAP_m(tin,CF)$

$$\Pi_3 = (\{A,B,A',B',C,D,C',D',X,Y,a\},\{a\},\#,[_1[_2]_2]_1,$$

$$\left\{\begin{matrix} aA \\ BZ \end{matrix}\right\}, \emptyset, \emptyset, R_1, R_2, 2),$$

$$R_1 = \left\{(1)\ (A\,\#\rightarrow a\,A',\{B\}), (2)\ \left(\begin{matrix} B \\ \# \end{matrix} \rightarrow \begin{matrix} a \\ B' \end{matrix}, \{A'\}\right), (3)\ (A' \rightarrow A, \{B'\}),\right.$$

$(4)\ (B' \to B, \{A\}),\ (5)\ (B'\# \to a\,C, \{A'\}),\ (6)\ (C' \to C, \{D'\}),$

$(7)\ (D' \to D, \{C\}),\ (8)\ \begin{pmatrix} A' \\ \# \end{pmatrix} \to \begin{matrix} a \\ D \end{matrix}, \{C\}\Big),\ (9)\ (C\# \to a\,C', \{D\}),$

$(10)\ \begin{pmatrix} D \\ \# \end{pmatrix} \to \begin{matrix} a \\ D' \end{matrix}, \{C'\}\Big),\ (11)\ (C\# \to a\,X, \{D\}),\ (12)\ \begin{pmatrix} D \\ \# \end{pmatrix} \to \begin{matrix} a \\ Y \end{matrix}, \{X\}\Big),$

$(13)\ (Z\# \to a\,Z, \{X\}),\ (14)\ \begin{pmatrix} Z \\ \# \end{pmatrix} \to \begin{matrix} a \\ Z \end{matrix}, \{X\}\Big),$

$15)\ (\#Z \to Z\,a, \{X\}),\ (16)\ \begin{pmatrix} \# \\ Z \end{pmatrix} \to \begin{matrix} Z \\ a \end{matrix}, \{X\}\Big)\Big\}$

$$R_2 = \{(17)X \to a,\ (18)\,D \to a,\ (19)\,Z \to a\}$$

The computation starts with the initial array in region 1. Rules (1) to (4) enable the top border and left border to grow equally, one step at a time until the rules(5) and (8) make the top border to turn down and the left border to turn right. The rules (6), (7), (9), (10) make the bottom border and right border to grow equally, one step at a time until the rule (11) is applied which makes in a correct computation the symbols D and E to meet. This makes the rule (12) not applicable which really is an indication of a correct computation. The remaining rules (13) to (16) enable filling up the interior in rows and columns, until no more rule is applicable. The application of the rules throughout the computation is guided by the permitting symbols. Due to type *tin* of the system, the array moves to region 2 where all the nonterminals are changed into the terminal a thus yielding a solid square of $a's$ in a halting computation. Note that any incorrect sequence of application of the rules will result in the symbol Y getting stuck in the array and thus not contributing anything to the language. $\square$

Remark 1. We note that the t − *communicating* array P system with context-free array rewriting rules generating the set of solid squares of $a's$ given in [17] involves four membranes whereas two membranes are enough when the system is endowed with the additional permitting feature, with regular array rewriting rules only.

4 Conclusion

We have considered here the features of permitting symbols in the rules and t−mode of communication in the regions of an array P system and examined the generative power of such a system. It is of interest to note that array P system with t−communication and permitting symbols and regular array rewriting rules generates solid squares of $a's$. It is possible to construct permitting t−communicating array P systems, as done for solid squares, to generate picture arrays representing other kinds of geometric figures such as hollow squares, solid and hollow rectangles and so on. Comparison with the array P systems considered in [19] can also be made. Also the techniques used here can be applied to construct corresponding *P* systems for triangle-tiled pictures [18].

Acknowledgements The authors are grateful to the referees for their time spent and for their very useful comments. The first and the second authors gratefully acknowledge support from a FRGS grant No. 203/PKOMP/6711267 of the Ministry of Higher education, Malaysia.

References

1. Ceterchi, R., Gramatovici, R., Jonoska, N.: Tiling rectangular pictures with P systems. Lecture Notes in Comp. Sci. **2933**, 263-269 (2004).
2. Ceterchi, R., Gramatovici, R., Jonoska, N., Subramanian, K.G.:Tissue-like P Systems with Active Membranes for Picture Generation. Fundam. Inform. **56**, 311-328 (2003).
3. Ceterchi, R., Mutyam, M., Păun, Gh., Subramanian, K.G.: Array - rewriting P systems. Natural Computing **2**, 229-249 (2003).
4. Csuhaj-Varjú, E., Dassow, J., Kelemen, J., Păun, G.: Grammar Systems: A Grammatical Approach to Distribution and Cooperation. Gordon and Breach Science Publishers, Topics in Computer Mathematics 5, Yverdon 1994).
5. Csuhaj-Varjú, T., Vaszil, G., Păun, G.: Grammar systems versus membrane computing: The case of CD grammar systems. Fundamenta Informaticae **76**, 271-292 (2007).
6. Dassow, J., Păun, G.: Regulated Rewriting in Formal Language Theory. Springer-Verlag, Berlin (1989).
7. Ewert, S., van der Walt, A.: Random Context Picture Grammars. Publicationes Mathematicae Debrecen **54**, 763-786 (1999).
8. Ewert, S., van der Walt, A.: Generating Pictures using Random Permitting Context. Int. J. Pattern Recogn. Artificial Intell. **13** 339-355 (1999).
9. Ewert, S., van der Walt, A.: Generating Pictures using Random Forbidding Context. Int. J. Pattern Recogn. Artificial Intell. **12**, 939-950 (1998).
10. Păun, G.: Computing with membranes. Journal of Computer and System Sciences **61**, 108-143 (2000).
11. Păun, G., Rozenberg, G., Salomaa, A. (Eds.): The Oxford Handbook of Membrane Computing. Oxford University Press, Inc., New York, NY, USA (2010)
12. Rosenfeld, A.: Picture Languages - Formal Models for Picture Recognition. Academic Press, New York, (1979).
13. Rosenfeld, A., Siromoney, R.; Picture languages - a survey, Languages of design. **1**, 229 - 245 (1993).
14. Rozenberg, G., Salomaa, A. (Eds.): Handbook of Formal Languages. Vol. 1-3, Springer, Berlin (1997).
15. Salomaa, A.: Formal languages. Academic Press, London (1973).
16. Subramanian, K.G.: P systems and picture languages. Lecture Notes in Comp. Sci., **4664**, 99-109 (2007).
17. Subramanian, K.G., Ali, R.M., Nagar, A.K., Margenstern, M.: Array P systems and t-communication. Fundam. Inform. **91**, 145-159 (2009).
18. Subramanian, K.G., Geethalakshmi, M., Nagar, A.K., Lee, S.K.: Triangle-tiled picture languages, In "Progress in Combinatorial Image Analysis" (Eds. P. Wiederhold, R.P. Barneva), Research Publishing Services, 165-180 (2010).
19. Subramanian, K.G., Pan, L., Lee, S.K., Nagar, A.K.: A P system model with pure context-free rules for picture array generation. Math. Comp. Modelling **52** 1901-1909 (2010).
20. Wang, P.S.P. (ed.): Array grammars, Patterns and recognizers, Series in Computer Science, Vol. 18, World Scientific, (1989).

21. Yamamoto, Y., Morita, K., Sugata, K.: Context-sensitivity of two-dimensional regular array grammars, In "Array Grammars, Patterns and Recognizers" (P.S.-P. Wang, ed.), WSP Series in Computer Science, 18, World Scientific Publ., Singapore 17- 41 (1989).

An ACO framework for Single Track Railway Scheduling Problem

Raghavendra G. S. and Prasanna Kumar N

Abstract This work focus on application of ant algorithms to railway scheduling problem. The railway scheduling problem especially on a single track is considered to be NP hard problem with respect to number of conflicts in the schedule. The train scheduling is expected to satisfy several operational constraints, thus making the problem more complex. The ant algorithms have evolved as more suitable option to solve the NP hard problem. In this paper, we propose a mathematical model to schedule the trains that fits into ACO framework.The solution construction mechanism is inspired by orienteering problem. The proposed methodology has the capability to explore the complex search space and provides the optimal solution in reasonable amount of time. The proposed model is robust in nature and flexible enough to handle additional constraints without any modification to the model. The model assumes that set of trains will be scheduled in a zone, that covers several cities and they are optimized with respect to number of conflicts.

Keywords: Ant, Optimization, Railway, Schedules, Train.

1 Introduction

The train timetable generation is a tedious and time consuming task. Traditionally, timetable is generated manually by trial and error method based on experience and information. The advent of computer aided tools have helped the planner to come up with the effective timetable [1,2] and to access the effectiveness in terms of robustness in routing [3], revenue profitablity etc. The aim of the train scheduling problem is to come with the ideal timetable that satisfies several objectives. The objectives can be maximizing the number of passengers, minimizing the number of conflicts,

Raghavendra G. S.
BITS-Pilani K. K. Birla Goa Campus, Goa e-mail: gsr@bits-goa.ac.in

Prasanna Kumar N
BITS-Pilani K. K. Birla Goa Campus, Goa e-mail: prasannak@bits-goa.ac.in

J. C. Bansal et al. (eds.), *Proceedings of Seventh International Conference on Bio-Inspired Computing: Theories and Applications (BIC-TA 2012)*, Advances in Intelligent Systems and Computing 201, DOI: 10.1007/978-81-322-1038-2_4, © Springer India 2013

waiting time of the passangers, revenue maximization and so on. Hence, scheduling
is a multi-objective optimization problem. In addition, timetable needs to satisfy the
set of constraints that can be grouped into three categories.

1. **User Requirements** - Some of the user requirements like passengers expection
   to have trains towards particular destination at particular interval of time or trains
   can delay at most δ time units in its overall journey time.
2. **Traffic Constraints** - It can be :

   a. **Journey time** - The scheduled train t will take certain fixed amount of time to
      travel from one station to another station.
   b. **Crossing** - Two trains travelling in opposite directions (example train t from
      city i to j and train t' from city j to i) cannot occupy same track at the same
      time.
   c. **Commercial stop** - Each train is expected to stop in station for C units of
      time.
   d. **Delay for unexpected stop** - If train t stops at station/ double line j to avoid
      conflict with train t', then a time delay need to be incorporated in the schedule
      of train t to reflect the delay in the arrival time at station $j+1/j-1$.

3. **Infrastructure Constraints** - It can be :

   a. **Finite Capacity of Stations** - A train can arrive at station, if atleast one track
      is available for stoppage.
   b. **Headway Time** - If train t and t' are travelling in the same direction, then Δ
      time difference need to be maintained in terms of arrival or departure of trains.

It is possible to deduce many more constraints to reflect the realistic railway schedules. The constraints described above are generic constraints that need to be satisi-
fied by idealistic timetable. The scope of this work is to device a suitable schedule
order for a set of trains on a single track with the objective to have a minimum
number of conflicts in there journey.

2 Literature Review

2.1 Railway Scheduling

The Rail transportation planning provides rich number of problems that can be modeled and solved using optimization techniques. The problems can be classified into
two groups [4] :

1. **Train Routing Problem** - The pre planned activities need to be completed in
   the yard inorder to schedule freights and trains. The routing problem is concerned with efficient use of available resource like tracks, crews in the yard to
   complete the pre planned activities. The other important issues being addressed

in this problem category are assignment of different kind of freight cars, crew management and planning the movement of freight cars on the tracks.

2. **Train Scheduling Problem** - It is concerned about the generation of train time table or scheduling the new train between pair of cities.

The above problems itself containing lot of subproblems and a mathematical model can be developed to tackle each of these problems. The current trend involves providing an integrated solution to entire rail transport inorder to have a efficient working system. Several optimization techniques like mathematical programming, mixed integer linear programming, branch and bound have been used to solve the train problems. Although they were able to provide consistently good solutions, but need large memory space and time to compute the solution. Heuristics techniques were employed to obtain results in reasonable amount of time. These techniques uses problem's domain knowledge to arrive at the solution. [5] proposed greedy heuristic approach to resolve the conflicts. The set of rules were deviced to determine the best way to resolve the conflicts. The results obtained using heuristics often deviates more from the optimal solution. [6] proposed a heuristics based on Local Heuristics Search (LSH) that tries to resolve the conflicts using previous schedules. The LSH technique replace the existing solution with better solution by searching in the neighbourhood region. The train conflicts that happens at sidings will be shifted by one position and solution will be accepted, if it results in minimum conflict delay. [7] proposed a methodology that combines LSH and tabu search to shift more than one conflicts at a time with the intention to have a reduced total conflict delay. [8] used the co-evolutionary approach to generate the automatic timetable. [9] used genetic algorithms to solve the train timetable problem. The chromosomes were encoded using activity list representation and genes were represented as a sequence of (train,section) pair. The genes were organised to satisfy the feasibilty constraint to ensures no two trains occupy the same track section at the same time. The problem involves generating a population of chromosomes that satisfy the feasibility constraints and selecting the best chromosome that results in minimum conflict delay.

2.2 Ant Colony Optimization

The tiny creatures ants exhibits a collective behaviour to solve the day to day problems. The collective behaviour is termed as swarm intelligence. Swarm intelligence techniques focus on collective behaviour of decentralized, self-organized agents that interact with each other and with the environment. These interactions leads to the evolution of single global pattern and it is unknown to the agents. The interactions that happens at local level leads to a generic global behaviour. The foraging activity of ants exhibits the collective behaviour. Ants share the experience of food hunting by laying the pheromone trial on the travelled path and this act will help the fellow ants to search for the food. Initially, there might be multiple paths between food source and the destination nests with varying amount of pheromone trial associated with these paths. These tiny creatures have ability to sense the quantity of pheromone trial and follow the path that has a better amount of pheromone trial. This process enables them to find the shortest path between food source and the nest. The pheromone trial, a chemical substance acts as a knowledge repository for the ants in making decision and being a volatile substance evoporates over a period of time. [10] pioneered the first ant algorithm based on this collective behaviour. Ant algorithms combines the greedy mechanism and heuristic information to arrive at the optimal solution. In literature, several variants of algorithm [11, 12, 13] have been proposed and each of them tries to balance the exploration of new regions and the exploitation of good solutions in the search space. ACO algorithms have been successfully applied to other problems like Quadratic Assignment Problem[14], Vehicle Routing Problem[15], Telecommunication Networks[16] and Graph Coloring[17] to name few.

3 A Model for Train Scheduling Problem

In this section, we will discuss the generally used terminology, set of referred notations and the input requirements of the model. The model is described by objective functions and the constraints that is supposed to be satisfied.

3.1 Definitions

1. **Sidings** - An unexpected stop that occurs on partially double track section for crossing or passing of trains.
2. **Conflict Delay** - The amount of time spent by train due to sidings.
3. **Minimum Headway** - The minimum length of time seperating two trains on a single track.
4. **Train Conflict** - It occurs under two circumstances on single track line:

   a. When two trains approach each other.
   b. When a fast train catches up the slow train.

5. **Resolving a Conflict** - If two trains are involved in conflict, then one of the train must be forced for sidings so that other train can cross or pass it.
6. **line time** - The time taken by the train to cover the line.
7. **Dwell time** - The waiting time of a train in a station.
8. **Station** - A place where passengers will board or get down from the train.

The Fig. 1 shows the single track divided into lines l. The partial double lines $\{S1, S2, S3\}$ repeats at every alternate line and sidings can be next to the station. The train $t0$ can conflict with the $t1$ and conflict can be resolved by siding one of the train, so that crossing can take place. The general assumption is UP trains move from left to right direction and the DOWN trains moves from right to left direction.

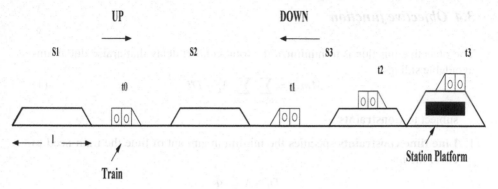

Fig. 1: Train Diagram

3.2 Notations

The following notations will be used in the model:

1. T : Set of trains $\{t_1, t_2 \ldots t_n\}$.
2. $T_{up} \subset T$: Set of $n/2$ trains moving in up directions.
3. $T_{down} \subset T$: Set of $n/2$ trians moving in down directions.
4. $T = T_{up} \cup T_{down}$ and $T_{up} \cap T_{down} = \emptyset$.
5. L : Set of lines $\{l_1, l_2 \ldots l_m\}$. Set of lines connecting a pair of stations.
6. S : Set of stations $\{ s_1, s_2 \ldots s_j \}$
   $j \ll m$ and $S \subset L$.
7. DL : Set of partial double lines. It can be obtained by
$$DL = L\%k$$
   k controls the re-occurance of partial double lines.
8. Φ : Time spent by the train to cover the line.
9. Δ : specifies the minimum headway time.

3.3 Variables

1. $A_i^{t_j}$: Represents the arrival of train t_j on line i.
2. $D_i^{t_j}$: Represents the departure of train t_j from line i.
3. $C_i^{t_j}$: Commercial stoppage or Dwell time of train t_j on line i.

3.4 Objective function

The objective function is to minimize the total conflict delay that araise due to un-
avoidable sidings.

$$Min_{cd} = \sum_{i \in T} \sum_{k \in DL} A_k^{l_i} - D_k^{l_i} \tag{1}$$

subject to **Constraints**:

1. **Line time constraints** specifies the minimum amount of time the train need to
   cover the line.

$$D_i - A_i = \Phi$$

2. **Headway constraints** specifies the minimum time difference need to be maintained between the departure of a train w.r.t arrival of another train in the same
   direction on the same line.

$$A_i^{l_m} - D_i^{l_n} = \Delta$$

3. **Train Dispatch constraint**

   a. Train scheduled in *up* direction will use the lines in the following order:

$$\{ l_i, l_{i+1}, l_{i+2}...l_k \}$$

   b. Train scheduled in *down* direction will use the lines in the following order:

$$\{ l_k, l_{k-1}, l_{k-2}...l_i \}$$

   where $i \geq 0$, $k \leq n$ and $i < k$.

4. **Stop time Constraints** specifies the minimum time, the train need to stop on a
   station line.

$$A_i^{l_m} - D_i^{l_m} = \Upsilon_i^{l_m}$$

3.5 Assumption

The following assumptions are made with regard to the model:

1. All the stations are identified w.r.t line numbers l_i and will have partial double
   line. These partial double lines are in addition to *DL*.
2. All the trains will travel with the same speed, will have same length, dwell time
   and have equal weightage, in the sense there will be only one type of train.
3. The train t_i that start its journey in *up* direction from source to destination station
   will come back again to source station in its *down* journey, but with the different
   train identifier say t_j.
4. All the trains will run on a daily basis.

5. All the lines are of equal length. The length of line will be more than the length of the train, which ensures crossing can be done without any problem.
6. A train cannot be rescheduled inorder to adjust the crossings w.r.t to unscheduled train.
7. At any time, line is occupied by only one train (To ensure operational safety).
8. The time measurement is expressed in minutes. The range of time spans from 1 - 1440 (24 X 60). We refer each minute as one time unit.

3.6 Input

The following input parameters are considered for the model:

1. Org_{t_i} = Origin station for train $t_i \in T$ expressed in terms of line number.
2. Dst_{t_i} = Destination station for train $t_i \in T$ expressed in terms of line number.
3. n = number of trains.
4. L = number of lines.
5. S = number of stations.
6. Δ = The Headway time is expressed in terms of time unit.
7. Φ = The line time is expressed in terms of time unit.
8. Υ = The Commercial waiting time in station.

4 ACO framework for Train Scheduling Problem

The proposed work is inspired by the Orienteering Problem (OP) [18]. The OP can be stated as follows: given a set of n nodes and scores for each node, the goal is to find the subset of nodes starting from vertex 1 to vertex n that maximize the total score with in the time T_{max}. The edges connecting the vertex is associated with time t and once vertex is covered it should not considered for further inclusion in journey. It should be noted that, all the vertices may not be covered due to timing constraints T_{max}. Hence, problem of finding the multiple paths from source to destination vertex with the timing constraints is a NP hard problem. The OP is comparable to Travelling Salesman Problem(TSP) with few exceptions:

- The OP involves finding a path between two distinct vertices where as TSP involves finding the tour(first and last vertex will be same).
- If T_{max} is relaxed or suffciently large, then OP reduces to generalized TSP with the possibility to cover all the vertices.

The train scheduling problem can be tranformed into OP problem , since it generalizes the TSP. We define a complete graph called Train Schedule Order Graph (TSOG) G=(V,E), where V is set of vertices $\{v_1, v_2 ... v_n\}$ and E is set of $\{e_1, e_2, ... e_m\}$. The V corresponds to set of trains and $e_i \in E$ connecting t_i with t_j depicts the schedule order (i.e., t_i followed by t_j or vice versa). Each edge is undirected, weighted,

symmetric and associated with pheromone trail that stores the goodness of selecting the edge in previous schedules. The ants will use TSOG representation to move
from one vertex to another vertex in order to construct the train schedule order. The
Fig. 2 shows the TSOG representation consisting of 5 trains. If ant has selected
the $t0$ as the starting train, then one of the possible schedule selection order can be
$t0 - t2 - t4 - t3$.

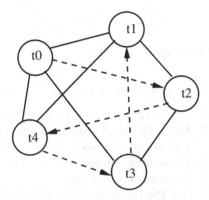

Fig. 2 Train Schedule Diagram

4.1 Tour Construction

The solution construction proceeds as follows: The edges in TSOG will be initialized with pheromone trail of quantity $\tau_0 = \frac{1}{n*\eta_0}$, where n is the number of trains
and η_0 is the smallest time difference between two scheduled trains. Initially, each
ant will be randomly assigned the starting train. The selection of next train to be
scheduled by ant k will be based on probabilistic funtion of next nearest schedule
time of the train and the amount of pheromone trial present on the connecting path.
The probabilistic function is given by the equation:

$$p_{t_i t_j}^k = \frac{[\tau_{t_i t_j}]^\alpha \cdot [\eta_{t_i t_j}]^\beta}{\sum_{k \notin tabulist} [\tau_{t_i t_j}]^\alpha \cdot [\eta_{t_i t_j}]^\beta} \tag{2}$$

$\tau_{t_i t_j}$ is the heuristic information, that specifies the amount of pheromone present
on the path between t_i and t_j and $\eta_{t_i t_j}$ is the visibilty factor, computed as $\frac{1}{D_{Org_{t_i}}^{t_i} - D_{Org_{t_j}}^{t_j}}$.

The two parameters α and β controls the importance of previous experience and the
visibilty factor. A *tabu list* is maintained for each ant k inorder to keep track of selected trains that ensures no train is selected more than once in a given iteration.
After the selection, train will be scheduled ensuring all the constraints are satisfied.
If there are n trains, then each ants will make $n-1$ train selections to come up with
a single train schedule order.

4.2 Pheromone Updation

The train during its journey experience the conflict delays in the form of crossing and passing that affects its total journey time. The individual ants construct the individual train schedule order and the selection experience can be expressed as *Total Conflict Time* (TCT). The TCT is defined as sum of all the conflict delays experienced by all the trains in a single schedule order. The total Conflict time will be used for trial reinforcement along the selection path of the trains. The trial reinforcement is done according to the following equation:

$$\tau_{t_i t_j} = \rho \cdot \tau_{t_i t_j} + \Delta \tau_{t_i t_j} \tag{3}$$

where $\rho \in [0,1]$ is pheromone coefficient such that $(1-\rho)$ represents the evoporation rate.

$$\Delta \tau_{t_i t_j} = \sum_{k=1}^{m} \Delta \tau_{t_i t_j}^k$$

where $\Delta \tau_{t_i t_j}^k$ is the amount of pheromone trial laid by the ant k on the edge(t_i, t_j) and it is given by equation.

$$\Delta \tau_{ij}^k = \begin{cases} Q/L_k & \text{if the k-th ant travel on } edge(t_i, t_j) \\ 0 & \text{otherwise} \end{cases} \tag{4}$$

where Q is a constant and L_k is the total conflict time of the k-th ant.

5 Experimental Study

5.1 Parameter Settings and Input to the Algorithm

The proposed model is extensively simulated to identify the schedule that has minimal conflict delay. The parameters relevent to ACO α, β were varied from 1 to 5, ρ was varied from 0.7 to 1.0 with incremental value of 0.5, and number of ants m were varied from 5 to 20. The parameters relevent to train scheduling problem were set as follows - number of trains $n = 20$, number of lines $L = 200$, number of stations $S = 15$, Headway time $\Delta = 6$, line time $\Phi = 3$, Commercial waiting time in station Υ = 2 and k that controls the repeatation of double line was set to 8. The line number $l = \{ 0, 17, 38, 47, 56, 72, 80, 95, 117, 126, 149, 163, 179, 190, 199 \}$, where $l_i \subseteq L$ and each l_i represents the line number associated with the station. The train origin station and destination station is expressed in line number. The train details is expressed in tuple $TP =< Orig_{t_i}, Dst_{t_i}, Dept_{Orig_{t_i}} >$, where $Orig_{t_i}, Dst_{t_i}$ represents the origin and destination station of train t_i and $Dept_{Orig_{t_i}}$ represents the departure time of train t_i from the origin station. The train details considered for the experimental purpose is $< \{38, 117, 330\}, \{95, 190, 680\}, \{56, 149, 740\}, \{126, 199, 1120\}$,

{17, 179, 445}, {72, 190, 870}, {80, 199, 1300}, {38, 126, 540}, {0, 199, 1430}, {56, 163, 990}, {117, 38, 1202}, {190, 95, 475}, {149, 56, 670}, {199, 126, 1400}, {179, 17, 800}, {190, 72, 100}, {199, 80, 1350}, {126, 38, 340}, {199, 0, 550}, {163, 56, 1190} >.

5.2 Result Analysis

The Table 1 shows the comparitive analysis for some of the ant algorithm variants available in the literature. The assessment was done with respect to TCT and better schedules will have smaller TCT. The Table 1 reports the TCT and observed parameters values for that TCT. The experiment was carried out for partial double lines and partial triple lines. For the partial double lines, best TCT is obtained for RA and for partial triple lines, ACS provides the best result. The MMAS+IB and MMAS+IB+PTS variants suffers from search stagnation and this may be due to 'limiting the pheromone strength' mechanism of algorithm. It can be concluded that limiting mechanism overcome the search stagnation for ACO algorithms that uses TSP as a benchmark program, but for OP problem, it leads to search stagnation. The result obtained by partial triple lines are better than the partial double lines demonstrating the better availability of lines that results in lesser waiting time due to sidings.

Table 1: Comparitive Analysis of Train Scheduling Problem for various Ant Variants.

Algorithms	pml=2		pml=3	
	TCT	Parameter details	TCT	Parameter details
AS	790	$\alpha=3\ \beta=3\ \rho=0.75$	757	$\alpha=4\ \beta=3\ \rho=0.75$
ACS	787	$\alpha=3\ \beta=4\ \rho=0.85$	**716**	$\alpha=4\ \beta=1\ \rho=0.85$
EA	792	$\alpha=4\ \beta=1\ \rho=0.75$	751	$\alpha=5\ \beta=1\ \rho=0.7$
RA	**783**	$\alpha=4\ \beta=2\ \rho=0.85$	722	$\alpha=3\ \beta=1\ \rho=0.9$
MMAS+IB	1023	$\alpha=1\ \beta=4\ \rho=0.70$	1025	$\alpha=2\ \beta=1\ \rho=0.8$
MMAS+IB+PTS	1023	$\alpha=1\ \beta=4\ \rho=0.85$	1025	$\alpha=4\ \beta=3\ \rho=0.9$

The Fig. 3 shows the comparitive results for partial double lines. The RA variant was analyzed, as it provides the best results compared to other variants. It can be observed from Fig 3(a) that better performance was obtained for smaller ant population and deterioates with the increase in number of ants. The best solution was obtained for $n=5$. The Fig 3(b) shows the variation in pheromone trial strength and the better result was obtained for trial strength of 0.8. Similarly, Fig 4 shows the comparitive results for partial triple lines. It can be observed from Fig 4(a) that, as the number of ants increases performance improvises and better results were obtained, when number of ants are around 15. Similarly, Fig 4(b) reveals that relatively better result were obtained for lower pheromone trial than the higher pheromone concentration and optimal result was obtained for trial strength of 0.85.

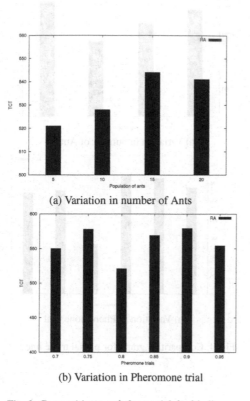

(a) Variation in number of Ants

(b) Variation in Pheromone trial

Fig. 3: Comparitive graph for partial double lines.

6 Conclusion and future direction

In this paper, we have presented the train scheduling model that fits into ACO framework. We have demonstrated that OP can be transformed into single track scheduling problem. The distributed approach followed by the ACO algorithm results in obtaining the optimal schedule order of train. The model can be made more realistic by adding few more constraints like specifying the time bound for the departure of a train to a particular destination, scheduling different type of trains that travels with different speed etc to name few. The future direction includes extending the model for multi track scheduling, incorporation of local search to improvize the solution and to compare the performance with other variants of ant algorithm.

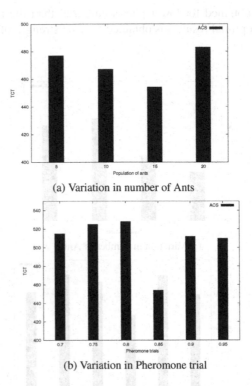

(a) Variation in number of Ants

(b) Variation in Pheromone trial

Fig. 4: Comparitive graph for partial triple lines.

References

1. Caprara, A.M., Kroon, L.G., Monaci, M., Peeters, M. & Toth, P. (2007). Passenger railway optimization. Transportation Science. Elservier, 14:129-187.
2. Kroon, L., Huisman, D., Abbink, E., Fioole, P., Fischetti, M., Maroti, G., Schrijver, A., Steenbeek, A. & Ybema, R. (2009). The new Dutch timetable: the OR revolution, Interfaces, 39(1): 6-17.
3. Zwaneveld, P.J., Kroon, L.G. & van Hoesel, S.P.M. (2001). Routing trains through a railway station based on a node packing model. European Journal of Operational Research, 128:14-33.
4. Cordeau, J., Toth, P. & Vigo, D. (1998). A survey of optimization models for train routing and scheduling. Transportation Science, Elservier, 32(4):380-404.
5. Cai, X. & Goh, C. J. (1994) A fast heuristic for the train scheduling problem, Computers and Operation Research, 21(5):499-510.
6. Kraay, D., Harker, P. & Chen, B. (1991), Optimal pacing of trains in freight railroads, Operation Research, 39(1):82-99.
7. Higgins, A., Kozan, E. & Ferreira, L. (1997), Heuristic techniques for single line train scheduling, Journal of Heuristics, 3:43-62.
8. Kwan, R. S. K. & Mistry, P (2003), A Co-evolutionary Algorithm for Train Timetabling, Research Report Series.13, School of Computing, University of Leeds, Leeds.

9. Tormos, P., Lova, A., Barber, F., Ingolotti, L., Abril, M.,& Salido, M. A. (2008), A Genetic Algorithm for Railway Scheduling Problems, Studies in Computational Intelligence, 128:255-276.
10. Dorigo, M., Maniezzo, V. & Colorni, A. (1996), Ant System: Optimization by a colony of cooperating agents. IEEE Transaction on Systems, Man and Cybernetics 26(1):29-41.
11. Bullnheimer, Hartl, R.F., & Strauss, C. (1999), A new rank based version of the Ant System: A computational study. Central European Journal for Operation Research and Economics 7(1):25-38
12. Blum, C., Roli, A. & Dorigo, M. (2004), HC-ACO: The Hypercube framework for Ant Colony Optimization. IEEE Transaction on Systems, Man and Cybernetics 34(2):1161-1172.
13. Stutzle, T. & Hoos, H.H. (2000), MAX - MIN Ant System. Future Generation Computer System (16)8:889-914.
14. Maniezzo, V. & Colorni, A. (1999), The Ant System Applied to the Quadratic Assignment Problem, IEEE Transactions on Knowledge and Data Engineering, 11(5):769-778.
15. Bullnheimer, B., Hartl, R.F. & Strauss, C. (1999), Applying the Ant System to the Vehicle Routing Problem, in S. Voss, S. Martello, I. H. Osman, and C. Roucairol (eds.), MetaHeuristics: Advances and Trends in Local Search Paradigms for Optimization, Kluwer, pp. 285-296.
16. Di Caro, G., & Dorigo, M. (1998), Ant Colonies for Adaptive Routing in Packet-Switched Communication Networks, Proceedings of the 5th International Conference on Parallel Problem Solving from Nature (PPSN V), pp. 673-682.
17. Costa, D. & Hertz, A. (1997), Ants Can Colour Graphs, Journal of the Operational Research Society, 48, pp. 295-305.
18. Vansteenwegen, P., Wouter, S. & Oudheusden, D.V. (2011), The Orienteering Problem: A Survey, European Journal of Operational Research, 209, pp. 1-10.

Bio-Inspired Soft-Computational Framework for Speech and Image Application

Dipjyoti Sarma and Kandarpa Kumar Sarma

Abstract Artificial Neural Network (ANN) based recognition systems show dependence on data and hardware for achieving better performance. The work here describes the use of DSP processors to design a bio-inspired soft-computational framework with which processing of speech and image inputs are carried out. Certain nonlinear activation function for implementation in DSP processor framework is also designed and configured appropriately to train a soft-computational tool like ANN. The results derived show that the capability of the ANN improves with the derived DSP processor framework. Its performance is further enhanced using the approximation of *tan-sigmoidal* nonlinear activation function. In terms of computational capability, the proposed approach shows around 12% improvement compared to a conventional framework. Similarly, improvement in recognition rate is around 4% with applications involving speech and image samples.

Keywords: ANN, Bio-inspired, Recognition.

1 Introduction

Application of digital signal processing and certain bio-inspired soft-computing tools such as Artificial neural Network (ANN) on speech and image signals demands high computing requirement. Computation of ANN resembles brain. As in brain, the ANN also employs many computational elements that works concurrently and finally achieves a brain like structure. ANN that performs speech recognition and synthesis, or pattern classification consist of large number of neurons and inputs. Every neuron computes a weighted sum of its inputs and applies a nonlinear

Dipjyoti Sarma
Department of ECT, Gauhati University, e-mail: dipsarma4u@gmail.com

Kandarpa Kumar Sarma
Department of ECT, Gauhati University, e-mail: kandarpaks@gmail.com

J. C. Bansal et al. (eds.), *Proceedings of Seventh International Conference on Bio-Inspired Computing: Theories and Applications (BIC-TA 2012)*, Advances in Intelligent Systems and Computing 201, DOI: 10.1007/978-81-322-1038-2_5, © Springer India 2013

function to its result [1]. The ANN recognizes patterns based on information and weights during training. However, the use of ANN classifier remains constrained due to the availability of powerful hardware to provide sufficient speed during training. The basic operation performed by a neuron during classification can be written as,

$$Y = f\sum_i x_i * w_i + b \qquad (1)$$

Thus for each classification the network must perform one multiplication and one addition for every connection which translates to a few billion multiply add operations per second. Only parallel implementations, in which several connections are evaluated concurrently, achieve such computational power [2]. General-purpose personal computers (GPPC) and workstations are the most popular computing platforms used by researchers to simulate ANN algorithms. They provide a convenient and flexible programming environment and technology advances have been rapidly increasing their performance and reducing their cost [3]. But ANN simulations for image and speech signals can still overwhelm the capabilities of even the most powerful GPPC.

Although, the use of super computer reduces the required CPU time however this is not a clever solution as it is expensive. A convenient solution of this constraining solution is derived using DSP processors. These are design wise parallel processing blocks with a host of features which make them suitable for real time signal processing.Bio-inspired processing shows all features of an advanced form of real-time signal processing with supportive cognitive capabilities. Therefore, any bio-inspired system design must combine real-time computation with cognition.

Such a setup designed using DSP processors has been proposed here. We specially focusses the implementation of certain ANN based applications involving image and speech inputs. The proposed architecture shows distinct advantage in terms of processing power and cognitive capability as compared to conventional approach of implementing a soft-computational framework like ANN for image and speech applications. In terms of computational capability, the proposed approach shows around 12% improvement compared to a conventional framework. Similarly, improvements in recognition rate is around 4% with applications involving speech and image samples.

This paper focuses on the design of a bio-inspired soft computational framework using DSP processors. The DSP processor's high throughput characteristic and capability of executing million instructions per second provides better computational result as the ANN by deign wise provides a parallel architecture. A prototype of the work is also reported in [4] using parallel processing. The role played by parallel computing environment in increasing the processing performance of real time applications involving speech and image processing is shown here. It provides certain insights into bio-inspired system design. Experimental results show that multicore CPU arrangement helps ANN to learn applied patterns better.

Section 1 provides a brief introduction of the bio-inspired tools and related things. In Section 2, certain important features of the DSP processors for Bio-Inspired design are discussed. A brief introduction to TMS320C6713 is also provided in this section. The system model of the work and experimental steps in detail are discussed in Section 3. Results of the experiments are provided in Section 4. Finally the work is concluded in Section 5.

2 Key Features of the DSP Processor for Bio-Inspired Design

The features like speed, cost-effectiveness, reprogram ability in the field, energy efficiency etc have made the DSP Processor suitable and advantageous for application in bio-inspired soft-computing tool design.

DSP's differ from ordinary microprocessors in that they are specifically designed to rapidly perform the sum of products operation required in many discrete-time signal processing algorithms. They contain parallel multipliers, and functions implemented by microcode in ordinary microprocessors are implemented by high speed hardware in DSP's. Since they do not have to perform some of the functions of a high end microprocessor like an Intel Pentium, a DSP can be streamlined to have a smaller size, use less power, and have a lower cost [5]. Most of these processors share various common features so as to support the high performance, repetitive, numeric intensive tasks and lowering the computational complexity. Some of the key advantages are specialize CPU architecture, Multiply and Accumulate units (MACs) and Multiple Execution Units, Efficient Memory Access, Circular Buffering, Dedicated Address Generation Unit, Specialized Instruction Sets etc.

As shown in eq. 1 the multiplication and addition of ANN is mostly performed by the MAC operation of DSP processor. The basic DSP arithmetic processing blocks are registers, multipliers, Arithmetic Logic Units (ALUs), shifters which work in parallel during the same clock cycle and thus optimizing MAC as well as other arithmetic operations for faster computation.

TMS320C6713 DSP Processor has been used here. This is a floating point DSP Processor of TMS320C6x(C6x) family manufactured by Texas Instruments (TI). During the training phase of an ANN, the resultant output of the nodes and the adaptively measured weights are generally floating point values. So the floating point processor, in this case, provides more reliable and precise results compared to a fixed point processor.

DSP processors such as the TMS320C6x (C6x) family of processors are Fast special-purpose microprocessors with a specialized type of architecture and an instruction set appropriate for signal processing. The C6x notation is used to designate a member of Texas Instruments (TI) TMS320C6000 family of DSP processors. The architecture of the C6x digital signal processor considered suitable for digital signal processing. Based on a very long instruction word (VLIW) architecture, the C6x is considered to be one of the TIs most powerful processor [6]. The TMS320C6713 DSK which has been used during the experimental work contains the TMS320C6713 digital signal Processor. TMS320C6713 is a high performance floating point DSP, its working frequency up to 225 MHz, the single instruction execution cycle is only 5 ns, with a strong fixed-point floating-point computing power generates a computational speed of up to 1.3 GFLOPS. TMS320C6713 processor consists of three main components: CPU core, memory and peripherals. The CPU contains eight functional units that can operate in parallel, has two sets of registers, address are 32 bit wide. On-chip program memory bus has a width of 256 bit. Peripherals including the expansion of the direct memory access (EDMA), low-power, external memory interface (EMIF), serial port, McBSP Interface, IIC interfaces and timers. The C6713 DSK is a low-cost standalone development platform that enables users to evaluate and develop applications for the TI C67xx DSP family. The figure 1 shows the functional block diagram of the DSK. It also serves as a hardware reference design for the TMS320C6713 DSP. Schematics, logic equations and application notes are available to ease hardware development and reduce time to market.

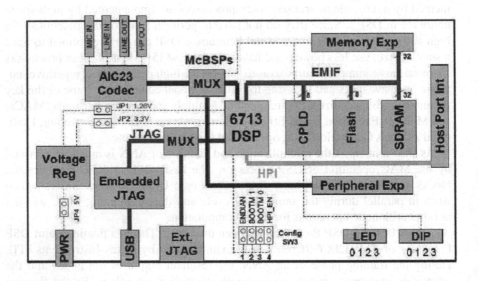

Fig. 1 Functional Block Diagram Of TMS320C6713 DSK.

3 System Model and Experimental Details

The ANN implementation to speech and image data is carried out using back propagation feed forward ANN algorithm. The samples, generated from different sources contain speech extracts and face captures. Some of the samples are mixed with noise. The sample sets thus generated consists of a sizeable number of data for use with the proposed system. Of these about 25% are categorized as training set, another 25% for validation and the rest taken for testing of the recognizer. Set of speech samples are recorded with variable sampling rates between 8 Kbps to 16 Kbps. The soft computational framework is designed with the DSP Processor and its performance is compared with INTEL Dual Core Processor. Figure 2 shows the process logic of the framework.

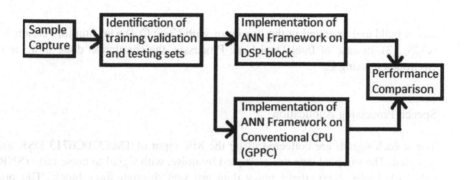

Fig. 2 Process Logic Diagram.

Table 1 Configuration of Intel duel core TMS320c6713.

Parameter	Intel duel core	TMS320c6713
Frequency	2 GHz	225 MHz
Memory	2 GB	256 KB

The configuration of both the processors are shown in table 1. The preprocessed image and speech data are applied to ANN, which is simulated in both the TMS320C6713 DSP Processor and Intel Dual Core Processor. The ANN is made as per the configuration as shown in table 2. Although the experiment is done using several sets of number of input neurons such as 10, 25, 50, 100, 150, and 200 however 50 input neurons is chosen as optimum value with a view to the memory available and the numbers of epochs required for training. The number of output number is 4 as we have recognized 4 different patterns.

The same ANN is also simulated with another set of transfer functions, where the *tanh* like non linear transfer function is used in hidden layer instead of *log sigmoid*. The simulation in TMS320C6713 DSP Processor is done using C language

Table 2 Configuration of Artificial Neural Network.

Parameter	Value
No of Input Neuron	50
No of Hidden Neuron	Varied between .5 times to 2.5 times of the No of input neurons
No of Output Neuron	4
No of Patterns	4
Transfer Functions	log sigmoid, log sigmoid, log sigmoid
Learning Rate	.5

and is build and executed in the processor with Code Compose Studio, version 3.3 (CCS 3.3). In case of Intel Dual Core Processor, the simulation done in a Linux environment using C.

Speech Processing Application

The speech signals are collected using the Mic input of TMS320C6713 DSK and sampled. The sampled data are corrupted by noise, with signal to noise ratio (SNR) value of $\pm 5dB$. Next, these noisy data are sent through filter block. The pre-emphasis filter is a digital filter, designed with adjusting components changing the filter coefficients so as to update the frequency characteristics [7]. For the speech signal, sets a transposed equiripple FIR filtering is used. Several filter structers are designed using TMS320C6713, transposed equiripple FIR structure are found to provide the minimum mean square error (MSE) and also less processing time [8].

Image Processing Application

Ten numbers of images are taken using the web cam of computer and each of clean and noise corrupted images are used for the experiment. Various preprocessing steps are implemented and then the image data are used for ANN training.

4 Experimental Results

For various number of hidden neurons, the number of epochs and processing time with TMS320c6713 and Intel Dual Core are shown in tables 3 and 4 using patterns as speech data and image data respectively.

As shown in the tables 3 and 4, for both the speech and image data the number of epochs and hence the processing time required to meet the MSE of 1×10^{-3} is less for TMS320C6713, compared to the INTEL Dual Core processor. For number of hidden layer neurons equal to ($2\times$ input layer) neurons the processing speed

Table 3 Comparison of Epochs and processing time for various number of hidden neurons with TMS320c6713 and Intel Dual Core using Speech data pattern.

Number of Hidden Layer	Number of Epochs required using	Number of Epochs required using	Total time required (In Seconds)	Total time required (In Seconds)
Neurons	TMS320C6713	INTEL Dual Core	TMS320C6713	INTEL Dual Core
25	492	612	3.67	4.07
50	417	506	2.88	3.71
75	276	390	1.73	2.66
100	205	314	1.34	2.10
125	360	522	2.27	3.43

Table 4 Comparison of Epochs and processing time for various number of hidden neurons with TMS320c6713 and Intel Dual Core using Image data pattern.

Number of Hidden Layer	Number of Epochs required using	Number of Epochs required using	Total time required (In Seconds)	Total time required (In Seconds)
Neurons	TMS320C6713	INTEL Dual Core	TMS320C6713	INTEL Dual Core
25	549	719	3.45	3.91
50	361	594	2.19	3.28
75	185	403	1.30	2.82
100	148	336	1.05	2.41
125	395	458	2.74	3.67

performance is found to better compared to other cases of hidden layer neurons. In this case, compared to INTEL Dual Core processor, the TMS320C6713 processor provides improvement in processing speed efficiency of around 43% and 37% for image data and speech data respectively. However, if we compare the processing time between the number of hidden layer neurons equal to ($1.5\times$ input layer) neurons and ($2.5\times$ input layer) neurons with reference to ($2\times$ input layer) neurons, we see that the processing time for ($2.5\times$ input layer) neurons is much higher than ($1.5\times$ input layer) neurons. So this can be concluded here that for number of hidden layer neurons between ($1.5\times$ input layer) neurons to around ($2\times$ input layer) neurons the soft-computational framework will have faster processing.

The figure 3, shows the plot of MSE versus number of Epochs. It clearly dictates that TMS320C6713 process the ANN faster than the INTEL dual Core processor. A plot of number of hidden layer neurons versus number of average epochs for both image and speech data required to have MSE of 10^{-3} is shown in figure 4.

Next we have processed the ANN for 2 second with both speech and image data. Table 5 and 6 shows this performance. It is observed from the tables 5 and 6, that with varying the number of hidden layer neurons apart from the variation in MSE and there is a difference in the recognition efficiency shown by the framework

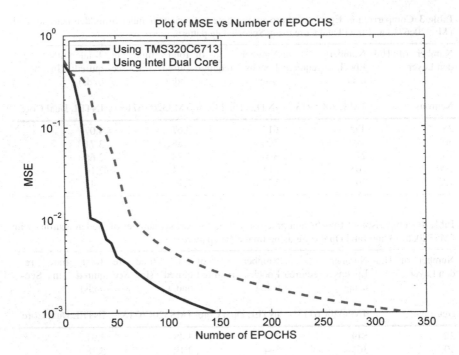

Fig. 3 Plot of MSE vs Number of Epochs for TMS320c6713 and Intel Dual Core.

formed by the processors. The recognition efficiency is better using TMS320C6713 DSP Processor compared to INTEL Dual Core Processor. The recognition efficiency difference in percentage between the processors is shown in the tables 5 and 6. The recognition efficiency found to increase by around 4% and 5% with TMS320C613 compared to INTEL Dual Core for 100 neurons in the hidden layer for image and speech data respectively.

Table 5 Performance of TMS320C613 and Intel Dual Core processor for 2 seconds with speech data.

Number of Hidden Layer Neurons	MSE attained using TMS320C6713	MSE attained using INTEL Dual Core	% Difference in Recognition Efficiency between TMS320C6713 and INTEL Dual Core processor
25	.103	.5549	1 %
50	.0621	.2811	3 %
75	.0006	.0239	5 %
100	.0002	.0077	5 %
125	.0337	.0926	2 %

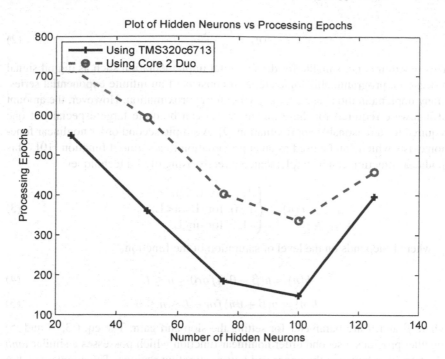

Fig. 4 Plot of Hidden Neurons vs processing Epochs for TMS320c6713 and Intel Dual Core

Table 6 Performance of TMS320C613 and Intel Dual Core processor for 2 seconds with image data.

Number of Hidden Layer Neurons	MSE attained using TMS320C6713	MSE attained using INTEL Dual Core	% Difference in Recognition Efficiency between TMS320C6713 and INTEL Dual Core processor
25	.0744	.0880	0 %
50	.0031	.0097	5 %
75	.0007	.0045	2 %
100	.0003	.0029	4 %
125	.0319	.0612	2 %

4.1 Evaluation of the Design and implementation of tanh activation function

The hyperbolic tangent (tanh) sigmoid function is a popular and one of the most frequently used activation function in backpropagation ANN applications. This activation function is (referred to as *"tansig"* in Matlab) provides at the output of a neuron, a non linear function that has tanh like transition between the lower and upper saturation regions and is given by eq. 2

$$f(n) = \frac{e^n - e^{-n}}{e^n + e^{-n}} \tag{2}$$

This function is not suitable for direct digital implementation such as digital signal processors, programmable logic etc, as it consists of an infinite exponential series. Many implementations use a lookup table for approximation. However, the amount of hardware required for these lookup tables can be quite large especially if one required for a reasonable approximation [9]. A simple second order nonlinear function exists which can be used as an approximation to a sigmoid function [10]. This nonlinear function can be implemented directly using digital techniques.

$$f(n) = \begin{cases} 1, & \text{for } L \leq n; \\ h(n), & \text{for } -L < n < L; \\ -1, & \text{for } -n \leq L. \end{cases} \tag{3}$$

where L depends on the level of saturation of the function.

$$h(n) = n(\beta - \theta n) \, for 0 \leq n \leq L \tag{4}$$

$$h(n) = n(\beta + \theta n) \, for -L \leq n \leq 0 \tag{5}$$

where β and θ are parameter for setting the slop and gain. The eq. (3, 4 and 5) together provides a second order nonlinear function which possesses a similar *tanh* like transition between the upper and lower saturation regions. The comparison between the sigmoid defined by eq. 2 and the hardware approximation defined by eq. (3, 4 and 5)is shown in figure 5. Tables 7 and 8 shows the comparison of epochs and processing time for various number of hidden neurons applying this activation function at the output of hidden layer to train the ANN using TMS320C6713 using speech and image data respectively.

Table 7 Comparison of Epochs and processing time for various number of hidden layer neurons using approximated transfer function of figure 1 at hidden layer output with speech data.

Number of Hidden Layer Neurons	Number of Epochs required	Total time required (In Seconds)
25	488	3.61
50	410	2.82
75	272	1.61
100	201	1.27
125	357	2.23

Comparing table (7 and 3) and table (8 and 4), we see that application of approximated transfer function of figure 1 decreases the number of epochs required and increases the processing speed. Thus the formulated DSP based framework for designing a bio-inspired soft-computation tool not only improves computational capability in terms of less number of processing cycles but also enhances the recog-

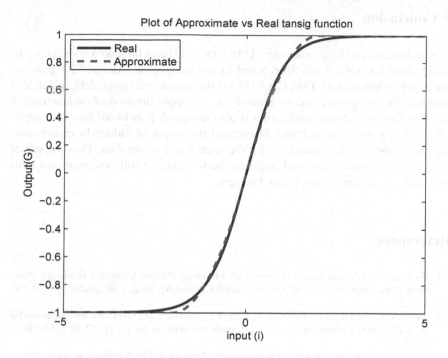

Fig. 5 Plot of approximation versus real *tansig* function.

Table 8 Comparison of Epochs and processing time for various number of hidden layer neurons using approximated transfer function of figure 1 at hidden layer output with image data.

Number of Hidden Layer Neurons	Number of Epochs required	Total time required (In Seconds)
25	541	3.41
50	357	2.11
75	179	1.23
100	141	.971
125	387	2.67

nition accuracy in both speech and image samples. This is obviously due to the support obtained from a specialized hardware like the DSP processor. It is thus obvious that use of specialized hardware framework improves performance of soft-computational tools and facilitate design of bio-inspired system.

Here, MLP has been used which is a feed forward ANN. It is trained using back propagation(BP) algorithm. Replacement training algorithms can be BP with adaptive learning (ALR) algorithm, BP with LevenbergMarquardt (LM) optimization etc. If MLP is replaced by recurrent ANNs, it will provide better computational efficiency. These are some likely future directions of the work.

5 Conclusion

The performance of DSP processor (TMS320C6713) and Intel duel core processor is examined for ANN based Speech and Image recognition. Almost 12% processing gain is achieved with TMS320c6713 for the speech and image data, which also increases the recognition rate by around 5%. An approximation of *tanh* activation function for direct digital application is also designed. It is found that with application of this approximated *tanh* function at the output of hidden layer increases the processing gain by around 5% for the speech and image data. Thus the use of specialized hardware framework improves performance of soft-computational tools and facilitates design of bio-inspired system.

References

1. Seiffert, U.: Artificial Neural Networks on Massively Parallel Computer Hardware, European Symposium on Artificial Neural Networks (ESANN), Bruges (Belgium), pp. 319-330, (2002).
2. Boser, B.E., Sackinger,E., Bromley, J., Cun, Y.L., Jackel, L.D.: Hardware Requirements for Neural Network Pattern Clasifiers, A Case Study and Implementation. pp. 32-40. IEEE Micro, (1992).
3. Asanovic. K.: Programmable Neurocomputing. Appears in *The Handbook of Brain Theory and Neural Networks*, 2nd edition, M.A. Arbib, Ed., Cambridge, MA: The MIT Press, (2002). Available via *www.eecs.berkeley.edu/ krste/papers/neurocomputing.pdf*.
4. Sarma, D., Sarma, K., K.: Multicore Parallel Processing Architecture for ANN based Speech and Image Processing Applications, accepted for publication in *Journal of Instrument Society of India*, IISC, Bangalore, India. (In Press). (2012).
5. Tretter, S.A.,: *Communication System Design Using DSP Algorithms, with Laboratory Experiments for the TMS320C6713 DSK*, Springer, (2008).
6. Nadiminti, K., Dias de Assunao. M., Buyya, R.: *Distributed Systems and Recent Innovations: Challenges and Benefits.*, available in, *www.cloudbus.org/papers/InfoNet-Article06.pdf*.
7. Hisashi, K., Mano, F., T.: *Patent application, Title: Filter Circuit*, mi.eng.cam.ac.uk / ajr / SA95/ node43.html
8. Sarma, D., Sarma, K., K.: Real Time Pre-Processing Filter Design for a Speech Processing System using TMS320C6713. IICAI, pp. 982-993, (2011)
9. Khalil. R. A.:Hardware Implementation of Backpropagation Neural Networks on Field programmable Gate Array (FPGA), *Al-Rafidain Engineering*, Vol.16, No.3, (2008).
10. Kwan , H.K. : Simple sigmoid like activation function suitable for digital hardware implementation. *Electronic Letters* Vol. 28 , pp. 1379 1380, (1992).

Leukocyte Classification in Skin Tissue Images

Mukesh Saraswat and K. V. Arya

Abstract : Automated leukocyte classification can assist histopathologist for quantifying inflammatory cells in microscopic images. Most of the work for classification of leukocytes have been done on blood smear or immunohistochemically (IHC) stained or immunofluroscence (IF) stained tissue section images. But rare work have been initiated till date to automate identification of inflammatory cells in the tissue section images stained with routinely used Hematoxylin & Eosin (H&E) staining. This is due to the coarse background and availability of different artifacts in the tissue section images. Therefore, in this paper, an automated method for classification of inflammatory cells into monomorphonuclear cells and polymorphonuclear cells for H&E stained skin tissue section images has been presented.

Key words: Leukocytes, Cell Segmentation, Feature Extraction, Cell Classification

1 Introduction

Inflammation is a complex protective reaction which responds to infection, irritation, injury, burns, wound etc. and it is characterized by pain, redness, swelling, and loss of function [1]. Inflammation process destroys, dilutes or walled-off the injurious agents. For many diseases, body tissues are incapable to fight with the cause and require drug supplementation. Discovery of drugs undergo preclinical screening by pathologist with microscope on laboratory animals before their use for humans. The quantification of inflammatory cells, also known as leukocytes, can give important

Mukesh Saraswat

ABV- Indian Institute of Information Technology & Management, Gwalior, India.
e-mail: saraswatmukesh@gmail.com

K. V. Arya

ABV- Indian Institute of Information Technology & Management, Gwalior, India.
e-mail: kvarya@iiitm.ac.in

J. C. Bansal et al. (eds.), *Proceedings of Seventh International Conference on Bio-Inspired Computing: Theories and Applications (BIC-TA 2012)*, Advances in Intelligent Systems and Computing 201, DOI: 10.1007/978-81-322-1038-2_6, © Springer India 2013

information about the efficacy of the discovered drug. The inflammatory cells can be categorized into monomorphonuclear cells and polymorphonuclear cells based on the number of nuclei present in the cell [1]. Polymorphonuclear cells consist of number of lobes while monomorphonuclear cells have only one nucleus.

The inflammatory cells are counted manually by histopathologist using microscope at 40x or higher magnification which is a time consuming process. Further, this observation is always biased in nature and depends highly on the experience and knowledge of the histopathologist. Therefore, to reduce the human workload and individual biasness, automation of the inflammatory cells is required. A number of successful algorithms have been designed for identification of leukocytes in blood smear images as background of these images are regular [2, 3, 4, 5, 6]. For the automated recognition of these cells in tissue section images, immunohistochemically (IHC) staining or immunofluorescence (IF) staining are used which are costly and generally used for specific pattern recognition such as cancer cell identification. Further, reported work, related to cell identification, used the images which were acquired at 100x or higher magnification. Rare work have been reported for identification of leukocytes in tissue section images stained with routinely used H&E staining and acquired at lower magnification due to their complex background along with presence of different noisy elements (artifacts).

Therefore, this paper introduces a method for classification of already cropped leukocytes into monomorphonuclear cells and polymorphonuclear cells available in the H&E stained tissue section images which are acquired at 40x magnification. Rest of the paper is organized as follows: A review of number of techniques on quantitative estimation of inflammation cells are discussed in Section 2. Proposed method is explained in Section 3. Experimental results are discussed in Section 4 and Section 5 concludes the paper.

2 Related Work

Automated classification of leukocytes into polymorphonuclear cell and monomorphonuclear cell is a four steps process; (i) normalization of the images, (ii) leukocyte segmentation, (iii) feature extraction, and (iv) cell classification. Chan et al. [7] proposed an automatic nuclei segmentation method and developed the method of counting the number of lobes in a cell nuclei. But in their work, only location of the lobes of particular leukocyte is undertaking, no counting of cells is given. Forero et al. [8] presented an image processing method to count automatically the number of mitotic glial cells labeled with anti-phospho-histone H3 and glial cells labeled with anti-Repo in Drosophila embryos. But they used immunohistochemistry to localize the mitotic glial cells by color detection only.

Ulrich et al. [9] combined whole slide scanning technology with object orientated image analysis to characterize inflammatory cells of lungs in mouse model for chronic asthma. But they also used immunohistochemistry to localize the cells. Chubb et al. [10] make a study on "BioVision", that can be trained quickly and effectively to classify and quantify user definable histological objects within single or double-labeled immunocytochemically stained sections. De Boer et al. [11] compared fully automated digital image analysis with interactive digital cell counting and semi-quantitative scoring of cytokine expression in 52 patients with mild to moderate atopic asthma. Immunohistochemistry was done to localize the cells.

From the previous research, it is found out that obtaining quantitative data from histological sections represents a tricky challenge. Rare efforts have been done to automate the classification and quantification of histological structures in various disease models. Most of the image analysis softwares for leukocyte classification on tissue section images were developed on immunohistochemically (IHC) staining or immunofluroscence (IF) staining which makes quantification easier due to monochromatic target specific staining. Therefore, in this paper, a method to classify the inflammatory cells into monomorphonuclear cells and polymorphonuclear cells in the images of inflamed mouse skin section stained using H&E staining has been proposed.

3 Proposed Method

The proposed method as shown in the Figure 1 consists of three steps: (i) cell segmentation, (ii) feature extraction, and (iii) cell classification. Each step is described in the following sections:

3.1 Cell Segmentation

In this paper, the cell locations are found out with the help of trained pathologist and these cell locations are cropped after preprocessing the input image. Image preprocessing consists of normalization followed by statistical threshold. Since the taken images are H&E stained skin tissue images, the hue of the nuclei appears bluish type as compare to other components. Therefore, the red component of the *RGB* images is taken to segment the nuclei as its value is lower than other components. For normalizing the illumination and staining variations of the input images, a reference image is taken with perfect illumination and staining properties and all the images are converted as per the histogram distribution of reference image using histogram specification [12]. Then, a threshold value is selected using statistical threshold to separate out the nuclei of the cells. The threshold value can be calculated for each image using mean & standard deviation of the histogram distribution and the distribution of reference image. It is experimentally found that best results are received for the following threshold value:

$$T_{nucleus} = \mu - 4 * \sigma \qquad (1)$$

where T is the threshold value, μ is mean value of the histogram, and σ is standard deviation. The Eq. 1 gives a binary image having nuclei pixels (Nuclei image). Since in this paper, leukocytes are identified using nuclei structure, sub-images are cropped from the nuclei image using the location information of the cells as given by expert pathologist. This process gives the set of sub-images each consisting of all the nuclei of one cell. These sub-images are used in further processing steps. The size of each cropped sub-image is taken as $N \times N$.

3.2 Feature Extraction and Cell Classification

For the feature extraction process, principal component analysis (PCA) [13] has been used. A set of M cropped cells from one class are selected randomly to generate the eigen cell similar to eigenface, proposed by Turk et al. [14]. But this process has a limitation that all the cells of one type must be aligned to one direction which can

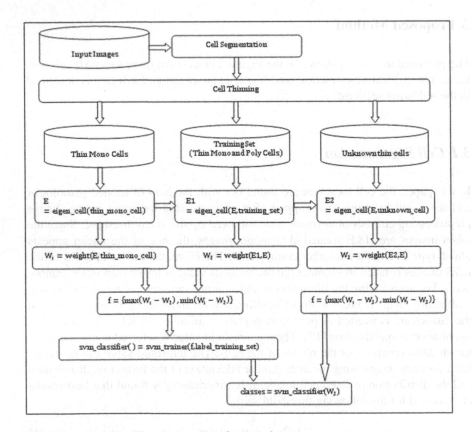

Fig. 1 Flow chart describing the overview of proposed system.

be achieved by rotation method. To overcome this limitation, all cropped nuclei are thinned followed by dilation operation. Thinning operation converts all monomorphonuclear cells into almost similar structure in comparison to polymorphonuclear cells whose structure are always random. The features of known monomorphonuclear cells will be treated as basis for classification of cells.

Each M thinned monomorphonuclear cell is converted to eigen cell and its weight is calculated. A data set consisting of monomorphonuclear and polymorphonuclear cells are collected for training and their thinned eigen cells using above mentioned M eigen cells are calculated followed by their weights. The Euclidian distance between these weights of the cells from training data set and previously calculated weights of monomorphonuclear cells are found out and a maximum and minimum values are taken as feature vector. This feature vector along with label of each cell is given to Support Vector Machine (SVM) trainer which generates the classifier. The output function of previous step is served as classifier for unknown cells. The unknown cells are processed for thinned eigen cells generation using base eigen cells and their weights are calculated. Then euclidian distance between these weights of the cells and weights of monomorphonuclear cells are find out and a maximum and minimum values are taken as feature vector to SVM classifier [15].

4 Experimental Results

To measure the performance of the proposed method, the experiments are carried out on 30 microscopic digital images of mouse skin sections stained with H&E staining and acquired at 40x magnification. All the images were acquired using DC500 camera (Leica, Germany) attached with DMLB microscope (Leica, Germany). These images were taken from archived animal studies that have a prior approval from the Institutional Animal Ethical Committee of Defence Research & Development Establishment, Gwalior, India. Images are taken by trained pathologist and the size of each image is 1300×1030. Images consist of both the monomorphonuclear cells and polymorphonuclear cells. Representative images from the taken database are shown in Figure 2.

As the database consists of heavy and low stained images, we have divided the whole image data set into three groups of similar size for result analysis: (i) heavy stained image, (ii) normal stained image, and (iii) low stained image. Figure 3 shows representative image from heavy and low stained image groups and the effect of histogram specification on these images. For separating the images into above groups, we have observed that heavy stained images have lower value of Luma while low stained images have higher value of Luma. Luma is given by Eq.2 [16]. Images are classified into above groups using average of log of Luma as per Eq. 3:

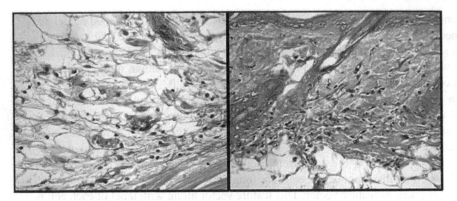

Fig. 2 Representative images of inflamed mice skin section stained with H&E staining and acquired at 40x magnification.

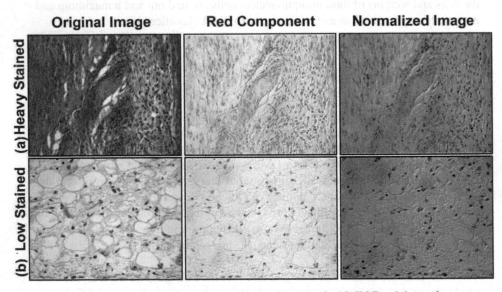

Fig. 3 Representative images of inflamed mice skin section stained with H&E staining and acquired at 40x magnification (a) Heavy stained image, its red component and result of histogram specification (b) Low stained image, its red component and result of histogram specification.

$$Y' = 0.299 * red + 0.587 * green + 0.114 * blue \qquad (2)$$

$$Image = \begin{cases} High\ stained & 0 < T < T_1 \\ Normal\ stained & if\ T_1 < T < T_2 \\ Low\ stained & if\ T > T_2 \end{cases} \qquad (3)$$

Table 1 Summary of total leukocytes marked by expert pathologist and segmented regions by proposed system from 30 images of inflamed mice skin sections

Image Data Set Set	Cells Marked by Expert Pathologist		Total Segmented Regions	Validation Results of Proposed System by expert on Segmented Regions		
	Monomorpho-nuclear Cells	Polymorpho-nuclear Cells		Monomorpho-nuclear Cells	Polymorpho-nuclear Cells	Noise
Heavy Stained	76	129	10962	76	129	10757
Low Stained	108	163	11824	108	163	11553
Normal Stained	124	128	7100	124	128	6848

After normalizing the images, the nuclei image is generated using statistical threshold. The resultant nuclei image of one representative image along with original image is shown in Figure 4. A comparison of segmented regions and cells information given by expert pathologist for all the 30 images is tabulated in Table 1. From the table, it is evident that the proposed segmentation method gives all required locations of actual cells which are to be cropped for subsequent steps of feature extraction. Figure 5 shows the cropped sub-images consisting of the cell nuclei and the resultant images after thinning and dilation operation. Figure 6 shows the generated thinned eigen cells of monomorphonuclear cells. The accuracy of SVM classifier is calculated using the Eq. 4.

$$accuracy = \frac{TP+TN}{TP+FP+FN+TN} \tag{4}$$

where, TP, FP, FN and TN are the true positive, false positive, false negative, and true negative respectively. For the selected database, 69.45% accuracy of classifier for leukocytes classification into monomorphonuclear and polymorphonuclear cell is observed.

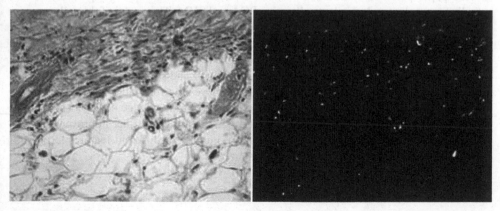

Fig. 4 The result of nuclei extraction (a). Original Image, (b). Nuclei image.

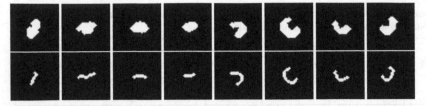

Fig. 5 First four images and last four images of first row are the cropped sub-images of the nuclei of monomorphonuclear cells and polymorphonuclear cells respectively. The images of second row represents the corresponding resultant images of first row after thinning and dilation operation.

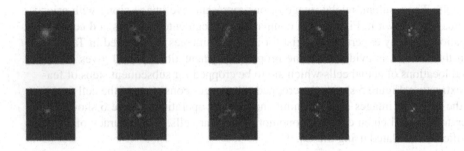

Fig. 6 Thin eigen cells of monomorphonuclear cells.

5 Conclusion

This paper presents a classification method to differentiate monomorphonuclear cells from polymorphonuclear cells using PCA and SVM. The nuclei of the cells are cropped using prior knowledge of its locations and some preprocessing steps like normalization and statistical threshold. The segmentation results show that along with the nuclei, a large number of noisy elements are also extracted out as depicted in Table 1. Classification of monomorphonuclear cells and polymorphonuclear cells is done with 69.45% accuracy. Further work involves the fully automated method for cell segmentation with noise reduction without prior knowledge of cell location. Since the obtained accuracy is not satisfactory and requires a further improvement, studies are to be carried out for selecting more number of different features to increase the classification accuracy of the system.

Acknowledgement

Authors are thankful to Defence Research and Development Establishment, Gwalior, India for funding a part of this work under the project (DRDE-P1-2011/Task-190). Authors are also thankful to Dr. S. C. Pant, Scientist 'F' at DRDE and Dr. Vinay Lomash, Ph.D., MVSc. for their valuable help in the analysis of microscopic images. Further, first author acknowledges ABV-Indian Institute of Information Technology & Management, Gwalior for providing research grant to carry out this work.

References

1. V. Kumar, A. K. Abbas, N. Fausto, and J. Aster, *Robbins and cotran pathologic basis of disease*. Elsevier Saunders, 2010.
2. C. Pan, D. Park, Y. Yang, and H. Yoo, "Leukocyte image segmentation by visual attention and extreme learning machine," *Neural Computing & Applications*, pp. 1–11, 2011.
3. B. C. Ko, J. Gim, and J. Nam, "Automatic white blood cell segmentation using stepwise merging rules and gradient vector flow snake," *Micron*, vol. 42, no. 7, pp. 695–705, 2011.
4. N. Theera-Umpon and S. Dhompongsa, "Morphological granulometric features of nucleus in automatic bone marrow white blood cell classification," *IEEE Transactions on Information Technology in Biomedicine*, vol. 11, no. 3, pp. 353–359, 2007.
5. J. Angulo and G. Flandrin, "Automated detection of working area of peripheral blood smears using mathematical morphology," *Analytical Cellular Pathology*, vol. 25, no. 1, pp. 37–49, 2003.
6. D. Wermser, G. Haussmann, and C. Liedtke, "Segmentation of blood smears by hierarchical thresholding," *Computer Vision, Graphics and Image Processing*, vol. 25, no. 2, pp. 151–168, 1984.
7. Y. Chan, M. Tsai, D. Huang, Z. Zheng, and K. Hung, "Leukocyte nucleus segmentation and nucleus lobe counting," *BMC Bioinformatics*, vol. 11, 2010.
8. M. G. Forero, A. R. Learte, S. Cartwright, and A. Hidalgo, "Deadeasy mito-glia: Automatic counting of mitotic cells and glial cells in drosophila," *PLoS ONE*, vol. 5, 2010.
9. C. Apfeldorfer, K. Ulrich, G. Jones, D. Goodwin, S. Collins, E. Schenck, and V. Richard, "Object orientated automated image analysis: Quantitative and qualitative estimation of inflammation in mouse lung," *Diagnostic Pathology*, 2008.
10. C. Chubb, Y. Inagaki, P. Sheu, B. Cummings, A. Wasserman, E. Head, and C. Cotman, "Biovision: an application for the automated image analysis of histological sections," *Neurobiol Aging*, 2006.
11. J. K. Sont, W. I. De Boer, W. A. A. M. Van Schadewijk, K. Grnberg, J. H. J. M. Van Krieken, P. S. Hiemstra, and P. J. Sterk, "Fully automated assessment of inflammatory cell counts and cytokine expression in bronchial tissue," *American Journal of Respiratory and Critical Care Medicine*, vol. 167, pp. 1496–1503, 2003.
12. R. C. Gonzalez and R. E. Woods, *Digital Image Processing*. Pearson Education, 2009.
13. T. Jolliffe, *Principal Component Analysis*, ser. Springer Series in Statistics. Springer New York, 2006.
14. M. A. Turk and A. P. Pentland, "Face recognition using eigenfaces," in *Computer Vision and Pattern Recognition, 1991. Proceedings CVPR '91., IEEE Computer Society Conference on*, jun 1991, pp. 586 –591.
15. C. Cortes and V. Vapnik, "Support-vector networks," in *Machine Learning*, 1995, pp. 273–297.
16. C. Poynton, *Digital video and HDTV algorithms and interfaces*. Morgan Kaufmann, An imprint of Elsevier Science, 2003.

Solving application oriented graph theoretical problems with DNA computing

Veronika Halász, László Hegedüs, István Hornyák, and Benedek Nagy

Abstract Social networks are represented by graphs. Important features of the network, e.g., length of shortest paths, centrality, are given by graph theoretical way. Bipartite graphs are used to represent various problems, for example, in medicine or in economy. The relations between customers and goods can be represented by bipartite graphs. Genes and various diseases can also form a bipartite graph, where a disease is connected to those genes that could cause it. In this paper DNA computing approach is presented for solving some graph theoretical problems. Since DNA computing uses a massively parallel approach, hard graph theoretical problems can be solved (at least in theory). Our main contribution is to present Projection algorithms for bipartite graphs; the molecular tube obtained by them can be used as a base for further processes.

Key words: DNA computing, Graph algorithms, Bipartite graphs, DNA algorithm, Networks, Social network, Bioinformatics

Veronika Halász
Faculty of Informatics, University of Debrecen, Debrecen, Hungary,
e-mail: veronika@macko.atomki.hu

László Hegedüs
Faculty of Informatics, University of Debrecen, Debrecen, Hungary,
e-mail: hegedus.laszlo@inf.unideb.hu

István Hornyák
Faculty of Informatics, University of Debrecen, Debrecen, Hungary,
e-mail: ihornyak@namafia.atomki.hu

Benedek Nagy
Department of Computer Science, Faculty of Informatics, University of Debrecen, Debrecen,
Hungary, e-mail: nbenedek@inf.unideb.hu

J. C. Bansal et al. (eds.), *Proceedings of Seventh International Conference on Bio-Inspired Computing: Theories and Applications (BIC-TA 2012)*, Advances in Intelligent Systems and Computing 201, DOI: 10.1007/978-81-322-1038-2_7, © Springer India 2013

1 Introduction

DNA computing is a relatively new branch of theoretical computer science. It has connections to logic, graph theory, etc. and it can be applied in various fields like sociology, economy, bioinformatics and thus in medical sciences. The advantages of a DNA computing lie in its massive parallelism.

In 1994, Adleman solved via DNA computing the Directed Hamiltonian Path problem and one year later, Lipton presented a DNA algorithm for the solution of the SAT problem [1, 8]. These two ground breaking discoveries have initiated the research in applications of DNA computing to graph theory, and pushed toward, especially to hard problems. For a survey in this topic, see [5].

While some problems of graph theory can be solved using DNA computing, the structure of molecules (including DNA strands) can also be presented by graphs. In [11], Uejima and Hagiya used bipartite graphs to indicate incompatibility of base pairs in DNA and RNA sequences. In molecular biology, gene expression data is used to represent the response of genes to certain conditions. Of course these data can be stored in a bipartite graph with all the genes in one partition and all the conditions in the other partition with the edges as responses. A biclustering algorithm of gene expression data is given in [10]. In [9], the interaction between amino acids and the nucleotides of DNA molecules are also analysed via bipartite graphs. A more mathematical application of bipartite graphs to tilings can be found in [6]. In this paper one of our focuses are bipartite graphs, but from the viewpoint of graph theoretical algorithms.

The structure of the paper is as follows. After some preliminaries (definitions, operations of our DNA algorithms) in Section 2, we present our results in two sections. In Section 3 applications related to (social) networks are presented; while in Section 4 practical (e.g., bioinformatics and marketing) problems represented by bipartite graphs. The paper is finished by a short summary and future thought.

2 Preliminaries

In this section we recall important (graph theoretical and DNA computing related) concepts that we are using in this paper.

A directed graph G is a tuple $G = (V, E)$, where V is an arbitrary set of *vertices* and E is a subset of $V \times V$, the set of *edges*. For a pair $(v_1, v_2) \in E$, we say, that an edge goes from v_1 to v_2 and it is also denoted by $v_1 \to v_2$. A *walk* is a series of edges $(v_1, v_2), (v_2, v_3), \ldots, (v_{n-2}, v_{n-1}), (v_{n-1}, v_n)$. We usually denote the vertices of V with natural numbers from 1 to n where $n = |V|$, and we write e_{ij} instead of $(v_i, v_j) \in E$. A *trail* is a walk, in which all the edges are distinct and a (cycle-free) *path* is a trail, in which all the vertices are distinct. A trail is called a *cycle* if $v_1 = v_n$ and all the other vertices are distinct. We call a graph bipartite if there exists a partition U, W of the set of vertices V such that for all $(v_1, v_2) \in E$, one of v_1 or v_2 is in U and the other vertex is in W.

A nonempty set Σ is called an *alphabet*, and the elements of Σ are called letters. A *word* w over Σ is a series of letters of Σ, i.e., $w = w_1 w_2 \cdots w_n$, where $w_i \in \Sigma$ for all $i = 1, 2, \ldots, n$. The length of such word w is n and is denoted by $|w|$. The *empty* word is a word which does not contain any letter and is denoted by λ. The set of all nonempty words over Σ is denoted by Σ^+, and we denote $\Sigma^+ \cup \{\lambda\}$ by Σ^*. The product, or concatenation of $u, v \in \Sigma^*$, $u = u_1 u_2 \cdots u_n$, $v = v_1 v_2 \cdots v_m$ is the word $uv = u_1 u_2 \cdots u_n v_1 v_2 \cdots v_m$. From now on, we may refer to the ith letter of the word w with w_i, where $i = 1, 2, \ldots, n$.

Any word over the alphabet $\Sigma = \{A, C, G, T\}$ is called a *single stranded DNA sequence*. These letters denote the molecules (or more precisely: nucleotides) *Adenine, Cytozine, Guanine, Thymine*, respectively. In nature, each single stranded DNA sequence has two ends, namely the $3'$ and $5'$ ends. Thus each strand has a direction $3' \rightarrow 5'$ caused by the underlying chemical bonds. In nature, DNA usually occurs in a double stranded form, which we denote by $\binom{u}{v}$, where u and v are single stranded sequences, with opposite directions and each pair u_i, v_i satisfies the reflexive relation $\rho = \{A - T, C - G\}$ called the *Watson-Crick complementarity relation*. We see, that u and v mutually determine each other, so in most cases we can write either u or v instead of $\binom{u}{v}$ noting that it is a double-stranded sequence. We call v the complement of u and denote it by $\bar{u}$. Note, that $\bar{v} = \bar{\bar{u}} = u$. We call a DNA sequence of length n (contains n nucleotides) an *n-mer*. We refer to containers of DNA strands as tubes. A tube can be empty, if it does not contain any strands.

There are several ways for formal descriptions of the usual DNA operations. In [5] NP-complete graph theoretical problems are solved in polynomial time with DNA computing. We will use the operations and notation described in [5] with some modifications (some operations are removed, since we do not need them).

1. *Extract*(T, s, T^+, T^-): Where T and s are a tube and a short single stranded DNA sequence respectively. The results will be in T^+ and T^-. The tube T^+ contains all molecules of T which contain s as a subsequence, and $T^- = T \setminus T^+$.
2. *Merge*$(T_0, T_1, \ldots, T_n)$: $T_0, T_1, T_2, \ldots, T_n$ are given. We will get the result in T_0 as the union of $T_0, T_1, \ldots, T_n$.
3. $Detect(T) = \begin{cases} True & \text{if } T \text{ contains at least one DNA molecule,} \\ False & \text{if } T \text{ is empty.} \end{cases}$
4. *Discard*(T): if tube T is not needed, we can discard it.
5. *Amplify*$(T_0, T_1, \ldots, T_n)$: $T_0 \neq \emptyset$ and $T_1, \ldots, T_n$ empty tubes are given. Tubes $T_1, \ldots, T_n$ will be identical copies of T_0. The tube T_0 will be empty at the end of the operation.
6. *Number*(T) = the number of different DNA strands in T.
7. *Selection*(T_1, l, T_2): given $T_1 \neq \emptyset$, l is a positive integer and T_2 is an empty tube. All strands of length l are removed from T_1 and put into T_2.
8. *Denaturation*(T): all double stranded sequences are disassociated to single stranded sequences in T.
9. *Annealing*(T): we put a large number of nucleotides and ligase enzyme in T, then cool the tube, thus in T all feasible double stranded DNA sequences are formed from the single stranded sequences, also the formed sequences are completed to whole double stranded sequences by nucleotides and ligation takes place.

We modified the *Annealing* operation for our convenience (including the ligation also in it).

In the next sections we will use these descriptions of the operations in our algorithms.

3 Social network related problems

In this section, as a counterpart of [5], we present an algorithm that computes all shortest paths between all pairs of nodes in a graph (that may represent a social network, or other network [3, 7]). Based on our algorithm we also give a method to help one who is interested to other features of the networks as centrality.

3.1 Shortest path between all vertices: the algorithm

We can obtain the shortest paths between all pairs of vertices with the following algorithm. We give an algorithm for graphs, where the weight of the edges are equal (i.e., for unweighted graphs). The algorithm can be seen on the next page.

In the most inner loop the algorithm checks whether there exists a path of length l containing v_i and v_j while gradually increasing l. If $Detect(P_{ij}^*)$ yields true, we know that there is a path and so, there is a shortest path between i and j.

This loop (the code in 3.1.4.) although seems like a very complex procedure, can be implemented in constant time with gel electrophoresis (i.e., a method used to determine the length of DNA strands). In this way the whole algorithm runs on quadratic time ($\sim n^2$ where n is the number of nodes) due to the two cycles (line 3 and line 3.1). However in laboratory the $\sim n^2$ cases can also be computed in a parallel manner of that degree that the (number of) equipments allow.

At the end of the algorithm, P^* contains the shortest path between all pairs of vertices v_i, v_j. One may ask how the paths are checked to contain only distinct vertices. Actually, if it is a shortest walk, then there is no repetition of any vertices (a shorter walk could be obtained by cutting the part between the repetition out), therefore it is also a shortest cycle-free path. From P^* we can easily obtain some centrality measures of the graph.

We note here that a shortest path contains shortest paths between all pair of vertices that are contained in the path. With this knowledge one may develop a faster algorithm to have all shortest paths. It is not necessary to run the loop for each pair of vertices.

Begin "Shortest Paths AV"
1. Let $G = (V, E)$ be a graph with $|V| = n$ vertices. For all $v_i \in V$ we assign $(2 \cdot x)$-mer strands where x depends on the size of G. We do this in a way that no strand forms a hairpin with itself (i.e., no strand is in the form $uv\bar{u}$) and no two strands have a common prefix/subword (with length comparable to x).

We construct the strands corresponding to the edges: $e_{ij} \sim \bar{B}_i \bar{A}_j$, where $v_i \sim A_i B_i$ and $v_j \sim A_j B_j$, and $|A_i| = |B_i| = |A_j| = |B_j| = x$. Therefore the edges are also coded by $(2 \cdot x)$-mer strands.

2. Let P include the strands $A_i B_i$, $A_j B_j$, for all $v_i, v_j \in V$. And let P contain $\bar{B}_k \bar{A}_\ell$ if $e_{k\ell} \in E$, $v_k, v_\ell \in V$. We assume that P contains multiple copies of every strand. Let all other tubes be empty. Then:

    2.a. Annealing(P)
    2.b. Denaturation(P)

After step 2.b. we have all paths in G represented in P.

    2.c. Amplify(P, P_{ij}) $(1 \le i, j \le n)$
3. for $i := 1$ to $n - 1$ do
    3.1 for $j := 1$ to n do
        if $i = j$ then endfor
            else Amplify(P_{ij}, P)
        endif
        3.1.1. Extract($P, \{A_i B_i\}, P_i, P_i'$)
        3.1.2. Discard(P)

        We empty the tube P. To get rid of excess molecules.

        3.1.3. Extract($P_i, \{A_j B_j\}, P_{ij}, P_{ij}'$)

        Strands in P_{ij} encode all paths that having vertices v_i and v_j.

        3.1.4. if Detect(P_{ij})

        We check wether there exist a path between v_i and v_j.

        then
            let $l := 4 \cdot x$
            while not Detect(P_{ij}^*) do
                Selection(P_{ij}, l, P_{ij}^*)
                let $l := l + 2 \cdot x$
            end while
            let $l := l + 2 \cdot x$
            Merge(P^*, P_{ij}^*)

            We need these strands, they represent the shortest paths.

        end if
    end for
end for
End "Shortest Paths AV"

3.2 Descriptors of a network

In addition to know the distance or a shortest path between vertices there are some other important features of a network:
- centrality of a vertex: the average distance of the vertex from other vertices. It is the average of the length of the shortest path from the vertex. (The length of a shortest path is also called the geodesic distance of the vertices.)

Actually, this measure can be computed during the algorithm (when the value of the loop variable is considered as fixed) we obtain the length l of the shortest paths, i.e., all the distances to other vertices. We need only compute the average of these values (divided by $2x$ and one needs to decrease the obtained value by 1 (since we have distance in the terms of number of nodes in the path).
- eccentricity of a vertex: is the greatest geodesic distance between the vertex and any other vertex.

This measure can easily be computed from the tube containing all shortest paths starting from the given vertex (with a Selection operation in a loop) by gel electrophoresis.

The diameter of a graph is the maximum eccentricity of any vertex in the graph.

It can be computed from the tube P^* obtained in the Algorithm with a Selection operation in a loop, i.e., by gel electrophoresis.
- centrality of an edge: the shortest path between two vertices, opposite to the Euclidean space, can be ambiguous. The centrality of an edge is the sum of the ratio of shortest paths containing the given edge. Formally it is computed for the edge (v_i, v_j):

$$C_{i,j} = \sum_{s \neq t \in V} \frac{|t:(v_i,v_j) \in t, t \in S_{s,t}|}{|S_{s,t}|}, \text{ where } S_{s,t} \text{ is the set of shortest paths between vertices } v_s$$

and v_t.

In our algorithm we obtained all the shortest paths for any pair of vertices. At the step 3.1.4. of the algorithm, before merging the newly found shortest paths $(P^*_{i,j})$ to the shortest paths given between other pairs of vertices, one could measure how many shortest paths are found (Number operation) and how many of them contain the given edge (Extract operation, and then Number operation). By these values and the given formula, the centrality of an edge is easily computable.

Having the tube of all shortest paths of the graph other measures of the graph also can be determined in a simple manner.

4 Problems related to bipartite graphs

Bipartite graphs are frequently applied, not only in computer science (e.g., Petri Nets), but in other sciences as well. In medical sciences bipartite graphs are used to denote the connection between diseases and causes, or genes and characteristics, etc. [4]. Thus scientists are interested in the direct paths from v_i to v_j, or all direct paths from v_i to other vertices, or all direct paths between any two vertices.

In this section we show a graph transformation algorithm: Starting from a bipartite graph we obtain its projection [12], i.e., a graph with labeled edges having only vertices from one of the sets of the bipartition. This is one of our main contributions.

Applications related to economy, e.g., marketing are also presented.

4.1 Projection algorithm

Using the notation we have fixed in Section 2 we give an algorithm for the projection of a bipartite graph $G = (V, E)$. (In several cases the bipartite graph is not directed, in this cases one can represent an undirected edge between v_i and v_j by adding both (v_i, v_j) and (v_j, v_i) to the representation. Let the partition of the vertices be $U_1, U_2 \subset V$. Our aim is to obtain (the tubes coding the) graphs with the following properties: let $G_1 = (U_1, E_1)$ be defined with $E_1 = \{(v_i, v_j) \mid v_i, v_j \in U_1$, there exists $v_k \in E_2$ such that $(v_i, v_k), (v_k, v_j) \in E\}$. Moreover we will code the type of the connection, i.e., the code of v_k, into the edge connecting v_i and v_j. Similarly $G_2 = (U_2, E_2)$ can be defined. The algorithm is related to our previous algorithm, but now, we want to obtain only paths with special lengths.

The algorithm:

Begin "Projection"
1. Let U_1, U_2 be the bipartition of V. For all $v_i \in U_1$ we assign $(2 \cdot x + r)$-mer strands and we assign $(2 \cdot x + p)$-mer strands to all $u_i \in U_2$ (x depends on the size of G and $r \neq p$). We do this in a way that no strand forms a hairpin with itself (i.e., no strand is in the form $uv\bar{u}$) and no two strands have a common prefix (with length comparable to x). This way, we can distinguish between vertices in U_1 and vertices in U_2.

We construct the strands corresponding to the edges: $e_{ij} \sim \bar{B}_i \bar{A}_j$, where $v_i \sim A_i s_i B_i$ and $v_j \sim A_j t_j B_j$, where $|A_i| = |B_i| = |A_j| = |B_j| = x$.

2. Let P_1 include the strands $A_i r_i B_i$ and P_2 include the strands $A_j p_j B_j$ for all $v_i \in U_1$ and $v_j \in U_2$. And let E' contain $\bar{B}_k \bar{A}_\ell$ if $e_{k\ell} \in E$, $v_k, v_\ell \in V$. Let P_1' contain the codes of the vertices of the first partition with doubled length, i.e., $A_i r_i B_i A_i r_i B_i$. We assume that all tubes contain multiple copies of every strand. Further, initially, let all other tubes be empty. Then:

   2.a. Merge(P,P₁,P₂,E')
   2.b. Annealing(P)
   2.c. Denaturation(P)

After step 2 we have all paths in G represented in P.

3. for $i := 1$ **to** $|V|$ **do**
  3.1 for $j := 1$ **to** $|V|$ **do**
    3.1.1. Extract($P, \{A_i r_i B_i\}, P_i, P_i'$**)**
    3.1.2. Discard(P**)**
    We empty the tube P. To get rid of excess molecules.
    3.1.3. Extract($P_i, \{A_j p_j B_j\}, P_{ij}, P_{ij}'$**)**
    Strands in P_{ij} encode all paths that start at v_i and end at v_j.
    3.1.4. Selection($P_{ij}, 4 \cdot x + 2 \cdot r + p, P_{ij}^*$**)**
    Strands in P_{ij}^* encode all the paths in the form $v_1 \to u \to v_j$, where
    $v_1, v_j \in U_1$ and $u \in U_2$.
    3.1.5. Merge(P^*, P_{ij}^***)**
    We need these strands as the labeled edges of our new graph.
  end for
end for
4. Merge(P, P_1', P^***)**
End "Projection"

At step 3.1.5. of the algorithm, P^* contains all the paths that connect two vertices from U_1 with a vertex from U_2 between them. And P^* is ready to further processing. Actually, we need those strands that contain the original part from the vertex in U_2. In this way these newly generated edges can be used with the vertices coded in P_1', and so, in P this new graph is coded: At the end of the algorithm P contains all vertices of U_1 with the newly generated labeled edges. With a similar method, we can construct paths in the form $u_i \to v \to u_j$, where $u_i, u_j \in U_2$ and $v \in U_1$ and also the projection graph having vertices of U_2 can be constructed. The time complexity of this algorithm is also quadratic ($\sim |V|^2$).

We show another algorithm for projection that produces only path of the desired length (and so it is faster and more efficient) using two different representations of each vertex of either partition depending on our aim. The algorithm is shown in the next page.

At the end of the run of the algorithm "Fast Projection" P^* contains only the paths in the form $v_i \to u_k \to v_\ell$, where $v_i, v_\ell \in U_1$ and $u_k \in U_2$. Similarly, if the role of U_2 and U_1 are switched, we can easily construct paths that start and end in U_2 and only contain a node from U_1 in between.

This fast algorithm could run on constant time assuming that the reaction time in the tubes does not depend on the number of molecules in the tube (i.e., it is independent of the size of the graph).

Begin "FastProjection"

1. Let U_1, U_2 be the bipartition of V. For all $v_i \in U_1$ we assign $(2 \cdot x + r)$-mer strands $A_i r_i (B_i)'$ and $(A_i)' r_i B_i$ (which are both variations of the strand $A_i r_i B_i$) where $(A_i)'$ and $(B_i)'$ are given in a way, that no molecule can bound to them from the left and from the right respectively. (If one wants only the projection graph without further DNA algorithm, then this aim can easily be achieved by using dideoxy nucleotides. When further actions are planned with the resulted projection graph we do not recommend to use dideoxy molecules.) So every node in U_1 is represented by two strands.

And we assign $(2 \cdot x + p)$-mer strands to all $u_i \in U_2$. In these operations x depends on the size of G and $r \neq p$. We do these assignments in a way that no strand forms a hairpin with itself (i.e., no strand is in the form $uv\overline{u}$) and no two strands have a common prefix (with length comparable to x). We construct the strands corresponding to the edges: $e_{ij} \sim \overline{B_i A_j}$, where $v_i \sim A_i s_i B_i$ (i.e., $A_i s_i (B_i)'$ and $(A_i)' s_i B_i$) and $v_j \sim A_j t_j B_j$, where $|A_i| = |B_i| = |A_j| = |B_j| = x$.

2. Let P_1 include the strands $(A_i)' r_i (B_i)$, $A_i r_i (B_i)'$ and P_2 include the strands $A_j p_j B_j$ for all $v_i \in U_1$ and $v_j \in U_2$. And let E' contain $\overline{B_k A_\ell}$ if $e_{k\ell} \in E$, $v_k, v_\ell \in V$. We assume that all tubes contain multiple copies of every strand. Further, initially, let all other tubes be empty. Then:

    2.a. Merge(P, P_1, P_2, E')

    2.b. Annealing(P)

    2.c. Denaturation(P)

After step 2 we have in P all paths in G that are in the form $v_i \to u_k \to v_\ell$, where $v_i, v_\ell \in U_1$ and $u_k \in U_2$. We may have some shorter paths in P, that can be removed with the following command.

3. Selection($P, 4x + 2r + p, P^*$)
End "FastProjection"

4.2 Applications in bioinformatics

As we already mentioned, in medical sciences bipartite graphs are used to denote the connection between, for instance, diseases and genes.

By our graph coding method, we can distinguish the vertices coding genes (U_1) and vertices coding diseases (U_2) by the length of the molecules. By considering the projected graph the role of related genes can be seen: two genes are connected in this new graph (represented in tube P at the end of the algorithm) if there is a disease that is caused by both of them (e.g., by gene error). This network of genes has important applications in bioinformatics to predict the probability of some diseases based on the known genes.

Also, considering the other projection of the original bipartite graph has some important information. In this graph, the relation of some diseases can be obtained. They are connected if there is a common gene that could cause them. These relations of diseases can be used for statistical and other tests for medical purposes.

There is another application of bipartite graphs in medicine: the drugs and their possible side effects can be represented in a similar way as genes and diseases.

4.3 Applications in marketing

In this subsection we show how our algorithm, i.e., projections of bipartite graphs can be applied in economy (market, business). Let a bipartite graph be given with two types of vertices: goods and customers. By our coding the vertices of goods U_1 and customers U_2 are easily separable by their length.

By projection one could obtain a graph where the customers are connected via the products they buy. In this way some partitions can be obtained by the type of the customers, and they could get semi-direct marketing. The customers who buy similar articles can get the same advertisements or coupons... (Moreover one could measure and use also the strength of the connections (i.e., how many goods connect two customers).

In the other hand, we have a graph where the goods are connected by edges of the customers. The goods that are connected by several edges are usually bought together, and so, this fact could be used in marketing. The price of one such product can be decreased with making large advertisement about this big sales; but at the same time the price of the other (strongly connected) products is increased by the amount is reasonable for the supermarket (of course, without making advertisement).

Other possibility is to extend this study to larger relations. Allowing shortest paths between goods up to a predefined length, from the projection graph one may obtain m-tuples of products that are frequently bought together, and so the supermarket may get a larger profit to play with a few goods together... It is also a good idea to select m-tuples which includes one of the most popular, most frequent-bought product and/or products that gives maximum profit to the supermarket, because probably the most of these products are purchased by most of the people.

5 Conclusions, future work

Some DNA algorithm are presented for various graph problems related to social sciences, networks, medical sciences and economy. We believe that DNA computing can effectively be used to solve some special problems. Our projection algorithm and so, the representation of the obtained graph is so natural that we could use it for further studies. The tube obtained by our algorithm can be used as a base of further algorithms such as the tube with all possible truth assignments is used for filtering/testing SAT formulae in Lipton's experiment [8].

Since we did not do any experiments it is a matter of future work to see, how the presented method could work in practice. Theoretically DNA algorithms (so the presented ones) are deterministic parallel algorithms, however in practice DNA-like algorithms have some upper limits. With very large graphs the result may not be correct.

Acknowledgements

The authors wish to thank to the reviewers for their valuable remarks. The work is supported by the TÁMOP 4.2.1/B-09/1/KONV-2010-0007 and TÁMOP 4.2.2/C-11/1/KONV-2012-0001 projects. The projects are implemented through the New Hungary Development Plan, co-financed by the European Social Fund and the European Regional Development Fund.

References

1. Leonard M. Adleman. Molecular computation of solutions to combinatorial problems. *Science* 266, 1994, pp. 1021–1024.
2. Armen S. Asratian, Tristan M. J. Denley, and Roland Häggkvist. Bipartite Graphs and their Applications. *Cambridge University Press*, New York, USA, 1998.
3. Albert-László Barabási. Linked: The New Science of Networks, Perseus Publishing, 2002.
4. Anna Bauer-Mehren, Michael Rautschka, Ferran Sanz, and Laura I. Furlong. DisGeNET: a Cytoscape plugin to visualize, integrate, search and analyze genedisease networks. Bioinformatics 26/22 (2010), 2924–2926.
5. Hossein Eghdami and Majid Darehmiraki. Application of DNA computing in graph theory. *Artificial Intelligence Review* 38, 2012, 223–235.
6. Jean-Claude Fournier. Combinatorics of perfect matchings in plane bipartite graphs and application to tilings. *Theoretical Computer Science* 303, 2003, 333–351.
7. L. C. Freeman. Centrality in social networks: Conceptual clarification. Social Networks, 1/3, (1979), 215–239.
8. Richard J. Lipton. DNA solution of HARD computational problems. *Science* 268, 1995, pp. 542–545.
9. R. Sathyapriya, M.S. Vijayabaskar, S. Vishveshwara. Insights into protein-DNA interactions through structure network analysis. *PLoS Comput Biol* 4, 2008
10. Amos Tanay, Roded Sharan, Ron Shamir. Discovering statistically significant biclusters in gene expression data. *Proceedings of the Tenth International Conference on Intelligent Systems for Molecular Biology*, 2002, pp. 136–144.
11. Hiroki Uejima, Masami Hagiya. Analyzing secondary structure transition paths of DNA/RNA molecules. *DNA Computing, 9th International Workshop on DNA Based Computers, Madison, USA* Springer-Verlag, Berlin, Heidelberg, 2004, pp. 86–90.
12. Katharina Anna Zweig, Michael Kaufmann. A systematic approach to the one-mode projection of bipartite graphs. Social Network Analysis and Mining 1 (2011) 187-218.

Human Identification using Heartbeat Interval Features and ECG Morphology

Yogendra Narain Singh and Sanjay Kumar Singh

Abstract This paper presents a novel method to characterize the ECG signal for human identification. The characterization process utilizes the analytical and appearance based techniques to analyze the ECG signal with an aim to make the measurements insensitive to noise and non-signal artifacts. We extract heartbeat interval features and interbeat interval features using analytical based technique and use them as a complementary information with the morphological features that are extracted using appearance based technique for improved identification accuracy. We perform identification using one-to-many comparisons based on match scores that are generated using statistical pattern matching technique. Results demonstrate that the proposed method for automated characterization of the ECG signal is efficiently used in identifying the normal as well as the arrhythmia subjects. In particular, the recognition accuracy for the subjects of MIT-BIH Arrhythmia database is reported to 87.37% whereas the subjects of our IIT(BHU) database are recognized with an accuracy of 92.88%.

Key words: Human identification, electrocardiogram, biometrics, signal processing and pattern recognition.

1 Introduction

Heartbeat is normally used in diagnosing intraventricular conduction disturbances and arrhythmia [1]. The heart is a muscular organ, which is electrically polarized.

Y. N. Singh
Department of Computer Science & Engineering, Institute of Engineering & Technology,
Guatam Buddh Technical University, Lucknow - 226 021, India. e-mail: singhyn@gmail.com

S. K. Singh
Department of Computer Engineering, Indian Institute of Technology
Banaras Hindu University, Varanasi - 225 021, India.

J. C. Bansal et al. (eds.), *Proceedings of Seventh International Conference on Bio-Inspired Computing: Theories and Applications (BIC-TA 2012)*, Advances in Intelligent Systems and Computing 201, DOI: 10.1007/978-81-322-1038-2_8, © Springer India 2013

The polarization of cardiac cells cause the heart to beat. The electrocardiogram (ECG) is used to record the electrical activity of the heart. It measures rate and regularity of the heartbeats. More formally, an ECG signal is a transthoracic interpretation of the electrical activity of the heart over a period of time. Recent, studies [2]-[10] have suggested that the ECG acquired from different individuals show heterogeneous characteristics. The heterogeneity has also been marked in the studies conducted for diagnosing arrhythmia present in the heart function [11]. Therefore, the ECG signal of an individual can be characterized by specific patterns that are sufficient to discriminate his/her identify from others.

The distinctiveness of the ECG signal among individuals is generally resulted due to the change in ionic potential, the time of ionic potential to spread from different parts of heart muscle, the plasma levels of electrolytes (e.g., potassium, calcium and magnesium etc.) and the rhythmic differences. The difference in heart structure such as, chest geometry, position, size, and physical condition can also manifest unique characteristic in the rhythm of an individual heartbeats. These distinctions are reflected in the change in morphology, difference in amplitudes and the variation in time intervals of the dominant fiducials in an individual heartbeats. The main advantage of using the ECG signal as a biometric is the robustness to circumvention, replay and obfuscation attacks, however these are prime concerns associated to the conventional biometrics [12], [13]. An ECG signal exploits the physiological features that exist in all (live) humans and as such, it is naturally secured. It has an inherent feature of vitality that signifies the life signs [14]. It is difficult to mimic, and hard to be copied or stolen. Therefore, the ECG has the strong credentials to successfully address the security and privacy issues of an individual. In a multibiometric system, an ECG signal can also be combined with other and independent biometric modalities as a complementary information to enable secure and efficient individual authentication [15]. The ECG signal as a biometric can solve the problem of identity theft and therefore, it can be used as a tool for information security [16].

The issue of using the ECG signal as a biometric includes the variation present in the signal that makes the data representation more difficult [17], [18]. The variations in the signal are resulted due to noise and non-signal artifacts such as 50/60 Hz power line interference, muscle contractions close to the electrodes and the motion artifacts. In addition, fluctuation in isoelectric line caused by respiration and motion of the subjects degrade the quality of the ECG signal significantly [19].

In this paper, we present a novel method to characterize heartbeat features for human identification that is insensitive to signal variations and non-signal artifacts. The method performs the ECG characterization using analytical and appearance based techniques. Using analytical technique, we extract clinically dominant fiducials from the selected heartbeats that include interval features and amplitude features while the appearance based technique extracts the morphological features from the beats. The advantage of using the analytic based features is that it captures local information of a heartbeat. The drawback of analytical based technique is that it is not robust in analyzing all types of ECG traces. In order to overcome from its limitation, we use an appearance based feature extraction technique that captures the heartbeat features in a holistic manner such that the complete information of a heart-

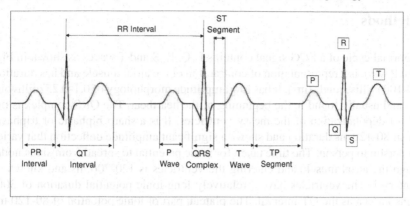

Fig. 1 A typical ECG signal that includes three successive heartbeats and information lying in P, Q, R, S and T waves.

beat can be preserved. We utilize the analytical based features as a complementary information and combined with the morphological features for improved recognition accuracy. In order to avoid the sudden changes in an ECG signal the aforementioned characterization selects a sequence of heartbeats such that the current beat can be analyzed only if its predecessor and the successor beats are segmented correctly.

At appearance level, the two stage procedure is employed to extract the ECG morphological features. In the first stage, segmented ECG morphological features are extracted from the beats that show consistent characteristics using the information of analytical based features. In the second stage, fixed interval ECG morphological features are extracted from the selected beats. In this stage the effect of noise and motion artifacts are minimized by scaling the signal using Pareto normalization [20]. For the identification experiment, the classification is performed using statistical pattern matching technique on the basis of match scores. The performance of the proposed method is evaluated on 73 ECG recordings selected from publically available MIT-BIH Arrhythmia database of PhysioBank [21] and our IIT(BHU) database. The identification results confirm the effectiveness of proposed characterization of the ECG signal and support the presence of distinct physiological features that can be used for human recognition.

The rest of the paper is organized as follows. Section 2 presents the methods used for ECG characterization and feature extraction. In section to follow the description of statistical technique that generates match scores from derived features of the ECG signal is presented. The description of a human identification system using the ECG signal is given in   Section 3. The experimental results that prove the efficacy of the proposed biometric system on the publically available database and on our database are presented in Section 4.   Finally, some conclusions are drawn in Section   5.

2 Methods

The normal cycle of a ECG signal contains P, Q, R, S and T waves as shown in Fig. 1. The P wave is a representation of contraction of the atrial muscle and has duration of 60-100 milliseconds (ms). It has low amplitude morphology of 0.1-0.25 millivolts (mV) and usually found in the beginning of the heartbeat. The QRS complex is the result of depolarization of the messy ventricles. It is a sharp biphasic or triphasic wave of 80-120 ms duration and shows a significant amplitude deflection that varies from person to person. The time taken for ionic potential to spread from sinus node, through the atrial muscle and entering the ventricles is 120-200 ms and known as PR interval. The ventricles have a relatively long ionic potential duration of 300-420 ms known as the QT interval. The plateau part of ionic potential of 80-120 ms after the QRS and known as the ST segment. The return of the ventricular muscle to its resting ionic state causes the T wave that has an amplitude of 0.1-0.5 mV and duration of 120-180 ms. The duration from resting of ventricles to the beginning of the next cycle of atrial contraction is known as TP segment which is a long plateau part of negligible elevation.

Prior to use the ECG signal in the subsequent processing of heartbeat segmentation and features extraction all signals are passed through a two-stage median filters of width 200 ms and 600 ms, respectively to remove the baseline wander. The first median filter suppressed the QRS complexes and P waves while the second median filter suppressed the T waves. The resulting signal is then subtracted from the original signal to produce the baseline corrected ECG signal [22].

2.1 Heartbeat Detection and Segmentation

The heartbeats are detected from the ECG signal using the QRS complex delineator. We employed the technique proposed by Pan and Tompkins [23] with some improvements. It uses digital analysis of slope, amplitude and width information of the ECG waveforms. The beginning and end of the QRS complex i.e., QRS_{onset} and QRS_{offset} time instances (fiducials), respectively are delineated according to the location and convexity of the R peak.

Once the heartbeat is detected, temporal time windows are defined heuristically before and after the QRS complex fiducials to seek for the P and the T waves. The technique proposed in [24] is used to determine the P_{onset} and the P_{offset} fiducials from the P wave, while the technique proposed in [25] is used to determine the T_{onset} and the T_{offset} fiducials from the T wave including their peak fiducials. The delineation results of the ECG characteristic waves obtained by the employed methods are shown in Fig. 2. From computed fiducials of a heartbeat, we derive three different classes of features such as, (1) heartbeat interval features, (2) interbeat interval features and (3) ECG morphological features.

Fig. 2 Delineation of ECG
characteristic waveforms and
their clinically dominant
fiducials.

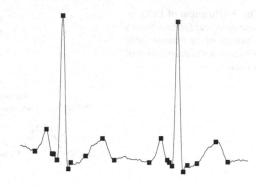

2.2 Feature Extraction

2.2.1 Heartbeat Interval Features

Five features related to the heartbeat intervals are computed after heartbeat segmentation. The QRS width is the duration between the QRS_{onset} and the QRS_{offset}, fiducials. The T wave duration is defined as the time interval from QRS_{offset} to T_{offset} fiducials. The PQ segment is defined as the time interval from P_{onset} to QRS_{onset} fiducials. The pre-TP segment is defined as the duration between the current beat P_{onset} and the previous beat T_{offset} fiducials. Similarly, the post-TP segment is defined as the duration between the current beat T_{offset} and the following beat P_{onset} fiducials.

2.2.2 Interbeat Interval Features

Ten features related to inter heartbeat intervals are computed after segmentation of the heartbeats. These features include PP, QQ, SS, TT and RR sequence that are extracted from successive heartbeats. The pre-PP (post-PP) interval is the duration between the P_{onset} of the current heartbeat and the P_{onset} of the previous (following) heartbeat. pre-QQ (post-QQ) interval is the duration between the Q_{Peak} of the current beat and the Q_{peak} of the previous (following) beat. The pre-SS (post-SS) interval is the duration between the S_{peak} of the current beat and the S_{peak} of the previous (following) beat. The pre-TT (post-TT) offset interval is the duration between the T_{offset} of the current beat and the T_{offset} of the previous (following) beat. Similarly, the pre-RR (post-RR) interval is defined as the RR interval between the current heartbeat and the previous (following) heartbeat.

Fig. 3 Extraction of ECG
morphological features from a
heartbeat where fiducial point
(FP) represents the position of
R peak.

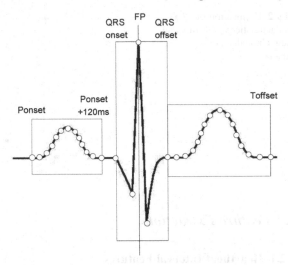

2.2.3 ECG Morphological Features

We divided the ECG morphological features into two groups where both groups
contained the amplitude values of the segmented heartbeats of an ECG trace. The
main distinction between the groups lie on the method used to extract the amplitude features. The first group extracts the morphological features from a heartbeat
using the segmented information derived by the analytical method within the sampling windows e.g., the P wave, the QRS complex and the T wave as shown in
Fig. 3. In total, thirty-three features are derived within the sampling windows. The
first window is set between the QRS_{onset} and the QRS_{offset} fiducials. Five features
are extracted corresponding to the QRS_{onset}, the Q_{peak}, the R_{peak}, the S_{peak} and the
QRS_{offset} fiducials. The boundaries of the second window is set as such so that it
approximately cover the P wave. It contains the portion of heartbeat between the
P_{onset} and the P_{onset} + 120 ms fiducials. Using linear interpolation method, thirteen
features are estimated uniformly within the sampling window. Similarly, the third
window is bounded between the QRS_{offset} and the T_{offset} fiducials. Fifteen amplitude features are derived uniformly within this window using linear interpolation.

The second group extracts the morphological features from a scaled ECG signal.
It contains twenty-eight features which are extracted from a heartbeat on fixed interval basis. In a scaled signal the amplitude difference from x_{nT} to the mean, μ is
measured in units of standard deviation σ such as,

$$x_{nT}' = \frac{x_{nT} - \mu}{\sqrt{\sigma}} \tag{1}$$

where x_{nT} represents the data sample of size n at discrete time instance T [20]. The
aim of scaling is to reduce the sensitivity of the signal, both to noise and non-signal
artifacts that are contaminated to them. In this group the sampling windows are

Fig. 4 Extraction of ECG morphological features from the scaled samples of a heartbeat.

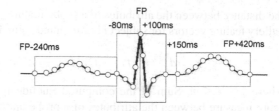

defined with respect to the location of the heartbeat fiducial point (FP) as shown in Fig. 4. The first window approximately covered the QRS complex and contained the portion of the ECG signal between $FP - 80$ ms and $FP + 100$ ms. Nine features are resulted from this window. The second window approximately covered the P wave and extended from $FP - 80$ ms to $FP - 240$ ms towards its left. Nine features are resulted within this window. The third window approximately covered the T wave and extended from $FP + 150$ ms to $FP + 420$ ms. Ten amplitude features are derived from this window. In the described windows the features are derived from uniformly distributed sample positions using linear interpolation method.

2.3 Generation of Match Scores

In order to generate the match scores from feature vectors of the gallery and the probe samples of the ECG signal, statistical framework technique is adopted [15]. Consider, the subject i has an ECG signal of length t unit of time. The m sub signals of length l unit of time ($l < t$) are arbitrarily selected from the complete trace of the ECG signal. For each sub signal, a vector of d-dimension is prepared from the successive occurrence of the selected beats by taking average of attributes of their feature vectors. Let $P^{(i)}$ be the pattern matrix consisting of m vectors of the subject i of size $m \times d$ can be defined as,

$$P^{(i)} = \begin{pmatrix} f_{1,1} & f_{1,2} & \cdots & f_{1,d} \\ f_{2,1} & f_{2,2} & \cdots & f_{2,d} \\ \vdots & \vdots & & \vdots \\ f_{m,1} & f_{m,2} & \cdots & f_{m,d} \end{pmatrix} \tag{2}$$

where element $f_{j,k}$ represents the kth feature of jth sub dataset. The values of m and d are set to 10 and 74, respectively in this experiment. The purpose of arbitrarily selection of subdata set is to analyze the signal statistically for the variation present in different heartbeats of an individual ECG. Consider, the population size is N, so there are N different ECG signals. Thus, N different pattern matrices $P^{(i)}$ are generated in the database where, $1 \leq i \leq N$.

Similarly, a probe sample, Q is prepared from the testing dataset of an individual ECG and a feature vector f', where $f' = \{f'_1, f'_2, \ldots, f'_d\}$ is generated. Statistically,

the distance between the attributes of a probe feature vector and the attributes of the
gallery feature vectors for subject i is computed using Euclidean distance as follows,

$$d_j^{(i)} = \left(|f_{j,1} - f_1'| \; |f_{j,2} - f_2'| \cdots |f_{j,d} - f_d'| \right) \tag{3}$$

where $1 \leq j \leq m$. Sum of the computed Euclidean distances return the distance
score measure between the attributes of a probe and the gallery feature vectors for
an individual i such as

$$s_j^{(i)} = \sum_{k=1}^{d} |f_{j,k} - f_k'| \tag{4}$$

In order to acknowledge the variation present in an ECG signal for the subject i, the
mean of the distance scores, $s^{(i)}$ can be computed and determined as follows

$$s^{(i)} = \frac{1}{m} \sum_{j=1}^{m} s_j^{(i)} \tag{5}$$

A smaller value of distance score indicates a good match while a higher value of
distance score indicates a poor match.

3 Identification System

The schematic representation of the proposed automatic system of ECG characterization for human identification is shown in Fig. 5. The identification process is an
outcome of the processing of three different stages such as, a preprocessing stage,
a data representation stage and a decision making stage. First, the ECG signal is
acquired from individuals and preprocessed. It utilizes a filtering unit that makes
necessary correction of the signal from noise and non-signal artifacts. The data representation stage consists of heartbeat detection and feature extraction modules. The
heartbeat detection module attempts to locate all heartbeats with heartbeat segmentation. The heartbeat segmentation includes the detection of the P, Q, R. S, and
T waves and determination of their end fiducials. The feature extraction includes
the determination of heartbeat interval features, interbeat interval features and ECG
morphological features from a set of heartbeats. From the derived features, a vector
of measurement is prepared as template and stored in the database. A similar process
is adopted for generating a vector of measurement from the probe ECG signal. Finally, the probe information is compared with the templates stored in the database.
The identification decision can be taken on the basis of the generated match scores
using $1 : N$ matching criterion under the predefined threshold.

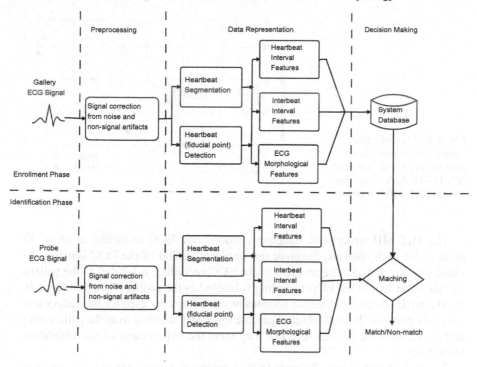

Fig. 5 Schematic representation of automated characterization of ECG for human identification.

4 Results

The efficacy of the proposed characterization of the ECG signal for human identification is tested on two different database. The first database is prepared from the publically available PhysioBank archives [21], in particular MIT-BIH Arrhythmia database is used. It contains 48 half-hour excerts of two-channel ECG recordings of the subjects aged 23 to 89 years whereas over 50% of the included subjects are suffered with clinically significant arrhythmia. Forty-four ECG recordings are randomly selected from this database in this study. The second database is prepared in our laboratory at the School of Biomedical Engineering, Indian Institute of Technology, IIT(BHU), using the PowerLab 4/25 of AD-Instruments. The total 29 volunteers aged 20 to 56 years participated in data enrollment process and the data is acquired in multiple sessions across a period of one year. Each session contains the ECG recording of five minutes for each subjects whereas the subjects do not have any known cardiac arrhythmia. We perform the data acquisition in a more simplistic manner, with subjects merely sitting on a wooden stool under relax condition and the clamp electrodes are fixed to both wrist and left ankles. The data are bandpass filtered at $0.3 - 50$ Hz and sampled at 1000 Hz.

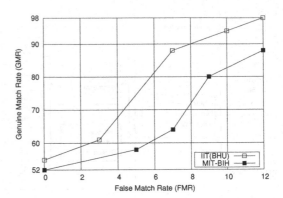

Fig. 6 ROC curve representing the identification performance for the subjects of MIT-BIH Arrhythmia database and IIT(BHU) database.

The MIT-BIH Arrhythmia database contains one ECG recording of about 30 minutes for each subjects, therefore the whole recording of the ECG signal is divided into two halves. The first half of the ECG recording is labeled as the gallery and the other half of the ECG recording is labeled as the probe, for each subject. In the IIT(BHU) database, different sessions of recordings are used for the gallery and the probe purpose. Ten sets of heartbeats are randomly selected from the gallery data and the features are derived such that they meet the requirement of the delineators in each set.

The aim of testing two different ECG database is to evaluate the effectiveness of the proposed method of ECG characterization and subsequently the performance of the aforementioned human identification system on the normal subjects and the subjects suffering with cardiac arrhythmia. The performance of the system is evaluated on equal error rate (EER) and receiver operating characteristic (ROC) curve [26]. An EER is the error rate where the likelihood of fraudulent matches (FMR) and the likelihood of non-matches of individuals who should be correctly verified (FNMR) assume the same value. The EER can be adopted as a unique measure for characterizing the security standard of a biometric system. The ROC curve is a two dimensional measure of classification performance that plots the likelihood of fraudulent matches against the likelihood of genuine matches (GMR).

The identification system reports the EER values of 7.12% and 12.63%, respectively for the subjects of IIT(BHU) database and MIT-BIH Arrhythmia database. The ROC curves representing the classification accuracy on both the database are shown in Fig. 6. The subjects of IIT(BHU) database achieves better classification accuracy of 92.88% than the subjects of MIT-BIH Arrhythmia database that are classified with an accuracy of 87.37%. Further, the subjects of IIT(BHU) database are genuinely matched with the rate of 55% when no fraudulent match is allowed by the system. The system achieves the GMR of 61% at the FMR of 3%. Subsequently, the system achieves higher GMR of 88%, 94% and 98%, respectively at the FMR of 7%, 10% and 12%. The subjects of MIT-BIH Arrhythmia database are genuinely matched with the rate of 54% when there is no fraudulent matched. Performance of the system raises further to the GMR of 64%, 80% and 88%, respectively at the

FMR of 7%, 9% and 12%. One of the reasons of reporting better identification performance for the subjects of IIT(BHU) database is because, it contained the ECG recordings of healthy subjects that are acquired under normal condition.

5 Conclusion

This study has proposed a novel method for automated characterization of the ECG signal for human identification. The characterization process has involved the analytical based method and the appearance based method that made the ECG analysis insensitive to noise and non-signal artifacts. The vector of measurement has derived from the analytical based features that include interval features and amplitude features while the appearance based features include morphological features derived from the successive heartbeats. The aforementioned system has performed the identification on the basis of match scores that are statistically generated using pattern matching technique through one-to-many comparisons. The system has performed efficiently for the subjects of test database, in particular the identification results on the normal ECG recordings of IIT(BHU) are found better. The results have also proved that there exists a number of heartbeat features that are remained consistent in each individual inspite that they are suffering with significant cardiac arrhythmia.

References

1. Kligfield P (2002) The centennial of the Einthoven electrocardiogram. J Electrocardiology 35:123129
2. Biel L, Pettersson O, Philipson L et al (2001) ECG analysis: a new approach in human identification. IEEE Trans on Instrumentation and Measurement 50(3):808-812
3. Shen TW, Tompkins WJ, Hu YH (2002) One-lead ECG for identity verification. Proc Second Joint EMBS/BMES Conf Houston, USA:62-63
4. Irvine JM, Israel SA, Scruggs WT et al (2008) eigenPulse: robust human identification for cardiovascular function. Pattern Recognition 41(11):3427-3435
5. Wang Y, Agrafioti F, Hatzinakos D et al (2008) Analysis of human electrocardiogram for biometric recognition. EURASIP Journal on Advances in Signal Processing, Article ID 148658, 2008:1-11
6. Singh YN, Gupta P (2008) ECG to individual identification. Proc Biometrics: Theory, Applications and Systems (BTAS'2008), Washington DC, USA: 1-8
7. Chan A, Hamdy M, Badre A (2008) Wavelet distance measure for person identification using electrocardiogram. IEEE Trans on Instrumentation and Measurement 57(2):248:253
8. Singh YN, Gupta P (2009) Biometric method for human identification using electrocardiogram. ICB 2009, Lecture Notes of Computer Science, Springer-Verlag Berlin Heidelberg, 5558, pp. 1277-1286.
9. Li M, Narayanan S (2010) Robust ECG biometrics by fusing temporal and cepstral information. Proc 20th Int'l Conf on Pattern Recognition (ICPR'2010) Istanbul, Turkey:1326-1329
10. Singh YN, Gupta P (2011) Correlation based classification of heartbeats for individual identification. J of Soft Computing 15(3):449-460
11. Hampton JR (2001) The ECG Made Easy. 5th edn. Churchill Livingstone, London

12. Singh YN and Singh SK (2012) A taxonomy of biometric system vulnerabilities and defenses. International Journal of Biometrics 5(2): pp. TBA [In Press]
13. Singh YN and Singh SK (2012) Challenges of biometrics: evaluation of system attacks and defences. Journal of Information Assurance & Security 7(3):207-221
14. Singh YN, Singh SK (2011) Vitality detection from biometrics: state-of-the-art. Proc. 2011 World Congress on Information and Communication Technologies (WICT), Mumbai, India:106-111
15. Singh YN, Singh SK, Gupta P (2012) Fusion of electrocardiogram with unobtrusive biometrics: An efficient individual authentication system. Pattern Recognition Letters 33(2012):1932-1941
16. Singh YN, Singh SK (2011) The State of Information Security. Proc AIATA 2011 Artificial Intelligence and Agents: Theory and Applications, Varanasi, India:363-367
17. Singh YN, Singh SK (2012) Bioelectrical signals as emerging biometrics: Issues and challenges. ISRN Signal Processing 2012 Article ID 712032:1-13 [doi:10.5402/2012/712032]
18. Singh YN, Singh SK (2012) Evaluation of electrocardiogram for biometric authentication. J of Information Security 3(1):39-48 [doi:10.4236/jis.2012.31005]
19. Friesen GM, Thomas CJ, Manal AJ et al (1990) A Comparison of the noise sensitivity of nine QRS detection algorithms. IEEE Trans on Biomedical Engineering 37(1):85-98
20. van den Berg RA, Hoefsloot HCJ, Westerhuis JA et al (2006) Centering, scaling, and transformations: improving the biological information content of metabolomics data. BMC Genomics 7(142):1-15
21. Physionet, PhysioBank archives. Massachusetts Institute of Technology Cambridge Available online at: http://www.physionet.org/physiobank/database/#ecg. Accessed on January 2011.
22. Chazal P, O'Dwyer M, Reilly RB (2004) Automatic classification of heartbeat using ECG morphology and heartbeat interval features. IEEE Trans on Biomedical Engineering 51(7):1196-1205
23. Pan J, Tompkins WJ (1985) A real time QRS detection algorithm. IEEE Trans on Biomedical Engineering 33(3):230-236
24. Singh YN, Gupta P (2009) A robust delineation approach of electrocardiographic P waves. Proc 2009 IEEE Symposium on Industrial Electronics and Applications (ISIEA'2009) 2:846-849
25. Singh YN, Gupta P (2009) A robust and efficient technique of T wave delineation from electrocardiogram. Proc Second Int'l Conf on Bio-inspired Systems and Signal Processing (BIOSIGNALS'2009) IEEE-EMB:146-154
26. Duda RO, Hart PE, Stork DG (2009) Pattern Classification. 2nd edn. Wiley, India

Improved Real-Time Discretize Network Intrusion Detection System

Heba F. Eid, Ahmad Taher Azar and Aboul Ella Hassanien

Abstract Intrusion detection systems (IDSs) is an essential key for network defense. Many classification algorithms have been proposed for the design of network IDS. Data preprocessing is a common phase to the classification learning algorithm, which leads to improve the network IDS performance. One of the important data preprocessing steps is discretization, where continuous features are converted into nominal ones. This paper addresses the impact of applying discretization on building network IDS. Furthermore, it explores the impact of the quality of the classification algorithms when combining discretization with genetic algorithm (GA) as a feature selection method for network IDS. In order to evaluate the performance of the introduced network IDS, several classifiers algorithms; rules based classifiers (Ridor, Decision table), trees classifiers (REPTree, C 4.5, Random Forest) and Naïve bays classifier are used. Several groups of experiments are conducted and demonstrated on the NSL-KDD dataset. Experiments show that discretization has a positive influence on the time to classify the test instances. Which is an important factor if real time network IDS is desired.

Keywords: Disretization; Real-time network intrusion detection; Feature selection; Network Intrusion Classification.

Heba F. Eid
Faculty of Science, Al-Azhar University,Cairo, Egypt e-mail: heba.fathy@yahoo.com

Ahmad Taher Azar
Faculty of Engineering, Misr University for Science & Technology
e-mail: ah-mad_T_azar@ieee.org

Aboul Ella Hassanien
Faculty of Computers and Information, Cairo University e-mail: aboitcairo@gmail.com

J. C. Bansal et al. (eds.), *Proceedings of Seventh International Conference on Bio-Inspired Computing: Theories and Applications (BIC-TA 2012)*, Advances in Intelligent Systems and Computing 201, DOI: 10.1007/978-81-322-1038-2_9, © Springer India 2013

1 Introduction

Anderson in 1980 [1] proposed the concept of Intrusion Detection System (IDS).
IDS is a major research problem in network security. The goal of network IDS is
to identify automatically unusual access or attack to secure the networks channels
[2, 3].

However, many issues need to be consider when building an IDS, such as data
collection, data preprocessing and classification accuracy. Several machine-learning
techniques have been proposed for the design of IDS. In particular, particular, these
techniques are developed to classify whether the incoming network trances are normal or intruder. The classifiers algorithms include rules based classifiers (Ridor,
Decision table [4]) , trees classifiers (REPTree, C 4.5 [5], Random Forest [6]) and
Naïve bays classifier [7] .

Classification is the prediction of a class label of an unknown instance. The instances are usually described by a set of features, which can be nominal or continuous features. Therefore, discretization is one of the important data preprocessing
steps. Discretization converts the continuous features into nominal ones. Many discretization methods have been proposed during the last two decades [8–10].

Another important research challenge for constructing high-performance IDS is
dealing with data containing a large number of features. Therefore, feature selection is required to deal with a large feature set. Feature selection reduces the computational complexity and removes information redundancy, which speeds up the
learning algorithm and increase the accuracy of the learning algorithm [11]. Different feature selection methods are proposed to enhance the performance of IDS [12].
Genetic algorithm (GA) [13, 14] is one of the successfully global search algorithms
used to solve the feature selection tasks.

This paper addresses the impact of applying discretization on building network
IDS using different classifiers algorithms; rules based classifiers (Ridor, Decision
table), trees classifiers (REPTree, C 4.5, Random Forest) and Naïve bays classifier.
Furthermore, it explores the impact of the quality of the different classifiers algorithms when combining discretization with GA feature selection on building the
network intrusion detection system.

The rest of this paper is organized as follows: Section 2 gives an overview of
Pre-Processing Approaches: Disretization, feature selection and Genetic algorithm.
Section 3 describes the proposed framework of the network intrusion detection system. The experimental results and conclusions are presented in Section 4 and 5
respectively.

2 Pre-Processing Approaches

2.1 Disretization

Discretization is a process of converting the continuous space of features into a nominal space [15]. The goal of the discretization process is to find a set of cut points, which split the range into a small number of intervals. Each cut-point is a real value within the range of the continuous values, which divides the range into two intervals one greater than the cut-point and other less than or equal to the cut-point value [16]. Discretization is usually performed as a pre-processing phase to the learning algorithm.

Discretization methods can be classified into five categories [17]:

1. Supervised vs. Unsupervised
2. Static vs. Dynamic
3. Global vs. Local
4. Top-down (splitting) vs. Bottom-up (merging)
5. Direct vs. Incremental

Supervised methods use the class labels during the discretization process. In contrast, Unsupervised methods do not use information about the class labels and generate discretization schemes based only on distribution of the values of the continuous attributes. Researches show that supervised methods are better than unsupervised methods [18]. Dynamic and static methods depends on whether the method considers the interdependence among the features into account or not [19]. Global methods use the entire value space of a numeric attribute for the discretization. While, Local methods use a subset of instances when deriving the discretization. Top-down(splitting) discretization methods start with one interval of all values of feature and split it into smaller intervals at the subsequent iterations. While, the bottom-up (merging) methods start with the maximal number of sub-intervals and merge these sub intervals until achieving a certain stopping criterion or optimal number of intervals [9]. Direct methods divide the range into equal-width of intervals, it requires the user to determine the number of intervals. Incremental methods begin with a simple discretization and go through an improvement process until reaching a stopping criterion to terminate the discretization process [20].

Fayyad et al. [21] proposed the Information Entropy Maximization (IEM) discretization method. It is a supervised, local, splitting and incremental discretization method. IEM algorithm criterions are based on information entropy, where the cut points should be set between points with different class labels.

2.2 Feature Selection

Feature selection (FS) is a preprocessing step before classification. Its purpose is to improve the classification performance through the removal of redundant or irrelevant features. FS methods generate a new set of features by selecting only a subset of the original features.

Based on the evaluation criteria feature selection methods fall into two categories: filter approach [22, 23] and wrapper approach [24, 25]. Filter approaches evaluate and select the new set of features depending on the general characteristics of the data without involving any machine algorithm. Frequently used filter methods include chi-square [26], information gain [27] and Pearson correlation coefficients [28]. Wrapper approaches use the classification performance of a predetermined machine algorithm as the evaluation criterion to select the new features set. Machine learning algorithms such as Genetic algorithm (GA) [29] ,ID3 [30] and Bayesian networks [7] are commonly used as induction algorithm for wrapper approaches.

2.3 Genetic Algorithm

Genetic algorithm (GA) is an adaptive search technique initially introduced by Holland [29]. It is computational model designed to simulate the evolutionary processes in the nature. GA includes three fundamental operators: selection, crossover and mutation within chromosomes.

1. **Selection:** A population is created with a group of randomly individuals. The individuals in the population are then evaluated by fitness function. Two individuals (offspring) are selected for the next generation based on their fitness.
2. **Crossover:** crossover randomly chooses a point in the two selected parents and exchanging the remaining segments of them to create the new individuals.
3. **Mutation:** mutation randomly changes one or more components of a selected individual. This process continues until a suitable solution has been found or a certain number of generations have passed [31].

Given a well bounded problem GAs can find a global optimum, which makes them well suited to feature selection processes.

3 A Framework of Real-Time Discretize Network Intrusion Detection

The framework for the proposed anomaly intrusion detection approach is shown in Fig 1. It is comprised of the following three fundamental building phases: (1) Data set Pre-processing by mapping and IEM discretization, (2) Data reduction by GA feature selection, and (3) Intrusion detection and classification of a new intrusion into five outcome.

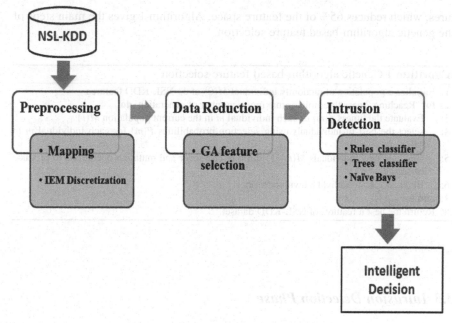

Fig. 1 Real-time Discretize Network Intrusion Detection Framework

3.1 Pre-processing Phase

The following two pre-processing stages has been done on NSL-KDD dataset:

1. **Mapping**:

   - symbolic features to numeric value.
   - Attack names to one of the five classes, 0 for *Normal*, 1 for DoS (Denial of Service), 2 for *U2R* (user-to-root: unauthorized access to root privileges), 3 for *R2L* (remote-to-local: unauthorized access to local from a remote machine), and 4 for *Probe* (probing:information gathering attacks.

2. **Discretization**: Features where discritized by Information Entropy Maximization (IEM) discretization method.

3.2 GA Feature Selection Phase

GA is applied as a feature selection method to reduce the dimensionality of the dataset. GA efficiently reduces the NSL-KDD dataset from 41 features to 14 features, which reduces 65% of the feature space. Algorithm 1 gives the main steps of the genetic algorithm-based feature selection.

Algorithm 1 Genetic algorithm-based feature selection

1: Initialize a population of randomly individual $M(0)$ of 41 NSL-KDD features.
2: **for** Reaching ?tness threshold or maximum number of generation **do**
3:    Evaluate the fitness $f(m)$ of each individual m in the current population $M(t)$
4:    select the best-fit individuals using selection probabilities $P(m)$ for each individual m in $M(t)$
5:    Generate new individuals $M(t+1)$ through crossover and mutation operations to produce offspring.
6:    Replace least-fit individual with new ones.
7: **end for**
8: Return the best n features of NSL-KDD dataset.

3.3 Intrusion Detection Phase

we evaluate the performance of the proposed high speed network intrusion detection framework on different set of classifier. The set of classifier includes rules based classifiers (Ridor, Decision table), trees classifiers (REPTree, C 4.5, Random Forest) and Naïve bays classifier.

4 Experiments and Analysis

4.1 Network Dataset Characteristics

NSL-KDD dataset [32] is a benchmark used for evaluating network intrusion detection systems. It consists of selected records of the complete KDD'99 dataset [33]. Each NSL-KDD connection record contains 41 features (e.g., protocol type, service, and flag) and is labeled as either normal or an attack. The training set contains a total of 22 training attack types, with additional to 17 types of attacks in the testing set. The attacks fall into four categories:

1. DoS denial of service e.g Neptune, Smurf, Pod and Teardrop.
2. R2L: unauthorized access to local from a remote machine e.g Guess-password, Ftp-write, Imap and Phf.
3. U2R: unauthorized access to root privileges e.g Buffer-overflow, Load-module, Perl and Spy.
4. Prob: collect information to helpful for make an attack in the future eg. Port-sweep, IP-sweep, Nmap and Satan.

4.2 Comparison Criteria

The Comparison Criteria to evaluate the proposed network intrusion detection system are: (1) the speed of the ID system and (2) the classification Accuracy.

Classification performance of ID system is measured in term of *precision, recall* and *F − measure*; which are calculated based on the confusion matrix given in Table 1. F-measure is a weighted mean that assesses the trade-off between precision and recall. An ID system should achieve a high recall without loss of precision.

Table 1 Confusion Matrix

	Predicted Class	
	Normal	Attake
Actual Class Normal	True positives (TP)	False positives (FP)
Attake	False negatives (FN)	True negatives (TN)

Where, TP and TN indicates that normal and attacks events are successfully labeled as normal and attacks, respectively. FP refer to normal events being predicted as attacks; while FN are attack events incorrectly predicted as normal [34].

$$Recall = \frac{TP}{TP+FN} \tag{1}$$

$$Precision = \frac{TP}{TP+FP} \tag{2}$$

$$F - measure = \frac{2*Recall*Precision}{Recall+Precision} \tag{3}$$

4.3 Results and Analysis

The proposed real time network intrusion detection system is evaluated using the NSL- KDD dataset, where 59586 records are randomly taken. All experiments have been performed using Intel Core 2 Duo 2.26 GHz processor with 2 GB of RAM.

We evaluate the proposed framework on different categories of classifiers; tree classifiers (REPTree, C 4.5, Random Forest), rule based classifiers (Ridor, Decision table) and Naïve bayes classifier.

Table 2 and 3 shows the F-measures and speed achieved for the different set of classifiers; without applying any preprocessing phase, applying IEM discritization and finally applying IEM discritization combined with GA feature selection (14 features). The comparison results are based on 10 fold cross-validation.

From table 2, applying IEM Discretization method leads to highly improve the speed of the systems especially for C4.5 classifier; which is very important for real

Table 2 Comparison of F-measures and speed for tree classifiers

	REPTree		C4.5		Random Forest	
Preprocess approach	F-measure	speed (sec.)	F-measure	speed(sec.)	F-measure	speed(sec.)
Non	98.3%	6.07	98.8%	43.46	99.2%	34.58
Discretization	98.1%	3.75	99.0%	3.05	99.1%	2.87
Discretization + GA	98.7%	1.20	98.8%	0.77	99.3%	1.76

time network intrusion detection systems. Also, the classification accuracy for REP-Tree and Random Forest classifier does not effect by discretization, while it is improved for C4.5 classifier. The systems speed shows another improvement when combining the discretization with GA faeture selection; which reduces the NSL-KDD dimentions from 41 features to 14 features.

Table 3 Comparison of F-measurs and speed for Rules based and Naïve bayes classifiers

	Ridor		Decision table		Naïve Bayes	
Preprocess approach	F-measure	speed(sec.)	F-measure	speed(sec.)	F-measure	speed(sec.)
Non	98.3%	435.57	96.4%	136.6	72.16%	4.21
Discretization	97.2%	129.16	96.3%	132.0	93.6%	0.21
Discretization + GA	97.8%	61.14	97.9%	25.5	94.5%	0.09

Table 3, gives the impact of applying discretization and applying discretization combined with GA feature selection. For Ridor classifier the system speed shows a good improvement. Also, it is clear that, discretization has a positive impact on the naïve bayes classifier, that is, it helps to highly improve the detection accuracy and speed.

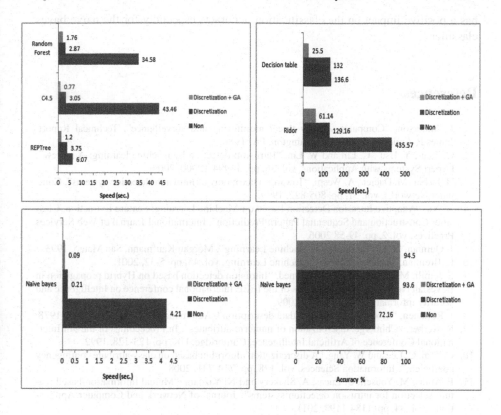

Fig. 2 An overall speed comparision of the proposed network ID framework.

Figures 2 shows the speed comparision of the proposed network ID framework for the different set of classifiers. Also, it gives the accuracy comparision for the Naïve bayes classifier.

5 Conclusions

In this study, a real time discritize network ID framework is proposed. We explore the impact of applying IEM discretization and GA feature selection on the performance of network IDS. Different classifiers algorithms; rules based classifiers (Ridor, Decision table), trees classifiers (REPTree, C4.5, Random Forest) and Naïve bays classifier are used to evaluate the classification time and accuracy of the introduced network ID framework. Experiments on the NSL-KDD dataset show that IEM discretization helps to highly improve the time to classify the test instances. Which is an important factor for real time network IDS. Also, IEM discretization has a positive impact on the classification accuracy, especially for the naïve bayes classifier.

References

1. J. Anderson, "Computer security threat monitoring and surveillance", Technical Report, James P. Anderson Co., Fort Washington, PA, 1980.
2. C. Tsai , Y. Hsu , C. Lin and W. Lin, "Intrusion detection by machine learning: A review", Expert Systems with Applications, vol. 36, pp. 11994-12000, 2009.
3. H. Debar ,M. Dacier , A. Wespi, "Towards a taxonomy of intrusion-detection systems", Computer Networks, vol. 31, pp. 805-822, 1999.
4. X. Xu, "Adaptive Intrusion Detection Based on Machine Learning: Feature Extraction, Classifier Construction and Sequential Pattern Prediction", International Journal of Web Services Practices, vol. 2, pp. 49-58 2006.
5. J. Quinlan, "C4.5: Programs for Machine Learning", Morgan Kaufmann, San Mateo, 1993.
6. L. Breiman, "Random Forests", Machine Learning, vol. 45, pp. 5-32, 2001.
7. F. Jemili, M. Zaghdoud and M. Ahmed, "Intrusion detection based on Hybrid propagation in Bayesian Networks," In proceedings of the IEEE international conference on Intelligence and security informatics, pp. 137-142, 2009.
8. J. Rissanen, "Modeling by shortest data description", Automatica, vol. 14, pp. 465471, 1978.
9. R. Kerber, "Chimerge: discretization of numeric attributes", In Proceedings of the 9th International Conference of Artificial Intelligence, Cambridge, UK, pp. 123-128, 1992.
10. C. Tsai, C. Lee and W. Yang, "A discretization algorithm based on class-attribute contingency coefficient", Information Sciences, vol. 178, pp. 714-731, 2008.
11. F. Amiri, M. Yousefi, C. Lucas, A. Shakery and N. Yazdani, "Mutual information-based feature selection for intrusion detection systems", Journal of Network and Computer Applications, vol. 34, pp. 1184-1199, 2011.
12. C. Tsang, S. Kwong and H. Wang, "Genetic-fuzzy rule mining approach and evaluation of feature selection techniques for anomaly intrusion detection", Pattern Recognition, vol. 40, pp. 2373-2391, 2007.
13. K. Chan, C. Kwong, Y. Tsim, M. Aydin and T. Fogarty,"A new orthogonal array based crossover, with analysis of gene interactions, for evolutionary algorithms and its application to car door design", Expert Systems with Applications, vol. 37, pp. 3853-3862, 2010.
14. Y. Li, S. Zhang and X. Zeng, "Research of multi-population agent genetic algorithm for feature selection", Expert Systems with Applications, vol. 36, pp. 11570-11581, 2009.
15. M. Mizianty, L. Kurgan and M. Ogiela, "Discretization as the enabling technique for the Naïve Bayes and semi-Naïve Bayes-based classification", The Knowledge Engineering Review, vol. 25, pp. 421-449, 2010.
16. S. Kotsiantis and D. Kanellopoulos, "Discretization Techniques: A recent survey",GESTS International Transactions on Computer Science and Engineering, vol.32, pp. 47-58, 2006.
17. H. Liu, F. Hussain, C. Tan and M. Dash, "Discretization: an enabling technique", Data Mining and Knowledge Discovery, vol. 6, pp. 393-423, 2002.
18. J. Dougherty, R. Kohavi and M. Sahami, "Supervised and unsupervised discretization of continuous features", In Proceedings of the 12th international conference on machine learning, San Francisco: Morgan Kaufmann; pp. 194-202, 1995.
19. H. Steck and T. Jaakkola, "Predictive discretization during model selection", In Proceedings of DAGM Symposium In Pattern Recognition, Tbingen, Germany, pp. 1-8, 2004.
20. J. Cerquides and R. Lopez, "Proposal and Empirical Comparison of a Parallelizable Distance Based Discretization Method". In Proceedings of the III International Conference on Knowledge Discovery and Data Mining (KDDM97). Newport Beach, California USA, pp. 139-142, 1997.

21. U. Fayyad and K. Irani, "Multi-interval discretization of continuous-valued attributes for classification learning", In Proceedings of the International Joint Conference on Uncertainty in AI. Morgan Kaufmann, San Francisco, CA, USA, pp. 1022-1027, 1993.
22. M. Dash, K. Choi, P. Scheuermann and H. Liu, "Feature selection for clustering-a filter solution", In Proceedings of the Second International Conference on Data Mining, pp. 115-122, 2002.
23. L. Yu and H. Liu, "Feature selection for high-dimensional data: a fast correlation-based filter solution," In Proceedings of the twentieth International Conference on Machine Learning, pp. 856-863, 2003.
24. R. Kohavi and G. John, "Wrappers for feature subset selection," Artificial Intelligence, vol. 2, pp. 273-324, 1997.
25. Y. Kim, W. Street and F. Menczer, "Feature selection for unsupervised learning via evolutionary search", In Proceedings of the Sixth ACM SIGKDD International Conference on Knowledge Discovery and Data Mining, pp. 365-369, 2000.
26. X. Jin, A. Xu, R. Bie and P. Guo, "Machine learning techniques and chi-square feature selection for cancer classification using SAGE gene expression profiles," Lecture Notes in Computer Science, 3916, DOI: 10.1007/1169173011, pp. 106-115, 2006.
27. M. Ben-Bassat, "Pattern recognition and reduction of dimensionality," Handbook of Statistics II, North-Holland, Amsterdam, vol. 1, 1982.
28. H. Peng, F. Long, C. Ding, "Feature selection based on mutual information criteria of max-dependency, max-relevance ,and min redundancy," IEEE Transactions on Pattern Analysis and Machine Intelligence, vol. 27, pp. 1226-1238, 2005.
29. J. Holland, "Adaptation in Natural and Artificial Systems. University of Michigan Press,Ann Arbor, MI, 1975.
30. J. R. Quinlan, "Induction of Decision Trees", Machine Learning, vol. 1, pp. 81-106, 1986.
31. B. Jiang , X. Ding , L. Ma , Y. He , T. Wang and W. Xie," A Hybrid Feature Selection Algorithm:Combination of Symmetrical Uncertainty and Genetic Algorithms", In Proceedings of the Second International Symposium on Optimization and Systems Biology OSB'08), China, pp. 152-157, 2008.
32. I. Cohen, Q. Tian, X. Zhou and T. Huang, "Feature Selection Using Principal Feature Analysis", In Proceedings of the 15th international conference on Multimedia, Augsburg, Germany, September pp. 25-29, 2007.
33. KDD'99 dataset, http://kdd.ics.uci.edu/databases, Irvine, CA, USA, July, 2010.
34. R. Duda, P. Hart and P. Stork, "Pattern Classification (Second Edition)", JohnWiley & Sons, USA, 2001.

Identification and Impact Assessment of High-Priority Field Failures in Passenger Vehicles using Evolutionary Optimization

Abhinav Gaur, Sunith Bandaru, Vineet Khare,
Rahul Chougule and Kalyanmoy Deb

Abstract This paper presents a method for prioritizing field failures in passenger vehicles based on their potential for improvement in the Customer Satisfaction Index (CSI_{QSR}). CSI_{QSR} refers to Customer Satisfaction Index pertaining to quality, service and reliability of the vehicle and is referred to as simply 'CSI' in this paper. A novel method for quantitative modeling of the CSI function using an evolutionary approach was presented in [3]. Such a CSI function can be used to capture individual customer's perception of a vehicle model as well as to compare overall CSI of multiple vehicle models. This work is firstly aimed at improving the previous modeling technique and validating it against Consumer Reports reliability ratings. More importantly, it presents a procedure for identifying high impact field failures based on their CSI Improvement Potential (CIP). These high priority field failures can then be further studied for root cause analysis.

Keywords: Customer Satisfaction Index, Quantitative modeling, Evolutionary optimization, Field failures

1 Introduction

Customer satisfaction measurement for most products, including consumer vehicles, has traditionally been performed through surveys conducted on a sample of customers. Automotive OEMs use reports published by firms such as the American Customer Satisfaction Index [1], J.D. Power [2] and Consumer Reports [4].

Abhinav Gaur, Sunith Bandaru and Kalyanmoy Deb
Kanpur Genetic Algorithms Laboratory, Indian Institute of Technology Kanpur, U.P., India
e-mail: {abnvgaur, sunithb, deb}@iitk.ac.in

Vineet Khare and Rahul Chougule
India Science Lab, General Motors Global R&D, Bangalore, India
e-mail: {vineet.khare, rahul.chougule}@gm.com

J. C. Bansal et al. (eds.), *Proceedings of Seventh International Conference on Bio-Inspired Computing: Theories and Applications (BIC-TA 2012)*, Advances in Intelligent Systems and Computing 201, DOI: 10.1007/978-81-322-1038-2_10, © Springer India 2013

The resulting rating is popularly known as the Customer Satisfaction Index or CSI. While such assessment is inevitable for many products, consumer vehicles (along with some other products) come with a warranty period during which customers can approach authorized dealers and service stations for repairs/service. In this way automobile companies can keep a record of all vehicle related problems faced by the customers. In service terminology, these problems are called field failures (see Table 1). Within automotive OEMs, quality (or warranty) data analysis is focused on number of failures in the field (example Incidences per Thousand Vehicle or IPTV [10], Problems per Hundred Vehicles or PPH) and limited emphasis is placed on the assessment of individual customers' (or consumers') perception and satisfaction.

In [3], the authors presented an evolutionary optimization based modeling approach that uses the recorded data to build a CSI distribution function, which can quantitatively predict the satisfaction rating for various vehicle models. The recorded data consists of two parts as shown in Table 2. The sales data for each vehicle is a one-time entry recorded at the point of sale. Service data is updated each time a customer visits a service station. Among other fields that are self-explanatory, it consists of the Vehicle Identification Number (VIN) and the repair code which is unique to each different type of field failure.

In this work, we first propose an improvement to the previous method for a more realistic modeling approach. With the improved methodology in place, datasets from five vehicle models currently available in the market are used to obtain ten different CSI distribution functions using all ($^5C_3 = 10$) possible three model combinations. Subsequently, these vehicle models are ordered based on their overall satisfaction rating as predicted by each CSI function. In each case, the resulting ordering is validated against published Consumer Reports [4] ratings. The best CSI function among the ten is chosen based on its performance in the optimization problem and is used for determining the priorities of different types of failures that can occur in these vehicle models. Table 1 shows a summary of the vehicle model datasets analysed in this work. This data was collected during the period January 2008 to August 2009 from various service stations in a given region. The total number of customers for all the models are represented as multiples of the same for Model-5.

Table 1: Total number of customers and unique field failures for the five datasets.

Vehicle Model & Segment	Total Custs.	Unique Field Failures
1 (Compact)	$19.61 \times C$	1026
2 (Mid-Size)	$13.35 \times C$	1084
3 (Luxury)	$1.93 \times C$	776
4 (Mid-Size)	$30.19 \times C$	1228
5 (Luxury)	$1.00 \times C$	606

Table 2: Data fields in the sales and service data of a vehicle and the notation used in this work

Vehicle Sales Data (One-time entry)		Vehicle Service Data (For i-th claim of a single visit)	
Identification no.	VIN	Identification no.	VIN
Sale date	d_0	Repair Start date	d_i
Mileage at sale	m_0	Repair End date	e_i
		Mileage at repair	m_i
		Repair cost	c_i
		Repair code	r_i
		Repair Severity rating	s_i

The rest of the paper is organised as follows. Sect. 2 briefly describes the original methodology [3] for obtaining the quantitative CSI model along with the improvement suggested in this paper. In Sect. 3 the validation results for the improved method are presented on different combinations of the vehicle models. Sect. 4 discusses the procedure for finding the improvement potential associated with different types of failures. It also presents and discusses a novel method of identifying important field failures which utilize the best CSI function obtained in Sect. 3. Sect. 5 concludes this work and gives future research directions.

2 Methodology

A recent study by J.D. Power and Associates [8] shows that close to 50% of the overall satisfaction of a customer is comprised of satisfaction related to vehicle quality, reliability and service. These same attributes are captured indirectly in the form of service data each time a customer visits the service station. The original methodology [3] extracts and works with those features from the combined sales and service data that influence the vehicle CSI the most. In the improved method, one of these extracted features, i.e. – *the waiting period for the customer while the vehicle is in service*, is changed in order to better reflect realistic waiting times. This section briefly explains the original methodology. The typical fields found in the vehicle sales and service data and the notations used in this work are shown in Table 2. The features $\mathbf{x}$ extracted from vehicle sales and service data can then be used for evaluating the CSI of the vehicle using,

$$CSI_{vehicle} = f(\mathbf{x}). \tag{1}$$

The function f is not known *a priori* and is built adaptively using a genetic algorithm representation. The optimization problem formulation is based on the assumption that most of the customers of a vehicle model should have similar overall view (satisfaction) of the vehicle model. In other words, the CSI distribution over all the customers should have a small variance resulting in a good bell shaped distribution. A narrower CSI distribution means better agreement among the customers thus making averaging more sensible. The following sections briefly describe the features extracted from the sales and service data and the optimization problem formulations. The reader is referred to [3] for a more detailed explanation.

2.1 Characteristic Features

A total of six features are extracted from the sales and service data for each customer. They are enumerated below along with the logical dependency of each with respect to the CSI.

1. x_1: *Number of visits made by the customer*. It is obtained by counting the number of times a particular VIN occurs in the service data. $CSI_{vehicle} \propto 1/x_1$.

2. x_2: *Sum of all repair times in days.* In the original methodology, this feature is the sum of all repair times in hours. Its purpose is to capture the total time for which the vehicle is unavailable to the customer while in service. The repair times of all repairs (of a visit) do not truly reflect this waiting time. Instead, it is better captured by the number of days from the day the vehicle came in for repairs/service till the day it was returned to the customer. This total waiting time for a customer over all the visits in a period could thus be given by $x_2 = \sum_{i=1}^{x_1}(e_i - d_i)$. Also, $CSI_{vehicle} \propto 1/x_2$.

3. x_3: *Sum of all service/repair costs.* It is given by $\sum_{i=1}^{x_1} c_i$. Also $CSI_{vehicle} \propto 1/x_3$.

4. x_4: *Average time interval between visits.* It is easy to see that $CSI_{vehicle} \propto x_4$. It is calculated using,

$$x_4 = \frac{1}{x_1}\left(d_1 - d_0 + \sum_{i=2}^{x_1}(d_i - d_{i-1})\right).$$

5. x_5: *Average miles run between visits.* Like x_4, $CSI_{vehicle} \propto x_5$. It is calculated using,

$$x_5 = \frac{1}{x_1}\left(m_1 - m_0 + \sum_{i=2}^{x_1}(m_i - m_{i-1})\right).$$

6. x_6: *Sum of problem severity ratings.* Each repair or field failure is associated with a repair code r_i which defines the type of service or repair performed. All repair codes are assigned a severity rating between 1 (for minor problems; e.g. oil change) and 5 (for major problem; e.g. engine replacement) by subject matter experts. Since severity rating has a negative impact on the CSI, we have $CSI_{vehicle} \propto 1/x_6$.

The extracted features are linearly normalized between 0 and 1 to avoid bias in the CSI function due to the scales of the features. Let x_i^{nr} be these normalized features. The normalized features are further transformed as follows for simplicity of notation,

$$X_i = \frac{1}{(1+x_i^{nr})}, \ \forall \ \{i = 1,2,3,6\},$$
$$X_i = (1+x_i^{nr}), \ \forall \ \{i = 4,5\}. \tag{2}$$

2.2 Functional form of CSI function

The proposed structure for the CSI model in Eq. (1) comprises of six terms,

$$CSI_{vehicle} = f(X_1,X_2,X_3,X_4,X_5,X_6) = T_1 \oplus T_2 \oplus T_3 \oplus T_4 \oplus T_5 \oplus T_6, \tag{3}$$

where,

$$T_l = \prod_{i=1}^{6} X_i^{\alpha_{il}\beta_{il}}, \ \forall \ l \in \{1,2,\ldots,6\}, \tag{4}$$

and '$\oplus$' can be either addition or multiplication operation. The model is adaptive in the sense that its exact functional form is not known *a priori* but is derived using a genetic algorithm (GA) for solving an optimization problem explained later in Sect. 2.4. The GA works with a population of CSI functions (f). Each of these f's is a function of the six transformed features X_i's, shown in Eq. (2). The exact functional form of each CSI function is dependent on a number of model parameters namely, α's, β's and γ's, which are randomly initialized by the GA. The parameters α_{il} and β_{il} decide the composition of the term 'T_l' as shown in Eq. (4). The Boolean parameter γ_i decides the operation '$\oplus$' between the terms T_i and T_{i+1} in Eq. (3). A summary of all the CSI model parameters is as follows:

$$0 \le \alpha_{il} \le 1 \quad \forall\, i,l \in \{1,2,\ldots,6\} : 36 \text{ real variables,}$$
$$\beta_{il} \in \{0,1\} \quad \forall\, i,l \in \{1,2,\ldots,6\} : 36 \text{ Boolean variables,} \tag{5}$$
$$\gamma_k \in \{0,1\} \quad \forall\, k \in \{1,2,\ldots,5\} : 5 \text{ Boolean variables.}$$

2.3 Bi-objective Optimization Problem

In order to obtain a customer level CSI function for each vehicle model the following bi-objective optimization problem is formulated,

$$\begin{aligned} &\text{Minimize} \quad \sigma, \\ &\text{Minimize} \quad |g|, \\ &\text{Subject to} \quad \textstyle\sum_l \beta_{il} \ge 1 \,\forall\, i \in \{1,2,\ldots,6\} : 6 \text{ constraints} \end{aligned} \tag{6}$$

The GA works with a population of CSI functions, obtained using randomly initialized CSI model parameters α's, β's and γ's given by Eq. (5). Referring to our earlier argument, the ideal CSI function should have a narrow distribution, i.e. small standard deviation σ, over the customers of the vehicle model under consideration. Further, to obtain non-trivial solutions to this problem, an additional objective of minimizing absolute skewness ($|g|$) of the CSI distribution has to be considered (refer [3] for justifications). The six constraints on the β parameters ensure that each X_i is used at least once in the CSI model. The non-dominated solutions to this bi-objective optimization problem are obtained by solving Eq. (6) for each of the five vehicle models in Table 1 separately using NSGA-II [6]. It is observed that the trade-off front for each vehicle model has a *knee*, a narrow region where a good trade-off between the objectives is seen. The knee point for each trade-off front is identified using the bend-angle approach [7].

2.4 Single Objective Optimization Problem

The bi-objective approach described above provides a different CSI model for each vehicle model. Thus, the relative CSI ranking between different vehicles cannot be established. However, a study similar to the one described in [3] gives us the following bounded region (in the objective space)

$$\sigma \leq 0.05 \text{ and } |g| \leq 1.0,$$

which contains the knees from all five trade-off fronts. This region encloses the set of all CSI functions which yield a low variance CSI distribution, without much asymmetry in distribution (low absolute skewness).

With the single objective formulation our aim is to obtain a single CSI model which can differentiate between two or more vehicle models as distinctly as possible. Assuming the mean μ of the CSI distributions as the index for overall satisfaction rating of a vehicle model, the optimum CSI model for two vehicle models m and n can be obtained by maximizing $|\mu_m - \mu_n|$. When more than two vehicle models are to be optimized, we use the following optimization formulation where this quantity is summed for all $\{m,n\}$ pairs,

$$\text{Maximize} \quad \sum_{\{m,n|m\neq n\}} |\mu_m - \mu_n|,$$
$$\text{Subject to} \quad \sigma_m \leq 0.05 \ \forall \ m, \tag{7}$$
$$|g_m| \leq 1.0 \ \forall \ m,$$
$$\textstyle\sum_l \beta_{il} \geq 1 \ \forall \ i \in \{1,2,\ldots,6\}.$$

Note how the bounded knee region obtained in Sect. 2.3 has been incorporated in the above formulation by converting the objectives of the bi-objective formulation into constraints to guarantee a low and low-skewness CSI distribution.

3 Validation Tests and Results

In this section, we apply the single objective optimization formulation to all ten ($^5C_3$) possible three-vehicle-model combinations obtained from the five vehicle models in Table 1. The results are summarized in Table 3. In each of the ten cases, the precedence of overall satisfaction of all vehicle models is established using the means of their CSI distributions obtained from the optimum CSI model for that three-vehicle-model combination. For example, $CSI_{\{1,2,3\}}$ represents the optimum CSI model for the combination of Models 1, 2 and 3. Using $CSI_{\{1,2,3\}}$ on the datasets of all five vehicle models gives five CSI distributions. The resulting order of their means gives, $CSI_5 \prec CSI_3 \prec CSI_1 \prec CSI_4 \prec CSI_2$ which means that Model 5 has the worst satisfaction rating according to $CSI_{\{1,2,3\}}$ and so on. The correct CSI precedence for the very same five vehicle models is available from the surveys conducted by Consumer Reports [4] during the same time period. According to [5], the five vehicle models are ranked as follows:

$$CSI_5 \prec CSI_3 \prec CSI_1 \prec CSI_4 \approx CSI_2 \tag{8}$$

Table 3: Overall CSI precedence obtained by solving Eq. (7) for all 3-model combinations. All predictions match that of Consumer Reports. $CSI_{\{2,4,5\}}$ (highlighted) gives the best objective value..

CSI Model	Predicted CSI Precedence Order	Objective Value (10^{-2})	CSI Form (simplified forms are omitted for conciseness)
$CSI_{\{1,2,3\}}$	$CSI_5 \prec CSI_3 \prec CSI_1 \prec CSI_4 \prec CSI_2$	3.980	$T_1 + T_2 \times T_3 \times T_4 + T_5 + T_6$
$CSI_{\{1,2,4\}}$	$CSI_5 \prec CSI_3 \prec CSI_1 \prec CSI_2 \prec CSI_4$	2.659	$T_1 + T_2 + T_3 + T_4 + T_5 + T_6$
$CSI_{\{1,2,5\}}$	$CSI_5 \prec CSI_3 \prec CSI_1 \prec CSI_2 \prec CSI_4$	6.580	$T_1 \times T_2 \times T_3 \times T_4 \times T_5 \times T_6$
$CSI_{\{1,3,4\}}$	$CSI_5 \prec CSI_3 \prec CSI_1 \prec CSI_4 \prec CSI_2$	4.654	$T_1 + T_2 + T_3 \times T_4 \times T_5 \times T_6$
$CSI_{\{1,3,5\}}$	$CSI_5 \prec CSI_3 \prec CSI_1 \prec CSI_4 \prec CSI_2$	3.652	$T_1 + T_2 + T_3 \times T_4 + T_5 + T_6$
$CSI_{\{1,4,5\}}$	$CSI_5 \prec CSI_3 \prec CSI_1 \prec CSI_2 \prec CSI_4$	8.015	$T_1 \times T_2 \times T_3 \times T_4 \times T_5 \times T_6$
$CSI_{\{2,3,4\}}$	$CSI_5 \prec CSI_3 \prec CSI_1 \prec CSI_4 \prec CSI_2$	4.706	$T_1 \times T_2 \times T_3 + T_4 + T_5 + T_6$
$CSI_{\{2,3,5\}}$	$CSI_5 \prec CSI_3 \prec CSI_1 \prec CSI_4 \prec CSI_2$	5.620	$T_1 + T_2 \times T_3 \times T_4 + T_5 + T_6$
$CSI_{\{2,4,5\}}$	$CSI_5 \prec CSI_3 \prec CSI_1 \prec CSI_2 \prec CSI_4$	8.458	$T_1 \times T_2 \times T_3 \times T_4 \times T_5 \times T_6$
$CSI_{\{3,4,5\}}$	$CSI_5 \prec CSI_3 \prec CSI_1 \prec CSI_2 \prec CSI_4$	5.801	$T_1 + T_2 \times T_3 \times T_4 + T_5 + T_6$

in terms of quality, reliability and service. Since the overall CSI ratings of Models 2 and 4 are very close, assuming both $CSI_4 \prec CSI_2$ and $CSI_2 \prec CSI_4$ to be acceptable, it can be seen in Table 3 that the predicted CSI precedence matches that of Consumer Reports in all ten cases. Additionally, it is seen that the combination $\{2, 4, 5\}$ attained the best objective value (8.458×10^{-2}) which means that the corresponding CSI model best differentiates between all vehicle models. The simplified functional form of $CSI_{\{2,4,5\}}$ is given by,

$$CSI_{\{2,4,5\}} = X_1^{0.0665} X_2^{1.9547} X_3^{5.9082} X_4^{0.5356} X_5^{2.6382} X_6^{5.9102}. \tag{9}$$

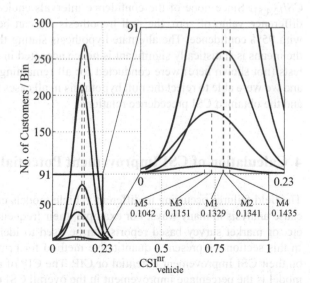

Fig. 1 Normal distribution fits to the CSI distributions obtained for the five vehicle models using $CSI_{\{2,4,5\}}$. The mean of each normal distribution fit is shown by a broken vertical line. The relative position of the means gives the relative CSI ordering of the vehicle models. Hypothesis testing enables us to say that the small difference between these means is statistically significant.

Table 4: Results of t-test for CSI function $CSI_{\{2,4,5\}}$. The second column $(M_i - M_j)$ represents the difference between the population means of CSI of the i-th and j-th vehicle model while the third column $(\mu_i - \mu_j)$ is the difference between the sample means of their CSI distributions. The implication of each test is also shown in the last column.

Model Pair $\{i,j\}$	95% Confidence Interval for $(M_i - M_j)$	$(\mu_i - \mu_j)$	Implication
$\{1,2\}$	(-0.001759,-0.000617)	-0.0012	$CSI_1 \prec CSI_2$
$\{1,3\}$	(0.016336, 0.019163)	0.0177	$CSI_1 \succ CSI_3$
$\{1,4\}$	(-0.011131,-0.010174)	-0.0107	$CSI_1 \prec CSI_4$
$\{1,5\}$	(0.026697, 0.030599)	0.0286	$CSI_1 \succ CSI_5$
$\{2,3\}$	(0.017503, 0.020373)	0.0189	$CSI_2 \succ CSI_3$
$\{2,4\}$	(-0.010004,-0.008925)	-0.0095	$CSI_2 \prec CSI_4$
$\{2,5\}$	(0.027869, 0.031803)	0.0298	$CSI_2 \succ CSI_5$
$\{3,4\}$	(-0.027002,-0.029803)	-0.0284	$CSI_3 \prec CSI_4$
$\{3,5\}$	(0.008545, 0.013251)	0.0109	$CSI_3 \succ CSI_5$
$\{4,5\}$	(0.037358, 0.041242)	0.0393	$CSI_4 \succ CSI_5$

The CSI distributions for the five vehicles obtained using $CSI_{\{2,4,5\}}$ are shown in Fig. 1. The positions of the means of individual distributions are shown at the bottom of the inset. The mean values of all the distributions are fairly close. In order to ascertain that the difference between these means is indeed statistically significant and can be used for determining the relative CSI ordering of the vehicle models, we conduct pairwise Welch t-tests as described in [3] for all ten ($^5C_2$) possible vehicle model pairs. For each test, consider the null hypothesis, $H_0 : M_i - M_j = 0$, where M_i and M_j are the population means for Model i and j respectively. [9] suggests the use of Welch's t-test for testing this hypothesis. Note that M_i is different from μ_i which is the mean of the CSI distribution of i-th vehicle model or the sample mean. Table 4 shows the results of the tests for the CSI distributions obtained using $CSI_{\{2,4,5\}}$. Since none of the confidence intervals enclose the hypothesized mean difference value of zero, the null hypothesis H_0 can be rejected in all ten cases with 95% confidence. The alternate hypothesis stating that the difference between the means is statistically significant is hence accepted in all cases. Similar pairwise tests (not shown here) were conducted for all remaining CSI functions of Table 3 and we were able to reject the null hypothesis in all cases thus validating the method and the obtained CSI precedence relations.

4 Calculation of CSI Improvement Potential

The field failures occurring in different vehicle models can be prioritized for a root cause analysis in many ways. For example, their frequency of occurrence, severity, etc. or market survey based reports can be used to identify high impact failures. In this section we present a quantitative method for prioritizing the failures based on their CSI Improvement Potential or CIP. The CIP of a failure type for a vehicle model is the percentage improvement in the overall CSI obtained for that model by completely eliminating all failures of that type. The procedure for finding the CIP of a field failure is described here. It assumes the knowledge of the quantitative CSI model obtained as shown in Sect. 3. The following notation will be used henceforth:

1. RD: It stands for Raw Data. It is a matrix comprising of nine fields (in columns) drawn from the combined sales and claims data. The fields are shown in Table 2. The RD matrix for a vehicle model contains as many rows as the number of claims made by all customers of that model in a given period.
2. FM: It stands for Feature Matrix. It has six columns corresponding to the six features mentioned in Sect. 2.1. It has as many rows as the number of customers. The i-th column of FM is shown as a vector x_i, where $i \in \{1,2,\ldots,6\}$ following the notation in Sect. 2.1.
3. r_c: It stands for the chosen repair code, representing the field failure for which CIP needs to be evaluated.
4. iCSI and dCSI: iCSI stands for individual CSI. It represents the CSI value for a single customer of a vehicle model. dCSI stands for CSI distribution and it is a column vector containing the iCSI values for all the customers of a vehicle model.
5. rd2fm: This function takes RD of a vehicle-model as input and returns its corresponding FM matrix as output.
6. fm2dcsi: This function takes FM as input and returns its corresponding dCSI vector as output.
7. normalize: This function takes a vector as input and normalizes each of its components linearly in the range $[0, 1]$. If the input to the function normalize is a matrix, then it applies to each column of the matrix separately.
8. min and max: The min and max functions take a vector as an input and give the minimum and maximum value amongst all the components of the vector as output.

The following steps describe the procedure to obtain the CIP for repair code r_c for the vehicle model under consideration:

1. From the original RD (RD_{orig}), remove all claim records (rows) having the repair code r_c. This gives RD_{new}.
2. Obtain $FM_{orig} = \mathrm{rd2fm}(RD_{orig})$ and $FM_{new} = \mathrm{rd2fm}(RD_{new})$.
3. The removal of records from RD_{orig} in the first step introduces certain aberrations in FM_{new} data. These aberrations and their possible remedies are as follows:

   a. Removal of records from RD_{orig} for any repair code r_c should always lead to an improvement in the satisfaction of every customer that claimed the corresponding repair, resulting in an overall increase in CSI of the model. However, this does not always happen with respect to features x_4 and x_5. For example, in case of x_4 (and similar explanation holds for x_5), as shown in Sect. 2.1,

$$x_4 = \frac{1}{x_1}\left(d_1 - d_0 + \sum_{i=2}^{x_1}(d_i - d_{i-1})\right) = \frac{1}{x_1}(d_L - d_0), \qquad (10)$$

where d_L is the last date of customer visit for the period of data and d_0 is the vehicle sale date. While deleting all claim records containing repair code r_c, if a particular record being removed was made on date d_L, then the numerator of x_4 decreases (numerically more than the decrease in denominator). This decrease is an aberration since the period for which the customer has been offered repair services should remain same even in the hypothetical case of zero r_c occurrences. To check this aberration, both x_4 and x_5 in FM_{new} are re-evaluated by extracting their numerators from RD_{orig} and the denominators from RD_{new}.

  b. There can be some customers, all of whose claim records in RD_{orig} contain only the repair code r_c. In such cases, the complete removal of r_c records from RD_{orig} completely removes such customers from FM_{new}. However, these customers must be among the most satisfied customers as they had no failures in the hypothetical case of complete r_c removal. Hence, post extraction, compensatory records are appended to FM_{new} representing these removed customers. Each compensatory record should evaluate to the highest CSI amongst rest of the customers. Hence for such a record,

$$x_i = \begin{cases} \min(\mathbf{x}_i), & \text{for } i \in \{1,2,3,6\}, \\ \max(\mathbf{x}_i), & \text{for } i \in \{4,5\}. \end{cases}$$

4. Normalize all vectors $(\mathbf{x}_1, \mathbf{x}_2, \ldots, \mathbf{x}_6)$ in FM_{orig} and FM_{new} together as shown below. It is important that FM_{orig} and FM_{new} are normalized together so as to avoid any bias in the CSI values obtained because of different numerical ranges.

$$\begin{bmatrix} \text{FM}_{orig}^{nr} \\ \text{FM}_{new}^{nr} \end{bmatrix} = \texttt{normalize}\left(\begin{bmatrix} \text{FM}_{orig} \\ \text{FM}_{new} \end{bmatrix} \right).$$

5. Evaluate dCSI_{orig} and dCSI_{new} vectors using,

$$\begin{bmatrix} \text{dCSI}_{orig} \\ \text{dCSI}_{new} \end{bmatrix} = \texttt{fm2dcsi}\left(\begin{bmatrix} \text{FM}_{orig}^{nr} \\ \text{FM}_{new}^{nr} \end{bmatrix} \right).$$

6. Next, for an unbiased comparison of the means of the original and the new CSI distributions, dCSI_{orig} and dCSI_{new} are normalized together as,

$$\begin{bmatrix} \text{dCSI}_{orig}^{nr} \\ \text{dCSI}_{new}^{nr} \end{bmatrix} = \texttt{normalize}\left(\begin{bmatrix} \text{dCSI}_{orig} \\ \text{dCSI}_{new} \end{bmatrix} \right).$$

7. The expected percentage improvement in the CSI of the model because of complete removal of repair code r_c can now be calculated as,

$$\text{CIP} = \frac{\text{mean}(\text{dCSI}_{new}^{nr}) - \text{mean}(\text{dCSI}_{orig}^{nr})}{\text{mean}(\text{dCSI}_{orig}^{nr})} \times 100\%. \tag{11}$$

Next, we present some results utilizing the CIP calculation method described here.

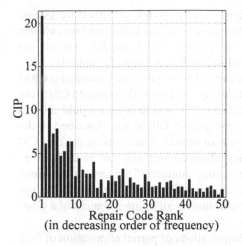

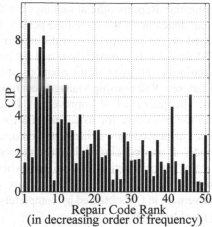

Fig. 2: CIP for top fifty most-frequent repair codes of Model 1.

Fig. 3: CIP for top fifty most-frequent repair codes of Model 5.

The CIP can be calculated for any desired number of repair codes. In our case of five vehicle models, we shall use $CSI_{\{2,4,5\}}$ for all CIP calculations. The reason being that this CSI model gives the best objective value among all CSI models shown in Table 9. Moreover, Welch t-tests on all pairwise model combinations as shown in Table 4 reveal that the CSI values obtained using $CSI_{\{2,4,5\}}$ are reliable.

Fig. 2 shows the CIP for top fifty most-frequent repair codes in Model 1. It shows that, for Model 1, the CIP and frequency of repair codes are positively correlated. However, the CIP does not fall monotonically with frequency of repair code. Similarly Fig. 3 shows the CIP for top fifty most-frequent repair codes in Model 5. It is clear here that a high frequency failure does not necessarily mean that rectifying the associated problem leads to proportional improvement in the CSI rating. Similar studies on other vehicle models were performed. In all cases it was observed that the degree of correlation between CIP and frequency of occurrence of the repair codes varies considerably. Hence in comparison to frequency, the quantitatively obtained CIP value is a better criterion for prioritizing repair codes for root cause analysis.

5 Conclusions

This paper has presented an improvement to an existing method for quantitative modeling of the CSI function for predicting the satisfaction of passenger vehicle owners. The methodology uses information regarding vehicle field failures gathered in the form of service data. Datasets from five vehicle models currently available in the market have been used and CSI models have been built for different combinations. The CSI rankings predicted by all CSI models were found to agree with the

Consumer Reports official rankings for these vehicle models published for assessment done during the same time period. Thereafter, the best CSI model was chosen based on the performance in differentiating different vehicle models. The chosen CSI model was statistically tested by conducting Welch t-tests for pairwise combinations of all five CSI distributions obtained using it. Next, we presented a CRM (Customer Relationship Management) specific application of the obtained CSI models. The quantity, CSI Improvement Potential (CIP), was defined for field failures and the procedure for its calculation for any specific failure was described in detail. The procedure was illustrated on datasets of Models 1 and 5 using the best CSI model. It was concluded that the frequency of occurrence may not be the best indicator for prioritizing field failures when a large number of the same are present. Instead, the CIP provides a better measure since it can quantitatively predict the improvement in CSI obtained by completely eliminating all future instances of a given type of field failure.

In near future, we wish to explore the implications of partial elimination of field failures. Other CRM specific studies can also be performed. The most immediate application being the utilization of the CSI function to identify the most dissatisfied and the most satisfied customer sets and then finding classification rules which can help identify potential customers for making buyback offers to. Such applications are potent tools in the hands of customer relation managers and they can help in improving the overall CSI immensely.

Acknowledgements The financial support and vehicle related data provided by India Science Lab, General Motors R&D are greatly appreciated. Authors thank Dr. Prakash G. Bharati, Dr. Pulak Bandyopadhyay and Dr. Pattada A. Kallappa for helpful discussions.

References

1. The American Customer Satisfaction Index. ACSI (2010). www.theacsi.org
2. JDPower.com. J.D. Power and Associates (2010). www.jdpower.com
3. Bandaru, S., Deb, K., Khare, V., Chougule, R.: Quantitative modeling of customer perception from service data using evolutionary optimization. In: Proceedings of the 13th annual conference on Genetic and evolutionary computation, pp. 1763–1770. ACM (2011)
4. ConsumerReports.org. Consumers Union of U.S., Inc. (2010). www.consumerreports.org
5. ConsumerReports. ConsumersReports.org (2010)
6. Deb, K., Agarwal, S., Pratap, A., Meyarivan, T.: A fast and elitist multi-objective genetic algorithm: NSGA-II. IEEE Transactions on Evolutionary Computation 6(2), 182–197 (2002)
7. Deb, K., Gupta, S.: Understanding knee points in bicriteria problems and their implications as preferred solution principles. Engineering optimization (2011)
8. J.D. Power/What Car? 2011 UK Vehicle Ownership Satisfaction Study (2011). www.jdpower.com
9. Lomax, R.: An introduction to statistical concepts for education and behavioral sciences. Lawrence Erlbaum (2001)
10. Robinson, J., Chukova, S.: Estimating mean cumulative functions from truncated automotive warranty data. In: Communications of the Fourth International Conference on Mathematical Methods in Reliability, Methodology and Practice, pp. CD–ROM (4 pages). Santa Fe, New Mexico, USA (2004)

Automatic Agricultural Leaves Recognition System

Meenakshi, Durga Puja, Mukesh Saraswat, K. V. Arya

Abstract : India is an agricultural country where large number of human beings are involved in cropping different plants for their living. But these plants may be affected by different diseases which are to be handled by the farmers within time to increase their productivity. An automatic plant disease identification system can be helpful for the farmers to identify the disease and their cures within time. Most of these diseases can be identified using the leaves of the plants. Therefore, an automatic classification of leaves would be the prior step for disease identification system. The leaf recognition system is a complex task due to the presence of large variations in the leaves. Therefore, this paper proposes a novel methodology for classification of agricultural leaves into their respective class based on principal component analysis (PCA) and support vector machine (SVM). The results show that the proposed method is accurate and fast enough to classify the leaves.

Keywords: Principal Component Analysis, Support Vector Machine, Leaf Segmentation, Leaf Classification

Meenakshi
BSA College of Engineering & Technology, Mathura, India.
e-mail: meenakshisaraswat6@gmail.com

Durga Puja
BSA College of Engineering & Technology, Mathura, India.
e-mail: durga.puja@bsacet.org

Mukesh Saraswat
ABV- Indian Institute of Information Technology & Management, Gwalior, India.
e-mail: saraswatmukesh@gmail.com

K. V. Arya
ABV- Indian Institute of Information Technology & Management, Gwalior, India.
e-mail: kvarya@iiitm.ac.in

J. C. Bansal et al. (eds.), *Proceedings of Seventh International Conference on Bio-Inspired Computing: Theories and Applications (BIC-TA 2012)*, Advances in Intelligent Systems and Computing 201, DOI: 10.1007/978-81-322-1038-2_11, © Springer India 2013

1 Introduction

India has a rich agricultural land and a large set of the population harvest different types of food plants. But due to the lack of technical facilities provided to farmers, many problems regarding the harvesting are faced by the farmers and every year a large area of crops are ruined. Lack of plants' disease knowledge is one of these problems which affects the farmers. There are very few experts who have a proper knowledge of these plants' diseases and may be approached by the farmers to increase the food productivity by asking their advice. Sometimes experts are at a long distance from farmers. Even though, if farmers cover these distances, expert may not be available at that time. Sometimes, the expert is also unaware of such type of diseases and could not help the farmers. For such situations, the expert advice becomes very expensive and time consuming. The crops can be saved from many diseases by providing appropriate and timely solutions to the farmers. Therefore, automated plant disease recognition system can provide the best solution to this problem. In the most of the cases, plants' diseases can be identified by seeing the leaves of the plants. Therefore, automated leaf recognition system is the prior step to solve this problem. Further, leaf recognition system can also be used to search important ayurvedic plants in the forests. By taking the photo of the leaf, the system can recognize the plant which can be very useful for human being.

Leaf recognition is a complicated task due to the large variations in the properties of leaves such as size, color, texture, shape, etc. Different strategies have been proposed to automate the recognition of the leaves. Most of the methods have been based on shape recognition techniques [1, 2, 3, 4, 5]. Sakai et al. [6] used geometrical parameters like area, perimeter, maximum length, maximum width, etc. to classify the four types of rice grains. Shape features give excellent results in recognizing the leaves having different shapes. For identifying the leaf having similar shapes, like beetle and pepper, texture and color properties are used [7]. Some researchers [8] proposed the combined texture and shape features to differentiate the leaves of the plant. But more number of features increase the computational complexity. Therefore, some methodologies are required which take less number of properties to make the recognition process fast and accurate.

In the proposed method, five different agricultural leaves were taken for classification. These are Cajanus cajan (Arhar), Chenopodium album (Bathua), Triticum aestivum (Wheat), Trifolium alexandrinum (Bearseem Clover), and Parthenium hysterophorus (Gajar Ghas). The leaves are extracted from the background image using Otsu's threshold method [9]. Further, a novel fully automated method for classification of leaves using principal component analysis (PCA) [10] and support vector machine (SVM) [11] is introduced which will use minimal set of features for classification. The overview of the whole systematic approach is shown in Figure 1.

Rest of the paper is organized as follows: Section 2 explains the proposed methodology. Experimental results are discussed in Section 3 and Section 4 concludes the paper.

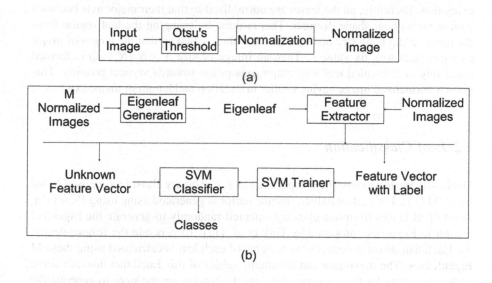

Fig. 1 Detailed flow graph of proposed method (a) Leaf segmentation followed by normalization, (b) Leaf classification.

2 Materials and Method

Database of 200 images of four different types of leaves (Arahar, Bathua, Wheat, and Bearseem Clover) was created with uniform background. Further, 36 images of Gajar Ghas were also taken as unwanted leaves. All the images were acquired using Nikon Coolpix S200 digital camera.

The proposed method as shown in the Figure 1 consists of two phases: leaf segmentation followed by classification. Phase-I segments the leaf from the background followed by shape normalization. Five cascaded binary classifiers using SVM are used for identification purpose. Detailed description of each phase is explained in the following subsections.

2.1 Leaf Segmentation

As all images have single leaf over uniform background, it is an easy task to segment the leaf from the background using threshold based methods. Therefore, to extract the leaf from the background, Otsu's threshold method [9] is used on gray level image which chooses the threshold to minimize the intra class variance of the leaf and background.

The varying shape, size, and orientation of the leaves impose criticality in feature extraction. Therefore, all the leaves are normalized so that their major axis becomes equivalent without shape disorder. This is done by cropping the leaf region from the image using boundary box property and then scaled so that the length of major axis increases to a fix value L. Then an image of size $N \times N$ ($N > L$) is formed consisting of the scaled leaf with major axis points towards vertical position. This gives a normalized image having similar orientation and length of major axis.

2.2 Leaf Classification

To classify the leaves into their respective class, five binary classifiers are used based on SVM [11]. For each classifier, feature vector is generated using using PCA [10]. A set of M leaves from one class are selected randomly to generate the Eigenleaf similar to Eigenface, proposed by Turk et al. [12]. To generate the feature vector, the Euclidian distance between the weights of each leaf is calculated using these M Eigenleaves. The maximum and minimum values of this Euclidian distance serve as feature vector for the corresponding leaf. Following are the steps to generate the features of the leaves.

1. A set S with M leaf images from one class has been formed as shown in Eq. 1.

$$S = \{A_1, A_2, ..., A_M\} \tag{1}$$

2. Mean image Φ from the set S using Eq. 2 has been found out.

$$\Phi = \frac{1}{M} \sum_{n=1}^{M} A_n \tag{2}$$

3. The difference θ has been calculated between the input and the mean image as shown in Eq. 3.

$$\theta = A_i - \phi \tag{3}$$

4. A set of M orthonormal vectors, u_n, have been found out describing the best distribution of the data. The eigenvectors, u_k, are chosen such that the eigenvalues, λ_k, as given by Eq. 4 is maximum.

$$\lambda_k = \frac{1}{M} \sum_{n=1}^{M} (u_k^T \theta_n)^2 \tag{4}$$

$$u_l^T u_k = \delta_{lk} = \left\{ \begin{array}{l} 1 \text{ if } l = k \\ 0 \text{ otherwise} \end{array} \right\} \tag{5}$$

5. The Eigenleaves are calculated using Eq. 6.

$$u_l = \sum_{k=1}^{M} u_{lk}\theta_k \; for \; l = 1,...,M \tag{6}$$

6. The input image is compared with the mean image. Their difference is multiplied with each eigenvector of the u_k. Each value would represent a weight, W_k.

$$W_k = u_k^T(A - \phi) \tag{7}$$

7. The weights of all images in set S are represented as W_1. Similarly weights of training set images and testing set images are represented as W_2 and W_3 respectively.
8. Feature sets f_1 and f_2 for learning and testing the classifier are calculated as shown in Eq. 8 and Eq. 9 respectively.

$$f_1 = \{max(W_1 - W_2), min(W_1 - W_2)\} \tag{8}$$

$$f_2 = \{max(W_1 - W_3), min(W_1 - W_3)\} \tag{9}$$

Above sets, f_1 and f_2, are used as feature sets for training and testing the classifier respectively. The feature vector, f_1, along with label of each leaf from training data set is given to SVM trainer which generates the classifier. The unknown leaves are processed similarly for feature extraction using M Eigenleaves and their weights are calculated. Then Euclidian distance between these weights of the leaves and weights of M Eigenleaves are find out and a maximum and minimum value are taken as feature vector to SVM classifier which classifies the input feature vector into its respective class. This process is repeated for all five classifiers which are further cascaded as a single system.

3 Experimental Results

Database consists of the 200 images of size 2304 x 3072 pixels from four different classes: (i) Chenopodium album (Bathua), (ii) Trifolium alexandrinum (Bearseem Clover), (iii) Cajanus cajan (Arhar), and (iv) Triticum aestivum (Wheat) each consisting of 50 images. 36 Images of Parthenium hysterophorus (Gajar Ghas) of similar size were also taken as unwanted leafs. Representative images of each class are shown in Figure 2. The data set is divided into two groups. First group consists of 118 images which are used for training the system to generate the SVM classifier. Other set, having 118 images, is used for testing the system for its accuracy.

To diminish the effect of variable shape, and varying orientation, all the images goes through normalization process after cropping of the leaves that results an image of size $N \times N$ with similar orientation and length of major axis of the leaf. One leaf after normalization from each class is shown in Figure 3. The feature vector for each leaf is calculated using eigen values. For the same, 20 Eigenleaves are generated for each class and one such set is represented in Figure 4.

Leaf Image	Scientific Name	Common Name
	Chenopodium album	Bathua
	Trifolium alexandrinum	Berseem Clover
	Cajanus cajan	Arhar
	Triticum aestivum	Wheat
	Parthenium hysterophorus	Gajar Ghas

Fig. 2 Representative images of individual class (a) Chenopodium album (Bathua), (b) Trifolium alexandrinum (Bearseem Clover), (c) Cajanus cajan (Arhar), (d) Triticum aestivum (Wheat), and (e) Parthenium hysterophorus (Gajar Ghas).

Fig. 3 Representative normalized images of individual class (a) Chenopodium album (Bathua), (b) Trifolium alexandrinum (Bearseem Clover), (c) Cajanus cajan (Arhar), (d) Triticum aestivum (Wheat), and (e) Parthenium hysterophorus (Gajar Ghas).

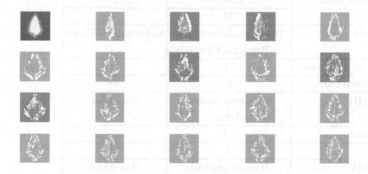

Fig. 4 Generated eigenleaves of Chenopodium album (Bathua) for feature extraction.

In the proposed method, five SVM based cascaded binary classifiers are used. The individual accuracy of each classifier for leaf identification from testing database is tabulated in Table 1. The confusion matrix for the same is given in Table 2. It is observed that classification accuracy decreases due to similarity in the leaf shape. Further, total accuracy of 77.9661% for whole system is achieved. From these results, it has been observed that accuracy degrades only in those images which have very much similarity in their structure.

Table 1 The individual accuracy of each classifier for leaf identification from testing database

	Accuracy
Classifier-I (Chenopodium album (Bathua))	79.6610
Classifier-II (Trifolium alexandrinum (Bearseem Clover))	88.1356
Classifier-III (Cajanus cajan (Arhar))	97.4576
Classifier-IV (Triticum aestivum (Wheat))	100
Classifier-V (Parthenium hysterophorus (Gajar Ghas))	90.6780

Table 2 The confusion matrix for each classifier for leaf identification from testing database

A	**Classifier-I**	Chenopodium album (Bathua)	Non-Bathua
	Chenopodium album	24	1
	Non-Bathua	23	70
B	**Classifier-II**	Trifolium alexandrinum (Bearseem Clover)	Non-Bearseem Clover
	Trifolium alexandrinum	11	14
	Non-Bearseem Clover	0	93
C	**Classifier-III**	Cajanus cajan (Arhar)	Non-Arhar
	Cajanus cajan	23	2
	Non-Arhar	1	92
D	**Classifier-IV**	Triticum aestivum (Wheat)	Non-Wheat
	Triticum aestivum	25	0
	Non-Wheat	0	93
E	**Classifier-V**	Parthenium hysterophorus (Gajar Ghas)	Non-Gajar Ghas
	Parthenium hysterophorus	9	9
	Non-Gajar Ghas	2	98

4 Conclusion

In this paper, we have developed a system to classify the agricultural leaves into their respective classes using PCA and SVM with an accuracy of 77.9661%. For most of the images with varying size and shape, the classification method described here is effective and computationally efficient. The classification method based on eigen values and SVM classifier gives perfect results in comparison to data taking different parameters as features. Further work will include more number of classes in the system and find out the best features to classify the leaves with state-of-the-art comparison.

References

1. Z. Wang, Z. Chi, D. Feng, and Q. Wang, "Leaf image retrieval with shape features," in *Advances in Visual Information Systems*, ser. Lecture Notes in Computer Science, R. Laurini, Ed. Springer Berlin / Heidelberg, 2000, vol. 1929, pp. 41–52.
2. Z. Wang, Z. Chi, and D. Feng, "Shape based leaf image retrieval," *Vision, Image and Signal Processing*, vol. 150, no. 1, pp. 34 – 43, feb 2003.
3. Y. Shen, C. Zhou, and K. Lin, "Leaf image retrieval using a shape based method," in *Artificial Intelligence Applications and Innovations*, ser. IFIP International Federation for Information Processing, D. Li and B. Wang, Eds. Springer Boston, 2005, vol. 187, pp. 711–719.
4. C.-L. Lee and S.-Y. Chen, "Classification of leaf images," *International Journal of Imaging Systems and Technology*, vol. 16, no. 1, pp. 15–23, 2006.
5. C. Caballero and M. C. Aranda, "Plant species identification using leaf image retrieval," in *Proc. of the ACM International Conference on Image and Video Retrieval*. New York, NY, USA: ACM, 2010.
6. N. Sakai, S. Yonekawa, A. Matsuzaki, and H. Morishima, "Two-dimensional image analysis of the shape of rice and its application to separating varieties," *Journal of Food Engineering*, vol. 27, no. 4, pp. 397–407, 1996.
7. A. R. Backes and O. M. Bruno, "Plant leaf identification using multi-scale fractal dimension," in *Proc. of the 15th International Conference on Image Analysis and Processing*. Berlin, Heidelberg: Springer-Verlag, 2009, pp. 143–150.
8. B. S. Bama, S. M. Valli, S. Raju, and V. A. Kumar, "Content based leaf image retrieval (cblir) using shape, color and texture features," *Indian Journal of Computer Science and Engineering*, vol. 2, no. 2, pp. 202–211, 2011.
9. N. Otsu, "A threshold selection method from gray-level histograms," *IEEE Transactions on Systems, Man and Cybernetics*, vol. 9, no. 1, pp. 62 –66, jan. 1979.
10. T. Jolliffe, *Principal Component Analysis*, ser. Springer Series in Statistics. Springer New York, 2006.
11. C. Cortes and V. Vapnik, "Support-vector networks," in *Machine Learning*, 1995, pp. 273–297.
12. M. A. Turk and A. P. Pentland, "Face recognition using eigenfaces," in *Proc. of IEEE Computer Society Conference on Computer Vision and Pattern Recognition*, jun 1991, pp. 586 –591.

Non-Uniform Mapping in
Binary-Coded Genetic Algorithms

Kalyanmoy Deb, Yashesh D. Dhebar, and N. V. R. Pavan

Kanpur Genetic Algorithms Laboratory (KanGAL)
Indian Institute of Technology Kanpur
PIN 208016, U.P., India
deb,yddhebar@iitk.ac.in,nvrpavan@yahoo.co.in

Abstract. Binary-coded genetic algorithms (BGAs) traditionally use a
uniform mapping to decode strings to corresponding real-parameter variable values. In this paper, we suggest a non-uniform mapping scheme
for creating solutions towards better regions in the search space, dictated by BGA's population statistics. Both variable-wise and vector-wise
non-uniform mapping schemes are suggested. Results on five standard
test problems reveal that the proposed non-uniform mapping BGA (or
NBGA) is much faster in converging close to the true optimum than the
usual uniformly mapped BGA. With the base-line results, an adaptive
NBGA approach is then suggested to make the algorithm parameter-
free. Results are promising and should encourage further attention to
non-uniform mapping strategies with binary coded GAs.

Keywords: Non-uniform mapping, binary-coded genetic algorithms, optimization, adaptive algorithm.

1 Introduction

Genetic algorithms (GAs) were originally started with a binary-coded representation scheme [1, 2], in which variables are first coded in binary strings comprising of Boolean variables (0 and 1). Since such a string of Boolean variables
resembled a natural chromosomal structure of multiple genes concatenating to
a chromosome, the developers of GAs thought of applying genetic operators (recombination and mutation) on to such binary strings – hence the name genetic
algorithms.

For handling problems having real-valued variables, every real-valued variable
x_i is represented using a binary substring $\mathbf{s}^i = (s_1^i, s_2^i, \ldots, s_{\ell_i}^i)$ where $s_j^i \in \{0, 1\}$.
All n variables are then represented by $\ell = \sum_{i=1}^{n} \ell_i$ bits. Early GA researchers
suggested a uniform mapping scheme to compute x_i from a substring $\mathbf{s}^i$:

$$x_i = x_i^{(L)} + \frac{x_i^{(U)} - x_i^{(L)}}{2^{\ell_i} - 1} DV(\mathbf{s}^i), \qquad (1)$$

where $\ell_i = |\mathbf{s}^i|$ is the number of Boolean variables (or bits) used to represent
variable x_i, $x_i^{(L)}$ and $x_i^{(U)}$ are lower and upper bounds of variable x_i and $DV(\mathbf{s}^i)$

J. C. Bansal et al. (eds.), *Proceedings of Seventh International Conference on Bio-Inspired
Computing: Theories and Applications (BIC-TA 2012)*, Advances in Intelligent Systems
and Computing 201, DOI: 10.1007/978-81-322-1038-2_12, © Springer India 2013

is the decoded value of string $\mathbf{s}^i$, given as follows: $DV(\mathbf{s}^i) = \sum_{j=1}^{\ell} 2^{j-1} s_j^i$. This mapping scheme is uniform in the range $[x_i^{(L)}, x_i^{(U)}]$, because there is a constant gap between two consecutive values of a variable and the gap is equal to $(x_i^{(U)} - x_i^{(L)})/(2^{\ell_i} - 1)$. The uniform mapping mentioned above covers the entire search space with uniformly distributed set of points. Without any knowledge about good and bad regions in the search space, this is a wise thing to do and early users of BGA have rightly used such a scheme. However, due to its uniformity, BGA requires a large number of iterations to converge to the requisite optimum. Moreover, in every generation, the GA population usually has the information about the current-best or best-so-far solution. This solution can be judiciously used to make a faster search.

In this paper, we suggest the possibility of a non-uniform mapping scheme in which the above equations are modified so that binary substrings map non-uniformly in the search space. If the biasing can be done to have more points near the good regions of the search space, such a non-uniform mapping may be beneficial in creating useful solutions quickly. The binary genetic algorithm (BGA) framework allows such a mapping and we investigate the effect of such a mapping in this paper.

In the remainder of the paper, we first describe the proposed non-uniform mapping scheme and then describe the overall algorithm (we call it as NBGA). Thereafter, we present simulation results of NBGA on a number of standard problems and compare its performance with original uniformly mapped BGA. The NBGA approach involves an additional parameter. Finally, we propose an adaptive scheme (ANBGA) that do not require the additional parameter and present simulation results with the adaptive NBGA approach as well. Conclusions of this extensive study are then made.

2   Past Efforts of Non-Uniform Mapping in BGA

Despite the suggestion of BGAs in early sixties, it is surprising that there does not exist too many studies related to non-uniform mapping schemes in coding binary substrings to real values. However, there are a few studies worth mentioning.

ARGOT [3] was an attempt to adaptively map fixed binary strings to the decoded variable space. The methodologies used in the study used several environmentally triggered operators to alter intermediated mappings. These intermediate mappings are based on internal measurements such as parameter convergence, parameter variance and parameter 'positioning' within a possible range of parameter values. The dynamic parameter encoding (DPE) approach [4] adjusts the accuracy of encoded parameters dynamically. In the beginning of a GA simulation, a string encodes only most significant bits of each parameter (say 4bits or so), and when GA begins to converge, most significant bits are recorded and dropped from the encoding. New bits are introduced for additional precision. In the delta coding approach [5], after every run, the population is reinitialized with the substring coding for each parameter representing distance or Δ value away from the corresponding parameter in best solution of the previous run, thereby forming a new hypercube around the best solution. The size of hypercube is controlled by adjusting number of bits used for encoding.

All the above were proposed in late eighties and early nineties, and have not been followed up adequately. Non-uniform mapping in BGA can produce faster convergence if some information about a good region in the search space is identified. Here, we suggest a simple procedure that uses the population statistics to create a non-uniform mapping in BGAs.

3   Proposed Non-Uniformly Mapped BGA (NBGA)

In NBGA, more solutions get created near the best-so-far solution $(\mathbf{x}^{b,t})$ at any generation t. We suggest a polynomial mapping function for this purpose with a user-defined parameter η. Let us consider Figure 1 in which the lower bound is a, upper bound is b of variable x_i and i-th variable value of best-so-far solution is $x_i^{b,t}$ at the current generation t. Let us also assume that the substring $\mathbf{s}^i$ decodes

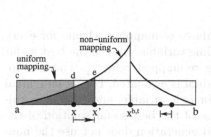

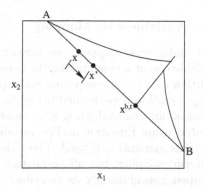

Fig. 1. Non-uniform mapping scheme is shown. Area acdxa is equal to area aex'a.

Fig. 2. Vector-wise mapping scheme is shown.

to the variable value x_i using Equation 1 with $a = x_i^{(L)}$ and $b = x_i^{(U)}$. The quantity in fraction varies between zero and one, as the decoded value $DV(\mathbf{s}^i)$ varies between zero and $(2^{\ell_i} - 1)$. Assuming the fraction $\zeta = DV(\mathbf{s}^i)/(2^{\ell_i} - 1)$, which lies in $[0, 1]$, we observe a linear behavior between x_i and ζ for uniform mapping used in BGA, $x_i = a + \zeta(b - a)$, or $\zeta = (x_i - a)/(b - a)$.

For NBGA, we re-map value x_i to obtain x_i' $(m : x_i \to x_i')$ in the non-dimensionalized x-space $(\zeta = (x-a)/(b-a))$, so that there is more concentration of points near $x_i^{b,t}$, as follows (with $\eta \geq 0$) and as shown in Figure 1:

$$m(\zeta) = k\zeta^{\eta}, \qquad (2)$$

where k is a constant, which is chosen in such a manner that the point $x_i = x_i^{b,t}$ remains at its own place, that is, $\int_0^{(x_i^{b,t}-a)/(b-a)} m(\zeta)d\zeta = (x_i^{b,t}-a)/(b-a)$. This condition gives us: $k = (\eta+1)(b-a)^\eta/(x_i^{b,t}-a)^\eta$. Substituting k to Equation 2 and equating area under the mapped curve ($\int_0^{(x_i'-a)/(b-a)} m(\zeta)d\zeta$) with area under uniform mapping ($(x_i - a)/(b - a)$), we have the following mapping:

$$x_i' = a + \left((x_i - a)(x_i^{b,t} - a)^\eta\right)^{1/(1+\eta)} . \tag{3}$$

Figure 1 shows how the points in between $a \leq x_i \leq x_i^{b,t}$ are re-mapped to move towards $x_i^{b,t}$ for $\eta > 0$. For a point in the range $x_i^{b,t} \leq x_i \leq b$, a mapping similar to the above will cause the solution to move towards $x_i^{b,t}$ as well, as shown in the figure. It is clear that depending on the location of $x_i^{b,t}$ at a generation, points towards left and right of it will get re-mapped towards the best-so-far point – thereby providing a non-uniform mapping of the binary substrings in the range $[a, b]$. Note that the discontinuity of mapping function at $x_i^{b,t}$ is not an important matter, as solutions on either side of $x_i^{b,t}$ is considered separately.

The above non-uniform mapping scheme can be implemented in two ways: (i) variable-wise and (ii) vector-wise. We describe them in the following subsections.

3.1 Variable-wise Mapping

In variable-wise mapping, we perform the above re-mapping scheme for every variable one at a time, taking the corresponding variable value of the best-so-far solution $(x_i^{b,t})$ and computing corresponding re-mapped value x_i'. This remapping operation is performed before an individual is evaluated. For an individual string $\mathbf{s}$, first the substring $\mathbf{s}^i$ is identified for i-th variable, and then x_i' is calculated using Equation 3. The variable value of the best-so-far solution of the previous generation is used. Thus, the initial generation does not use the non-uniform mapping, but all population members from generation 1 onwards are mapped non-uniformly as described. It is important to note that the original string $\mathbf{s}$ is unaltered. The individual is then evaluated using variable vector $\mathbf{x}'$. The selection, recombination and mutation operators are performed on the string $\mathbf{s}$ as usual.

3.2 Vector-wise Mapping

In the vector-wise mapping, we coordinate the mapping of all variables in a systematic manner. The mapping operation is explained through Figure 2 for a two-variable case. For any individual $\mathbf{s}$, its location $(\mathbf{x})$ in the n-dimensional variable space can be identified using the original uniform mapping. Also, the location of the best-so-far $(\mathbf{x}^{b,t})$ solution in the n-dimensional space is located next. Thereafter, these two points are joined by a straight line and the line is extended to find the extreme points ($\mathbf{A}$ and $\mathbf{B}$) of the bounded search space. Parameterizing, $\mathbf{x}' = \mathbf{A} + d(\mathbf{x} - \mathbf{A})$, and substituting $a = 0$, $x_i = 1$ and $x_i^{b,t} =$

$\|\mathbf{x}^{b,t} - \mathbf{A}\|/\|\mathbf{x} - \mathbf{A}\|$ in Equation 3, we compute the re-mapped point x_i' and save it as d'. Thereafter, the re-mapped variable vector can be computed as follows:

$$\mathbf{x}' = \mathbf{A} + d'(\mathbf{x} - \mathbf{A}). \tag{4}$$

The re-mapped vector will be created along the line shown in the figure and will be placed in between $\mathbf{x}$ and $\mathbf{x}^{b,t}$. The vector-wise re-mapping is done for every population member using the best-so-far solution and new solutions are created. They are then evaluated and used in the selection operator. Recombination and mutation operators are used as usual. Binary strings are unaltered. The vector-wise re-mapping is likely to work well on problems having a linkage among variables. We present the results obtained from both methods in the following section.

4 Results

We use the binary tournament selection, a multi-variable crossover, in which a single-point crossover is used for each substring representing a variable, and the usual bit-wise mutation operator. Variable-wise mapping is used in this section.

We consider six standard unconstrained objective functions, which are given below:

$$\text{Sphere: } f(\mathbf{x}) = \sum_{i=1}^{n} x_i^2, \tag{5}$$

$$\text{Ellipsoidal: } f(\mathbf{x}) = \sum_{i=1}^{n} i x_i^2, \tag{6}$$

$$\text{Ackley: } f(\mathbf{x}) = -20 \exp\left(-0.2\sqrt{\frac{1}{n}\sum_{i=1}^{n} x_i^2}\right) - \exp\left(\frac{1}{n}\sum_{i=1}^{n}\cos(2\pi x_i)\right) + 20 + e, \tag{7}$$

$$\text{Schwefel: } f(\mathbf{x}) = \sum_{i=1}^{n}\left(\sum_{j=1}^{i} x_i\right)^2, \tag{8}$$

$$\text{Rosenbrock: } f(\mathbf{x}) = \sum_{i=1}^{n-1}\left[100(x_{i+1} - x_i^2)^2 + (1 - x_i^2)\right]. \tag{9}$$

All problems are considered for $n = 20$ variables. Following parameter values are fixed for all runs:

- Population size = 100,
- String length of each variable = 15,
- Lower Bound = −8, Upper Bound = 10,
- Selection type: Tournament Selection,
- Crossover: Variable wise crossover with crossover probability = 0.9,
- Mutation: Bitwise mutation with mutation probability = 0.0033.

An algorithm terminates when any of the following two scenarios take place: (i) a maximum generation of $t_{max} = 3,000$ is reached, or (ii) the difference between the best objective function value and the corresponding optimal function value of 0.01 or less is achieved. For each function, 50 runs are performed and the number of function evaluations for termination of each run is recorded. Thereafter, minimum, median and maximum number of function evaluations (FE) are reported. Note that a run is called success (S), if a solution with a difference in objective function value from the optimal function value of 0.01 is achieved, otherwise the run is called a failure (F). In the case of failure runs, the obtained objective function values (f) are reported.

4.1   Linear Increase in η:

Figure 1 suggests that a large value of η will bring solutions close to the current best solution. During initial generations, the use of a large η may not be a good strategy. Thus, first, we use an η-update scheme in which η is increased from zero linearly with generations to a user-specified maximum value η_{max} at the maximum generation t_{max}:

$$\eta(t) = \frac{t}{t_{max}} \eta_{max}. \tag{10}$$

We use different values of $\eta_{max} = 10, 100, 1000, 5000, 10000,$ and $15000,$ for variable-wise mapping case and compare results with the usual uniformly-mapped binary coded GA. The results are tabulated for sphere function in Table 1. The

Table 1. Performance of linearly increasing η schemes are shown for the sphere function.

Method	FE or f	S or F	min	median	max
BGA	FE	S = 17	9,701	13,201	23,501
	f	F = 33	0.0517	0.0557	0.157
NBGA ($\eta_{max} = 10$)	FE	S = 50	10,501	12,901	176,701
NBGA ($\eta_{max} = 100$)	FE	S = 50	7,401	10,101	87,901
NBGA ($\eta_{max} = 1,000$)	FE	S = 50	3,901	5,901	19,501
NBGA ($\eta_{max} = 5,000$)	FE	S = 50	2,301	2,901	3,501
NBGA ($\eta_{max} = 10,000$)	FE	S = 50	**1,801**	**2,301**	**2,801**
NBGA ($\eta_{max} = 15,000$)	FE	S= 50	1,901	2,301	3,101

first aspect to note is that the original BGA (with uniform mapping) is not able to solve the sphere problem in 33 out of 50 runs in 3,000 generations, whereas all NBGA runs with $\eta_{max} \geq 10$ are able to find the desired solution in all 50 runs. This is a remarkable performance depicted by the proposed NBGA.

The best NBGA performance is observed with $\eta_{max} = 10,000$. Note that although such a high η_{max} value is set, the median run with 2,301 function evaluations required 23 generations only. At this generation, η value set was $\eta =$

$(23/3000) \times 10000$ or 76.67. The important matter in setting up $\eta_{max} = 10,000$ is that it sets up an appropriate rate of increase of η with generations (about 3.33 per generation) for the overall algorithm to work the best. An increase of η from 1.67 per generation to 3.33 per generation seems to work well for the Sphere function, however, an increase of η by 5 per generation (equivalent to $\eta_{max} = 15,000$) is found to be not so good for this problem. Since the sphere function is a unimodal problem, a large effective η maps the solution closer to the best-so-far solution. But a too close a solution to the best-so-far solution (achieved with a large η) may cause a premature convergence, hence a deterioration in performance is observed.

Figure 3 shows the variation in function value of the population-best solution with generation for the sphere function of the best run. It is clear that NBGA is much faster compared to BGA.

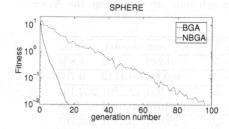 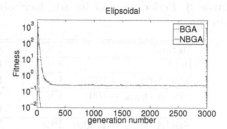

Fig. 3. Population-best objective function value with generation for the sphere function.

Fig. 4. Population-best objective function value with generation for the ellipsoidal function.

Next, we consider the ellipsoidal function and results are shown in Table 2. For this problem, $\eta_{max} = 15,000$ turns out to be the best option. This corre-

Table 2. Performance of linearly increasing η schemes are shown for the ellipsoidal function.

Method	FE or f	S or F	min	median	max
BGA	f	F = 50	0.2374	4.5797	29.6476
NBGA ($\eta_{max} = 10$)	FE	S = 11	40,001	236,501	284,201
	f	F = 39	0.01	0.0564	3.3013
NBGA ($\eta_{max} = 100$)	FE	S = 46	24,501	78,001	198,101
	f	F = 04	0.0137	0.01986	0.0246
NBGA ($\eta_{max} = 1,000$)	FE	S = 50	12,001	31,401	125,201
NBGA ($\eta_{max} = 5,000$)	FE	S = 50	7,301	21,601	48,601
NBGA ($\eta_{max} = 10,000$)	FE	S = 50	6,401	8,501	29,801
NBGA ($\eta_{max} = 15,000$)	FE	S = 50	**5,701**	**7,701**	**21,501**

sponds to an increase of η of 5 per generation. At the final generation of the best run (having minimum function evaluations), an effective $\eta = (57/3000) \times 15000$ or 285 was used. This problem is also unimodal; hence a larger η produced a faster convergence. Notice here that the original BGA (with uniform mapping of variables) is not able to find the desired solution (close to the true optimum) in all 50 runs in a maximum of 3,000 generations, whereas all runs with $\eta_{max} \geq 1,000$ are able to solve the problem in all 50 runs with a small number of function evaluations. Figure 4 shows the variation in function value of the population-best solution with generation for the ellipsoidal function of the best run. Again, NBGA is much faster compared to BGA.

The performance of BGA and NBGA approaches on Ackley's function are shown in Table 3. The best performance is observed with $\eta_{max} = 10,000$ with

Table 3. Performance of linearly increasing η schemes are shown for the Ackley's function.

Method	FE or f	S or F	min	median	max
BGA	f	F = 50	0.249	1.02	1.805
NBGA ($\eta_{max} = 10$)	f	F = 50	0.0208	0.1113	0.703
NBGA ($\eta_{max} = 100$)	FE	S = 33	74,901	122,801	232,101
	f	F = 17	0.01073	0.0203	0.08
NBGA ($\eta_{max} = 1,000$)	FE	S = 50	20,601	32,401	203,701
NBGA ($\eta_{max} = 5,000$)	FE	S = 50	11,001	40,101	110,701
NBGA ($\eta_{max} = 10,000$)	FE	**S = 50**	**10,301**	**31,601**	**66,701**
NBGA ($\eta_{max} = 15,000$)	FE	S = 36	9,201	25,701	51,301
	f	F = 14	1.1551	1.1551	2.0133

100% successful runs. Here, $\eta \approx 1,054$ is reached at the final generation of a median NBGA's run. Importantly, a rate of increase of 3.33 per generation seems to be the best choice for Ackley's function. The BGA is found to fail in all 50 runs. Figure 5 shows the objective function value of the population-best solution with generation for Ackley's function. The results from the median run is shown. The proposed NBGA approach is found to be much faster compared to BGA.

Next, we consider Schwefel's function. Table 4 tabulates the results. Here, $\eta_{max} = 15,000$ performs the best and the original BGA fails to make any of its 50 runs successful. Figure 6 shows the objective function value of the population-best solution with generation for Schwefel's function. The proposed NBGA approach is found to be much faster compared to BGA.

Next, we consider Rosenbrock's function. Table 5 shows the results. This function is considered to be a difficult function to solve for the optimum. BGA could not find the desired optimum in all 50 runs. Also, NBGAs with small η_{max} values could not solve the problem. But, NBGA with $\eta_{max} = 10,000$ is able to find the desired solution in 27 out of 50 runs.

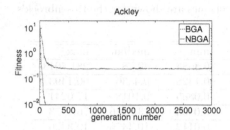
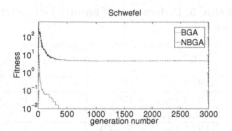

Fig. 5. Population-best objective function value with generation for Ackley's function.

Fig. 6. Population-best objective function value with generation for Schwefel's function.

Table 4. Performance of linearly increasing η schemes are shown for the Schwefel's function.

Method	FE or f	S or F	min	median	max
BGA	f	F = 50	4.7183	17.417	37.6447
NBGA($\eta_{max} = 10$)	f	F = 50	2.129	6.2925	16.093
NBGA ($\eta_{max} = 100$)	f	F = 50	1.1814	2.3364	3.8168
NBGA ($\eta_{max} = 1,000$)	FE	S = 19	73,501	251,001	298,501
	f	F = 31	0.0107	0.0436	0.1954
NBGA ($\eta_{max} = 5,000$)	FE	S = 50	27,601	144,401	292,801
NBGA ($\eta_{max} = 10,000$)	FE	S = 50	35,501	104,901	167,301
NBGA ($\eta_{max} = 15,000$)	FE	S = 50	**25,701**	**74,401**	**124,301**

It is clear from the above section that (i) The proposed non-uniform mapping method (NBGA) performs much better than the usual BGA, and (ii) In general, a high value of η_{max} produces a better performance.

5 Adaptive NBGA

An increasing η with a rate of about 3 to 5 per generation seems to work well, but the optimal performance depends on the problem. In this section, we propose an adaptive approach in which, instead of a predefined increase in η with generation, the algorithm will decide whether to increase or decrease η after every few generations.

In the adaptive NBGA, we initialize η to be zero at the initial generation. Thereafter, we keep track of the current-best [1] objective function value ($f(\mathbf{x}^{cb,t})$) at every five generations, and re-calculate η_{max}, as follows:

$$\eta_{max} = \eta_{max} - 100\frac{f(\mathbf{x}^{cb,t}) - f(\mathbf{x}^{cb,t-5})}{f(\mathbf{x}^{cb,t-5})}. \tag{11}$$

---

[1] The current-best solution is different from the best-so-far solution, as the former is the population-best solution at the current generation.

Table 5. Performance of linearly increasing η schemes are shown for the Rosenbrock's function.

Method	FE or f	S or F	min	median	max
BGA	f	F = 50	0.3686	42.78	281.7013
NBGA ($\eta_{max} = 10$)	f	F = 50	0.1328	0.4836	90.1493
NBGA ($\eta_{max} = 100$)	f	F = 50	0.0865	0.1918	17.0241
NBGA ($\eta_{max} = 1,000$)	f	F = 50	0.0505	13.2236	76.1873
NBGA ($\eta_{max} = 5,000$)	f	F = 50	0.0112	10.4476	19.3426
NBGA ($\eta_{max} = 10,000$)	FE	**S = 27**	200,301	247,001	288,101
	f	F = 23	0.012	0.3443	19.0727
NBGA($\eta_{max} = 15,000$)	FE	S = 24	109,001	183,701	213,801
	f	F = 26	0.0112	3.9888	69.2418

The above equation suggests that if the current-best is better than the previous current-best solution (five generations ago), η_{max} is increased by an amount proportional to percentage decrease in objective function value. Since an increase in η_{max} causes decoded values to move closer to the best-so-far solution, an improvement in current-best solution is rewarded by a further movement towards the best-so-far solution. On the other hand, if the new current-best is worse than the previous current-best solution, η_{max} is reduced. In effect, an effort is being made to increase diversity of the population by not moving the solutions much towards the best-so-far solution.

When the change in the current-best function value is less than 0.001 in three consecutive updates of η_{max}, it is likely that the population diversity is depleted and in such a case we set $\eta_{max} = 0$ and also set mutation probability to 0.5 to introduce diversity in the population. From the next generation on, the mutation probability is set back to 0.0033 as usual, but η_{max} is set free to get updated using Equation 11. After updating η_{max} after every five generations, η is varied from $\eta_{max}/5$ to η_{max} for the next five generations in uniform steps.

5.1 Results with Adaptive NBGA Approach

All five problems are chosen and identical parameter settings as those used in Section 4 are used here. Table 6 compares the performance of the adaptive NBGA approach with both variable-wise and vector-wise non-uniform mappings. In both cases, identical adaptation scheme (Equation 11) is used. The table also compares the adaptive NBGA results with the original variable-wise NBGA approach discussed earlier.

Figure 7 shows variation of η_{max} and population-best objective value with generation on three problems. It can be observed that both variable-wise and vector-wise non-uniform mapping approaches perform well. However, the fact that adaptive approach does not require a pre-defined η_{max} value and is still able to perform almost similar to the pre-defined increasing η-scheme, we recommend the use of the adaptive NBGA approach.

Table 6. Comparison of performance between adaptive NBGA and original NBGA approaches.

Function	Method	S or F	FE/f	min	median	max
Sphere	Variable-wise	S = 50	FE	2,001	3,401	4,501
	Vector-wise	S = 50	FE	2,001	2,601	3,501
	orig. NBGA (η_{max} = 10000)	S = 50	FE	**1,801**	**2,301**	**2,801**
Ellipsoidal	Variable-wise	S = 50	FE	5,401	8,501	**14,601**
	Vector-wise	S = 50	FE	**5,101**	**7,501**	17,101
	orig. NBGA (η_{max} = 15000)	S = 50	FE	5,701	7,701	21,501
Ackley	Variable-wise	S = 50	FE	**8,901**	**29,901**	69,001
	Vector-wise	S = 49	FE	9,101	25,101	241,001
		F = 01	f	1.841	1.841	1.841
	orig. NBGA (η_{max} = 10000)	S = 50	FE	10,301	31,601	**66,701**
Schwefel	Variable-wise	S = 50	FE	24,501	**58,001**	**103,601**
	Vector-wise	S = 50	FE	**18,801**	75,601	113,901
	orig. NBGA (η_{max} = 15, 000)	S = 50	FE	25,701	74,401	124,301
Rosenbrock	Variable-wise	F = 50	f	0.019	0.092	12.413
	Vector-wise	F = 50	f	0.023	0.083	12.888
	orig. NBGA (η_{max} = 10000)	**S = 27**	FE	200,301	247,001	288,101
		F = 23	f	0.012	0.3443	19.0727

6 Conclusions

In this paper, we have suggested two non-uniform mapping schemes in binary-coded genetic algorithms (BGA). Using the best-so-far solution information at every generation, the variable-wise and vector-wise non-uniformly coded BGAs (or NBGAs) map binary substrings to a variable value that is somewhat closer to the best-so-far solution than its original decoded value. The non-uniform mapping requires a parameter η_{max}, which takes a non-negative value. The variable-wise approach with an increasing η_{max}-update is applied to five standard test problems and the performance of the proposed NBGA has been found to be much better than the BGA approach that uses a uniform mapping scheme.

Thereafter, the NBGA approach has been made adaptive by iteratively updating the η_{max} parameter associated with the NBGA approach. Starting with η_{max} = 0 (uniform mapping) at the initial generation, the adaptive NBGA approach has been found to update the η_{max} parameter adaptively so as to produce similar results as in the case of increasing η_{max}-update strategy. Both variable-wise and vector-wise non-uniform approaches have been found to perform equally well.

The research in evolutionary algorithms originally started with binary-coded GAs, but recently their use has been limited due to the advent of real-parameter GAs and other EC methodologies. Instead of BGA's usual choice of uniform mapping, the effective use of a non-uniformly mapped representation scheme demonstrated in this paper should cause a resurrection of BGAs in the near future.

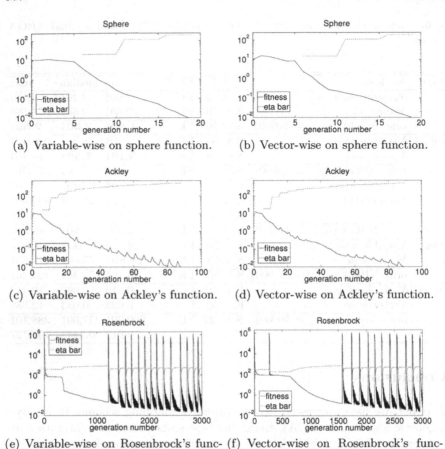

(a) Variable-wise on sphere function.   (b) Vector-wise on sphere function.

(c) Variable-wise on Ackley's function.   (d) Vector-wise on Ackley's function.

(e) Variable-wise on Rosenbrock's func-   (f) Vector-wise on Rosenbrock's function.   tion.

Fig. 7. Variations in $\eta_{\max}$ with generation number for variable-wise and vector-wise non-uniform mapping are shown.

References

1. J. H. Holland. *Adaptation in Natural and Artificial Systems.* Ann Arbor, MI: MIT Press, 1975.
2. D. E. Goldberg. *Genetic Algorithms for Search, Optimization, and Machine Learning.* Reading, MA: Addison-Wesley, 1989.
3. C. G. Shaefer. The argot strategy: Adaptive representation genetic optimizer technique. In *Proceedings of the Second International Conference on Genetic Algorithms,* pages 50–58, 1987.
4. N. N. Schraudolph and R. K. Belew. Dynamic parameter encoding for genetic algorithms. Technical Report LAUR90-2795, Los Alamos: Los Alamos National Laboratory, 1990.
5. D. Whitley, K. Mathias, and P. Fitzhorn. Delta coding: An iterative search strategy for genetic algorithms. In *Proceedings of the Fourth International Conference on Genetic Algorithms,* pages 77–84. San Mateo, CA: Morgan Kaufmann, 1991.

Control Words of Transition P Systems

Ajeesh Ramanujan and Kamala Krithivasan

Abstract A new way of associating a language with the computation of a P system is considered. A label is assigned to every rule in a P system, where the labels are chosen from a finite alphabet or λ. We associate a string, called control word, that is obtained by concatenating the labels of the rules in the transition sequence corresponding to a computation. We study the generative capacity of such control languages comparing them with family of languages such as regular, context-free, context-sensitive and recursively enumerable languages of Chomskian hierarchy.

Key words: P System; Finite language; Regular language; Context-free language; Context-sensitive language; Recursively enumerable language

1 Introduction

P systems introduced in [1] by Gh. Păun, are computing models inspired from the structure and functioning of the living cells. A P system consists of a finite number of membranes, each of which contains a multiset of objects from a finite alphabet. The membranes are organized as a Venn diagram or a tree structure where membranes may contain other membranes. The dynamics of the system is governed by a set of rules associated with each membrane. Each rule specifies how objects evolve and move into neighboring membranes. It has been introduced as a computing model which abstracts from the way live cells process chemical compounds in their compartmental (membrane) structure. Various models of P systems have been

Ajeesh Ramanujan
IITM, Indian Institute of Technology, Madras, Chennai-36, India, e-mail: ajeeshramanu-jan@gmail.com

Kamala Krithivasan
IITM, Indian Institute of Technology, Madras, Chennai-36, India e-mail: kamala@iitm.ac.in

J. C. Bansal et al. (eds.), *Proceedings of Seventh International Conference on Bio-Inspired Computing: Theories and Applications (BIC-TA 2012)*, Advances in Intelligent Systems and Computing 201, DOI: 10.1007/978-81-322-1038-2_13, © Springer India 2013

shown to be equivalent to Turing machines in computing power. For recent developments see [10].

Different ways of associating languages with P system were considered so far like external output, introduced in [3], where the objects leaving the system is collected and a language is obtained by arranging the symbols in a sequence (if more than one symbol exit the system every permutation is considered), P automata introduced in [5] (for more details see [4]), considered as a language acceptor, similar to external output, by arranging the symbols entering the system in a sequence, and trace introduced in [6], where an object is marked and its path is followed across membranes and the sequence of membrane labels visited by that object is considered as a string. Recently, another way of association language called control word with spiking neural P systems is considered in [9]. In [8], control words for other variants P systems is proposed but not investigated in detail. In this paper we extend the concept of control words introduced in [9, 8] to transition P system and we study the language generating capacity of such devices.

The paper is organized as follows. In Section 2, we review some automata theory basics and the basic definition and working of a register machine required for this paper. In Section 3, we give the definition of a system as defined in [1]. In Section 4 we introduce and define the control language associated with P systems and study the power of language generated with respect to a control language associated with P systems.

2 Basic Definition

In this section we give some definitions and notations related to automata theory.

Let Σ be a finite set of symbols called an alphabet. A string w over Σ is a sequence of symbols from Σ. λ denotes the empty string. The set of all strings over Σ is denoted by Σ^*. The length of a string $w \in \Sigma^*$ is denoted by $|w|$. A language L over Σ is a set of strings over Σ. The family of finite, regular, context-free, context-sensitive and recursively enumerable languages is denoted by FIN, REG, CF, CS and RE respectively.

A language $L \subseteq \Sigma^*$ is said to be regular if there is a regular expression E over Σ such that $L(E) = L$. The regular expressions are defined using the following rules. (i.) $\emptyset, \lambda$ and each $a \in \Sigma$ are regular expressions representing the regular set $\emptyset, \{\lambda\}$ and $\{a\}$ respectively. (ii.) If E_1 and E_2 are regular expressions over Σ representing the regular set R_1 and R_2, then $E_1 + E_2, E_1 E_2$ and E_1^* are regular expressions over Σ representing the regular sets $R_1 \cup R_2, R_1 R_2$ and R_1^* respectively and (iii.) nothing else is a regular expression over Σ. With each regular expression E, we associate a language $L(E)$.

For a set U, a multiset over U is a mapping $M : U \to N$, where N is the set of nonnegative integers. For $a \in U, M(a)$ is the multiplicity of a in M. If the set U is finite, $U = \{a_1, a_2, \cdots, a_n\}$, then the multiset M can be explicitly given in the form $\{(a_1, M(a_1)), (a_2, M(a_2)), \cdots, (a_n, M(a_n))\}$, thus specifying for each element of U its multiplicity in M. In membrane computing, the usual way to represent a multiset $M = \{(a_1, M(a_1)), (a_2, M(a_2)), \cdots, (a_n, M(a_n))\}$ over a finite set U is by using strings $w = a_1^{M(a_1)} a_2^{M(a_2)}, \cdots a_n^{M(a_n)}$ and all permutations of w represents M and the empty multiset is represented by λ. For example, the multiset $\{a, a, a, a, a, b, b, b, b, b, b, c\}$ is represented as $a^5 b^6 c$.

A register machine is a construct $M = (m, H, l_0, l_h, I)$, where m is the number of registers, H is the set of instruction labels, l_0 is the start label (labeling an *ADD* instruction), l_h is the halt label, (assigned to instruction halt), and I is the set of instructions labeled in a one-to-one manner by the labels from H. The instructions are of the following forms:

- $l_i : (ADD(r), l_j)$ (add 1 to register r and then go to the instruction with label l_j),
- $l_i : (SUB(r), l_j, l_k)$ (if register r is non-empty, then subtract 1 from it and go to the instruction with label l_j, otherwise go to the instruction with label l_k,
- $l_h : HALT$ (the halt instruction).

A register machine M accepts a number n in the following way: we start with number n in a specified register r_0 and all other registers being empty (i.e., storing the number 0), we first apply the instruction with label l_0 and we proceed to apply instructions as indicated by the labels (and made possible by the contents of the registers); if we reach the halt instruction, then the number n is said to be accepted by M. The set of all numbers accepted by M is denoted by $N(M)$. It is known (see, e.g., [7]) that register machines (even with only three registers, but this detail is not relevant in what follows) accepts all sets of numbers which are Turing computable.

In this paper all the vectors are row vectors and written in boldface letter. For example $\mathbf{u}$ represents a row vector u. The jth component of a vector $\mathbf{u}$ is denoted by $\mathbf{u}_j$.

3 P Systems

In this section we give the basic definition of a P system as given in [2]. For more details refer [1, 2].

Definition 1. A P system (with multiset rewriting rule) of degree $m \geq 1$ is a construct $\Pi = (O, H, \mu, w_1, w_2, \cdots, w_m, R_1, R_2, \cdots, R_m, i_0)$, where:

1. O is the alphabet of objects;
2. H is the alphabet of membrane labels;
3. μ is a membrane structure of degree m;
4. $w_1, w_2, \cdots, w_m \in O^*$ are multisets of objects associated with the m regions of Π;
5. $R_i, 1 \leq i \leq m$, are finite sets of multiset rewriting rules of the form $u \to v$, where u is a string over u, v is a string over $\{(a, here), (a, out), (a, in) | a \in O\}$ associated with the m regions of μ;
6. $i_0 \in H \cup \{e\}$ specifies the input/output region of Π, where e is a reserved symbol not in H.

In this paper we write a rule $a^2bc^3 \rightarrow (b, here)(a, here)(c, here)(a, here)(d, out)$ $(a, out)(c, in)(a, in)$ in a more compact way as $a^2bc^3 \rightarrow ba^2c(da, out)(ca, in)$. Here we consider the system as a language generator. So we do not specify any input/output region and in all the constructions that follows, we set $i_0 = \emptyset$.

A membrane system Π works as follows: It starts with the multiset of objects specified by the strings $w_1, w_2, \cdots, w_m$ in the corresponding membranes. A global clock is assumed that ticks at every time step. At each step, in each membrane, a multiset of objects and a multiset of rules is chosen and assignment of objects to rules is made. This is done in a non-deterministic maximally parallel manner. That is to say, no more rule can be added to the multiset of rules, because of the lack of objects and if there is a conflict of two rules for same object, then they are chosen non-deterministically. For example, if $w_i = a^5b^6c$ and the rules in R_i are $aab \rightarrow a(b, out)(c^2, in)$ and $aa \rightarrow (a, out)b$, then two copies of a can be assigned to one copy of first rule. Next two copies could be either assigned to another copy of first rule or to the second rule. And this choice is non-deterministic. Maximally parallel means that once four a's are consumed, no other rule can be applied to remaining objects. Thus non-deterministically, the ith membrane will contain a^3b^5c or a^2b^6c or ab^8c in the next step.

In essence we discuss the application of a single rule as follows. Assume the existence of a global clock that ticks during every step. When a rule $u \rightarrow v$ in membrane i is used, it acts only on objects placed in i (i.e. objects specified in u are available in i in required multiplicities). Then the following happens in a single step. The objects present in u in the specified multiplicities are removed from i, the objects present in v in the specified multiplicities are produced and the produced objects are placed either in i or in the neighbors of i according to the target indications in v. For example, if i is not an elementary membrane, the rule $a^2bc^3 \rightarrow ba^2c(da, out)(ca, in)$ is applicable if i contains at least two objects of a, one object b, and three objects c. The application of this rule removes from i two objects a, one object b, and three objects c and produces one object b, four objects a, two objects c and one object d and the produced object b remains in i, two produced objects a remain in i, one produced object c remains in i, the produced object d and one produced object a are sent to the upper neighbor, one produced object a is sent to a (non-deterministically chosen) lower neighbor, and one produced object c is sent to a (non-deterministically chosen) lower neighbor. If i is the skin membrane, then the produced object d and one produced object a are sent to the environment. Since there are no evolution rules placed in the environment, the skin region never receives objects from the environment and so the objects that leave the membrane structure can never return.

The output of a successful computation is the number of objects present in the output membrane in a configuration such that no further evolution is possible. (i.e. no rule applicable in any membrane). Such a configuration is called halting configuration. A vector of the number of objects of each kind (e.g, no. of a's, no. of b's, etc) can also be taken as the output.

4 Control Words of P Systems

In this section we introduce and define control languages for P systems and we consider the power of language generated with respect to a control word associated with a computation of a P system. Here we assign a label to every rule where the labels are chosen from a finite alphabet, say Σ or the labels can be λ. All the rules used in a computation step should have the same label, or they can also be labeled with λ.

Consider a P system $\Pi = (O, H, \mu, w_1, w_2, \cdots, w_m, R_1, R_2, \cdots, R_m, i_0)$ of degree m. Assume a total ordering on the rule. Let $R = R_1 \cup R_2 \cup \cdots \cup R_m$ and Σ be a finite alphabet. Define a function $l : R \rightarrow \Sigma \cup \{\lambda\}$ called the labeling function that assigns a label to every rule in R. We extend the labeling function for a label sequence $S = d_1 d_2 \cdots d_k \in R^*$ as follows: $l(\lambda) = \lambda$ and $l(S) = l(d_1 d_2 \cdots d_k) = l(d_1)l(d_2 d_3 \cdots d_k)$. Given two configurations c, c', we consider only transition $c \overset{b}{\Rightarrow} c', b \in \Sigma$ between configurations which use only rules with the same label b and rules labeled with λ. We say that such a transition is label restricted. With a label restricted transition, we associate the symbol b with the transition if at least one rule with label b is used. If all the used rules have the label λ, then we associate λ with the transition. The label restricted transitions which cannot use only rules with label λ are called λ-label restricted. So with any computation in Π starting from the initial configuration and using label restricted transitions, we associate a string called control word. The set of all strings associated with all λ-label restricted halting computations in Π is denoted by $L(\Pi)$. If we allow λ transitions, then the language is denoted by $L_\lambda(\Pi)$. The family of languages $L(\Pi)$ associated with P systems with at most m membranes is denoted by LP_m. If we allow λ transitions, then the language family is denoted by $L_\lambda P_m$. If the number of membranes is unbounded, then we replace m with $*$.

Remark 1. In the case of λ-restricted computation, we can see that the length of the control word corresponding to the computation has the same length as that of the number of steps used in the computation, whereas in the λ transition allowed case, the number of steps in the computation can be greater than or equal to the length of the control word.

We shall next prove some results. Even though Theorem 1 will follow from Theorem 2, we state it and give the proof to show the simple construction involved.

Theorem 1. $FIN \subset LP_1$.

Proof. Let $L \in FIN$ over an input alphabet Σ. Let $L = \{w_1, w_2, \cdots w_k\}$. Let $l_i = |w_i|$ and $l = l_1 + l_2 + \cdots + l_k$. Let $w_i = b_{i1} b_{i2} \cdots b_{il_i}$. Define a function $f(b_{ij}) = \sum_{r=1}^{i-1} l_r + j$ that maps the jth symbol of the ith string to an integer between 1 and l. We construct a P system $\Pi = (\{1, 2, \cdots, l, s, \$\}, \{1\}, [_1]_1, s, R_1, \emptyset)$ generating L with one membrane as follows:

$$R_1 = \{b_{i1} : s \rightarrow f(b_{i1}) | 1 \leq i \leq k\} \cup \{b_{ij} : f(b_{ij}) \rightarrow f(b_{i(j+1)}) | 1 \leq i \leq k, 1 \leq j \leq l_i - 1\} \cup \{b_{il_i} : f(b_{il_i}) \rightarrow \$ | 1 \leq i \leq k\}$$

The P system Π constructed in Theorem 1 works as follows: The system initially contains the object s which is used to guess the string to be generated. Suppose that the system generates the ith string $w_i = b_{i1}b_{i2}\cdots b_{il_i}$. In the first step the system generates b_{i1} using the rule $b_{i1} : s \to f(b_{i1})$. Then in the next $l_1 - 1$ steps, the system generates the next $b_{ij}, 2 \le j \le l_i - 1$ symbols in the ith string using the second type of rules. In the last step, the system generates the last symbol b_{il_i} by using the rule $b_{il_i} : f(b_{il_i}) \to \$$ thereby introducing the terminating symbol $\$$ into the system, halting the computation.

Theorem 2. $REG \subset LP_1$.

Proof. Let $G = (N, T, P, S)$ be a regular grammar that generates R. Assume that all productions are of the form $A \to bB, A \ne B$ or $A \to b$ or $S \to \lambda$, where $A, B \in N, b \in T$. If there are productions, of the form $A \to bA$, we replace $A \to bA$ with a set of productions $A \to bA'$ and $A' \to bA$ by defining a new variable A' and also add rules $A' \to cB$ for every rule $A \to cB \in P$. Let $G' = (N', T, P', S)$ be the modified grammar. We can see that the new grammar G' generates the same language as G. Let v be the number of variables in G'. Rename the variables as $A_i, 1 \le i \le v$, such that $A_1 = S$ and redefine the production rules using the renamed variables. Using G', we construct a P system with one membrane $\Pi = (N' \cup \{\$\}, \{1\}, [_1]_1, A_1, R_1, \emptyset)$ as follow:

$$R_1 = \{b : A_i \to A_j | A_i \to bA_j \in P'\} \cup \{b : A_i \to \$ | A_i \to b \in P'\}$$

The P system Π constructed from the regular grammar G' is diagrammatically shown in Figure 1. The P system Π constructed in Theorem 2 works as follows:

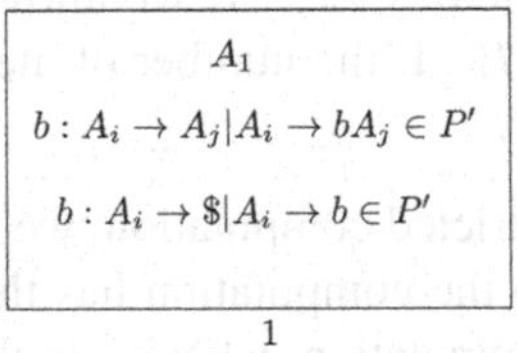

$$
\boxed{
\begin{array}{c}
A_1 \\
b : A_i \to A_j | A_i \to bA_j \in P' \\
b : A_i \to \$ | A_i \to b \in P'
\end{array}
}
$$

1

Fig. 1 P system construction from Theorem 2.

The system starts with the object A_1 which corresponds to the start symbol in the membrane. The use of a rule $b : A_i \to A_j$ simulates the use of the production $A_i \to bA_j \in P'$ and generates the symbol b. The system halts when the system uses the rule $b : A_i \to \$$ corresponding to $A_i \to b \in P'$ introducing the terminating symbol $\$$ in the membrane.

Theorem 3. $(CF - REG) \cap LP_* \ne \emptyset$.

Proof. Let L_2 be a context-free language $\{a^n b^n | n \ge 1\}$. We construct a P system with one membrane $\Pi_2 = (\{a_1, b_1, c_1\}, \{1\}, [_1]_1, a_1, \lambda, R_1, \emptyset)$, where,

$$R_1 = \{a : a_1 \to a_1 b_1, a : a_1 \to b_1 c_1, b : b_1 c_1 \to c_1\}$$

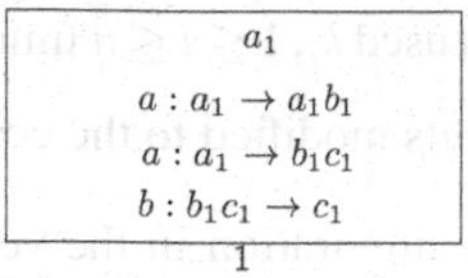

$$a_1$$
$$a : a_1 \rightarrow a_1 b_1$$
$$a : a_1 \rightarrow b_1 c_1$$
$$b : b_1 c_1 \rightarrow c_1$$

1

Fig. 2 P system for the context-free language $\{a^n b^n | n \geq 1\}$.

The P system Π constructed for the context-free language L_2 grammar is diagrammatically shown in Figure 2. The P system Π constructed in Theorem 3 works as follows: The system starts with the object a_1 in the membrane. By using the rule $a : a_1 \rightarrow a_1 b_1$ labeled with a, $n - 1$ times, the system generates the string a^{n-1} and introduces $n - 1$ b_1's into the system. Then it uses the a labeled rule $a : a_1 \rightarrow b_1 c_1$ once, generating one more a, removing the object a_1 and introducing one more b_1 and one c_1 into the system. So after this step, the the multiplicity of object b_1 is n. Then in the next n steps, the system uses the b labeled rule $b : b_1 c_1 \rightarrow c_1$, generating n b's and removing the b_1's from the system. After this step, the system contains the terminating object c_1 and the system halts the computation generating the string $a^n b^n$.

Theorem 4. $CF - LP_* \neq \emptyset$.

Proof. Consider the context free language $L = \{ww^R | w \in \{a,b\}^*\}$. Assume that there exists a P system $\Pi = (O, H, \mu, w_1, w_2, \cdots, w_m, R_1, R_2, \cdots, R_m, i_0)$ with m membranes that accepts L. Consider a string $uu^R \in L$ and let l be the length of u. After reading l symbols of u, Π must be able to reach as many different configurations as there are strings of length l. This must hold since Π has to remember the first half of the string uu^R in order to compare it with the second half. Since the alphabet size is two (the argument is applicable to any finite alphabet set of cardinality greater than 1), Π has to reach at least 2^l different configurations after reading l symbols. If Π cannot reach that many configurations, there are two different strings u and u', where the length of u' is strictly less than u, that leads Π to the same configuration. So it is required to prove that for sufficiently large l, only less than 2^l configurations are reachable. The proof is as follows: Let $O = \{a_1, a_2, \cdots, a_k\}$. Let $R = R_1 \cup R_2 \cup \cdots \cup R_m$ and $|R| = n$. Assume a total order on R. Let the ith configuration be represented by a row vector $c_i = (a_1^1, a_2^1, \cdots, a_k^1, a_1^2, a_2^2, \cdots, a_k^2, \cdots, a_1^m, a_2^m, \cdots, a_k^m)$ of size mk, where $a_i^j \in O$ represent the the multiplicity of the object a_i in membrane j. Every rule in Π is of the form $a_1^{p_1} a_2^{p_2} \cdots a_k^{p_k} \rightarrow a_1^{q_1}(a_1^{s_1}, in)(a_1^{t_1}, out)a_2^{q_2}(a_2^{s_2}, in)(a_2^{t_2}, out) \cdots a_k^{q_k}(a_k^{s_k}, in)$ $(a_k^{t_k}, out)$, $p_i, q_i, s_i, t_i \geq 0$ and at least one of $p_i \neq 0$. Application of a rule in membrane i takes away p_j a_j's from the membrane i and adds q_j, s_j, t_j a_j's to the membranes indicated in the target field of the rule. Associated with each rule $r_k, 1 \leq k \leq n$, we define a row vector called modification vector $\mathbf{v_k} = (b_1^1, b_2^1, \cdots, b_k^1, b_1^2, b_2^2, \cdots, b_k^2, \cdots, b_1^m, b_2^m, \cdots, b_k^m)$ of size mk, where $b_i^j \in O$ represent the change in the multiplicity of the object a_i in membrane j on the application of the rule. Application of each rule modifies the configuration by adding a vector $\mathbf{v_k}$ corresponding to rule k. Suppose that the rule $r_i, 1 \leq i \leq n$ is used $k_i, 1 \leq i \leq n$ times during the computation. The configuration of the P system gets modified to the configuration $c_0 + \sum_{i=1}^{n} k_i.\mathbf{v_i}$ where $\sum_i k_i = l$ and c_0 is the initial configuration in the vector form. Hence with n rules we can reach at most as many configurations as there are such tuples $(k_1, k_2, \cdots, k_n)$. These n numbers add exactly up to l and therefore $0 \leq k_i \leq l$ for all $i \in \{1, 2, \cdots, n\}$. So there are at most $(l+1)^n$ such tuples. Therefore, for sufficiently large l there are less than 2^l different configurations that are reachable by a P system that generates L. So Π is not able to distinguish between u and u' and so if it accepts uu^R then it also accepts $u'u^R$. But $u'u^R \notin L$. So we get a contradiction.

Theorem 5. $LP_* - CF \neq \emptyset$.

Proof. We can extend the construction of P system used in Theorem 3 to generate the context-sensitive language such as $\{a^n b^n c^n | n \geq 1\}$.

Theorem 6. $CS - LP_* \neq \emptyset$.

Proof. Consider the context-sensitive language $\{ww | w \in \{a, b\}^*\}$. We can prove that it is not possible to design a P system that generates the language in the same way we proved Theorem 4.

Theorem 7. $LP_* \subset CS$.

Proof. We show how to recognize a control word generated by a P system with a linear bounded automaton. In order to do this, we simulate a computation of P system by remembering the number of objects in each membrane after the generation of each symbol in the control word and show that the total number of objects in the system is bounded on the length of the control word.

Consider a control language L of a P system Π. Let $w = b_1 b_2 \cdots b_k, k \geq 0$ be a control word in L. Let the number of membranes be m and the total number of rules in all the membranes be n. We build a multi track nondeterministic LBA B which simulates Π. In order for B to simulate Π, it has to keep track of the number of objects in each membrane after generating each symbol. So B has a track assigned to every rule of Π, a track for each symbol-membrane pair $(a_i, j) \in O \times \{1, 2, \cdots, m\}$, and a track for each triple $(a_i, j, k) \in O \times \{1, 2, \cdots, m\}^2$. B keeps track of the configurations of Π by writing a positive integer on each track assigned to the symbol-membrane pair (a_i, j), denoting the number of objects a_i in membrane j. A single step of the computation of B is as follows: Based on the current configuration and the next symbol to be generated, B choses a set of rules that are to be applied in the next step by writing an integer on the track corresponding to the rules which indicates the number of times that a particular rule is to be applied. Then for each triple (a_i, j, k), B examines the chosen rule set and writes the number of objects a_i leaving from membrane j to membrane k on the corresponding track, decreasing the number on the track for (a_i, j) accordingly. Then it creates the next configuration by adding the values written on the track for each (a_i, j, k) to the number stored on the track for (a_i, k). We can see that in any step of the computation, the tracks contain integers bounded by the number of objects inside Π during the corresponding computation step. So if the number of objects inside the P system in a configuration c during a computation is bounded by $S(i)$, where i is the number of symbols generated, then the space used by B to record the configurations and to calculate the configuration change of Π is bounded by $t \times log_b(S(i))$, where b denotes the base of the track alphabet of B and t denotes the number of tracks used. Finally, B checks whether any further rules can be applied. If not, it accepts the string, else it rejects. So the total number of objects in the system is bounded on the input length and so the generated language is context-sensitive.

Theorem 8. $L_\lambda P_1 = RE$.

Proof. Let $L \subseteq \Sigma^*$ be a recursively enumerable language. Let $\Sigma = \{b_1, b_2, \cdots b_l\}$. Define an encoding $e : \Sigma \mapsto \{1, 2, \cdots, l\}$ such that $e(b_i) = i$. We extend the encoding for a string $w = c_1 c_2 \cdots c_k$ as follows: $e(w) = c_1 * (l+1)^{(k-1)} + \cdots + c_{(k-1)} * (l+1)^1 + c_k * (l+1)^0$. We use $l+1$ as the base in-order to avoid the digit 0 at the left end of the string.

For any L, there exists a deterministic register machine $M = (m, H, q_0, h, I)$ which halts after processing the input i_0 placed in its input register if and only if $i_0 = e(w)$ for some $w \in L$. So, it is sufficient to show how to generate the encoding $e(w)$, and simulate the instructions of a register machine with a P system. The value of register r is represented by the multiplicity of the object a_r in the membrane and the label l_i is represented by an object l_i.

The instructions of a register machine are simulated by a P system as follows:

- Add instruction $l_i : (ADD(r), l_j)$ is simulated by the instruction $l_i \to a_r l_j$. Removes the object l_i and introduces the objects a_r and l_j.
- Subtract instruction $l_i : (SUB(r), l_j, l_k)$ is simulated by the instructions

$$l_i \to l_i' l_i''$$
$$a_r l_i' \to l_i'''$$
$$l_i'' \to l_i''''$$
$$l_i''' l_i'''' \to l_j$$
$$l_i'''' l_i' \to l_k$$

The object l_i is replaced by two objects l_i', l_i'' by using the rule $l_i \to l_i' l_i''$. If an object a_r is present in the system, the number of it gets decreased by one by using the rule $a_r l_i' \to l_i'''$, which also introduces the object l_i''' in the next step. If no a_r is present, l_i' remains but l_i'''' is introduced into the system by using the rule $l_i'' \to l_i''''$. In the next step, either object l_j or l_k get introduced into the system depending on the objects present in the system by using the rule $l_i''' l_i'''' \to l_j$ or $l_i'''' l_i' \to l_k$.

- For the halt instruction $l_h : HALT$, nothing to do. The system halts when the object l_h is introduced into the system.

We construct a P system $\Pi = (O, \{1\}, [_1]_1, s, R_1, \emptyset)$, where

$O = \{s\} \cup \{l_i, l_i', l_i'', l_i''', l_i'''' | l_i \in H\} \cup \{a_r | 1 \le r \le m\} \cup \{l_g, l_{g1}, l_g', l_{gi}, l_{g(l+1)}, l_{a1} | 1 \le i \le l\}$

$R_1 = \{b_i : s \to l_g a_1^i a_3^i | b_i \in \Sigma, 1 \le i \le l\} \cup$ set of all rules corresponding to the register machine instructions (M and generating encoding in Step 2) labeled with λ.

with one membrane performing the following operations (a_1 and a_2 are two distinguished objects of Π, where multiplicity of a_1 represents the encoding corresponding to the symbol generated in each step and the multiplicity of a_2 represents the encoding of the generated string up to a particular step).

1. For some $1 \le i \le l$, generating symbol $b_i \in \Sigma$, is performed by using a rule $b_i : s \to l_g a_1^i a_3^i$, labeled with b_i, that introduces the object l_g, which is the label of the first instruction for generating the encoding in Step 2 and the objects a_1, a_3 with multiplicity i.

2. Perform the computation $e(ua) = (l+1) * e(u) + e(a), u \in \Sigma^*, a \in \Sigma$. Assume that the encoding of u is represented by the multiplicity of object a_2. The encoding of ua is performed by the following register machine sub-program.

   $l_g : (SUB(r_1), l_{g1}, l_g')$

   $l_{gi} : (ADD(r2), l_{g(i+1)}), 1 \le i \le l$

   $l_{g(l+1)} : (ADD(r_2), l_g)$

   $l_g' : (SUB(r_3), l_{a1}, s)$

   $l_{a1} : (ADD(r_2), l_g')$

   The instructions of the sub-program can be translated to the P system rules as shown in the beginning of the proof. The multiplicity of objects a_1, a_2 and a_3 corresponds to the content of registers r_1, r_2 and r_3.

3. Repeat from step 1, or, non-deterministically, stop the increase in the multiplicity of object a_2 by using a λ labeled rule $\lambda : s \to q_0$, where q_0 is an object that corresponds to the label of the first instruction of the register machine M in Step 4.

4. Multiplicity of a_2 is equal to $e(w)$ for some $w \in \Sigma^+$. We now start to simulate the working of the register machine M in recognizing the number $e(w)$. If the machine halts, by introducing the object h corresponding to the halt instruction in M, then $w \in L$, otherwise the machine goes into an infinite loop.

So, we can see that the computation halts after generating a string w if and only if $w \in L$.

5 Conclusion

In this paper we introduced and defined the control language of P systems and studied the relationship of the control language with other language families such as regular, context-free, context-sensitive and recursive enumerable languages under label restricted (without allowing the label of all the rules applied in a single step having λ label) and label unrestricted ways of computation. In the label restricted case we find that every regular language, some context-free and some context-sensitive languages are control languages of some P systems. There exists some context free languages such as $ww^R, w \in \{b,c\}^*$ and some context sensitive languages such as $ww, w \in \{b,c\}^*$ that cannot be a control language of any P system. But in the label unrestricted case, we find that, the system generates the family of recursively enumerable language.

References

1. G. Păun, Computing with membranes, *Journal of Computer and System Science*, 61(1),pp.108-143,2000.
2. G. Păun, Membrane Computing - An introduction, Springer-Verlag, Berlin, 2002.
3. G. Păun, G. Rozenberg, A. Salomaa, Membrane computing with an external output, *Fundamenta Informaticae*, 41(3),pp.313-340,2000.
4. M. Oswald, P Automata, PhD Thesis, TU Viena, 2003.
5. E. Csuhaj-Varjú, and G. Vaszil, P Automata or Purely Communicating Accepting P Systems, *Membrane Computing, International Workshop, WMC-CdeA, Curtea de Arges, Romania, August 19-23, 2002.*
6. M. Ionescu, C. Martin-Vide, G. Păun, P systems with symport/antiport rules: The traces of objects, *Grammars*, 5, pp.65-79,2002.
7. M. Minsky, Computation - Finite and infinite Machines, Prentice Hall, Englewood Cliffs, NJ, 1967.
8. K. Krithivasan, G. Păun and A. Ramanujan, Control Words Associated with P Systems, *Frontiers of Membrane Computing: Open Problems and Research Topics* by M. Gheorghe, G. Păun and M. J. Pérez-Jiménez - editors published in the second volume of the proceedings of 10th Brainstorming Week on Membrane Computing, Sevilla, 2012, 171-250.
9. A. Ramanujan and K. Krithivasan, Control Languages of Spiking Neural P Systems, Submitted.
10. The P System Web Page: http://ppage.psystems.eu

Iso-Array Splicing Grammar System

D.K. Sheena Christy and V. Masilamani and D.G. Thomas

Abstract Splicing systems were introduced by Head [4] on biological considerations to model certain recombinant behaviour of DNA molecules and are of current interest and study [5]. The splicing systems make use of a new operation called splicing on arrays of symbols [9]. Parallel splicing operation has been recently applied to rectangular arrays [6]. Further Dassow and Mitrana [2] introduced a new type of grammar system called splicing grammar system in which communication is done by splicing of strings. In this paper, we propose a new model called iso-picture splicing grammar system to generate iso-pictures. We compare the iso-picture splicing grammar system with iso-array grammars such as regular iso-array grammar and context free iso-array grammar.

Keywords DNA computing, Splicing system, Iso-picture languages, Splicing grammar system.

1 Introduction

The recombinant behaviour of DNA molecules has been modeled by splicing operation, and based on this operation, a computability model called splicing system is defined. The H splicing system was introduced by Head [4] for generating string

D.K. Sheena Christy
Department of Mathematics, SRM University, Kattankulathur, Chennai - 603 203, India, e-mail: sheena.lesley@gmail.com

V. Masilamani
Department of Computer Science and Engineering, IIITD&M Kanchipuram, Chennai - 600 036, India, e-mail: masila@iiitdm.ac.in

D.G. Thomas
Department of Mathematics, Madras Christian College, Tambaram, Chennai - 600 059, India, e-mail: dgthomasmcc@yahoo.com

J. C. Bansal et al. (eds.), *Proceedings of Seventh International Conference on Bio-Inspired Computing: Theories and Applications (BIC-TA 2012)*, Advances in Intelligent Systems and Computing 201, DOI: 10.1007/978-81-322-1038-2_14, © Springer India 2013

languages. To generate array languages, the splicing operation has been used [9]. A simple and effective parallel splicing on rectangular arrays has been introduced [6].

The theory of Grammar Systems [1] is an intensively investigated area of Formal Language Theory providing an effective grammatical framework for capturing several characteristics of multi-agent systems such as cooperation, distribution, communication, parallelism etc. The basic idea in a grammar system is to consider several usual grammars and to make them cooperate in order to generate a common language.

The splicing operation is a basic operation in DNA recombination under the influence of restriction enzymes and ligases [4]. The parallel and communicating grammar system is a very powerful mechanism, leading to a new characterization of recursively enumerable languages.

On the other hand, in the syntactic approach of pattern recognition, there have been several studies on theoretical models in the last few decades for generating or recognizing two dimensional objects, pictures and picture languages. Motivated by the study of recognizable rectangular picture languages using rectangular tiles [3], a new concept of recognizability has been introduced and studied for a class of picture languages called iso-picture languages through iso-triangular tiling systems (ITS) [7, 8].

Iso-arrays are made up of isosceles right angled triangles and an iso-picture is a picture, formed by catenating iso-arrays of same size [7]. By making use of iso-picture languages, one can generate variety of picture languages that cannot be generated by earlier models available in the literature [3, 5]. The hexagonal picture languages, rectangular picture languages, languages of rhombuses and triangles are some of the examples of iso-picture languages. One interesting study of iso-picture language is the generation of kolam patterns and it also includes the study of tiling the area of rectangular plane [7].

The present paper combines the concept of splicing operations on iso-arrays [10] with another interesting concept called parallel communicating grammar systems [2, 11, 12]. An iso-array splicing grammar system can be viewed as a set of grammar working in parallel on their own sentential forms (exactly as in parallel communicating grammar system) and from time to time, exchanging to each other segments of their sentential forms, determined by given splicing rules.

This paper gives the comparative study of iso-array splicing grammar system (IASGS) with regular iso-array grammar (RIAG) and context free iso-array grammar (CFIAG).

2 Basic Definitions

For basic notions of iso-picture language one can refer [7].

Definition 1. Let $\Sigma = \left\{ {}^{S_1}\!\!\triangle\!{}^{S_3}_{S_2}, {}_{S_3}\!\!\overset{S_2}{\nabla}\!{}^{S_1}, {}_{S_3}\!\!\overset{S_1}{\triangleleft}\!{}^{} c|s_2, s_2|D\overset{S_3}{\underset{S_1}{\triangleright}} \right\}$. The sides of each tile in Σ are of length $\frac{1}{\sqrt{2}}, 1, \frac{1}{\sqrt{2}}$.

An iso-array is an arrangement of isosceles right angled triangles of tiles from the set Σ.

An U-iso-array of size m is formed exclusively by m number of $\triangle$ tiles on side S_2 and it is denoted by U_m. It will have m^2 tiles in total (including the m number of A tiles on S_2). Similarly D-iso-array, L-iso-array and R-iso-array are formed exclusively by B-tile, C-tile and D-tile on side S_2 respectively.

Example 1. The following are the iso-arrays of size 3

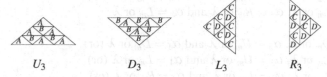

U_3 $\qquad\qquad$ D_3 $\qquad$ L_3 $\qquad\qquad$ R_3

Fig. 1

Iso-arrays of same-size can be catenated using the following four types of catenations of iso-arrays.
(i) Horizontal Catenation ($\ominus$):
The only possible catenation is $U \ominus D$.
(ii) Vertical Catenation ($\oslash$):
$L \oslash R$ is the only possible catenation of iso-arrays.
(iii) Right Catenation ($\oslash$):
The following catenations are possible under right catenation:
$D \oslash U, R \oslash U, D \oslash L, R \oslash L$.
(iv) Left Catenation ($\oslash$):
The following catenations are possible under left catenation:
$U \oslash D, U \oslash L, L \oslash R, R \oslash D$.
The catenation can be defined between any two gluable iso-arrays of same size.

The set of all iso-pictures over Σ is denoted by Σ_I^{**}. An iso-picture language L over Σ is a subset of Σ_I^{**}.

3 H Iso-Array Splicing System

The following definitions enumerate the basic concepts of iso-array splicing rules and the main notion of H iso-array splicing systems [10].

Definition 2. Let Σ be an alphabet, # and \$ are two special symbols not in Σ. An iso-array over Σ is an isosceles triangular arrangement of tiles ◁A◁ , ◁B▽ , ◁C◁ and ◁D▷ .

The splicing rule α_1 # α_2 \$$_\ominus$ α_3 # α_4 over Σ is called the
(i) horizontal splicing rule if $\alpha_1 = U_m$ or λ, $\alpha_2 = D_m$ or λ, $\alpha_3 = U_m$ or λ and $\alpha_4 = D_m$ or λ
(ii) vertical splicing rule if $\alpha_1 = L_m$ or λ, $\alpha_2 = R_m$ or λ, $\alpha_3 = L_m$ or λ and $\alpha_4 = R_m$ or λ
(iii) right splicing rule
(a) $\alpha_1 = D_m$ or λ, $\alpha_2 = U_m$ or λ, $\alpha_3 = D_m$ or λ and $\alpha_4 = U_m$ or λ (or)
(b) $\alpha_1 = R_m$ or λ, $\alpha_2 = U_m$ or λ, $\alpha_3 = R_m$ or λ and $\alpha_4 = U_m$ or λ (or)
(c) $\alpha_1 = D_m$ or λ, $\alpha_2 = L_m$ or λ, $\alpha_3 = D_m$ or λ and $\alpha_4 = L_m$ or λ (or)
(d) $\alpha_1 = R_m$ or λ, $\alpha_2 = L_m$ or λ, $\alpha_3 = R_m$ or λ and $\alpha_4 = L_m$ or λ
(iv) left splicing rule if
(a) $\alpha_1 = U_m$ or λ, $\alpha_2 = D_m$ or λ, $\alpha_3 = U_m$ or λ and $\alpha_4 = D_m$ or λ (or)
(b) $\alpha_1 = U_m$ or λ, $\alpha_2 = L_m$ or λ, $\alpha_3 = U_m$ or λ and $\alpha_4 = L_m$ or λ (or)
(c) $\alpha_1 = L_m$ or λ, $\alpha_2 = R_m$ or λ, $\alpha_3 = L_m$ or λ and $\alpha_4 = R_m$ or λ (or)
(d) $\alpha_1 = R_m$ or λ, $\alpha_2 = D_m$ or λ, $\alpha_3 = R_m$ or λ and $\alpha_4 = D_m$ or λ.
The set of all horizontal, vertical, left and right splicing rules over Σ are denoted by $R_\ominus, R_\oplus, R_\oslash, R_\oslash$ respectively.

Definition 3. An H iso-array scheme is a tuple $\sigma = (\Sigma, R_\ominus, R_\oplus, R_\oslash, R_\oslash)$ where Σ is an alphabet, $R_\ominus$ is a finite set of horizontal splicing rules. Similarly $R_\oplus, R_\oslash, R_\oslash$ are finite sets of vertical, right and left splicing rules.

An H iso-array splicing system is defined by $S = (\sigma, A)$ where A is finite subset of Σ_I^{**}. We define $\sigma(L)$ as

$$
\sigma(L) = \left\{ p \in \Sigma_I^{**} \left/ \begin{array}{c} (p_1, p_2) \overset{\ominus}{\to} p \ (\text{or}) \\ (p_1, p_2) \overset{\oplus}{\to} p \ (\text{or}) \\ (p_1, p_2) \overset{\oslash}{\to} p \ (\text{or}) \\ (p_1, p_2) \overset{\oslash}{\to} p \end{array} \right. \right\}
$$

for some $p_1, p_2 \in L$.

$\sigma^*(L)$ is defined iteratively as follows:
$\sigma^0(L) = L$
$\sigma^{i+1}(L) = \sigma^i(L) \cup \sigma(\sigma^i(L))$, for $i \geq 0$
$\sigma^*(L) = \cup_{i=0}^{\infty} \sigma^i(L)$.

The family of iso-picture languages generated by these splicing systems is denoted by FHIA.

4 Iso-Array Grammar

We now review the notion of iso-array grammar [7].

Definition 4. A Regular Iso-Array Grammar (RIAG) is a structure $G = (N, T, P, S)$ where

$N = \{\,\triangle A, \triangledown B, \triangleleft C, \triangleright D\,\}$ and $T = \{\,\triangle a, \triangledown b, \triangleleft c, \triangleright d\,\}$ are finite sets of symbols called nonterminals and terminals; $N \cap T = \phi$. $S \in N$ is the start symbol or the axiom. P consists of rules of the following forms:

(1) $\begin{smallmatrix}A\\\#\end{smallmatrix} \longrightarrow \begin{smallmatrix}a\\B\end{smallmatrix}$ (2) $\overline{\#A} \longrightarrow \overline{B\,a}$ (3) $\underline{A\,\#} \longrightarrow \underline{a\,B}$

(4) $\begin{smallmatrix}\#\\A\end{smallmatrix} \longrightarrow \begin{smallmatrix}D\\a\end{smallmatrix}$ (5) $\begin{smallmatrix}\#\\A\end{smallmatrix} \longrightarrow \begin{smallmatrix}C\\a\end{smallmatrix}$ (6) $\underline{A} \longrightarrow \underline{a}$

Similar rules can be given for the other tiles $\triangledown B$, $\triangleleft C$ and $\triangleright D$.

The regular iso-array language (RIAL) generated by G is defined by $\{W \mid S \Rightarrow_G^* W, W$ is a finite connected array over $T\}$ and is denoted by $L(G)$.

Definition 5. A Context-Free Iso-Array Grammar (CFIAG) is a structure $G = (N, T, P, S)$ where $N = \{\,\triangle A, \triangledown B, \triangleleft C, \triangleright D\,\}$ and $T = \{\,\triangle a, \triangledown b, \triangleleft c, \triangleright d\,\}$ are finite nonempty set of symbols called nonterminals and terminals, $N \cap T = \phi$. $S \in N$ is the start symbol or the axiom. P consists of rules of the form $\alpha \to \beta$, where α and β are finite connected array of one or more triangular tiles over $N \cup T \cup \{\,\triangle{\#A}, \triangledown{\#B}, \triangleleft{\#C}, \triangleright{\#D}\,\}$ and satisfy the following conditions:

1. The shapes of α and β are identical.
2. α contains exactly one nonterminal and possibly one or more #'s.
3. Terminals in α are not rewritten.
4. The application of the rule $\alpha \to \beta$ preserves the connectedness of the host array (that is, the application of the rule to a connected array results in a connected array).

The rule $\alpha \to \beta$ is applicable to a finite connected array γ over $N \cup T \cup \{\,\triangle{\#A}, \triangledown{\#B}, \triangleleft{\#C}, \triangleright{\#D}\,\}$, if α is a subarray of γ and in a direct derivation step, one of the occurrences of α is replaced by β, yielding a finite connected array δ. We write $\gamma \Rightarrow_G \delta$. The reflexive transitive closure of $\Rightarrow_G$ is denoted by $\Rightarrow_G^*$.

The context free iso-array language (CFIAL) generated by G is defined by $\{\delta : S \Rightarrow_G^* \delta, \delta$ is a finite connected array over $T\}$ and is denoted by $L(G)$.

5 Iso-Array Splicing Grammar System

In this section we introduce an iso-array splicing grammar system and obtain an interesting results.

Definition 6. An Iso-Array Splicing Grammar System (IASGS) is a structure

$$\mathscr{G} = (N, T, (S_1, P_1), (S_2, P_2), \ldots, (S_n, P_n), M)$$

where $N = \{\;\triangle\!A,\;\triangle\!B,\;\triangleleft\!C,\;\triangleright\!D\;\}$, called nonterminals and $T = \{\;\triangle\!a,\;\triangle\!b,\;\triangleleft\!c,\;\triangleright\!d\;\}$, called terminals; $N \cap T = \phi$. $S_i \in N$ is the start symbol or axiom. P_i, $1 \leq i \leq n$ is the finite set of production rules and it can be regular iso-array grammar (RIAG) (or) context-free iso-array grammar(CFIAG).

M is a finite set of horizontal or vertical or left or right domino splicing rules of the form $m = \alpha_1 \# \alpha_2 \$ \alpha_3 \# \alpha_4$ where

$$\alpha_i \in \left\{ \begin{array}{c} \triangle\!\!\!^A_B \end{array},\; \triangle\!B\!A,\; \triangle\!\!^B_C,\; \triangleright\!\!^B_D,\; \triangle\!A\,B,\; \triangleleft\!C\,D,\; \triangleright\!D\!\!\triangle_A,\; \triangle\!\!\triangleleft_A^C \right\},$$

for $i = 1, 2, 3, 4$ and $A, B, C, D \in N \cup T \cup \{\lambda\}$.

By a configuration, we mean an n-tuple consisting of iso-arrays over $N \cup T$.

For two configurations
$x = (x_1, x_2, \ldots, x_n)$, $x_i \in (N \cup T)_I^{**}$
$y = (y_1, y_2, \ldots, y_n)$, $y_i \in (N \cup T)_I^{**}$, $1 \leq i \leq n$.
i.e., x_i's and y_i's are iso-pictures with iso-arrays from $(N \cup T)_I^{**}$.

We define $x \underset{\mathscr{G}}{\Rightarrow} y$ iff one of the following two conditions holds:

(i) For each $1 \leq i \leq n$, $x_i \underset{P_i}{\Rightarrow} y_i$.

(ii) For $x_i = x_1 u_1 u_2 x_2$ and $x_j = y_1 v_1 v_2 y_2$ and for $u_1 \# u_2 \$ v_1 \# v_2 \in M$, we have $y_i = x_1 u_1 v_2 y_2$ and $y_j = y_1 v_1 u_2 x_2$ and in any other components (other than y_i and y_j) the iso-arrays generated at this instant will remain unchanged during this splicing process. Here there is no priority between the conditions (i) and (ii).

The language generated by i^{th} component is defined by

$$L_i(\mathscr{G}) = \{p_i \in T_I^{**} / (S_1, S_2, \ldots, S_n) \Rightarrow^* (p_1, p_2, \ldots, p_n), p_j \in (N \cup T)_I^{**}, j \neq i\}$$

where $\Rightarrow^*$ is the reflexive and transitive closure of the relation $\Rightarrow$.

The language $L_i(\mathscr{G})$ generated by single component will be called the individual language of the system.

The second associated language will be the total language, namely,
$$L_t(\mathscr{G}) = \bigcup_{i=1}^{n} L_i(\mathscr{G}).$$

Definition 7. $I_{iasgs}L_n(X)$ - The family of individual languages generated from first component by iso-array splicing grammar systems with n components of type X.

$T_{iasgs}L_n(X)$ - The family of total languages generated by splicing grammar system with n components (degree) of type X where $X \in \{IAREG, IACFG\}$.

In this paper we consider the individual languages. The same results can be applied to total languages as well.

Example 2. Let $\mathscr{G} = (N, T, (S, P), (S, P), (S, P), M)$ where $N = \{ \text{A}, \text{B}, \text{C}, \text{D} \}$,

$T = \{ \text{a}, \text{b}, \text{c}, \text{d} \}$ and $S = \text{c}$.

$P = \{$ $\begin{smallmatrix} S \\ \# \end{smallmatrix} \to \begin{smallmatrix} c \\ D \end{smallmatrix}$, $\begin{smallmatrix} D \\ \# \end{smallmatrix} \to \begin{smallmatrix} d \\ C \end{smallmatrix}$, $\begin{smallmatrix} C \\ \# \end{smallmatrix} \to \begin{smallmatrix} c \\ D \end{smallmatrix}$,

$\begin{smallmatrix} D \\ \# \end{smallmatrix} \to \begin{smallmatrix} d \\ A \end{smallmatrix}$, $\text{A}\# \to \text{a B}$, $\text{B}\# \to \text{b A}$, $\text{B} \to \text{b} \}$.

$M = \{B \# \lambda \$ d \# c\}$.

The production rules in a component generate L-token of all sizes of the form

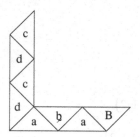

Fig. 2 L-token

before rewriting is terminated. In other words

$$\begin{smallmatrix} S \\ \# \end{smallmatrix} \Rightarrow \begin{smallmatrix} c \\ D \end{smallmatrix} \Rightarrow \begin{smallmatrix} c \\ d \\ C \end{smallmatrix} \Rightarrow \begin{smallmatrix} c \\ d \\ c \\ D \end{smallmatrix} \Rightarrow \begin{smallmatrix} c \\ d \\ c \\ d \\ A \end{smallmatrix}$$

$$\Rightarrow \begin{smallmatrix} c \\ d \\ c \\ d\; a\; B \end{smallmatrix} \Rightarrow \begin{smallmatrix} c \\ d \\ c \\ d\; a\; b\; A \end{smallmatrix} \Rightarrow \begin{smallmatrix} c \\ d \\ c \\ d\; a\; b\; a\; B \end{smallmatrix}$$

At this stage a right domino splicing rule in a component of iso-picture with another component of iso-picture can take place with terminating production rule. For example, using the splicing rule M the two identical iso-pictures of Fig. 2 yield the new staircase iso-picture Fig. 3.

Any iso-picture generated in the individual language of this iso-array splicing grammar system will generate the staircase iso-picture having only tiles $\triangle a$ and $\triangledown b$. The length and height of staircase iso-picture need not be the same.

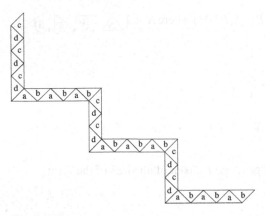

Fig. 3 Staircase

Example 3. Let $\mathscr{G} = (N, T, (S, P), (S, P), (S, P), M)$ where

$$N = \{\triangle A, \triangledown B, \triangleleft C, \triangleright D, \triangle A_i, \triangledown B_i, \triangleleft C_i, \triangleright D_i\} \text{ for } i = \{0, 1, 2\}$$

$$T = \{\triangle a, \triangledown b, \triangleleft c, \triangleright d, \triangle a_i, \triangledown b_i, \triangleleft c_i, \triangleright d_i\} \text{ for } i = \{0, 1, 2\}.$$

$S = \triangleright D$ and the production rules of CFIAG is given by

$$P = \left\{ \begin{array}{l} \text{(production rules shown as iso-picture tile diagrams)} \end{array} \right\}.$$

$M = \{c_2 \# D \$ c \# d_1; \; c_0 \# d \$ c \# d_0\}.$

In each component, the production rules generate iso-pictures of the form $xu^p v^p x$

where $x = \left(\boxed{\begin{smallmatrix} & b & \\ d & \times & c \\ & a & \end{smallmatrix}} \right)_r$, $u = \boxed{\begin{smallmatrix} & b_1 & \\ d_1 & \times & c_1 \\ & a_1 & \end{smallmatrix}} \ominus \left(\boxed{\begin{smallmatrix} & b_0 & \\ d_0 & \times & c_0 \\ & a_0 & \end{smallmatrix}} \right)_m \ominus \boxed{\begin{smallmatrix} & b_1 & \\ d_1 & \times & c_1 \\ & a_1 & \end{smallmatrix}}$ and

$v = \boxed{\begin{smallmatrix} & b_2 & \\ d_2 & \times & c_2 \\ & a_2 & \end{smallmatrix}} \ominus \left(\boxed{\begin{smallmatrix} & b_0 & \\ d_0 & \times & c_0 \\ & a_0 & \end{smallmatrix}} \right)_m \ominus \boxed{\begin{smallmatrix} & b_2 & \\ d_2 & \times & c_2 \\ & a_2 & \end{smallmatrix}}$ where $p, m \geq 1$; $r = m + 2$. Therefore we have

$$L_t(\mathcal{G}) = L_i(\mathcal{G}) = \{(xu^p v^p x)^i / m, p \geq 1, i = 1, 2, 3\}.$$

The splicing step is possible only if the symbol $\boxed{\begin{smallmatrix} \\ D \end{smallmatrix}\!\!\triangleright}$ occurs in one of the three sentential forms. On the other hand, after a splicing step, exactly one more usual rewriting step can be performed. One member of the above generated picture is given by

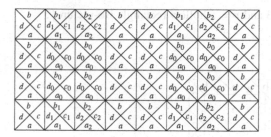

Fig. 4

Example 4. Let $\mathcal{G} = (N, T, (S, P), (S, P), (S, P), M)$ where

$N = \{ \boxed{A}, \boxed{B}, \boxed{C}, \boxed{D} \}$

$T = \{ \boxed{a}, \boxed{b}, \boxed{c}, \boxed{d} \}$

$S = \boxed{a}$

$P = \left\{ \boxed{\#/S} \rightarrow \boxed{B/a}, \; \boxed{/\#B} \rightarrow \boxed{A\,b}, \; \boxed{\#/A} \rightarrow \boxed{B/a}, \right.$

$\boxed{\#/A} \rightarrow \boxed{D/a}, \; \boxed{\#/D/d} \rightarrow \boxed{C/c}, \; \boxed{\#/C} \rightarrow \boxed{D/c},$

$\boxed{D/d} \rightarrow \boxed{/B/d}, \; \boxed{B/\#} \rightarrow \boxed{b/A}, \; \boxed{A\#} \rightarrow \boxed{a\,B},$

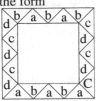

$M = \{c \, \# \, \lambda \, \$ \, \lambda \, \# \, d; \; C \, \# \, \lambda \, \$ \, \lambda \, \# \, d\}.$

The production rules in a component generate an iso-picture of hollow rectangles of the form

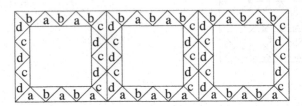

Before rewriting is terminated, a vertical domino splicing rule in a component with another component of iso-picture can take place with terminating production rule. Hence we have

$$L_t(\mathscr{G}) = L_i(\mathscr{G}) = \{p^i / i = 1, 2, 3 \text{ and } p = \text{a hollow rectangle iso-picture}\}.$$

One such iso-picture is given by

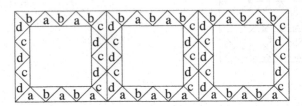

Fig. 5

Theorem 1. *For* $Y \in \{I, T\}$,

1. $RIAL = Y_{iasgs}L_1(RIAG)$
2. $RIAL \subseteq Y_{iasgs}L_3(RIAG)$
3. $CFIAL = Y_{iasgs}L_1(CFIAG)$
4. $CFIAL \subseteq Y_{iasgs}L_2(CFIAG)$

Proof. Statements (1) and (3) immediately follow from the definition. By Definition 7, $RIAL \subseteq Y_{iasgs}L_3(RIAG)$ and $CFIAL \subseteq Y_{iasgs}L_2(CFIAG)$. From Example 4, we observe that $RIAL \subset Y_{iasgs}L_3(RIAG)$, since the picture in Fig. 5 cannot be generated by RIAG. But it can be generated by $Y_{iasgs}L_3(RIAG)$. In statement (4), to hold the strict inclusion $CFIAL \subset Y_{iasgs}L_2(CFIAG)$ it is sufficient to consider only 2 components in Example 3. Clearly the iso-picture given in Fig. 4 is in the language generated by $Y_{iasgs}L_2(CFIAG)$. But, the rules of CFIAG cannot generate iso-pictures like Fig. 4.

Theorem 2. *For* $Y \in \{I, T\}$.

1. $RIAL = Y_{iasgs}L_1(RIAG) \subset Y_{iasgs}L_2(RIAG) \subset \cdots \subset Y_{iasgs}L_n(RIAG) \subset \ldots$
2. $CFIAL = Y_{iasgs}L_1(CFIAG) \subset Y_{iasgs}L_2(CFIAG) \subset \cdots \subset Y_{iasgs}L_n(CFIAG) \subset \ldots$

The first statement immediately follows from the Definition 7 and from Example 4. Similarly we can show statement (2) also.

Conclusion

In this paper we have introduced a new notion of iso-array splicing grammar system which is a combination of iso-array splicing system with parallel communicating grammar system. We made a comparison of iso-picture splicing grammar system with iso-array grammars such as regular iso-array grammar and context free iso-array grammar. It is a powerful system to describe and generate various iso-picture languages. We consider that these models deserve further comparisons and investigations.

References

1. E. Csuhaj-Varju, J. Dassow, J. Kelemen and Gh. Paun, Grammar systems: A grammatical approach to distribution and cooperation, Gordon and Breach Science Publishers, 1994.
2. J. Dassow and V. Mitrana, Splicing grammar systems, Computers and Artificial Intelligence, 15 (1996), 109–122.
3. D. Giammarresi and A. Restivo, Two-dimensional languages, in *Handbook of Formal Languages*, eds. A. Salomaa and G. Rozenberg, Vol. 3 (Springer-Verlag, 1997), 215–267.
4. T. Head, Formal language theory and DNA: an analysis of the generative capacity of specific recombinant behaviours, *Bull. Math. Biol.* 49 (1987), 735–759.
5. T. Head, Gh. Paun and D. Pixton, Language theory and molecular genetics: generative mechanisms suggested by DNA recomination, in *Handbook of Formal Languages*, eds. G. Rozenberg and A. Salomaa, Vol. 2, Ch. 7 (Springer-Verlag, 1997), 296–358.
6. P. Helen Chandra, K.G. Subramanian, D.G. Thomas and D.L. Van, A note on parallel splicing on images, *Electronic Notes in Theoretical Computer Science*, 46 (2001), 255–268.
7. T. Kalyani, A study on iso-picture languages, Ph.D. Thesis, University of Madras, (2006).
8. T. Kalyani, V.R. Dare and D.G. Thomas, Local and recognizable iso-picture languages, Lecture Notes in Computer Science, 3316 (2004), 738–743.
9. K. Krithivasan, V.T. Chakaravarthy and R. Rama, Array splicing systems, in *Computing with Bio-molecules: Theory and Experiments*, ed. Gh. Paun, (Springer-Verlag, 1998).
10. V. Masilamani, D.K. Sheena Christy and D.G. Thomas, Parallel splicing on iso-arrays, IEEE - International Conference on Bio-Inspired Computing: Theories and Application (2010), 1535–1542.
11. G.H. Paun and Santean L. Kari, Parallel communicating grammar systems: The regular case, Ann. Univ. Buc., See Matem-Inform, 38 (1989), 55–63.
12. A. Roslin Sagaya Mary, K.G. Subramanian and K.S. Dersanambika, Image splicing grammar systems, Proceedings of Grammar Systems, E. Csuhaj-Varju and Gy. Vaszil (Eds.), MTA SZTAKI, Budapest, (2004), 276–286.

GA based Dimension Reduction for enhancing performance of k-Means and Fuzzy k-Means: A Case Study for Categorization of Medical Dataset

Asha Gowda Karegowda, M.A. Jayaram, A.S. Manjunath, Vidya T, Shama

Abstract

Medical Data mining is the process of extracting hidden patterns from medical data. Among the several clustering algorithms, k-means is the one of most extensively used clustering techniques in addition to fuzzy k-means clustering. The performance of both k-means and fuzzy k-means clustering is influenced by the initial cluster centers and might converge to local optimum. In addition, the performance of any data mining algorithm is influenced by the significant feature subset. This paper attempts to augment the performance of both k-means and fuzzy k-means clustering using two stages. As part of first stage, this paper investigates the use of wrapper approach of feature selection for clustering, where Genetic algorithm (GA) is used as a random search technique for subset generation, wrapped with k-means clustering. In the second stage of projected work, GA and Entropy based fuzzy clustering (EFC) are used to find the initial centroids for both k-means and fuzzy k-means clustering. Investigations have been directed using standard medical dataset namely Pima Indians Diabetes Dataset (PIDD). Experimental results confirm markable decline of almost 7% in the classification error of both k-means and fuzzy k-means clustering with GA nominated significant features and GA identified initial centroids when compared to randomly selected centroids with all features.

keywords: k-means clustering, fuzzy k-means clustering, Genetic algorithm, feature selection, cluster center initialization, entropy based fuzzy clustering, Diabetic dataset

Asha Gowda Karegowda, M.A. Jayaram, A.S. Manjunath, Vidya T, Shama
Siddaganga Institute of Technology, Tumkur, India, e-mail: {ashagksit, jayaramdps, asmanju, vidu.tr, shama.ammu}@gmail.com

J. C. Bansal et al. (eds.), *Proceedings of Seventh International Conference on Bio-Inspired Computing: Theories and Applications (BIC-TA 2012)*, Advances in Intelligent Systems and Computing 201, DOI: 10.1007/978-81-322-1038-2_15, © Springer India 2013

1 Introduction

Huge data repositories, especially in medical domains, contain enormous amounts of data. Medical data mining has enormous potential for exploring the hidden patterns in the data sets of the medical domain. These patterns can be employed for clinical diagnosis. Clustering is the process of grouping the data into classes or clusters so that objects within a cluster have high similarity in comparison to one another, but are very dissimilar to objects in other clusters [9]. K-means and fuzzy k-means are the most commonly used clustering methods. The performance of both k-means and fuzzy k-means mainly hinges on the initial cluster centers and might converge to a local optimum [17, 22, 4, 18, 15]. However their performance can be upgraded by identifying the significant feature subset and the initial cluster centroids. The objective of feature selection for unsupervised learning is to find the smallest feature subset that best uncovers clusters from data according to the preferred criterion[10]. Feature selection in unsupervised learning is much tougher problem, due to the absence of class labels [13]. This paper investigates the use of GA [7] to augment the performance of both k-means and fuzzy k-means clustering by (i) identifying significant features and (ii) selection of initial cluster centroids for standard medical dataset namely Pima Indians Diabetes Dataset (PIDD)[16]. Performance of the projected work is compared with Entropy based fuzzy clustering (EFC)[21] identified k-means initial centroids. Section 2 elaborates on related work on feature selection for clustering and different ways to initialize the k-means cluster centroids. Section 3 and Section 4 briefs about the working of k-means and fuzzy k-means clustering respectively. Section 5 describes working of both binary and integer encoded GA for feature selection. Section 6 describes the functioning of both binary and real encoded GA and EFC method for initializing both k-means and fuzzy k-means centroids. Experimental results and conclusions are discussed in Section 7 and Section 8 respectively.

2 Related Works

Feature selection for unsupervised learning can be subdivided into filter methods and wrapper methods. Filter methods in unsupervised learning is defined as using some intrinsic property of the data to select features without utilizing the clustering algorithm [10].Volker [19]proposed a wrapper method where Gaussian mixture model combines a clustering method with a Bayesian inference mechanism for automatically selecting relevant features. Hao-Jun Sun [8] has used GA for feature subspace selection of k-means clustering with artificial datasets. They have used binary encoding to represent both the feature subspace and the cluster centers. The k-means clustering does not guarantee unique clustering since it engenders diverse results with randomly selected initial clusters for different runs of k-means. Several methods have been proposed to solve the cluster initialization for k-means algorithm. Bashar Al-Shbour [2]and Jimenez [11] have used GA to initialize the k-means cluster centers and discover the number of clusters [11] using binary encoding with artificial dataset. Vidyut Dey [20] has used EFC [21] to initialize the fuzzy c-mean initial cluster centers. Maulik [18] have applied GA to find the k-means cluster centers using floating point representation for clustering iris, crude oil ,vowels and artificial datasets. Several methods have been proposed to solve the cluster initialization for k-means algorithm using Pima diabetic dataset which contains 768 instances with eight input attributes and class with tested positive / tested negative value. Mohammad F [15] has used statistical information from the data set to initialize the k-means prototypes for iris, wine, Pima and synthetic datasets. Artificial bee colony [4] and Artificial fish swarm algorithm [22] have been used to find the initial centroids for fuzzy k-means and k-means respectively for clustering of Pima diabetic dataset. Maximum distance method has been adopted to find the initial centroids for k-means clustering using significant attributes identified by PCA [17] for Pima diabetic dataset. Around 376 records of PIMA dataset have missing values, represented as zero, which is biologically impossible. Most of the clustering work carried out on Pima diabetic dataset in the literature have neither removed the missing data nor identified significant attributes. Authors have investigated the proposed work with 392 cases (130 tested positive cases and 262 tested negative) after removal of instances with missing data and have used GA & EFC identified centroids with GA identified significant attributes.

3 K-means Clustering

K-means [12] takes the input parameter, k as number of clusters and partitions a dataset of n objects into k clusters, so that the resulting objects of one cluster are dissimilar to that of other cluster and similar to objects of the same cluster. K-means algorithm targets at minimizing an objective function namely sum of squared error (SSE). SSE is defined as

$$E = \sum_{i=1}^{k} \sum_{p \in C_i} \|p - m_i\|^2 \tag{1}$$

where E is sum of the square error of objects with cluster means for k clusters, p is the object belong to a cluster Ci and mi is the mean of cluster C_i.
K-means partitioning algorithm: Input is k is the number of clusters, D is input data set. Output is k clusters.

1. Randomly choose k objects from D as the initial cluster centers.
2. Repeat
3. Assign each object from D to one of k clusters to which the object is most similar based on the mean value of the objects in the cluster.
4. Update the cluster means by taking the mean value of the objects for each of k cluster.
5. Until no change in cluster means/ min error E is reached.

4 Fuzzy k-means clustering

Fuzzy K-means [3, 5] is an extension of K-Means clustering. The major difference between the fuzzy k-means and k-means is that the later discovers hard clusters where a particular sample can belong to only one cluster while the former discovers soft clusters where a particular sample can belong to more than one cluster with certain probability. This belongingness of a data sample to the cluster is represented using membership values. Let there be N data points each of L dimensions. The object X_{ij} represents i^{th} object with j^{th} dimension with i = 1 to N; j = 1 to L. The working of fuzzy k-means algorithm is discussed using following steps.

1. Input the number of clusters and appropriate level of cluster fuzziness $g > 1$. Initialize the N x C sized membership matrix $[\mu]$ at random, such that

$$\mu_{ij} \in [0.0, 1.0] \, and \sum_{j=1}^{C} \mu_{ij} = 1.0 \, for \; each \; i, for \; i \; = \; 1 \; to \; N, j \; = \; 1 \; to \; C \quad (2)$$

2. Compute k^{th} dimension of the j_{th} cluster center CC_{jk} using the equation (3)

$$CC_{jk} = \frac{\sum_{i=1}^{N} \mu_{ij}^{g} x_{ik}}{\sum_{i=1}^{N} \mu_{ij}^{g}} \quad (3)$$

3. Compute the Euclidean distance d_{ij} $d_{ij} = \|CC_j - x_i\|$ for i_{th} sample and j_{th} cluster center.
4. Update fuzzy membership matrix $[\mu]$ using equation 4 depending on $d_{\cdot ij}$ If $d_{ij} \geq 0$, then

$$\mu_{ij} = \frac{1}{\sum_{m=1}^{C} \left(\frac{d_{ij}}{d_{im}} \right)^{\frac{2}{g-1}}} \quad (4)$$

If $d_{ij} = 0$ then the sample coincides with j^{th} cluster center CC_j and it will have full membership value that is $\mu_{ij} = 1$.
5. Repeat steps 2-4 until the change in matrix $[\mu]$ is less than some user specified value.

The performance of fuzzy k-means depends on the randomly initialized matrix $[\mu]$. This paper proposes modified fuzzy k-means clustering algorithm by first initializing the centroids using 4 different methods: Random, EFC and real and binary encoded GA as mentioned in section 6. The Euclidean distance between the samples and the initial centroids identified by the proposed method is used to compute membership matrix $[\mu]$. The matrix $[\mu]$ in turn is used to compute new centroids and the process is repeated till user specified minimum change in the matrix $[\mu]$ is achieved.

5 Genetic Algorithm based feature selection

5.1 Genetic algorithm

GA [7] is an optimization techniques motivated by natural selection and natural genetics. GA is a stochastic general search method, proficient of effectively exploring large search spaces. GA is mainly composed of three operators: reproduction, crossover, and mutation. As a first step of GA, an initial population of individuals is generated at random or heuristically. In the selection process, the high fitness chromosomes are used to eliminate low fitness chromosomes. But selection alone does not produce any new individuals into the population. Hence selection is followed by crossover and mutation operations. Crossover is the process by which two selected chromosome with high fitness values exchange part of the genes to generate new pair of chromosomes. The crossover tends to facilitate the evolutionary process to progress towards the potential regions of the solution space. Mutation is the random change of the value of a gene, which is used to prevent premature convergence to local optima. The new population generated undergoes the further selection, crossover and mutation till the termination criterion is not satisfied. Convergence of the genetic algorithm is governed by the various criterions like fitness value achieved or the maximum number of generations[7, 6]. GA has been used in this paper to (i)identify the significant feature subset for k-means and fuzzy k-means clustering and (ii)to find k-means and fuzzy k-means initial cluster centers. The working of both binary and integer encoded GA for finding significant features is elucidated in section 5.2 and Section 5.3 respectively. The working of binary and real encoded GA for identifying k-means and fuzzy k-means initial cluster centroids is explained in Section 6.1 and Section 6.2 respectively.

5.2 Feature selection using Binary encoded GA

Genetic algorithm (GA) (binary and real encoded) is used as a random search technique for subset generation, wrapped with k-means clustering. The functioning of binary encoded chromosome for identifying significant features is briefed as follows. The length of the chromosome is set equal to total number of features say F. The ones and zeros of the binary encoded GA represent the significant and non-significant features. The GA is experimented with different values of MaxF, where MaxF is the maximum number of significant features selected in the range of [1 to F-1]. Each of the chromosomes has exactly MaxF number of ones. For example, the diabetic dataset has a total of 8 features (F = 8 features), hence the length of chromosome is equal to 8 with each gene value of one or zero (representing selection or dropping of an attribute). With MaxF = 4 and F= 8, the binary encoded chromosome: 10011001 represent 1st, 4th, 5th and 8th features as significant features and 2nd, 3rd, 6th and 7th as non-significant features for k-means clustering. The working of GA with binary encoded chromosomes for learning the significant feature subset for k-means is briefed below:

1. Initialize the chromosome population randomly using binary encoding (Each chromosome length is equal to total number of features F, where number of ones is equal to MaxF as explained above)
2. Repeat the steps (a-d) till terminating condition is reached

    a. Apply k-means clustering using significant features represented by each of the binary encoded chromosome.
    b. Select the chromosome with least SSE as the fittest chromosome. Replace the low fit chromosomes by highest fit chromosome.
    c. Select any two chromosomes randomly and apply crossover operation
    d. Apply mutation operation by randomly changing the bit 1 to 0 or bit 0 to 1 of the randomly selected chromosome. After mutation and crossover operation the number of ones in the chromosomes must be checked for not exceeding the number of required features: MaxF.

3. The positions of bit 1 in the fittest chromosome represent the significant features subset.

5.3 Feature selection using Integer encoded GA

The functioning of integer encoded chromosome used for feature selection is briefed as follows. With integer encoding, let F represent the total number of features. The length of the chromosome is equal to MaxF, where MaxF is the maximum number of significant features to be selected, which is in the range of 1 to F-1. Each gene in integer encoding may take a value in the range of 1 to F. For example with diabetic data set F = 8 features. With integer encoding of chromosomes, each gene represents the feature selected as part of feature subset selection. For example with MaxF of value 4, the integer encoded chromosome: 1457, represents a chromosome with 1st, $4^{th}, 5^{th}$ and 7th features as significant features and the $2^{nd}, 3^{rd}, 6^{th}$ and 8th features as non-significant features for k-means clustering. Experiments are conducted using different values of MaxF in the range of 1 to F-1. The working of GA for judgment of the significant feature subset for k-means using Integer encoded chromosomes:

1. Initialize the chromosome population randomly using integer encoding. Each chromosome length is equal to MaxF, where each gene is an integer number in the range of 1to F(no two genes in the chromosomes should have the same integer value).
2. Repeat the steps (a-d) till terminating condition is reached.

    a. Apply k -means clustering using significant features represented by each of the integer encoded chromosome.
    b. Select the chromosome with least SSE as the fittest chromosome. Replace the low fit chromosomes by highest fit chromosome.

    c. Select any two chromosomes randomly and apply crossover operation.

    d. Apply mutation operation by randomly selecting any one chromosome and randomly change gene by an integer value in the range of 1-F. (After mutation and crossover operation the chromosomes must be checked for not having the same feature selected more than once)

3. The integer numbers in the best-fit chromosome represent the significant features subset.

6 Methods used to initialize k-means and fuzzy k-means centroids

6.1 *GA (Binary encoded) based initial cluster centroids*

As a part of second stage of our projected work, the GA is further used to recognize the k-means and fuzzy k-means initial cluster centroids. Chromosomes are encoded using binary encoding where 1 represents the sample selected as initial cluster center and 0 represents the sample is not selected as initial cluster center [1]. The length of the chromosome is equal to total number of samples. After mutation and crossover operation the number of ones in the chromosomes must be checked for not exceeding the number of required clusters. More details can be found in [1].

6.2 *GA (Real encoded) based initial cluster centroids*

With real encoding, the user needs to specify two parameters: number of clusters: MaxC and number of dimensions for the input dataset: MaxD. The length of the chromosome is equal to MaxC x MaxD genes. For example with MaxC = 2, Max D = 3 and let the 5th sample (4.5, 6.7, 4.9) and 100th sample (2.2, 9.2, 1.5) be randomly selected as initial centroids. Then the chromosome using 5th and 100th sample as centroids, is represented using 2 X 3 = 6 genes as (4.5, 6.7, 4.9, 2.2, 9.2, 1.5). The population undergoes the selection, crossover and mutation process till the terminating condition is satisfied. Finally the fittest chromosomes (with the minimum SSE) represent the initial centroids which are decoded from left to right. The first MaxD genes represent the MaxD dimensions of first centroid, followed by next MaxD genes as dimension of second centroid and so on. With binary and real encoded GA, both one point and two-point crossovers resulted in almost the same centroids. The GA was investigated with population's size of 50 – 110 chromosomes, 15 to 30 generations and with both one point and two-point crossover. The terminating condition used for the experiment is that, 80% of the chromosomes represent the identical initial cluster centers. Once the terminating condition is reached, the highest fittest chromosomes represent the initial centroids for both k-means and fuzzy k-means clusters.

6.3 EFC recognized initial cluster centroids

Yao [21] introduced Entropy based fuzzy clustering (EFC), which identifies the number of clusters and initial cluster prototypes by itself. The entropy is calculated for each sample using equation (5).

$$E_i = -\sum_{\substack{j \in x \\ j \neq i}} (S_{ij} log_2 S_{ij} + (1 - S_{ij}) log_2(1 - S_{ij})) \tag{5}$$

where $S_{ij} = e^{-\alpha d_{ij}}$ is the similarity between two data points (i, j) and d_{ij} is the Euclidean distance between points (i, j) The algorithm for entropy based fuzzy clustering is as follows. The inputs for the algorithm are dataset D with N samples, β the threshold value, which can be viewed as a threshold of similarity among the data points in the same cluster, and constant α, which is computed as $\ln(0.5)/\overline{D}$, where $\overline{D}$ is the mean distance among the pairs of data points in a hyper-space and is usually set to 0.5.

1. Compute entropy E_i for each sample x_i from dataset D for i = 1 to N.
2. Identify x_i that has the minimum Ei value as the cluster centre.
3. Remove x_i and data points having similarity x_i > some threshold β from D.
4. If D is not empty then go to step b. The k centroids identified by EFC, are selected as k-means and fuzzy k-means initial cluster centres.

7 Experimental Results

As part of the first stage of proposed work, GA is investigated using both binary and integer encoded chromosome for feature selection as mentioned in section 5. Both one point and two point crossover were experimented. The terminating condition is that 80% of the chromosomes represent the same feature subset. Once the terminating condition is reached, the highest fittest chromosome (with the least sum of square error (SSE)) represents the significant feature subset. For both binary and integer encoding, the total number of features F is 8 for diabetic data-set, and MaxF is experimented with the values in the range of 4 − 6. With binary encoding the experiments were conducted with 50 − 100 population size, 15 − 30 generations and for MaxF in the range of 4 − 6. With integer encoding, experiments were conducted with 10 − 20 population size, 10 − 70 generations and for MaxF in the range of 4 − 6. Table 1.1 indicates that, for diabetic dataset the SSE is least with MaxF = 6. The following significant attributes were identified: Plasma glucose, pedigree, BMI, Age, Insulin and diastolic blood pressure with MaxF = 6. The significant attributes sets identified by both binary and integer encoded chromosomes are same with MaxF value of 5 and 6. For MaxF = 4, both integer and binary encoded chromosomes identified different sets of significant attributes.

Table 1 Performance of K-means clustering with random centroids for different values of MaxF using Binary/Integer encoded GA

Attributes selection using Binary /Integer Encoding	# Features	TP	FP	TN	FN	# Iterations	SSE	Error
—	8	74	56	197	65	7	0.0087	30.86
Binary/Integer	6	92	38	198	64	7	0.0027	26.02
Binary/Integer	5	89	41	196	66	7	0.0099	27.29
Binary	4	91	39	196	66	5	0.0076	26.78
Integer	4	69	61	205	57	5	0.0058	30.14

As a part of second stage of our proposed work, experiments were conducted by initializing the k-means and fuzzy k-means with random, GA (both real and binary encoding) and EFC identified centroids.

Table 2 Performance of k-means clustering with centroids initialized using random , GA and EFC random centroids for different values of MaxF with all and GA identified attributes

Method used to initialize k-means centroids	# Features	TP	FP	TN	FN	# Iterations	Sensitivity	Specificity	F-measure	Precision
Random	8	74	56	197	65	7	0.53	0.78	0.55	0.57
Random	6	92	38	198	64	7	0.59	0.84	0.64	0.71
G A (Binary)	8	86	44	197	65	5	0.57	0.82	0.61	0.66
G GA(Binary)	6	92	38	206	56	5	0.62	0.84	0.66	0.70
G A(Real)	8	78	56	206	52	4	0.60	0.78	0.59	0.58
G A(Real)	6	84	52	210	46	4	0.64	0.80	0.63	0.62
EFC	8	89	41	192	70	5	0.56	0.82	0.62	0.68
EFC	6	92	38	198	64	4	0.59	0.84	0.64	0.71

Table 2 and Table 3 illustrates the comparative performance of k-means and fuzzy k-means with all attributes and GA identified attributes using centroids identified by random, both real and binary encoded GA and EFC in terms of number of iterations, Sensitivity, Specificity, Precision, and F measure. Figure 1 and Figure 2 elucidates the enhanced classification accuracy of k-means and fuzzy k-means with all attributes and GA identified attributes using centroids identified by random, both real and binary encoded GA and EFC.

The hike in the accuracy of k-means and fuzzy k-means is the consequence of not only GA identified attributes but also because of the initial centroids identified by

Table 3 Performance of Fuzzy k-means clustering with centroids initialized using random , GA and EFC random centroids for different values of MaxF with all and GA identified attributes

Method used to initialize fuzzy k-means centroids	Fuzzifier value	# Features	TP	FP	TN	FN	# Iterations	Sensitivity	Specificity	F-measure	Precision
Random	1.15	8	94	36	179	83	6	0.53	0.83	0.61	0.72
Random	1.03	6	93	37	198	64	5	0.59	0.84	0.65	0.72
GA(Binary)	1.07	8	88	42	210	52	5	0.63	0.83	0.65	0.68
G A(Binary)	1.11	6	92	38	207	55	4	0.63	0.84	0.66	0.71
G A (Real)	1.03	8	84	46	209	53	4	0.61	0.82	0.63	0.65
G A(Real)	1.07	6	92	38	203	59	4	0.61	0.84	0.65	0.71
EFC	1.09	8	93	37	198	64	4	0.59	0.84	0.65	0.72
EFC	1.11	6	91	39	206	56	3	0.62	0.84	0.66	0.70

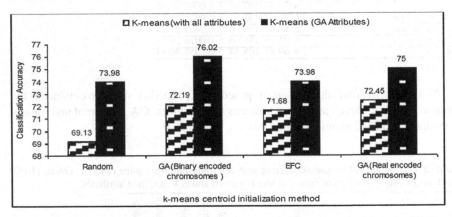

Fig. 1 k-means accuracy by initializing cluster centroids using random, GA (Binary and Real encoded chromosomes) and EFC method with all attributes and with significant attributes identified by GA for Diabeteic dataset.

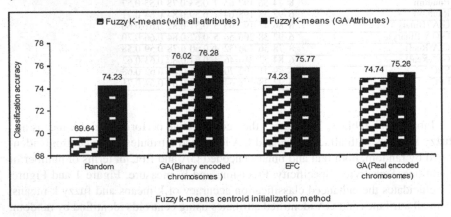

Fig. 2 Fuzzy k-means accuracy by initializing cluster centroids using random, GA (Binary and Real encoded chromosomes) and EFC method with all attributes and with significant attributes identified by GA for Diabeteic dataset.

GA and EFC. For the medical dataset experimented, the GA centroids proved to be better compared to EFC centroids. The performance of proposed work is compared with the some of the existing methods as depicted in Table 4.

Table 4 Comparison of classification accuracy of Proposed method with other methods for Pima diabetic dataset

Method Adopted	Classification Accuracy	Reference
Maximum distance method for initial centroids 768 instances with PCA for dimension reduction	74.00%	Rajashree et.al (2010)
Artificial Bee Colony for Fuzzy k-means initial centroids 768 instances with all attributes	68.95%	Karboga et.al (2010)
Statistical method for initial centroids 768 instances with all attributes	66.32%	Mohammed et.al (2011)
Artificial fish swarm algorithm for k-means for initial centroids 768 instances with all attributes	62.19%	Yazdani et.al (2010)
GA + k-means, GA + Fuzzy k-means for initial centroids (Binary encoded chromosome) EFC+ k-means, EFC + Fuzzy k-means for initial cen-troids	76.02% 76.28% 73.98% 75.77%	Proposed scheme
GA +k-means, GA + Fuzzy k-means for initial centroids (Real Encoded chromosome) 392 instances with GA identified significant attributes	75.00 % 75.26%	

8 Conclusions

This paper illustrates the upgraded performance of both k-means and fuzzy k-means clustering with centroids identified by both GA and EFC using GA identified attributes. The GA identified centroids outperformed the performance of k-means and fuzzy k-means when compared to EFC identified centroids. The performance of projected work is tested successfully for PIMA Indian diabetic dataset. In the future study, it is aimed to improve the performance of k-mediods and fuzzy k-mediods using GA and EFC identified mediods with GA identified attributes.

References

1. Asha Gowda Karegowda, S., Shama, T.R., Vidya, M.A. Jayaram, A.S. Manjunath. Improving performance of K-means clustering by initializing cluster centers using Genetic algorithm and Entropy based Fuzzy clustering for categorization of diabetic patients. In Proceedings of International Conference on Advances in Computing, MSRIT, Bangalore, (2012). (Advances in Intelligent Systems and Computing(Springer), Volume 174, 2013, pp.899-904)
2. BasharAl-Shbour, S.-H., Myaeng, Initializing K-means using Genetic Algorithm World Academy of Science, Engineering and Technlogy, (2009). 54: pp.114-118.

3. Bezdek J.C.(1973). Fuzzy mathematics in pattern classification, Ph.D thesis, Applied Mathematics Center,Ithca:Cornell University.

4. Dervis Karboga and Celal Ozturk, Fuzzy clustering with artificial bee colony algorithm, (18 July 2012), Scientific Research and Essays Vol 5(14), pp.1899-1902.

5. Dunn J.C. (1973). A fuzzy relative of the ISODATA process and its used in detecting compact well-separated clusters. Journal of Cybernetics, vol 3.pp.32-57.

6. Eduardo R.H., Ricardo Campello, Alex A., Freitas, Andre C.P, (2009), A Survey of Evolutionary Algorithm for Clustering. IEEE Transanctions on Systems, Man and Cybernetics, 39(2).

7. Goldberg, D., Genetic Algorithms in Search, Optimization , and Machine learning (1989): Addison Wesley.

8. Hao-Jun , Genetic Algorithm-based High-dimensional Data Clustering Technique, , in Sixth International Conference on Fuzzy systems and Knowledge Discovery pp.485-489, (2009).

9. Han, J., Kamber, M., Data Mining: Concepts and Techniques (2001), San Francisco, CA: Morgan Kaufmann Publishers.

10. Jennifer. G. Dy, C.E., Brodley, Feature Selection for Unsupervised Learning. Journal of Machine Learning Research, 5: pp.845-889, (2004).

11. Jimenez, J.F., Cuevas, F.J., Carpio, J.M. (2007). Genetic Algorithms applied to Clustering Problem and Data Mining. in 7th WSEAS International Conference on Simulation, Modeling and Optimization.

12. Mac Queen, (1967), J. Some methods for the classification and analysis of multivariate observations in Fifth Berkeley Symposium on Mathematical Statistics and Probability. Berkeley: University of California Press.

13. Manoranjan Dash, K.C. Peter Scheuermann, Huan Liu,.(2002), Feature Selection for Clustering - A Filter Solution. in Second International Conference on Data Mining.

14. Manoranjan Dash, H Liu, Feature selection for classification. Intelligent Data Analysis, (1998). 1: pp.131-156.

15. Mohammad F. Eltibi, W.M., Ashour, Initializing K-means Clustering Algorithm using Statistical Information International Journal of Computer Applications, (2011). 29: pp.51-55.

16. Newman, D.J., Hettich, S., Blake, C. L., Merz, C. J. , UCI repository of machine learning databases, (1998): University of California, Irvine, Dept. of Information and Computer Sciences.

17. Rajashree Dash, Debahuti Mishra, Amiya Kumar Rath, Milu Achrua. "A hybridized K-means clustering approach for high dimensional dataset", International Journal of Engineering , Science and Technology, vol2(2), (2010), pp.59-66

18. U. Maulik , S., Bandopadhyay, Genetic Algorithm-Based Clustering Technique. Pattern Recognition, (1999). 33: pp.1455-1465.

19. Volker Roth, Tilman Lange, Feature Selection in Clustering Problems, in In Advances in Neural Information Processing Systems (NIPS),(2003).

20. Vidyut Dey, D.K., Pratihar, Gauranga Lal Datta, (2011) Genetic algorithm-tuned entropy-based fuzzy C-means algorithm for obtaining distinct and compact clusters Fuzzy Optimization Decision Making 10: pp.153-166.

21. Yao.J, M., Dash, S.T, Tan, Liu.,H.,(2000) Entropy based fuzzy clustering and fuzzy modeling Fuzzy Sets and Systems. 113: pp.381-388.

22. Yazdani, D., Golyari, S., Meybodi, M.R., "A new hybrid approach for data clustering", 5th International Symposium on Telecommunications, (4-6 Dec. 2010) pp.914-919.

A Computational Intelligence based Approach to Telecom Customer Classification for Value Added Services

Abhay Bhadani[1], Ravi Shankar[2], and D. Vijay Rao[3]

[1]Bharti School of Telecommunication Technology and
Management, Indian Institute of Technology Delhi,
abhay_bhadani@dbst.iitd.ac.in,
http://web.iitd.ac.in/~bsz098503/abhay.html
[2]Department of Management Studies, Indian Institute of
Technology Delhi, ravi1@dms.iitd.ac.in,
http://web.iitd.ac.in/~ravi1
[3]Institute for Systems Studies and Analyses, Defence Research and
Development Organization (DRDO), Metcalfe House, Delhi,
doctor.rao.cs@gmail.com

Abstract

Customer classification is an imperative task for any organization catering to different market segments. In telecom industry it becomes even more important to identify which value added services(VAS) would be successful with a given customer segment. VAS provide a flexible revenue model that can be customized to different customer segments based on several attributes such as usage and preferences. Selecting and customizing VAS provides a wide canvas to the operators for maximizing their returns on the customer portfolio. Computational intelligence techniques such as Artificial Neural Network (ANN) and Support Vector Machine (SVM) have been successfully used for data mining and machine learning. These techniques provide a mathematical framework for identifying customers profiles and patterns in large datasets that representing the customers' data and their preferences. In this paper, we propose a methodology using SVM and ANN techniques to classify telecom customer data and identify the VAS best suited for the customer segment. We test our results with the SVM yielding high prediction accuracy for the unknown public test data with Radial Basis Function(RBF) Kernel using grid search technique.

Keywords: *Customer data classification, Support Vector Machine, Artificial Neural Network, Value Added Services*

J. C. Bansal et al. (eds.), *Proceedings of Seventh International Conference on Bio-Inspired Computing: Theories and Applications (BIC-TA 2012)*, Advances in Intelligent Systems and Computing 201, DOI: 10.1007/978-81-322-1038-2_16, © Springer India 2013

1   Introduction

With a fierce competition being faced by the telecom operators, they are left
with little scope for making profit with simple basic services like Voice or Short
Message Service(SMS) and people registering for Do Not Disturb kind of services. VAS provides a flexible revenue model that can be customized for different
customer classes based on their usage, preferences and likings. Telecommunication companies generate a tremendous amount of data continously. These
include details related to call data, which describes the calls that traverse the
networks, network data, describing the state of the hardware and software components used in the network, and static data like customer data, describing
the telecommunication customers, demographic information, describing the personal details of the customer including name, location, birth date. Computational Intelligence based Machine learning algorithms can classify large datasets
and find interesting patterns. Several techniques have been proposed in literature for the classification of large data sets such as decision trees, ANN, SVM,
Linear Discriminant Analysis, Linear Regression, Polynomial Regression, etc.
In this paper, we use ANN and SVM techniques for classifying the telecom customer data. A brief outline of these techniques has been discussed in section 2,
3.

SVM works on the principle of constructing a hyperplane[1] as the decision
surface such that the separation margin between positive and negative examples
is maximized, see Figure 1. ANN is a technique which tries to mimic the functioning of brain, See Figure 2. It was originally inspired from the functioning
of human brains ability to take decisions once it learns which is retained in the
memory and unknowingly helps in taking decision. Both these techniques can
be used for classification and prediction to determine business intelligence in
large sets of telecom customer's data. Customers use mobile phones for communication, entertainment and m-commerce applications. It is thus necessary
for telecom companies to identify the specific requirement and preferences of
customers and thus customize their VAS with a goal of maximizing the utility
to the customers and revenue to the operators.

In this paper, we propose a methodology using SVM and ANN techniques
to classify telecom customer data and identify the VAS Plan best suited for the
given customer segment. From these classified customer segments we derive the
classification rules where attributes are considered to be fuzzy.

2   Background and Literature Review

Decision making becomes a complex activity especially with large datasets being generated continuously as is the case of telecommunication industry (Weiss,
2005). Customer classification is a powerful tool being used in decision making and developing appropriate strategies. Broadly, classification techniques

---

[1]A hyperplane is a geometric generalization of the plane into a different number of dimensions.

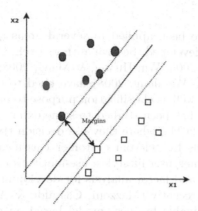

Figure 1: Support Vector Machine Model

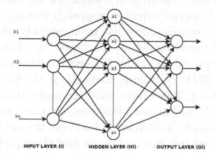

Figure 2: Artificial Neural Network Model

can be treated as parametric and non-parametric functions. There are several techniques available in the literature for regression and classification. Linear probability models, Generalized Linear models, Logistic Regression, Discriminant Analysis, Neural Networks, Genetic Algorithms, K-Nearest Neighbors, Decision Trees, Neural Networks, Perceptrons(single layer and multi-layered), Naive Bayes classifiers, Bayesian Networks, Instance based learning, Support Vector Machine (Large margin classifier) are some fo the popular techniques of computational intelligence used for making decisons with large input datasets.

Modeling complex, N-dimensional problems precisely is a challenging task. Highly predictive models continue to play an increasingly important role in 21st century marketing applications, particularly in areas such as automated modeling, mass-produced models, intelligent software agents, and data mining (Cui & Curry, 2005). Predictive accuracy is considered to be the standard way for measuring the model quality (Politz & Deming, 1953). The SVM avoids over reliance on particular structural assumptions and implicitly automates the model identification process. By doing so it enters the parameter estimation phase with a family of structural possibilities. Using kernel transformations cleverly, the SVM solves a nonlinear problem with a linear model (Cui & Curry,

2005).

ANN and SVM have been applied in several areas and a wide variety of applications have been developed. Several authors (Yeh, Chi, & Hsu, 2010; Lee, 2007; Lovell & Walder, 2006; Wu, Huang, & Meng, 2008; J.-J. Huang, Tzeng, & Ong, 2007; Guo-en & Wei-dong, 2008) have tried to use SVM for building models for regression as well as classification purpose to name a few. Cross selling from mobile phones has been studied to some extent by (Ahn et al., 2010; Ahn, Ahn, Oh, & Kim, 2011), where they use decision tree, ANN and Logistic Regression based models for telecom customer classification. Consumer behaviour based on consumer/user lifestyles, use-motivations and product/service attributes using cluster and factor analysis for developing their strategic plans have also been studied recently (Mazzoni, Castaldi, & Addeo, 2007). Reuver and Haaker propose a viable business model based on context aware mobile services from practitioners and experts in this field (Reuver & Haaker, 2009) and Constantiou proposes a theoretical framework for adoption of Mobile TV (Constantiou, 2009). A service oriented approach towards intelligence in telecom industry based on questionnaires has been recently proposed in (Ishaya & Folarin, 2012). A score based model was proposed by (Ahn, Ahn, Byun, & Oh, 2011) to study the likings and adoption of VAS by telecom user. Several authors (Sharma & Panigrahi, 2011; B. Huang, Kechadi, & Buckley, 2012; Pendharkar, 2009; Guo-en & Wei-dong, 2008) have tried to model telecom customer churn based on ANN and SVM techniques. However, all these studies lacked mathematical approach to predict the likings and success factors to generate business intelligence to understand telecom customers and their preferences.

3 Computational Intelligence Techniques for classification

Several techniques such as Bayesian network, decision trees and time series modeling have been proposed in literature classification of large data sets. The uncertainty associated with the data, the immense size of the data to deal with and the diversity of the data type and the associated rules and scales are important factors to rely on unconventional mathematical tools such as Computational intelligence based techniques in the domain of soft computing such as Artificial Neural Network (ANN) and Support Vector Machine (SVM), Logistic Regression, Fuzzy Reasoning, evolutionary computing for data analysis and interpretation have been successfully used for data mining and machine learning. These techniques provide a mathematical framework for identifying customers profiles and patterns in large datasets representing the customers' data and their preferences.

3.1 Artificial Neural Network

Neural Networks concept in computer science discipline was inspired by the architecture as well as functioning of the human brain. It exhibits certain features such as the ability to learn complex patterns of the information and then generalize the knowledge gained (Rosenblatt, 1962; Venugopal & Baets, 1994). Artificial Neural Networks (ANNs) attempt to model the architecture of biological neural networks. Biological neural networks are made of simple, tightly interconnected processing elements called *neurons*. The interconnections are made by the outgoing branches, the *"axons"*, which again form several connections *("Synapse")* with the other neurons. When a neuron receives a number of stimuli, and when the sum of the received stimuli exceeds a certain threshold value, it will fire and transmit the stimulus to adjacent neurons. The biggest advantage of neural network methods is that they can handle problems with large number of parameters, and still are able to classify objects well even when the distribution of objects in the N-dimensional parameter space. ANN has the ability to give multiple outputs from the same source of information. It classifies the output in multiple classes as desired by the analyst. The neural network design process primarily involves the following sequence of steps: *Collecting data, Creating the network, Configuring the network, Initializing the weights and biases, Training the network, Validating the network, Finally, Using the network for prediction purpose.*

Artificial Neural Network are have been designed in three stages/layers : *Input Layer(I), Hidden Layers(H_i) and Output Layer (O)* . This notation has been used in Table 3. Input layers can have n-inputs which is primarily decided based on the number of attributes to be used. Depending on the application it can have multiple outputs used for Multi-Class classification or single-class (also called as Regression) in the output layer. Hidden layers can be single or multiple depending on the need for designing the network. Usually one hidden layer is sufficient for most of the applications depending on the problem and network design. However, more than one hidden layers can be used to enhance the accuracy and robustness of the model. In this work, we have used one, two hidden layers for our experiments.

3.2 Support Vector Machine

SVM is a relatively new supervised machine learning techniques with a strong mathematical foundation being used for solving a variety of problems and resulting in high performance due to it's semi-parametric approach. Support Vector Regression(SVR) and SVM are based on statistical learning theory, Vapnik Chervonenkis(VC) theory, developed over the last several decades (Vapnik, 1995, 1998). The SVM, developed by Vapnik and others in 1995, is used for many machine learning tasks such as pattern recognition, object classification, and in the case of time series prediction, regression analysis. The SVM has been successful as a high performance classifier in several domains including pattern recognition, data mining and bioinformatics (Lovell & Walder, 2006; Hsu, Chang, & Lin, 2010). SVR, is the methodology by which a function is estimated using observed data which in turn "trains" the SVM.

To carry out the non-linear regression using SVR, it is necessary to map the input space $x(i)$ into a (possibly) higher dimension feature space $\varphi(x(i))$.

$$k(x, y) = \langle \phi(x), \phi(y) \rangle \tag{1}$$

$k(x, y)$ is some kernel function,
x denotes the feature space taken as input parameters,
y denotes the target or output vector.

The kernel functions used in our work are Polynomial, RBF (also sometimes termed as Gaussian) and Sigmoid (Hyperbolic Tangent) kernels which have been described below:

Polynomial Kernel

$$k(x, y) = \left(\alpha x^T y + c)^d\right) \tag{2}$$

RBF (Gaussian) Kernel

$$k(x, y) = exp\left(-\frac{||x - y||^2}{2\sigma^2}\right) \tag{3}$$

Sigmoid Kernel

$$k(x, y) = tanh(\alpha x^T y + c) \tag{4}$$

Kernel methods are a class of algorithms for pattern recognition and identifying patterns, whose best known element is the SVM. The use of kernels is the key in SVM/SVR applications. The kernel trick is a mathematical tool which can be applied to any algorithm which solely depends on the dot product of two vectors. Wherever a dot product is used, it can be replaced by a kernel function. When properly applied, linear algorithms are transformed into a non-linear algorithms (sometimes with little effort or reformulation). The use of a positive definite kernel ensures that the optimization problem will be convex and solution will be unique. SVMs have also been proven to outperform other non-linear techniques including ANN, Logistic Regression (Sapankevych & Sankar, 2009). These learning algorithms have also been applied to general regression analysis: the estimation of a function by fitting a curve to a set of data points. More details about kernel functions can be found here (Souza, 2010).

4 Classification of Telecom customer data

Telecom companies generate huge data based on customer profiles. Some of the data are static (such as, demographic data, socio-economic factors, gender and preferences) that are collected when the mobile numbers are issued; whereas others (such as call usage profile, packs activated, services used) are dynamically recorded in a spatio-temporal database. This data would prove to be an asset to the telecom company provided it can dig out reasonable amount of information about the customer inorder to identify and deliver appropriate VAS and customize them for the desired customer segment.

Based on the sample data from telecom companies, we have identified twenty-one attributes from the literature (Weiss, 2005; Ahn, Ahn, Oh, & Kim, 2011; Cui & Curry, 2005; J.-J. Huang et al., 2007; Mazzoni et al., 2007; Constantiou, 2009; Wang & Wang, 2009). The attributes used for the problem formulation are: *Age(age)*, *Gender(gen)*, *Religion(rel)*, *Qualification(qual)*, *Occupation(occ)*, *State(st)*, *Circle(ci)*, *Plan Opted(pl-opt)*, *Opting Any VAS(ot-vas)*, *Roaming Duration(roam)*, *Family Income(inc)*, *Operating Any Banks Transactions(bank)*, *Number of Family Members(f-mem)*, *Marital Status(mar)*, *Total Day Minutes(d-min)*, *Number of Total Day Calls(d-call)*, *Total Night Minutes(n-min)*, *Number of Total Night Calls(n-call)*, *Total Minutes per Day(m-day)*, *Average Duration per Call(dur)*, *Avg. Monthly Minutes consumed(avg-min)*.

4.1 Rules

Out of 21 attributes, ten important attributes have been considered for the classification of the customers inorder to identify their needs and calling patterns. This is presented in the Table 2.

Consultation with subject matter experts from various telecom companies helped us in identifying the needs of the customers and accordingly different schemes or products are designed as part of the targeted marketing after segementing the customers.

Several cost models have been explored and a linear cost model has been applied with the various plans associated with different customer segments. We are refining the cost model and the attributes and the plans from the company so that this forms a Decision Support System for formulating the plans to optimize the portfolio and maximize the company's revenue. In our cost model, we have considered the distribution of customers and considering the cost associated with the calls made to a local network, roaming network, Same Operator's Local or STD number, Other Operator's Local or STD number, Night Calls, SMS charges Local or STD, Data Usage(Internet), and ISD Calls to obtain the expected revenue from each plan.

5 Results and Analysis

The ANN and SVM have been trained to give outputs belonging to 13 classes labeled A to M. A test data set was generated using statistical distributions such as, uniform, normal and Bernoulli distributions based on the attribute having close resemblance to real data set. However, targets were decided based on different criteria like age, monthly income, mobile usage time and duration, gender, education and call usage.

The data was normalized between -1 to 1 before training and testing so that no attribute dominates in making decision. The ANN and SVM both were trained with 3000 data sets with 5 fold cross-validation checks. Both models were tested on another 3000 samples. We conducted extensive experiments using MATLAB for ANNs and LIBSVM (Chang & Lin, 2001) for SVM with

Rule based Classification		
S.No.	Rule Applied	Label
1	age(LOWER-YOUNG-AGE) and occ(Student, Unemployed) and mar(Unmarried) and avg-min($\geq$ 2000)	A - Single Young Unemployed Normal Phone users
2	age(LOWER-MIDDLE-AGE) and occ(Student, Unemployed) and mar(Unmarried) and avg-min(1500-2000)	B - Single Young Unemployed Cautious Phone users
3	age(MIDDLE-AGE) and qual(Graduate) and occ(Unemployed) and avg-min($\leq$ 3500)	C - Young unemployed graduate with high usage
4	roam(More than two weeks) and occ(Business) and avg-min($\geq$ 2000) and bank(Yes) and inc(Less than Rs. 10K)	D - Low income businessmen with high romaing activity
5	roam(More than a week) and occ(Business) and avg-min($\geq$ 2000) and inc(50K-100K)	E - High income businessmen heavy mobile users
6	age(MIDDLE-AGE) and roam(Less than a week) and occ(Professional) and avg-min($\geq$ 1000) and ci(A)	F - Young professional less phone usage residing in good economic zone
7	avg-min($\leq$ 1500) and cir(C)	G - Cautious user in low economic zone
8	avg-min($\geq$ 1500) and cir(C)	H - Heavy mobile user in low economic zone
9	avg-min($\leq$ 1500) and cir(A,B)	I - Cautious user in moderate economic zone
10	occ(Agriculture) and avg-min($\leq$ 2000) and ci(A,B,C)	J - Cautious Farmers
11	age(OLD-AGED)	K - Old Age people
12	age(UPPER-MIDDLE-AGE)	L - Middle Aged people
13	Others	M - All Others

Table 1: Associating rule applied for labeling the customer preferences

different kernels to make sure the robustness of the model developed by both the approaches.

We used Levenberg-Marquardt (LM) Technique to train Feed Forward Neural Network whereas three different kernels (Polynomial, RBF and Sigmoid) were used for SVM. Grid search technique was used to find the range which helped us to identify the proper values of the kernel parameters to achieve better accuracy or classification.

The classification accuracy achieved by both the techniques are presented in Table 3, 4, 5, 6:

We found SVM-RBF Kernel a very powerful kernel technique for the classification of telecom customers when compared with different kernel methods as well as ANN experimented over number of iterations. To find the best value for the given problem we used grid search technique to find different values for γ,

Customers' attributes based classification		
1.	(age = LOWER_YOUNG ∧ occ = STUDENT ∨ UNEMPLOYED ∧ mar = UNMARRIED ∧ incoming = HIGH ∧ outgoing = HIGH ∧ SMS = HIGH) ↦ Plan 1	(STUDENT)
2.	(age = LOWER_YOUNG ∧ occ = STUDENT ∨ UNEMPLOYED ∧ avg-min = AVERAGE ∧ incoming = LOW ∧ outgoing = HIGH ∧ SMS = LOW) ↦ Plan 2	(STUDENT or JOB SEEKER)
3.	(age = YOUNG ∧ qual = GRADUATE ∧ occ = UNEMPLOYED ∧ avg-min = VERY_HIGH) ↦ Plan 3	(JOB SEEKER)
4.	(age = YOUNG ∧ occ = SELF ∧ incoming = HIGH ∧ avg-min = HIGH ∧ SMS = LOW ∧ inc = HIGH) ↦ Plan 4	(AGGRESSIVE TALKER)
5.	(age = YOUNG ∧ occ = SELF ∧ incoming = HIGH ∧ avg-min = AVERAGE ∧ inc = BELOW_AVERAGE) ↦ Plan 5	(PURPOSEFUL TALKER)
6.	(age = YOUNG ∧ roam = HIGH ∧ incoming = AVERAGE ∧ occ = PROFESSIONAL ∧ avg-min = HIGH ∧ cir = A ∨ B) ↦ Plan 6	(FREQUENT TRAVELLER)
7.	(age = YOUNG ∧ roam = LESS ∧ incoming = LOW ∧ occ = PROFESSIONAL ∧ avg-min = LOW ∧ cir = B ∨ C ∧ ot-vas = YES) ↦ Plan 7	(RESERVED USER)
8.	(age = OLD ∧ avg-min = LOW ∧ cir = B ∨ C) ↦ Plan 8	(OLD PEOPLE)
9.	(age = YOUNG ∧ n-min = HIGH ∧ avg-min = VERY_HIGH ∧ cir = B ∨ C) ↦ Plan 9	(NIGHT PLANS)
10.	(age = YOUNG ∧ occ = AGRICULTURE ∧ n-min = LOW ∧ avg-min = HIGH) ↦ Plan 10	(AGRI PROMOTE PLAN)
11.	NO RULE ↦ Plan 11	(STANDARD RATES)

Table 2: Plan based classification as per customer preferences

Artificial Neural Network Technique			
S.No.	Architecture ($I{:}H_i{:}O$)	Mean Square Error	Accuracy
1	21:10:10:13	0.0219	77.7%
2	21:15:15:13	0.026	76.3%
3	21:10:13	0.0273	77.6%
4	21:15:13	0.0265	74.8%

Table 3: Neural Network Performance

C and d. We found that ANN struggled to cross 77.7% accuracy whereas SVM with sigmoid kernel staggered at around 74% and polynomial kernel resulted in giving accuracy of 78.16%. RBF kernel performed better than all the discussed technique with a performance yielding around 80.13% with $C = 1024$, $\gamma = 0.000976563$ or $log(\gamma) = -10.0$. With the grid search approach we can conclude that the performance of SVM may increase if C lies between 1024 and 2048 whereas γ may be set accurately ranging somewhere around $log10$.

SVM with Polynomial Kernel				
S.No	Degree(d)	γ	$\log(\gamma)$	Accuracy
1	2	0.0625	-4.0	78.0667%
2	**3**	**0.03125**	**-5.0**	**78.1667%**
3	4	0.03125	-5.0	77.9667%

Table 4: SVM Performance using Polynomial Kernel

SVM with Sigmoid/tanh Kernel			
S.No	γ	$\log(\gamma)$	Accuracy
1	0.0625	-4.0	59.1667%
2	**0.03125**	**-5.0**	**74.5667%**
3	0.015625	-6.0	62.4333%

Table 5: SVM Performance using Sigmoid/Tanh Kernel

SVM with RBF Kernel					
S.No	C	$\log(C)$	γ	$\log(\gamma)$	Accuracy
1	0.03125	-5	0.0625	-4	44.13 %
2	0.0625	-4	0.125	-3	51.40 %
3	0.125	-3	0.0625	-4	68.43 %
4	0.25	-2	0.0625	-4	75.27 %
5	0.5	-1	0.0625	-4	77.20 %
6	1	0	0.03125	-5	77.67 %
7	2	1	0.03125	-5	78.17 %
8	4	2	0.03125	-5	78.27 %
9	8	3	0.015625	-6	78.40 %
10	16	4	0.015625	-6	78.60 %
11	32	5	0.0078125	-7	78.80 %
12	64	6	0.0078125	-7	79.23 %
13	128	7	0.00390625	-8	79.57 %
14	256	8	0.00390625	-8	79.80 %
15	512	9	0.001953125	-9	79.97 %
16	**1024**	**10**	**0.000976563**	**-10.00**	**80.13%**
17	2048	11	0.000976563	-10.00	80.03 %
18	4096	12	0.000976563	-10.00	79.87 %
19	8192	13	0.000976563	-10.00	79.70 %
20	16384	14	0.000976563	-10.00	79.13 %
21	32768	15	0.000976563	-10.00	78.93 %
22	65536	16	0.000976563	-10.00	78.77 %

Table 6: SVM Performance using RBF Kernel

6   Conclusions and Future work

In this work we found SVM as a better alternative to ANN for classification
of customers in the telecom domain which can be used effectively for target

marketing and promoting services in the form of VAS to the increase the utility for the customers and a better revenue earning for the telecom operators. By using this approach, the marketing and the telecom companies would be in a better position to yield good returns. In future work, sensitivity analysis of different features is to be explored to see the effect on classification accuracy. Different bio-inspired optimization techniques also needs to be explored to find the global optimum value for the kernel parameters and designing of suitable plans for the right segment. Feature selection algorithms might also play crucial role in determining the accuracy of the model. Evaluation of different kernels which might be best suitable would be an interesting area to be explored. Cost models related to telecom services are also being explored.

References

Ahn, H., Ahn, J. J., Byun, H. W., & Oh, K. J. (2011, September). A novel customer scoring model to encourage the use of mobile value added services. *Expert Syst. Appl.*, *38*(9), 11693–11700.

Ahn, H., Ahn, J. J., Oh, K. J., & Kim, D. H. (2011, May). Facilitating cross-selling in a mobile telecom market to develop customer classification model based on hybrid data mining techniques. *Expert Syst. Appl.*, *38*(5), 5005–5012.

Ahn, H., Song, C., Ahn, J. J., Lee, H. Y., Kim, T. Y., & Oh, K. J. (2010, jan.). Using hybrid data mining techniques for facilitating cross-selling of a mobile telecom market to develop customer classification model. In *System sciences (hicss), 2010 43rd hawaii international conference on* (p. 1 -10).

Chang, C. C., & Lin, C. (2001). Libsvm: A library for support vector machines. Retrieved from http://www.csie.ntu.edu.tw/ cjlin/libsvm

Constantiou, I. D. (2009, August). Consumer behaviour in the mobile telecommunications' market: The individual's adoption decision of innovative services. *Telemat. Inf.*, *26*(3), 270–281.

Cui, D., & Curry, D. (2005, October). Prediction in marketing using the support vector machine. *Marketing Science*, *24*, 595–615.

Guo-en, X., & Wei-dong, J. (2008). Model of customer churn prediction on support vector machine. *Systems Engineering - Theory & Practice*, *28*(1), 71 - 77.

Hsu, C.-w., Chang, C.-c., & Lin, C.-j. (2010). A practical guide to support vector classification. *Bioinformatics*, *1*(1), 1–16.

Huang, B., Kechadi, M. T., & Buckley, B. (2012, January). Customer churn prediction in telecommunications. *Expert Syst. Appl.*, *39*(1), 1414–1425.

Huang, J.-J., Tzeng, G.-H., & Ong, C.-S. (2007). Marketing segmentation using support vector clustering. *Expert Syst. Appl.*, *32*(2), 313-317.

Ishaya, T., & Folarin, M. (2012, August). A service oriented approach to business intelligence in telecoms industry. *Telemat. Inf.*, *29*(3), 273–285.

Lee, Y.-C. (2007). Application of support vector machines to corporate credit rating prediction. *Expert Systems with Applications*, *33*(1), 67 - 74.

Lovell, B. C., & Walder, C. J. (2006). Business applications and computational intelligence. In K. Voges & N. Pope (Eds.), (pp. 267–290). Hershey, PA., U.S.A.: Idea Group.

Mazzoni, C., Castaldi, L., & Addeo, F. (2007, November). Consumer behavior in the italian mobile telecommunication market. *Telecommun. Policy*, *31*(10-11), 632–647.

Pendharkar, P. C. (2009, April). Genetic algorithm based neural network approaches for predicting churn in cellular wireless network services. *Expert Syst. Appl.*, *36*(3), 6714–6720.

Politz, A., & Deming, W. E. (1953). On the necessity to present consumer preferences as predictions. *Journal of Marketing*, *18*(1), pp. 1-5.

Reuver, M. de, & Haaker, T. (2009, August). Designing viable business models for context-aware mobile services. *Telemat. Inf.*, *26*(3), 240–248.

Rosenblatt, F. (1962). *Principal of neurodynamics*. Spartan Books.

Sapankevych, N., & Sankar, R. (2009, may). Time series prediction using support vector machines: A survey. *Computational Intelligence Magazine, IEEE*, *4*(2), 24 -38.

Sharma, A., & Panigrahi, D. P. K. (2011, August). A neural network based approach for predicting customer churn in cellular network services. *International Journal of Computer Applications*, *27*(11), 26-31.

Souza, C. R. (2010, 17 March). *Kernel functions for machine learning applications*. Retrieved from http://crsouza.blogspot.com/2010/03/kernel-functions-for-machine-learning.html

Vapnik, V. N. (1995). *The nature of statistical learning theory*. New York, NY, USA: Springer-Verlag New York, Inc.

Vapnik, V. N. (1998). *Statistical learning theory*. New York: Wiley.

Venugopal, V., & Baets, W. (1994). Neural networks and statistical techniques in marketing research: A conceptual comparison. *Marketing Intelligence Planning*, *12*(7), 30–38.

Wang, C., & Wang, Y.-Q. (2009, june). Segmentation of consumer's purchase behavior based on neural network. In *Cognitive informatics, 2009. icci '09. 8th ieee international conference on* (p. 552 -556).

Weiss, G. M. (2005). Data mining in telecommunications. In *The data mining and knowledge discovery handbook* (p. 1189-1201).

Wu, T.-K., Huang, S.-C., & Meng, Y.-R. (2008). Evaluation of ann and svm classifiers as predictors to the diagnosis of students with learning disabilities. *Expert Systems with Applications*, *34*(3), 1846 - 1856.

Yeh, C.-C., Chi, D.-J., & Hsu, M.-F. (2010). A hybrid approach of dea, rough set and support vector machines for business failure prediction. *Expert Systems with Applications*, *37*(2), 1535 - 1541.

An Efficient Approach on Rare Association Rule Mining

N. Hoque, B. Nath and D. K. Bhattacharyya

Abstract :Traditional association mining techniques are based on support-confidence framework, which enable us to generate frequent rules based on frequent itemsets identified on a market basket dataset with reference to two user defined threshold *minsup* and *minconf*. However, the infrequent itemsets referred here as rare itemsets ignored by those techniques often carry useful information in certain real life applications. This paper presents an effective method to generate frequent as well as rare itemsets and also consequently the rules. The effectiveness of the proposed method is established over several synthetic and real life datasets. To address the limitations of support-confidence based frequent and rare itemsets generation technique, a multi-objective rule generation method also has been introduced. The method has been found to perform satisfactory over several real life datasets.

Keywords: Support, confidence, frequent rule, rare rule, MOGA.

1 Introduction

Association rule mining problem was introduced by Agrawal[1] for any binary dataset, referred as market-basket dataset, where each row represents a transaction and each column or attribute represents an item. Association mining aims to extract interesting association among the attributes/attribute values of any market basket dataset.

N. Hoque
Tezpur University, Napam, Sonitpur, Assam-784028, e-mail: tonazrul@gmail.com

B. Nath,
Tezpur University, Napam, Sonitpur, Assam-784028, e-mail: bnath@tezu.ernet.in

D. K. Bhattacharyya
Tezpur University, Napam, Sonitpur, Assam-784028 e-mail: dkb@tezu.ernet.in

J. C. Bansal et al. (eds.), *Proceedings of Seventh International Conference on Bio-Inspired Computing: Theories and Applications (BIC-TA 2012)*, Advances in Intelligent Systems and Computing 201, DOI: 10.1007/978-81-322-1038-2_17, © Springer India 2013

1.1 Frequent rule mining: support-confidence framework

In the past couple of years, several novel works have been evolved to handle the association rule mining problem [2][4][8][13]. These algorithms work in two major steps using a support-confidence framework: frequent itemset generation and rule generation. The first step explains the concept of support to derive the frequent itemsets. Support of an itemset can be defined as the proportion of transactions in the dataset which contain the itemset. The confidence of a rule $X \Rightarrow Y$ is defined as Conf($X \Rightarrow Y$)=sup($X \cup Y$)/sup(X). A major limitation of these frequent itemset generation techniques is that they can extract only those itemsets which are frequent with respect to a given user threshold i.e support-count. However, in practical scenario, there can be some itemsets which have significance and can provide useful knowledge [10] but their support counts are relatively less. To extract the relationship among those itemsets having less frequency of occurrences, rare association rule mining came into existence.

1.2 Rare Rule: Definition

A rare rule r can be defined as an association rule using the support confidence framework , where $supp(r) < minsup$, i.e. the minimum support threshold, and $conf(r) >= minconf$ i.e. the minimum confidence threshold. Rare association rules are usually required to satisfy a user specified minimum support and a user specified minimum confidence at the same time.

1.3 Rare-rule generation: Issues and motivation

A major research issue related to rare rule generation is that with a single *minsup* value it may not be possible to generate all the rare rules. If *minsup* is set too high, we miss the frequent itemsets involving rare itemsets because rare items fail to satisfy high *minsup*. To find frequent itemsets consisting of both frequent and rare items, we have to set *minsup* very low. However, this may cause combinatorial explosion [9] and produce too many frequent itemsets. To address this issue multiple minsup values are used [6] but still it suffers from the same problem and also dropped some of the rare rules. This has motivated us to develop a rare association rule mining techniques to extract all the rare rules without generating uninteresting rules.

1.4 Organization

Section 2 provides a discussion on some popular rare association mining methods. In section 3 we defined the problem. A solution based on support-confidence framework is reported in section 4, followed by a detail discussion on experimental results in section 5. In section 6 we introduce a multi-objective rule generation method, followed by experimental results. Concluding remarks are given in section 7.

2 Rare Association Rule Generation

Rare association mining has been gaining importance in the recent days. Several significant developments have been evolved to extract interesting rare rules [6][7][14][15][16]. In this section we report some of the popularly known rare association mining techniques and analyze their pros and cons.

[A]Apriori-Rare[15]: It generates frequent as well as rare itemsets. It uses a subroutine called *supportcount* to find the support count of a given itemset. The main advantage of this algorithm is that it restores all the *minimal* rare itemsets. However, it fails to *find* all the rare itemsets.

[B]Apriori-Inverse [7]: It attempts to find all those sporadic association rules, which satisfy the minimum confidence threshold but fail to satisfy maximum support. The main advantage is that it can find the sporadic itemsets more quickly than apriori. However, a major limitation is that it is incapable of finding all the rare itemsets.

[C]MSapriori(Multiple Support Apriori) [6]: It is a variant of Apriori, which uses to determine the rare itemsets. It attempts to overcome the rare item problem [9] by altering the definition of minimum support. In their extended model, each item in the dataset can have a minimum item support (MIS) given by the user. This way, the user expresses different support requirements for different rules. This method tries to solve the rare item problem using MIS value but still it suffers from several serious limitations. The extraction of rare itemsets depends on the value of user defined threshold β rather than the support-count f. Though it can find rare itemsets, but high dependency on β limits the flexibility of the algorithm.

[D]RSAA algorithm (Relative Support Apriori Algorithm) [16]: The RSAA algorith generates rules by increasing the support threshold for items that have low frequency and by decreasing the support threshold for items that have high frequency of occurrences. So, it spends a significant amount of time looking for the rules which are not rare. If the minimum permissible relative support count is set close to zero, then RSAA takes a similar amount of time to that taken by Apriori to generate low support rules. RSAA can find all the rules, but it requires significant amount of time looking for rules which are not sporadic. Also, RSAA is dependent on two thresholds, *minsup* and *maxsup*, and results are highly sensitive to these thresholds.

[E] ARIMA (Another Rare Itemset Miner Algorithm) [14]: ARIMA exploits

Apriori-Rare to generate the minimal rare itemsets (MRIs), based on which it generates the rare itemsets. To reduce the search space, it uses the concept of zero generator. ARIMA is advantageous because dependency on two user defined thresholds i.e. *minsup* and *maxsup*, limits the flexibility of ARIMA.

2.1 Discussion and Motivation

Based on our study, it has been observed that none of the existing algorithms are totally free from the limitations. Except ARIMA and MSapriori, none of the algorithms is capable of generating all the rare itemsets. However, ARIMA takes the output of Apriori-rare while generating the rare itemsets. On the other hand MSapriori generates all the rare itemsets, however, all the rules generated by it are not interesting. To address these limitations, this paper presents an effective rare as well as frequent itemset generation technique. Also to overcome the limitation of single objective based (support-confidence framework) rule generation technique. This paper introduces a multi-objective technique.

3 Problem Formulation

The problem is to find all the frequent and rare itemsets for a given dataset D with reference to a user-defined threshold *minsup*, and consequently to find all the interesting and comprehensive rare rules, which have $confidence \geq minconf$, a user defined threshold.

4 The proposed Frequent Rare Itemset Mining Algorithm:FRIMA

FRIMA works on similar principle of Apriori, however it saves the computational overhead significantly. It uses bottom-up approach and attempts to find the frequent as well as rare itemsets based on the downward closure property of Apriori. FRIMA takes D and *minsup* as input and finds frequent and rare itemsets as outputs. FRIMA initially scans the dataset once and finds all the frequent and rare single itemsets. Based on the support count it categorizes the single itemsets as *zero itemsets* having support count zero, *frequent itemsets* having support count greater than *minsup* and rare itemsets having support count less than *minsup*. Next, it generates *three* candidate lists. The *first* candidate list is generated from the frequent single-itemset, *second* candidate list is generated from the rare itemset and the *third* list is generated by combining frequent and rare itemsets. Now, these three lists are combined to make one *single list* and make one database scan to find the *zero*, *frequent* and

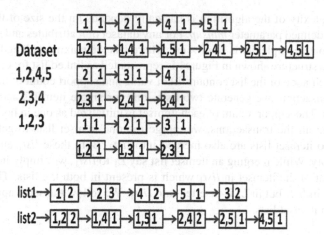

Fig. 1 One and Two level itemsets generation using single Database scan

rare itemsets of size 2. These frequent and rare 1 and 2-itemsets are maintained in a dynamic structure as shown in Figure1 for further consultation towards generation of frequent and rare itemsets greater than 2. This procedure continues for itemsets of large size until no more frequent or rare itemsets are produced. But, before scanning the database for k itemsets where $k > 2$, this algorithm first generates $(k-1)$ subsets of the candidate k itemsets. If any $(k-1)$ subset belongs to the $(k-1)$ *zero list* then k itemset is put into the *zero k* list. The steps of the method are given next. The

---

Algorithm 1: FRIMA

D =Dataset

$i=1$

Scan the database for once and construct one and two level frequent and rare itemsets.

Store frequent itemsets in L_i & L_{i+1}, rare itemsets in R_i & R_{i+1} and zero itemsets in MZG lists.

$i=2$

while (L_i *or* R_i *is not Null*), **do**

    L_{i+1}=CandidateGen(L_i)

    R_{i+1}=CandidateGen(R_i)

    LR_{i+1}=CandidateGen(L_i,R_i)

    $i = i+1$

    $C_i=L_i + R_i + LR_i$

    Supportcount(C_i)

    $L_i=i$ itemsets having supportcount larger minsup

    $R_i=i$ itemsets having supportcount smaller minsup

    $MZG=i$ itemsets having supportcount zero

end

L=Union of all large itemsets

R=Union of rare itemsets

---

overall complexity of the algorithm is basically depends on the size of the dataset and the user defined parameter *minsup*. For any dataset of n attributes and m records, the approximate complexity is $O(n^k \times m)$, where k is the maximum length itemset. The dynamic structure shown in Figure 1 is represented as linked list for one and two itemsets. Each node of the list contains item value and support count of the itemset. For each transaction, we generate two lists- one for single itemsets and other for two itemsets. The support count of each itemset is initialized as one in both the lists. After reading all the transactions, we merge the one itemset lists to get $list_1$ and similarly two itemset lists are also merged into $list_2$. Both these $list_1$ and $list_2$ are initially empty. While merging an itemset list say L_1 to $list_1$, we simply increase the support count of the itemset in $list_1$ which is present in both the lists. The itemset which is not in $list_1$ but in L_1 is inserted into $list_1$. We follow the same approach for merging two itemset lists.

4.1 Proof of Correctness

In this section we establish that *FRIMA* is correct in generating both frequent and rare itemsets. Following lemma provides the proof of correctness of our *FRIMA*.
Lemma 1: *FRIMA* is correct i.e the itemsets generated by the algorithm are either frequent or rare with reference to the user defined threshold *minsup*.
Proof: The correctness of *FRIMA* can be established from the fact that it generates the final list of frequent and rare itemsets based on three candidate lists i.e *zero*, *frequent* and *rare* with reference a user defined threshold i.e *minsup*. An itemset is put in the final list of frequent and rare itemset iff it satisfies the *minsup* condition, hence the proof.

5 Experimental Results

To implement the method we used C in a Linux environment on a 32-bit workstation having 2.94 Ghz core2 Due processor, 4GB RAM and 360GB Secondary storage. To evaluate the performance of the proposed *FRIMA*, we used three synthetic and five real-life benchmark UCI datasets with various dimensionality and number of instances. The characteristics of the datasets used are reported in Table 1.

5.1 Results

The performance of the proposed method was compared with *Apriori* and two other well known rare itemset finding techniques, viz *Apriori-Rare* and *ARIMA* and the results are reported for each dataset in Tables 2 and 3.

Table 1 Datasets used for evaluation

SI No	Dataset	Type	Attributes	Records
1	cancer	Real	4	32
2	Monk1	Real	5	423
3	Monk3	Real	5	423
4	Mushrooms	Real	128	8413
5	Chess	Real	37	3196
6	T20D10000k	Synthetic	5	20
7	T100D10000k	Synthetic	10	100
8	T1000D10000k	Synthetic	12	1000

Table 2 Results on Datasets 2 & 3 for Minsup 80% and Datasets 1,4 & 5 for Minsup 50%

Algorithm	Monk1		Monk3		Cancer		Mushrooms		Chess	
	FIs	RIs	FIs	RIs	FIs	RIs	FIs	RIs	FIs	RIs
Apriori	4	0	4	0	7	0	163	0	27,724	0
Apriori-Rare	4	5	4	5	7	0	163	147	27,724	6989
ARIMA	0	28	0	28	0	0	0	43,907	0	4,639
Proposed	4	6	4	6	7	0	163	47,767	27,724	7052

Table 3 Results on Synthetic datasets 6-8 for Minsup 50%

Algorithm	T20D10000K		T100D10000K		T1000D10000K	
	FIs	RIs	FIs	RIs	FIs	RIs
Apriori	31	0	511	0	15	0
Apriori-Rare	31	0	511	60	15	62
ARIMA	0	92	0	10,976	0	844092
Proposed	31	40	511	38,143	15	903768

5.2 Discussion

From the experimental results it can be observed that

- FRIMA is capable of generating both frequent and rare itemsets without any loss.
- It generates all those frequent itemsets generated by apriori.
- Except cancer dataset, in case of all other datasets FRIMA generates more number of rare itemsets than Apriori-rare. However, ARIMA performs better in some of the datasets.

6 Rare rule generation using Multi-Objective approach

The traditional approaches of rule generation either uses bottom-up or top-down or hybridization of both to find the frequent rules based on the support-confidence framework. However, a common limitation of those methods is that it generates rules with only one item in the consequent part. It was resolved by the Srikant's first algorithm [12] that practically it may contain any number of items in the consequent part. Again most methods check some candidate rules unnecessarily that waste significant amount of time. Srikant's second algorithm [12] overcomes this problem by eliminating unnecessary checking of candidate rules. But in rare rule generation we cann't directly use these algorithms because rare itemsets may have zero support value. So, specifically to address this rare rule generation problem using single objective based support-confidence framework may not be the right choice. So, here we introduce a multi-objective pareto based rule generation method to generate the rare rules.

Once the frequent or rare itemsets are found, the next step is to generate the rules either by using bottom-up, top-down or hybridization of both based on a support-confidence framework. Rules generated based on these approaches basically attempts to optimize only one objective function i.e. confidence. However, often it has been found that many rules generated by such approaches mayn't be interesting or non-comprehensive[3][11][17]. To address these limitations, this paper also presents a rule generation algorithm called Rare Association Rule Mining Algorithm(RARMA) which works based on the output generated by FRIMA, and attempts to optimize these measures i.e. confidence, comprehensibility and interestingness. From the rare itemsets rare rules using three objectives: confidence, comprehensibility and interestingness based on pareto optimal solution were generated using the multi-objective method. A conceptual framework of the rule generation method is shown in Figure2 where obj1, obj2 and obj3 represents confidence, comprehensibility and interestingness respectively.

The motivation for developing a multi-objective genetic algorithm(GA) for rule

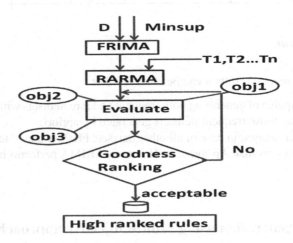

Fig. 2 Framework for multi-objective rule generation

generation[5]was that (i) GAs are a robust search method, capable of effectively exploring the large search spaces often associated with attribute selection problem; (ii) GAs perform a global search, (iii) GAs already work with a population of candidate solutions, which make them naturally suitable for multi-objective problem solving where the search algorithm is required to consider a set optimal solutions at each iteration.

6.1 Experimental results

To implement the proposed RARMA based method for rare rule generation, we used the similar programming platform and environment as reported in *section 5*. To evaluate the performance of our method we used four benchmark UCI datasets, namely *Monk1, Monk3, Mushrooms* and *Chess* dataset, and the results are reported for various *minsup* values in Tables 4 through 7. Rules generated from different datasets are shown below.

Table 4 Rules generated for *Monk1* Dataset and their effectiveness

Rule	Minsup in %	Confidence	Comprehensibility	Interestingness
6,7 →3	25	0.333333	0.682606	0.069959
1,4 →8	50	0.500000	0.682606	0.051988
2 →8	50	0.666666	0.792481	0.185969
2,5 →8	50	0.510638	0.682606	0.053094
6,7 →2	25	0.333333	0.682606	0.069959

Table 5 Rules generated for *Monk3* Dataset and their effectiveness

Rule	Minsup in %	Confidence	Comprehensibility	Interestingness
6,7 →3	25	0.333333	0.682606	0.069959
1,4 →8	50	0.500000	0.682606	0.051988
2 →8	50	0.666666	0.792481	0.185969
2,5 →8	50	0.510638	0.682606	0.053094
6,7 →2	25	0.333333	0.682606	0.069959

Table 6 Rules generated for *Mushrooms* Dataset and their effectiveness

Rule	Minsup in %	Confidence	Comprehensibility	Interestingness
2,65 →95	8	0.300000	0.682606	0.299049
21,117 →10	4	0.500000	0.682606	0.499762
36,80 →4	2	0.250000	0.682606	0.124970
4 →10	1	0.250000	0.792481	0.062493

Table 7 Rules generated for *Chess* Dataset and their effectiveness

Rule	Minsup in %	Confidence	Comprehensibility	Interestingness
21,65 →24	12	1.000000	0.682606	0.316667
23 →51	18	0.428571	0.792481	0.364286
25,47 →20	22	1.000000	0.682606	0.900000
56,63 →57	22	1.000000	0.682606	0.316667

6.2 *Discussion*

Association rule mining problem can be viewed as a multi-objective problem rather than single objective one. Using confidence, comprehensibility and interestingness as the various objective measures, rare rules are generated from the rare itemsets. Similarly, frequent rules are also generated from the frequent itemsets using the same objective measures. The *confidence* of a rule determines the reliability of the rule. *Comprehensibility* determines the significance of a rule based on number of items present in both the antecedent and consequent part. *Interestingness* helps in knowledge gathering. Based on these measures an attempt has been shown to determine the best set of rules. However, along with the meaningful rare rules, generation of redundant rules is a limitation of our method.

7 Conclusion and Future work

An effective frequent and rare itemset finding algorithm is present and established in this paper. A multi-objective rule generation method has been reported for rare rules extraction. The effectiveness of this method has been shown over several publicly available UCI datasets. Work is going on further enhancement of our RARMA based on rare association mining method by considering measures other than confidence, comprehensibility and interestingness, and for datasets with large number of high-dimensional instances to help finding only meaningful rare association rules. Attempt is going on to apply these two algorithms for rare class attack detection in intrusion detection system.

References

1. R. Agrawal, T. Imielinski, and A. Swami. Mining association rules between sets of items in large databases. pages 206–216. In ACM SIGMOD International Conference on Management of Data, 1993.
2. S. Brin, R.Motwani, J.D.Ullman, and S.Tsur. Dynamic itemset counting and implication rules for market basket data. volume 26, pages 255–268. in Proc. of the 1997 ACM SIGMOD Int'n Conf. on Management of data, 1997.
3. C.A.C Coello. A comprehensive survey of evolutionary-based multi-objective optimization technique. pages 269–308. Knowledge and information systems, 1999.

4. J. Han, J. Pei, Y. Yin, and R. Mao. Mining frequent patterns without candidate generation:a frequent-pattern tree approach. volume 8, pages 53–87. Data Mining and Knowledge Discovery, 2004.
5. N. Hoque, B. Nath, and D. K. Bhattacharyya. A new approach on rare association rule mining. volume 53. International Journal of Computer Applications (0975 - 8887), 2012.
6. R. U. Kiran and P. K. Reddy. Mining rare association rules in the datasets with widely varying items' frequencies. The 15th International Conference on Database Systems for Advanced Applications Tsukuba, Japan, April 1-4,, 2010.
7. Y. S. Koh and N. Rountree. Finding sporadic rules using apriori-inverse. pages 97–106. Springer-Verlag Berlin Heidelberg, 2005.
8. D. I. Lin and Z. M. Kedem. Pincer-search: an efficient algorithm for discovering the maximal frequent set. pages 105–219. In Proc. Of 6th European Conference on Extending DB Tech, 1998.
9. B. Liu, W. Hsu, and Y. Ma. Mining association rules with multiple minimum supports. pages 337–341. ACM Special Interest Group on Knowledge Discovery and Data Mining Explorations, 1999.
10. H. Mannila. Methods and problems in data mining. pages 41–55, 1997.
11. B. Nath and A. Ghosh. Multi-objective rule mining using genetic algorithm. pages 123–133. Information Science 163, 2004.
12. R.Srikant and R. Agrawala. Mining generalized association rules. pages 407–419. Proceedings of the 21st VLDB Conference Zurich, Swizerland, 1995.
13. A. Savesere, E. Omiecinski, and S. Navathe. An effective algorithm for mining asociation rules in large database. pages 432–443. In proceedings of International Conference on VLDB95, 1995.
14. L. Szathmary and P. Valtchev. Towards rare itemset mining. Soutenue publiquement le.
15. L. Szathmary, P. Valtchev, and A. Napoli. Generating rare association rules using the minimal rare itemsets family. volume 4, pages 219–238. International Journal on Software Informatics, 2010.
16. H. Yun, D. Ha, B. Hwang, and K. H. Ryu. Mining association rules on significant rare data using relative support. volume 67, pages 181–191. The Journal of Systems and Software, 2003.
17. E. Zitzler, K. Dev, and L. Thiele. Comparision of multi-objective evolutionary algorithms: empirical results. pages 125–148. Evolutionary Computation 8, 2000.

A Hybrid Multiobjective Particle Swarm Optimization Approach for Non-redundant Gene Marker Selection

Anirban Mukhopadhyay and Monalisa Mandal

Department of Computer Science and Engineering
University of Kalyani
Kalyani-741235, West Bengal, India
{anirban,monalisa}@klyuniv.ac.in

Abstract. The gene markers or biological markers indicate change in expression or state of protein that correlates with the risk or progression of a disease, or with the susceptibility of the disease to a given treatment. There are many approaches for detecting these informative genes from high dimensional microarray data. But in practice, for most of the cases a set of redundant marker genes are identified. Motivated by this fact a hybrid multiobjective optimization method has been proposed which can find small set of non-redundant disease related genes. In this article the optimization problem has been modeled as multiobjective problem which is based on the framework of particle swarm optimization. As the wrapper approaches depend on a specific classifier evaluation, hence artificial neural network classifier is used as evaluation criteria. Using the real life datasets, performance of proposed algorithm has been compared with other different techniques.

Keywords: Multiobjective Optimization, Particle Swarm Optimization, Biomarker, Non-redundant, Artificial Neural Network.

1 Introduction

Microarray is a 2D array where the rows represent samples or experimental condition and columns represent genes. Originally in raw data, besides the genes, one extra column can be viewed which corresponds to the class labels of the samples. For example, a microarray gene expression dataset consisting of g genes and s tissue samples is typically organized in a 2D matrix $E = [e_{ij}]$ of size $s \times g$. Each element e_{ij} represents the expression level of the jth gene for the ith tissue sample. Most of the time this data matrix can have missing value, noise and large number of genes which needs to be preprocessed. Furthermore dimensionality of the microarray data poses a big problem for processing.

Genes with significantly different expression in two different classes (normal and tumor or two different subtypes of cancer) are called differentially expressed genes. Therefore these genes are used as indicator or marker of disease and are

J. C. Bansal et al. (eds.), *Proceedings of Seventh International Conference on Bio-Inspired Computing: Theories and Applications (BIC-TA 2012)*, Advances in Intelligent Systems and Computing 201, DOI: 10.1007/978-81-322-1038-2_18, © Springer India 2013

also called biomarkers. There are already many supervised and unsupervised classification techniques available which have been adopted for classifying or clustering tissue samples, respectively. Most of the cases existing approaches result some top ranked genes or features which are often redundant. Motivated by this fact, in this article a variable length PSO based method has been applied to multiobjective optimization for selecting non-redundant gene markers. Then with the help of non-dominated sorting [1] and Crowding Distance measure [2], a small set of non-redundant informative genes have been identified.

Particle Swarm Optimization is a popular optimization technique which optimizes a problem by iteratively trying to improve its candidate solution with regard to given fitness measure. Previously a variety of PSO method has been introduced in [3] and [4]. In PSO, candidate solutions are called particles and a population of these particles is called a swarm. These particles move around in the search space. The position and velocity of each particle are updated according to a few formulae. Additionally, instead of fixed length particle, here PSO has been designed for encoding variable length particles.

The single objective optimization yields a single best solution but multiobjective optimization (MOO) produces a set of solutions which contains a number of non-dominated solutions, none of which can be further improved on any one objective without degrading it in another. Moreover multiobjective optimization problem has been modeled by applying PSO [5]– [6] in which fitness comparison takes Pareto dominance [7] into account when moving the particles and non-dominated solution are stored in an archive to approximate the Pareto front [8].

We have proposed a variable length multiobjective PSO based approach which has been modeled using the framework of non-dominated sorting [1] and crowding distance sorting [2]. Two objectives, sensitivity and specificity are simultaneously evaluated by the ANN classifier to get the non-redundant informative genes which should have maximum sensitivity and specificity at the same time. Non-redundancy is accomplished by clustering the genes while computing the fitness of a particle and choosing a representative gene (cluster prototype) from each gene cluster. The performance of the proposed method is compared with that of SFS, T-test and Ranksum based on three real-life gene expression data sets of various types of cancers.

2   Multiobjective Optimization (MOO)

Multiobjective Optimization problems have multiple goals or objectives, therefore the objectives may estimate different aspect of solutions which are partially or wholly in conflict. There are various methods for solving multiobjective optimization problems as illustrated in [2] and [9]. Find the vector $\overrightarrow{x}^* = [x_1^*, x_2^*, \ldots, x_n^*]^T$ of decision variables which satisfies m inequality constraints and p equality constraints:

$$g_i(\overrightarrow{x}) \geq 0, \quad i = 1, 2, \ldots, m; \qquad h_i(\overrightarrow{x}) = 0, \quad i = 1, 2, \ldots, p, \qquad (1)$$

and optimizes the vector function

$$\vec{f}(\vec{x}) = [f_1(x), f_2(x), \ldots, f_k(x)]^T \tag{2}$$

The constraints given in Eqn.1 define the feasible region $\mathcal{F}$ which contains all the admissible solutions. The vector $\vec{x}^*$ denotes an optimal solution in $\mathcal{F}$. In the context of multiobjective optimization, the difficulty lies in the definition of optimality, since it is only rarely that we will find a situation where a single vector $\vec{x}^*$ represents the optimum solution to all the objective functions. However the meaning of optimization in MOO can be defined through Pareto optimality [2]. Pareto optimal set of solutions consist of all those that it is impossible to improve any objective without simultaneous worsening in some other objective. It can be said that a vector of decision variables $\vec{x}^* \in \mathcal{F}$ is Pareto optimal if there does not exist another $\vec{x}^*$ such that $f_i(\vec{x}) \leq f_i(\vec{x}^*)$ for all $i = 1, \ldots, k$ and $f_j(x) < f_j(\vec{x}^*)$ for at least one j when the problem is minimizing one. Here, $\mathcal{F}$ denotes the feasible region of the problem (i.e., where the constraints are satisfied). Pareto optimal set generally contains more than one solution because there exists different 'trade-off' solutions to the problem with respect to different objectives. The set of solutions contained by Pareto optimal set are called non-dominated solutions. The plot of the objective functions whose non-dominated vectors are in the Pareto optimal set is called the Pareto front. Specifically MOO is a process of generating the whole Pareto front or an approximation to it.

3   Proposed Hybrid MOPSO-ANN Method

3.1   Particle Representation

Each particle has two parts; the first part contain n padding cells and next part contain n cluster centers or genes. The first n padding cells contain values between 0 and 1, and the last n cells of the particle consist of n genes from the dataset. Basically each gene represents the center of a gene-cluster. Initially the genes are chosen randomly from dataset $\{1, 2, 3, \ldots, g\}$ where g is the number of genes present in the data matrix. Each gene contains s sample values. Since each gene has the dimension s, the size of last the part of the particle is $n \times s$. Thus the total length of a particle is $(n + n \times s)$. If the value of ith padding cell is greater than a specific threshold then the gene represented by the ith cell of the second part of the particle is selected for fitness Computation. The particle encoding scheme has been demonstrated in Fig. 1(a) where a full particle is shown and in Fig. 1(b) where only the last n part of the particle is depicted.

3.2   Swarm Initialization

Initially the first part of the candidate solutions are randomly chosen values between 0 and 1, and the next part of the candidate solutions are randomly selected genes from the data set. After the initial particles are generated randomly, their corresponding fitness values are calculated. The fitness calculation is described

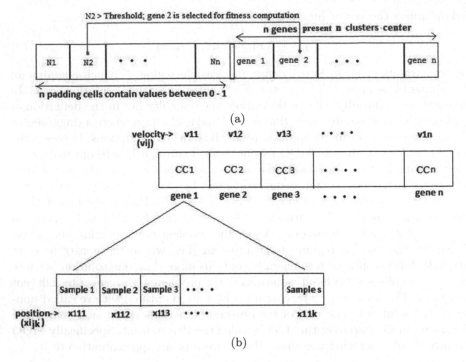

Fig. 1. Particle encoding: (a) A particle with $2n$ number of cells. First n cells contain values between 0 to 1 and remaining n cells represent n cluster centers or genes. The red line indicates if padding $cell(i) > threshold$ then ith gene is selected for fitness computation. (b) The second part of the candidate solution with n cluster centers (CC). Each cluster consists of one gene representing the cluster center. V_{ij} represents velocity of i-th candidate solution and j-th cluster. A gene contains s number of samples. X_{ijk} corresponds to position of i-th particle j-th cluster and k-th sample

later. Then the velocity of each cell and cluster are initialized to zero. In the second part, instead of velocity of each particle dimension, velocity of each gene-cluster is taken for moving a gene-cluster of a particle to same direction in the search space as shown in Fig. 1(b). The inputs of the proposed technique are swarm size=15, upper bound of gene cluster size=15, lower bound of the gene cluster size=3 and weighting factors c1 and c2 which are cognitive and social parameters respectively are set to 2. The number of iterations is taken as 50 for each dataset and the threshold of the padding cell is taken as 0.5.

3.3 Fitness Computation

In this article sensitivity and specificity are the objective functions which are to be maximized. Because of Particle Swarm Optimization algorithm has been designed as minimization problem, so fitness values are computed as $(1 - sensitivity)$ and $(1 - specificity)$. The first n padding cells are iterated as usual PSO evaluation. For the next part first the cluster centers are calculated for each candidate solution. Now fitness values are calculated for those genes which are chosen according to the values contained by first n padding cells. Therefore if a padding cell of the first part contains a value greater than a specific threshold (0.5 is taken here) value then the corresponding gene is selected for fitness computation. Thus the first part decides the gene participation for calculating fitness values. After that these candidate solutions are shuffled for k-fold cross validation with ANN classifier [10]- [11] as wrapper. Then these new class labels of the candidate solutions are compared with original class labels to figure out the number of false positives (fp), true negatives (tn), false negatives (fn) and true positives (tp). Thereafter using these four terms, sensitivity and specificity are determined as sensitivity $= \frac{tp}{tp+fn}$, specificity $= \frac{tn}{tn+fp}$.

3.4 Updating Position and Velocity

The positions of the padding cells contain values between 0 and 1, and velocity of each padding cell is initialized to zero. It is evident from Fig. 1(b) that each candidate solution contains n number of clusters and here velocity of each cluster has been considered instead of velocity of each particle dimension. At the beginning, the velocity of each cluster is set to zero. Using the information obtained from the previous step the position of each particle and velocity of each cluster are updated [3]- [4]. Each particle keeps track of the best position it has achieved so far in the history. This best position is also called *pbest* or local best. And the best position among all the particles is called global best or *gbest*. In multiobjective perspective, that position is chosen for *pbest* for which fitness of that particle dominates other fitnesses acquired by that particle in the history, if there is no such fitness then random choices have been done between current and previous position of that particle. For the *gbest*, random posion is chosen from the archive. Actually whenever a particle moves to a new position with a velocity, its position and velocity are changed according to the Equations 3 and 4 given below [5]:

$$v_{ij}(t+1) = v_{ij}(t) + c_1 * r_1 * (pbest_{ij}(t) - x_{ij}(t)) + c_2 * r_2 * (gbest_{ij}(t) - x_{ij}(t)) \tag{3}$$

$$x_{ij}(t+1) = x_{ij}(t) + v_{ij}(t+1) \tag{4}$$

Here t is the time stamp and the i-th particle and the j-th cluster have been considered. In Equation 3 new velocity $(v_{ij}(t+1))$ is acquired using the velocity of previous time $(v_{ij}(t))$, *pbest* and *gbest*. Then the new position $(x_{ij}(t+1))$ is obtained by adding new velocity with current position $(x_{ij}(t))$ as shown in Equation 4. r_1 and r_2 are two random values in the range from 0 to 1.

3.5 Updating Non-dominated Archive

For updating the archive, the current swarm is merged with the next generation swarm and then non-dominated solutions are yielded for next generation. The repository where the non-dominated population in the history is kept called archive. First the archive is initialized with non-dominated population, then next generation population is added, and finally again non-dominated sorting [1] and crowded distance sorting [2] are also applied for this combined population to improve the adaptive fit of a population of candidate solutions to a Pareto front constrained by a set of objective functions and to get better diversity of the Pareto optimal front respectively.

3.6 Updating Gene-cluster Centers

In the previous section it has been discussed that a particle has two parts; the first part contains n numbers between 0 and 1, and the second part contains n genes or features. Each gene is a representative of a cluster. In other way, we can say that the second part contains n clusters. To find the cluster center or representative gene the given steps are followed: 1)Clusters of a particle are separated to form an initial cluster-center matrix of size $c \times s$ where c is the number of genes or cluster centers and s is the number of samples. 2)A distance matrix is generated by calculating the Euclidean distance between each cluster center (representative gene) and each gene contained by the dataset. Thus a $c \times g$ distance matrix is formed, where g is the number of genes in the dataset. 3)Next a gene is assigned to a cluster for which it has the minimum distance value in the distance matrix. Thus each cluster contains some genes. 4)Then the genes belonging to a cluster and the initial value of that cluster are added and averaged with respect to each dimension to get the coordinate mean. Thus a $c \times s$ matrix is formed. 5)Again the Euclidean distance matrix of size $c \times g$ is calculated as described in second step. 6)Now for each cluster, the gene for which the sum of distances to other genes of that cluster is minimum, is selected as the new representative of that gene cluster.

3.7 Artificial Neural Network Classifier

The number of types of ANNs and their uses is very high. Starting from the first neural model by McCulloch and Pitts [10] hundreds of different models of ANNs have been developed. Here, we will present only an ANN which learns using the back propagation algorithm [11] for learning the appropriate weights, since it is one of the most common models used in ANNs, and many others are based on it. In this work we used feed-forward artificial neural network. This is the most widely used neural network model, and its design consists of one input layer, at least one hidden layer, and one output layer. Each layer is made up of non-linear processing units called neurons, and the connections between neurons in successive layers carry associated weights. Connections are directed and allowed only in the forward direction, e.g. from input to hidden, or from hidden layer to a subsequent hidden or output layer. Back propagation is a gradient-descent algorithm that minimizes the error between the output of the training input/output pairs and the actual network output. Therefore, a set of input/output pairs is repeatedly presented to the network and the error is propagated from the output back to the input layer. The weights on the backward path through the network are updated according to an update rule and a learning rule. ANNs are solely specified by the characteristics of their processing units and the selected training or learning rule. The network topology, i.e., the number of hidden layers, the number of units, and their interconnections, also has an influence on classifier performance. In our proposed method, one hidden layer with 10 neurons each having hyperbolic tangent activation function is used. The output layer neurons have linear activation function. Also feedforward back propagation algorithm is applied for learning the samples and gradient descent is used for adjusting the weights as training methodology.

3.8 Proposed MOPSO Algorithm

The proposed MOPSO technique is illustrated in Algorithm 1. The population is initialized by randomly chosen features from the data matrix and population fitness values are calculated using k-fold cross validation. The archive A is initialized by the population after non-dominated sorting of the initial population. Velocity and position are updated using Equations 3 and 4 respectively. Thereafter two boundary constrains $BoundaryConstraintsCell()$ and $BoundaryConstraintsDimension()$ function has been designed to set the cell value in between (0 - 1) for the first part and genes values between sample threshold i.e $Sample_{min}(g_i) \leq g_i \leq Sample_{max}(g_i)$ for the second part.

After updating the position, again the entire population are passed to a function $CenterSelection()$ for updating the representative genes for each gene cluster as discussed in Section 3.6. Then using non-dominated sorting [1] and crowding distance [2] sorting the archive is updated. These steps are repeated for specified number of iterations.

4 Datasets and Preprocessing

In this article three real-life gene expression datasets are used which are publicly available from the following website: www.biolab.si/supp/bi-cancer/projections/info/.

Prostate: Gene expression measurements for samples of prostate tumors and adjacent prostate tissue not containing tumor built this classification model. It contains 50 normal tissue and 52 prostate tumor sample. The expression matrix consists of 12533 number of genes and 102 number of samples.

DLBCL: Diffuse large B-cell lymphomas (DLBCL) and follicular lymphomas (FL) are two B-cell lineage malignancies that have very different clinical presentations and response to therapy. Total 7070 genes are there in the dataset. The number of samples of type DLBCL is 58 and type FL is 19.

GSE412 (Child-ALL): The childhood ALL dataset(GSE412) includes genes expression information on 110 childhood acute lymphoblastic leukemia samples. The dataset has 50 examples of type before therapy and 60 examples of type after therapy. The number of genes is 8280.

Algorithm 1 Algorithm : MOPSO-ANN

Input : data matrix dt, C=number of cluster center, N= number of particle, S=Number
of samples, threshold $thr = 0.5$.
Output: archive A
1: $[x_n, v_n, G_n, P_n]_{n=1}^N :=$ initialize(dt) Random locations and velocities
2: $A := x_n$ (if $x_n \not\succ u, \forall u \in A$) //Initialize archive A by first non-dominated x_n
3: $C1 = C$
4: $C2 = 2 \times C$
5: $vt = v$
6: **for** $n := 1 : N$ **do**
7:     **for** $d := 1 : C1$ **do**
8:         $v_{nd} := w.v_{nd} + r1.(P_{nd} - x_{nd}) + r2.(G_{nd} - x_{nd})$
9:         $x_{nd} := x_{nd} + v_{nd}$
10:         $x_{nd} := BoundaryConstraintsCell(x_{nd})$
11:         **if** $x_{nd} \geq thr$ **then**
12:             obj := d // obj contains id of cells having value greater than threshold
13:         **end if**
14:     **end for**
15:     **for** $d := C1 : C2$ **do**
16:         **for** $s := 1 : S$ **do**
17:             $vt_{nds} := w.vt_{nds} + r1.(P_{nd} - x_{nds}) + r2.(G_{nd} - x_{nds})$
18:         **end for**
19:         $V_{nd} := \sum vt_{nds}/S$
20:         **for** $s := 1 : S$ **do**
21:             $x_{nds} := x_{nds} + v_{nd}$
22:             $x_{nds} := BoundaryConstraintsDimension(x_{nds})$
23:             $xn_{ns} := x_{nds}$ // gene cluster (2nd part of the particle) is kept separately
        in xn_{ns}
24:         **end for**
25:     **end for**
26: **end for**
27: **for** $n := 1 : N$ **do**
28:     $x_n := CenterSelection(x_n)$
29:     $x_{obj} := xn_n(obj)$
30:     $y_n := f(x_{obj})$ // Evaluate objectives
31:     $A := A \cup x_n obj$ // Add $x_n obj$ to A, the whole particle ($C1 + C2$) saved to A
32:     $A := x_n obj$ if $x_n obj \not\succ u, \forall u \in A$ // Non-dominated sorting is applied to the
    updated archive
33:     CrowdingSort(A) // crowding distance sorting for archive
34:     **for** $n := 1 : N$ **do**
35:         **if** $x_n < P_n(fitnesses(x_n) \not\succ fitnesses(P_n))$ **then** // Update personal best
36:             $P_n := x_n$
37:             **if** Non-dominated fitnesses **then**
38:                 Random-choice $[(fitnesses(x_n), fitnesses(P_n)]$
39:             **end if**
40:         **end if**
41:         $G_n := selectGuide(x_n, A)$
42:     **end for**
43: **end for**

The above described two-class datasets can be obtained as matrix format whose columns are genes and rows are samples and preprocessed by signal to noise ratio (SNR) for each gene (column). For measuring SNR, mean and standard deviation (S.D.) of class1 and class2 are computed first. The $|SNR|$ of each gene can be defined in Equation 5. Next the genes (column) of the data matrix are sorted according to the decreasing order of obtained $|SNR|$.

$$|SNR| = \left| \frac{Mean(class1) - Mean(class2)}{S.D.(class1) + S.D.(class2)} \right| \tag{5}$$

Lastly from the data matrix top 100 genes are taken. After that the data matrix is normalized to set each gene expression value in the range from 0 to 1.

4.1 Results and Discussion

Here we first describe the performance metrics followed by the results of different algorithms. Performance is evaluated using sensitivity, specificity and accuracy. The total dataset is divided into two different sets namely training and testing set. The proposed approach is applied on the training data therefore a set of non-dominated candidate solutions are obtained. Next the F-score measure is computed for selecting one from these non-dominated candidate solutions. Typically the F-score measure (Equation 6) combines precision and recall through the harmonic mean of precision and recall:

$$F = \frac{2 * Precision * Recall}{Precision + Recall}, \quad Precision = \frac{tp}{tp + fp}. \tag{6}$$

In information retrieval positive predictive value is called precision defined in Equation 6 and sensitivity is called recall. The maximum F-score generating candidate solution is taken as the most informative solution. Three real-life datasets are taken for evaluation. Results are demonstrated in the tables below.

Table 1. Performance metric scores for Prostate Data Set

Algorithm	Sensitivity (sd)	Specificity (sd)	Accuracy (sd)	Fscore (sd)
Proposed	0.8846 (0.0942)	0.82 (0.118)	0.8529 (0.0345)	0.8594 (0.0338)
SFS	0.7393 (0.1272)	0.7864 (0.2313)	0.7950 (0.1047)	0.7126 (0.0847)
T-test	0.7778 (0.1462)	0.8244 (0.0615)	0.8497 (0.074)	0.8336 (0.1052)
Rank-Sum test	0.8547 (0.0976)	0.8375 (0.0825)	0.8768 (0.0399)	0.8522 (0.0317)

Score Analysis: The proposed algorithm has been compared with sequential forward search and other two statistical tests like T-test and Wilcoxon Rank-Sum test. The datasets are randomly divided into two set: training set and test set. This process is repeated 10 times and we got 10 train sets and their corresponding 10 test sets. Each of the algorithm is executed for each file (train and test) individually. Thus for each algorithm we have got 10 sensitivity, 10 specificity,

Table 2. Performance metric scores for DLBCL Data Set

Algorithm	Sensitivity (sd)	Specificity (sd)	Accuracy (sd)	Fscore (sd)
Proposed	0.9136 (0.0926)	0.8123 (0.16)	0.8663 (0.116)	0.7487 (0.1293)
SFS	0.5714 (0.2437)	0.7783 (0.1318)	0.7293 (0.0921)	0.5997 (0.1558)
T-test	0.7284 (0.3148)	0.9119 (0.0881)	0.8486 (0.1036)	0.7052 (0.268)
Rank-Sum test	0.7654 (0.2747)	0.8945 (0.0622)	0.8621 (0.0668)	0.7327 (0.1991)

Table 3. Performance metric scores for GSE412(Child-ALL) Data Set

Algorithm	Sensitivity (sd)	Specificity (sd)	Accuracy (sd)	Fscore (sd)
Proposed	0.6680 (0.1223)	0.8567 (0.1007)	0.6709 (0.0501)	0.7224 (0.0656)
SFS	0.46 (0.1908)	0.8556 (0.1089)	0.6758 (0.0656)	0.5402 (0.1766)
T-test	0.4960 (0.1943)	0.68 (0.3719)	0.5964 (0.1596)	0.5184 (0.1728)
Rank-Sum test	0.4640 (0.2296)	0.87 (0.2755)	0.6855 (0.1108)	0.5506 (0.1901)

10 accuracy and 10 F-score values. Now we have taken the average of these 10 values for each performance metric with standard deviation. For Prostate data it is evident from Table 1 that average sensitivity and F-score for the proposed method are 0.8846 and 0.8894 respectively which are better than other methods. For average specificity and accuracy the results differ slightly. Furthermore, from Table 5 it is evident that the Area Under ROC Curve(AUC) value produced by the proposed technique is 0.8573 which is the maximum among all the methods. Also from the same table as the average correlation value for the selected gene markers by the proposed technique is less compared to that for the other methods, it is evident that the proposed technique identifies small set of non-redundant genes as markers. For DLBCL data, Table 2 shows that with respect to average sensitivity, average accuracy and F-score our proposed technique (0.9136, 0.8663, 0.7484) outperforms any other algorithms presented in this paper. Moreover, with respect to average specificity the score is better than that produced by SFS. Table 5 shows that average correlation value produced by our proposed technique is very much lower than SFS, T-test and Ranksum test and also our method is superior than other methods with respect to the AUC value. For Child-ALL data it is clear from Table 3 that with respect to average sensitivity and F-score, the proposed method result (0.668 and 0.7224 respectively) which are better than other methods. With respect to average accuracy the outcome (0.6709) is slightly lower than SFS but better than T-test and Ranksum test. Again with respect to other objectives like average correlation and AUC the proposed method uniformly yields better values which proves the superiority of our proposed technique.

5 Gene Marker Analysis

After executing the proposed technique 10 times we got 10 feature sets. Thereafter we took those genes as maker which appears 8 times in the 10 feature sets. Table 4 shows that gene 32598_at and 36780_at for Prostate data, the gene $M63835\_at$ and $AB002409\_at$ for DLBCL data and the genes 38447_at, 37969_at and 34966_at for Child ALL data are selected as marker genes. Their symbol and description are also given in the same table. Among the above gene markers, many of those have already been validated to be associated with the respective cancer classes in different existing literature. Such as gene 32598_at of for Prostate Cancer data has been reported in [12] and [13]. Also the gene $AB002409\_at$ has also been reported in [14]. Again in [15], the gene 38447_at of Child-ALL data is addressed. Interestingly the gene 32598_at, $AB002409\_at$ and 38447_at are reposited in Bioinformatics Laboratory www.biolab.si/supp/ bi-cancer/projections/info/.

Table 4. Gene Markers Identified by the Proposed Method for various data set.

Data set	Gene ID	Symbol	Description
Prostate Cancer	32598_at	NELL2	NEL-like 2 (chicken)
	36780_at	CLU	clusterin
DLBCL	$M63835\_at$	FCGR1A	hifh affinity immunoglobulin gamma FC
	$AB002409\_at$	CCL21	chemokine (C-C motif) ligand 21
Child-ALL	38447_at	ADRBK1	adrenergic, beta, receptor kinase1
	37969_at	PTGS1	prostaglandin-endoperoxide synthase1
	34966_at	T	T brachyury homolog

Table 5. Average Correlation and AUC produced by Proposed method, SFS, T-test and Ranksum test.

Algorithm	Prostate		DLBCL		Child-ALL	
	AUC	AvgCorr	AUC	AvgCorr	AUC	AvgCorr
Proposed	0.8462	0.3803	0.8888	0.1709	0.8573	0.7274
SFS	0.7308	0.5514	0.4760	0.6767	0.84	0.7166
T-test	0.7817	0.4434	0.6672	0.6667	0.8253	0.9014
Rank-Sum test	0.8311	0.5177	0.4252	0.778	0.8114	0.9223

6   Conclusion

In this article, a PSO based multiobjective optimization approach has been proposed for identifying non-redundant marker genes. PSO has been modeled using the framework of non-dominated sorting and crowding distance sorting where objectives are evaluated by ANN classifier. Two objectives, sensitivity and specificity have been simultaneously optimized. Three real-life datasets have been used for performance analysis. The comparative study between the proposed technique, sequential forward search, T-test and Rank-Sum test has been done. As a future scope, we plan to incorporate Fuzzy based gene marker selection process guided by Gene Ontology to our proposed method.

Acknowledgement

The work is partially supported by DST PURSE scheme at University of Kalyani, India.

References

1. K. Deb, A. Pratap, S. Agrawal, and T. Meyarivan. A fast and elitist multiobjective genetic algorithm: NSGA-II. *IEEE Transactions on Evolutionary Computation*, 6:182–197, 2002.
2. K. Deb. Multi-objective optimization using evolutionary algorithms. *England: John Wiley and Sons, Ltd*, 2001.
3. X. Cui and T. E. Potok. Document clustering using particle swarm optimization. *IEEE Swarm Intelligence Symposium, Pasadena, California*, 2005.
4. X. Cui and T. E. Potok. Document clustering analysis based on hybrid pso+k-means algorithm. *Journal of Computer Sciences*, Special Issue:27–33, 2005.
5. K. E. Parsopoulos. Particle swarm optimization and intelligence: Advances and applications. *Information science reference, Hershey, New York*, 2010.
6. M. R. Sierra and C. A. Coello Coello. Multi-objective particle swarm optimizers: A survey of the state-of-the-art. *International Journal of Computational Intelligence Research*, 2(3):287–308, 2006.
7. M. H. Cheok, W. Yang, C-H. Pui, J. R. Downing, C. Cheng, C. W. Naeve, M. V. Relling, and W. E. Evans. Characterization of pareto dominance. *Operations Research Letters*, 31, Issue 1:711, 2003.
8. U. Maulik, A. Mukhopadhyay, and S. Bandyopadhyay. Combining pareto-optimal clusters using supervised learning for identifying co-expressed genes. *BMC Bioinformatics*, 10(27), 2009.
9. C. A. Coello Coello. A comprehensive survey of evolutionary-based multiobjective optimization techniques. *Knowledge and Information Systems*, 1(3):129–156, 1999.
10. W. McCulloch and W. Pitts. A logical calculus of the ideas immanent in nervous activity. *Bulletin of Mathematical Biophysics,*, 5:115133, 1943.
11. D. Rumelhart and J. McClelland. *Parallel Distributed Processing*. MIT Press,, Cambridge, Mass.
12. X. Wang and O. Gotoh. Accurate molecular classification of cancer using simple rules. *BMC Medical Genomics*, 2009.
13. J. Li, H. Liu, S-K. Ng, and L. Wong. Discovery of significant rules for classifying cancer diagnosis data. *Bioinformatics*, 19, 2003.
14. M. A. Shipp, K. N. Ross, P. Tamayo, A. P. Weng, J. L. Kutok, R.C.T. Aguiar, M. Gaasenbee, M. Angelo, M. Reich, G. S. Pinkus, T. S. Ray, M. A. Koval, K. W. Last, A. Norton, T. A. Lister, J. Mesirov, D. S. Neuberg, E. S. Lander, J. C. Aster, and T. R. Golub. Diffuse large b-cell lymphoma outcome prediction by geneexpression profiling and supervised machine learning. *Nature Medicine*, 8, 2002.
15. J-S. Zhang, Q. Liu, Y-M. Li, S. H. Hall, F. S. French, and Y-L. Zhang. Genome-wide profiling of segmental-regulated transcriptomes in human epididymis using oligo microarray. *Molecular and Cellular Endocrinology*, 2006.

Application of High Quality Amino Acid Indices to AMS 3.0: A Update Note

Indrajit Saha[1], Ujjwal Maulik[1] and Dariusz Plewczynski[2]

[1] Department of Computer Science and Engineering, Jadavpur University, Jadavpur-700032, West Bengal, India.
[2] Interdisciplinary Centre for Mathematical and Computational Modelling, University of Warsaw, 02-106 Warsaw, Poland
{indra,darman}@icm.edu.pl and umaulik@cse.jdvu.ac.in

Abstract. In this article, we are showing the application of high quality indices of amino acids to improve the performance of AutoMotif Server (AMS) for prediction of phosphorylation sites in proteins. The latest version of AMS 3.0 is developed using artificial neural network (ANN) method. The query protein sequence is dissected into overlapping short sequence segments and then represented it by ten different amino acid indices, which are various physicochemical and biochemical properties of amino acids. However, the selection of amino acid indices has done based on literature survey. Hence, this fact motivated us to use the recently proposed high quality amino acid indices for AMS 3.0. High quality amino acid indices have been developed after analyzing the AAindex database using fuzzy clustering methods. The significant differences in the performance are observed by boosting the precision and recall values of four major protein kinase families like CDK, CK2, PKA and PKC in comparison with the currently available state-of-the-art methods.

Keywords: AutoMotif Server, Artificial Neural Network, Amino Acid, High quality indices, Machine learning, Phosphorylation, Swiss-Prot database.

1 Introduction

Protein synthesis is the process in which cells build proteins. The term is sometimes used to refer only to protein translation but more often it refers to a multi-step process, beginning with amino acid (AA) synthesis and transcription of nuclear DNA into messenger RNA, which is then used as input to translation. After translation, the *Post-Translational Modification* (PTM) of amino acids extends the range of functions of the protein by attaching to it other biochemical functional groups such as acetate, phosphate, various lipids and carbohydrates, by changing the chemical nature of an amino acid (e.g. citrullination) or by making structural changes, like the formation of disulfide bridges. These modifications predominantly occur in protein linear functional motifs (LFMs), but only a small fraction of modified sites has been experimentally identified [1]. By

J. C. Bansal et al. (eds.), *Proceedings of Seventh International Conference on Bio-Inspired Computing: Theories and Applications (BIC-TA 2012)*, Advances in Intelligent Systems and Computing 201, DOI: 10.1007/978-81-322-1038-2_19, © Springer India 2013

the prediction of PTM sites from a protein sequence, we can obtain valuable biological information that can form the basis for further research.

Moreover, in the advent of massive (complex and time-consuming) sequencing experiments, the availability of whole proteomes requires accurate computational techniques for investigation of protein modification sites in the high-throughput scale. In order to address these needs, Basu *et al.* has developed the improved predictor for PTM based on local sequence information of proteins, called AMS 3.0 [2]. This predictor was trained by the Swiss-Prot dataset[3] and Phospho.ELM dataset[4]. During the training/testing of AMS 3.0, Basu *et al.* has encoded the short fragment of protein sequences by amino acid indices from AAindex database[5]. The AAindex database contains numerical indices representing various physicochemical and biochemical properties of amino acids and pairs of amino acids, which have been investigated through a large number of experiments and theoretical studies. Each of these amino acid properties can be represented by a vector of 20 numerical values, and we refer to it as an amino acid index [3,4]. Currently, 544 amino acid indices are released in AAindex database [5–7], which are categorized into six groups using hierarchical clustering technique. However, these indices are overlapping in nature. Hence, we performed consensus fuzzy clustering technique [8] to analyze the database, which gives us eight different clusters. For the purpose of consensus fuzzy clustering, we have used two automatic clustering methods, known as automatic differential evolution based fuzzy clustering (ADEFC) [9] and variable length genetic algorithm [10] based fuzzy clustering (VGAFC) [11], and three cluster define methods like fuzzy c-medoids (FCMdd) [12], differential evolution based fuzzy c-medoids (DEFCMdd) [13,14] clustering, and genetic algorithm based fuzzy c-medoids (GAFCMdd) [13–15] clustering. Thereafter, we computed three different sets of High Quality Inidces (HQIs) of amino acids, known as HQI8, HQI24 and HQI40 [8]. As the AMS 3.0 used short fragment of protein sequences, which are encoded by exhaustively selected amino acid indices, thus, this fact motived us to use the HQIs to see the effectiveness for four major protein kinase families like CDK, CK2, PKA and PKC.

In silico, automatic prediction of PTM sites is now very important area of interest for the bioinformatics research community. It shows by various types of PTM tools, such as AMS [16,17,2], GPS [18], NetPhosK [19], KinasePhos [20], PPSP [21], Scansite [22], PredPhospho [23], MetaPredictor [24], etc. It has been shown in [2] that the AMS 3.0 out performs better against all the available PTM tools. Thus, we confined ourselves to compare the new versions of AMS 3.0 (AMS3$_{HQI8}$, AMS3$_{HQI24}$ and AMS3$_{HQI40}$) with the latest AMS 3.0 and other PTMs prediction tools. For this purpose, we used Phospho.ELM dataset version 8.2 (April 2009) for four unbiased phosphorylation sites, namely the four major protein kinase families: CDK, CK2, PKA and PKC. Phospho.ELM version 8.2 contains 4687 substrate proteins covering 2217 tyrosine, 14518 serine and

---

[3] http://www.uniprot.org/

[4] http://phospho.elm.eu.org/dataset.html

[5] http://www.genome.jp/aaindex/

2914 threonine instances. Effectiveness of the different versions of AMS 3.0 are demonstrated by measuring the precision, recall/sensitivity, specificity, accuracy and AUC.

The rest of the article is organized as follows: the next section discusses the material and method. Section 3 presents the experimental results conducted on AMS 3.0 by using HQIs. Finally Section 4 concludes the article.

2 Material and Method

This section describes high quality indices of amino acid and the details of AMS 3.0.

2.1 High Quality Indices

The different set of HQIs is computed after performing a consensus fuzzy clustering (CFC) on AAindex database. It is to be noted that after analyzing the AAindex datadase, we have found eight clusters. The procedure of CFC method and details of each cluster have been described in [8]. After performing the CFS, we have computed HQI8 (considering only medoid of 8 clusters), HQI24 (considering medoid and two farthest points of medoid, in total 3 points, from each of 8 clusters, that means $8 \times 3 = 24$ points) and HQI40 (considering medoid, two farthest points of medoid and two closest point of medoid, in total 5 points, from each of 8 clusters, that means $8 \times 5 = 40$ points). The detail procedure of computing HQIs is also described in [8]. All of these high quality indices HQI8, HQI24 and HQI40 are separately given in the supplementary with their amino acid values at http://sysbio.icm.edu.pl/aaindex.

2.2 Description of AMS 3.0 with HQIs

The latest version of AMS 3.0 prediction tool [2] uses an efficiently designed Multi Layer Perceptron (MLP) pattern classifier. In this tool, the query protein sequences are dissected into overlapping short segments. Ten different physico-chemical features represent each amino acid from a sequence segment, hence the nine amino acids segment is represented as the point in a 90 dimensional abstract space of protein characteristics. The MLP used in this work, special Artificial Neural Network (ANN) algorithm, is developed to replicate learning and generalization abilities of human's behavior with an attempt to model the functions of biological neural networks of the human brain. The MLP architecture is build from a feed-forward layered network of artificial neurons, where each artificial neuron in the MLP computes a sigmoid function of the weighted sum of all its inputs. The MLP based ANNs are observed to be capable of classifying highly complex and nonlinear biological sequence patterns, where correlations between amino acid positions are important. Unlike earlier attempts, in AMS3, neural network has been implemented three different network models for each of the four major PTM types, by independently optimizing the network weights

Table 1. The best performance values of different PTMs prediction tools for the Kinases CDK, CK2, PKA and PKC, using 10-fold Cross-validation.

Protein kinase	Method	Precision	Recall/Sensitivity	Specificity	Accuracy	AUC
	$AMS3_{HQI8}$	70.68	97.09	95.89	95.25	**0.9488**
	$AMS3_{HQI24}$	66.63	94.27	93.63	93.72	0.9202
	$AMS3_{HQI40}$	69.81	97.01	95.08	94.66	0.9416
	AMS 3.0	67.35	95.65	94.00	94.16	0.9277
	GPS	81.95	90.80	80.00	84.40	0.8761
CDK	KinasePhos	79.07	79.15	79.05	79.10	0.8713
	NetPhosK	57.61	56.80	58.20	57.63	0.7767
	PPSP	62.87	64.73	61.77	62.93	0.8721
	PredPhospho	83.53	89.80	82.30	85.30	0.8670
	Scansite	88.96	40.83	94.93	73.27	0.7584
	MetaPredictor	70.62	91.20	83.20	86.40	0.8946
	$AMS3_{HQI8}$	69.83	78.26	97.08	87.96	0.8607
	$AMS3_{HQI24}$	62.46	70.04	93.63	81.58	0.8261
	$AMS3_{HQI40}$	69.88	78.58	97.82	85.86	0.8603
	AMS 3.0	64.36	73.33	95.00	92.17	0.8397
	GPS	86.94	69.90	89.50	81.60	0.8130
CK2	KinasePhos	88.06	47.93	93.50	75.25	0.7508
	NetPhosK	78.82	80.37	90.43	92.60	**0.9307**
	PPSP	67.19	57.33	72.00	66.17	0.8767
	PredPhospho	93.54	59.40	95.90	78.30	0.7791
	Scansite	98.20	36.37	99.33	74.13	0.7734
	MetaPredictor	65.02	87.80	90.40	89.30	0.9047
	$AMS3_{HQI8}$	66.64	92.86	91.15	91.46	**0.8966**
	$AMS3_{HQI24}$	60.75	89.42	86.63	87.88	0.8646
	$AMS3_{HQI40}$	66.08	91.61	90.08	90.74	0.8901
	AMS 3.0	62.99	91.34	89.31	89.63	0.8847
	GPS	81.70	81.70	80.90	81.20	0.8446
PKA	KinasePhos	62.70	62.70	87.15	77.35	0.8234
	NetPhosK	68.50	68.50	85.23	78.57	0.7581
	PPSP	60.83	60.83	67.83	65.03	0.8860
	PredPhospho	80.80	80.80	83.90	82.70	0.8537
	Scansite	40.80	40.80	96.30	74.13	0.7656
	MetaPredictor	63.81	88.30	82.80	85.00	0.8906
	$AMS3_{HQI8}$	78.93	84.66	87.42	88.72	**0.8542**
	$AMS3_{HQI24}$	73.57	80.73	82.08	84.38	0.8363
	$AMS3_{HQI40}$	78.03	84.08	86.88	87.14	0.8502
	AMS 3.0	76.39	82.02	86.74	85.36	0.8403
	GPS	71.80	71.80	75.30	73.90	0.7574
PKC	KinasePhos	48.63	48.63	83.80	69.73	0.7440
	NetPhosK	51.33	51.33	82.37	69.93	0.7581
	PPSP	56.70	56.70	67.23	63.03	0.7994
	PredPhospho	59.80	59.80	80.50	72.20	0.7149
	Scansite	21.67	21.67	93.53	64.80	0.6397
	MetaPredictor	64.07	77.30	79.10	78.40	0.8213

for three factors: optimum recall (sensitivity), precision, and maximizing the receiver operating characteristics of the prediction model. The consensus build by those three ANNs for each type of PTM gives an additional advantage in comparison to the previously reported ANN based PTM prediction models.

In the current version of AMS3, the three different sets of high quality indices are used separately during the encoding of short AA fragments, whereas the other configurations of AMS3 are kept intact. Hence, the sequence length of 9 amino acid is 72, 216 and 360 for $AMS3_{HQI8}$, $AMS3_{HQI24}$ and $AMS3_{HQI40}$, respectively. The developed software tools execute in two phases. In the first part, a Sequence Generator program reads any number of protein sequences in FASTA format, written in an input file and generates 9 amino acid long overlapping sequences for prediction. The Predictor program reads these short sequences and generates output. The user can specify the type of PTM for specific prediction and the nature of optimization required (based on AUC area, Recall or Precision values). The output is generated in the following format both as a binary decision (whether the sequence qualifies for a potential PTM site or not) and as a probabilistic confidence measure [2] (C_{ij}, for the ith query sequence in the decision class) for each short amino acid sequence, given as an input to the Predictor program. For example:

TRTYLGICH 0 0.9703

RTYLGICHI 0 0.9862

TYLGICHIT 0 0.7903

YLGICHITG 1 0.9622

LGICHITGR 0 0.9202

GICHITGRH 0 0.9633,

where, 0 and 1 signify potential negative and positive sequences respectively.

3   Results

The performance of the $AMS3_{HQI8}$, $AMS3_{HQI24}$ and $AMS3_{HQI40}$ are evaluated on k-fold cross-validation with $k = 4$, 6, 8 and 10, for four major protein kinase families. However, only the results of 10-fold cross-validation are shown in this article, whereas similar results are obtained in other cases. During training of the $AMS3_{HQI8}$, $AMS3_{HQI24}$ and $AMS3_{HQI40}$, the learning rate (η) and and acceleration factors (δ) for the feed-forward neural network with back-propagation learning algorithm are taken to be same as of AMS 3.0 [2]. The three sets of results (corresponding to the AUC, Recall and Precision optimizations) for each protein kinase family corresponds to the training of the network based on the optimized AUC area, recall and precision values for the random test dataset under consideration. Each network is trained with all possible variations of hidden neurons from 2 to 20 (with step size 2). Also the cross-validation has performed 50 times. Table 1 reports the performance measure of $AMS3_{HQI8}$, $AMS3_{HQI24}$ and $AMS3_{HQI40}$ in comparison with other existing PTMs tools. It is observed that the high recall, precision value (more than 80%) and AUC (more than 0.85) is

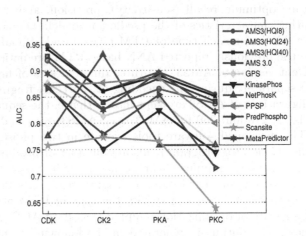

Fig. 1. Comparison of scope of AUC best values for the kinase families PKA, PKC, CDK and CK2, using the different PTM tools.

achieved on independent test datasets for four major protein kinase families. The scope of AUC values for four main kinases is shown in Fig. 1. For $AMS3_{HQI8}$, AUCs range from 0.8542 to 0.9488, for $AMS3_{HQI24}$, AUCs range from 0.8261 to 0.9202, while for $AMS3_{HQI40}$, AUCs range from 0.8502 to 0.9416. However, it has been observed that the performance of AMS 3.0 is lesser than $AMS3_{HQI24}$. The performance of the $AMS3_{HQI8}$ and $AMS3_{HQI40}$ is significantly better for all three major kinases of phosphorylation sites in comparision with previous version of AMS and other PTMs prediction tools. However, for CK2, NetPhosK is better for among the others.

Average execution times for the $AMS3_{HQI8}$ of software is around 23 ms for 100 entries of short amino acid sequences. However, other two versions are taking a bit more time than $AMS3_{HQI8}$ and AMS 3.0. Each such short sequence contains nine amino acids, extracted from a complete FASTA formatted protein sequence. The experiment is conducted on a moderately powerful desktop with 1.6 GHz processor and 768 MB primary memory in Linux based operating environment.

4   Conclusion

Summarizing, the existing AMS 3.0 tool has improved by using recently developed high quality amino acid indices for four major protein kinase families, such as CDK, CK2, PKA and PKC. The high quality indices have been designed after analyzing the AAindex database using fuzzy clustering techniques. The basic structure of AMS 3.0 is the multi layer perceptron based artificial neural network and the use of appropriate amino acid indices. The multi layer perceptron based design is much better suited to highly unbalanced ratio between positives and negatives in the training dataset, and also the use of HQIs for encoding the short protein sequences are therefore enhanced the performance of AMS 3.0 in comparison to other PTMs prediction tools.

As a scope of further research, We would like to make an online application for all different types of PTMs [2]. Also we would like to design a scheme for making a consensus among different versions of AMS 3.0. Moreover, the application of HQIs for protein subcellular localization [25–28], immunogenicity of MHC class I binding peptides [29, 30], protein SUMO modification site prediction [31] and protein-protein interaction prediction from local sequences [32] are one of our future interest too. Authors are working in this direction.

ACKNOWLEDGEMENTS

Mr. Saha is grateful to the All India Council for Technical Education (AICTE) for providing Notational Doctoral Fellowship (NDF) and Polish Ministry of Education and Science (N301 159735 and others) to support the work. Mr. Saha also thanks to Dr. Subhadip Basu for his valuable suggestions.

References

1. Pawson, T.: Specificity in signal transduction: from phosphotyrosine-SH2 domain interactions to complex cellular systems. Cell **116** (2004) 191–203
2. Basu, S., Plewczynski, D.: AMS 3.0: prediction of post-translational modifications. BMC Bioinformatics **11:210** (2010)
3. Nakai, K., Kidera, A., Kanehisa, M.: Cluster analysis of amino acid indices for prediction of protein structure and function. Protein Engineering **2** (1988) 93–100
4. Tomii, K., Kanehisa, M.: Analysis of amino acid indices and mutation matrices for sequence comparison and structure prediction of proteins. Protein Engineering **9** (1996) 27–36
5. Kawashima, S., Ogata, H., Kanehisa, M.: AAindex: amino acid index database. Nucleic Acids Research **27** (1999) 368–369
6. Kawashima, S., Pokarowski, P., Pokarowska, M., Kolinski, A., Katayama, T., Kanehisa, M.: AAindex: amino acid index database, progress report 2008. Nucleic Acids Research **36** (2008) D202–D205
7. Kawashima, S., Kanehisa, M.: AAindex: amino acid index database. Nucleic Acids Research **28** (2000) 374
8. Saha, I., Maulik, U., Bandyopadhyay, S., Plewczynski, D.: Fuzzy clustering of physicochemical and biochemical properties of amino acids. Amino Acids (**DOI 10.1007/s00726-011-1106-9**) (2011)
9. Maulik, U., Saha, I.: Automatic fuzzy clustering using modified differential evolution for image classification. IEEE Transactions on Geoscience and Remote Sensing **48**(9) (2010) 3503–3510
10. Bandyopadhyay, S., Pal, S.K.: Pixel classification using variable string genetic algorithms with chromosome differentiation. IEEE Transactions on Geoscience and Remote Sensing **39**(2) (2001) 303–308

11. Maulik, U., Bandyopadhyay, S.: Fuzzy partitioning using a real-coded variable-length genetic algorithm for pixel classification. IEEE Transactions on Geoscience and Remote Sensing **41**(5) (2003) 1075–1081

12. Krishnapuram, R., Joshi, A., Yi, L.: A fuzzy relative of the k-medoids algorithm with application to web document and snippet clustering. in Proceedings of IEEE International Conference Fuzzy Systems - FUZZ-IEEE 99 (1999) 1281–1286

13. Maulik, U., Bandyopadhyay, S., Saha, I.: Integrating clustering and supervised learning for categorical data analysis. IEEE Transactions on Systems, Man and Cybernetics Part-A **40**(4) (2010) 664–675

14. Maulik, U., Saha, I.: Modified differential evolution based fuzzy clustering for pixel classification in remote sensing imagery. Pattern Recognition **42**(9) (2009) 2135–2149

15. Maulik, U., Bandyopadhyay, S.: Genetic algorithm based clustering technique. Pattern Recognition **33** (2000) 1455–1465

16. Plewczynski, D., Tkacz, A., Wyrwicz, L.S., Rychlewski, L.: AutoMotif Server: prediction of single residue post-translational modifications in proteins. Bioinformatics Applications Note **21**(10) (2005) 2525–2527

17. Plewczynski, D., Tkacz, A., Wyrwicz, L.S., Rychlewski, L.: AutoMotif Server for prediction of phosphorylation sites in proteins using support vector machine: 2007 update. Journal of Molecular Modeling **14** (2008) 69–76

18. Xue, Y., Zhou, F., Zhu, M., Ahmed, K., Chen, G., Yao, X.: GPS: a comprehensive www server for phosphorylation sites prediction. Nucleic acids research (2005) W184–187

19. Hjerrild, M., Stensballe, A., Rasmussen, T.E., Kofoed, C.B., Blom, N., Sicheritz-Ponten, T., Larsen, M.R., Brunak, S., Jensen, O.N., Gammeltoft, S.: Identification of phosphorylation sites in protein kinase a substrates using artificial neural networks and mass spectrometry. Journal of Proteome Research **3** (2004) 426–433

20. Wong, Y.H., Lee, T.Y., Liang, H.K., Huang, C.M., Wang, T.Y., Yang, Y.H., Chu, C.H., Huang, H.D., Ko, M.T., Hwang, J.K.: KinasePhos 2.0: a web server for identifying protein kinase-specific phosphorylation sites based on sequences and coupling patterns. Nucleic acids research (2007) W588–W594

21. Xue, Y., Li, A., Wang, L., Feng, H., Yao, X.: PPSP: prediction of PK-specific phosphorylation site with bayesian decision theory. BMC Bioinformatics **7:163** (2006)

22. Yaffe, M.B., Leparc, G.G., Lai, J., Obata, T., Volinia, S., Cantley, L.C.: A motif-based profile scanning approach for genome-wide prediction of signaling pathways. Nature Biotechnology **19**(4) (2001) 348–353

23. Kim, J.H., Lee, J., Oh, B., Kimm, K., Koh, I.: Prediction of phosphorylation sites using SVMs. Bioinformatics **20**(17) (2004) 3179–3184

24. Wan, J., Kang, S., Tang, C., Yan, J., Ren, Y., Liu, J., Gao, X., Banerjee, A., Ellis, L.B., Li, T.: Meta-prediction of phosphorylation sites with weighted voting and restricted grid search parameter selection. Nucleic acids research **36**(4) (2008) e22

25. Huanga, W.L., Tung, C.W., Huangc, H.L., Hwang, S.F., Hob, S.Y.: ProLoc: Prediction of protein subnuclear localization using SVM with automatic selection from physicochemical composition features. BioSystems **90** (2007) 573–581

26. Tantoso, E., Li, K.B.: AAIndexLoc: predicting subcellular localization of proteins based on a new representation of sequences using amino acid indices. Amino Acids **35**(2) (2008) 345–353

27. Liao, B., Liao, B., Sun, X., Zeng, Q.: A novel method for similarity analysis and protein subcellular localization prediction. Bioinformatics **26**(21) (2010) 2678–2683

28. Laurila, K., Vihinen, M.: PROlocalizer: integrated web service for protein subcel-
    lular localization prediction. Amino Acids (2010, PMID: 20811800.)
29. Tung, W.C., Ho, Y.S.: POPI: predicting immunogenicity of MHC class I binding
    peptides by mining informative physicochemical properties. Bioinformatics **23**
    (2007) 942–949
30. Tian, F., Yang, L., Lv, F., Yang, Q., Zhou, P.: In silico quantitative prediction
    of peptides binding affinity to human MHC molecule: an intuitive quantitative
    structure-activity relationship approach. Amino Acids **36**(3) (2009) 535–554
31. Lu, L., Shi, X.H., Li, S.J., Xie, Z.Q., Feng, Y.L., Lu, W.C., Li, Y.X., Li, H., Cai,
    Y.D.: Protein sumoylation sites prediction based on two-stage feature selection.
    Molecular Diversity **14** (2010) 81–86
32. Klingström, T., Plewczynski, D.:   Protein-protein interaction and pathway
    databases, a graphical review. Briefings in Bioinformatics **12** (2010) 702–713

Constructive Solid Geometry based Topology Optimization using Evolutionary Algorithm

Faez Ahmed, Bishakh Bhattacharya and Kalyanmoy Deb

Abstract Over the past two decades, structural optimization has been performed extensively by researchers across the world. Most recent investigations have focused on increasing the efficiency and robustness of gradient based optimization techniques and extending them to multidisciplinary objective functions. The existing global optimization techniques suffer with requirement of enormous computational effort due to large number of variables used in grid discretization of problem domain. The paper proposes a novel methodology named as *Constructive Geometry Topology Optimization Method (CG-TOM)* for topology optimization problems. It utilizes a set of nodes and overlapping primitives to obtain the geometry. A novel graph based repair operator is used to ensure consistent design and real parameter genetic algorithm is used for optimization. Results for standard benchmark problems for compliance minimization have been found to give better results than existing methods in literature. The method is generic and can be extended to any two or three dimensional topology optimization problem using different primitives.

Keywords: Structural Optimization, Toplogy Optimazation, Genetic Algorithm, Constructive geometry

1 Introduction

Structural optimization is the determination of the topology, shape and size of the mechanism starting with a domain of material to which the external loads and supports are applied [9]. Topology optimization of compliant mechanism can be considered as determination of material connectivity among different ports such as input,

Bishakh Bhattacharya
Indian Institute of Technology, Kanpur, e-mail: bishakh@iitk.ac.in

Kalyanmoy Deb
Indian Institute of Technology, Kanpur, e-mail: deb@iitk.ac.in

J. C. Bansal et al. (eds.), *Proceedings of Seventh International Conference on Bio-Inspired Computing: Theories and Applications (BIC-TA 2012)*, Advances in Intelligent Systems and Computing 201, DOI: 10.1007/978-81-322-1038-2_20, © Springer India 2013

output and support ports (boundary conditions). These special ports and other material intersection ports can be termed as nodes and the topology defines the compliant mechanism skeleton and connection between such nodes. It provides the flexibility where both the shape of the exterior boundary and configuration of interior boundaries can be optimized simultaneously.

Most of the existing methods in literature use a grid approach for domain discretization. They use variables proportional to number of grid cells, and utilize gradient based methods to search for optimum topology. Hence the number of variables used are large. Problems like point flexure and mesh dependency are found in such methods and are often dealt with using different filtering techniques. Mesh refinement further increases the number of variables.

The objective function is often the compliance, that is, the flexibility of the structure under the given loads, subject to a volume constraint. The optimum distribution of material is measured in terms of the overall stiffness of the structure such that the higher the stiffness the more optimal the distribution of the allotted material in the domain. Two major methods existing in topology optimization field are homogenization and the solid isotropic material with penalization (SIMP) methods. Comprehensive details of SIMP and related methods can be found in [1]. For a continuum structure represented by a domain of finite elements and associated boundary conditions, the compliance minimization topology optimization problem in SIMP can be expressed mathematically as: Find the optimal distribution of solid and void elements that would

$$min f(\rho) = u^T K u = \sum_{j=1}^{N_e} u_j^T K_j u_j \qquad (1)$$

such that $\sum_{j=1}^{N_e} \rho_j V_j \leq V_0$
where $0 < \rho_{min} \leq \rho \leq 1$

here $f(\rho)$ represents the objective function which is total strain energy, ρ is the design variable vector of non-dimensional element densities, u the vector of global nodal displacements and K is the global stiffness matrix. u_j, K_j, ρ_j and V_j are the jth elements displacement vector, stiffness matrix, non-dimensional density and volume respectively. N_e is the number of elements and V_0 is the material available.

Genetic algorithms were used by Jakiela [5] for the optimal topology search of continuum structures. The design space was discretized into small elements with all of the finite elements forming a binary-coded bit-string chromosome, 0 and 1 for absence and presence of an element in the structure, respectively. To facilitate the transmission of topology and shape characteristics across generations, Tai et al. [10] utilized spline based arrangements of skeleton and flesh surrounding the bones to represent structural geometry. Other recent works in non-gradient methods include simulated biological growth (SBG) [7], bidirectional ESO (BESO) [8] and cellular automata [2, 6]. The usage of global optimization methods are generally not found to perform at par with the local optimization methods [9].

Constructive Solid Geometry (CSG) is a technique widely used in solid modelling. It uses Boolean operators to combine simple objects called solids or primitives, constructed according to geometric rules, and form complex two or three dimensional geometries. Simple shapes like rectangle, circle, ellipse or a generic polygon can be used as a CSG primitives in 2-D. The boolean operations can be summarized as Union, Intersection and Difference as shown in Figure 1 (a).

The idea of utilizing CSG primitives for topology optimization provided the motivation for the proposed technique. Using CSG, union of many primitives (rectangular bars) can be taken to form complex shape segments. The material where bar segments do not appear is left out as holes. This idea provides the backbone of CG-TOM method.

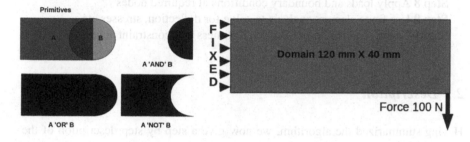

Fig. 1 (a) CSG Boolean operations on 2-D primitives (b) Problem domain for cantilever with end loading

2 Methodology

The key idea to the methodology is to utilize a small set of points (nodes) and there connectivity to generate the compliant mechanism or structure. A generic topology can be seen as comprising of few joints connected by segments. The number of joints and the number of segments connecting them may vary from one mechanism to other. The synthesis of compliant mechanisms has traditionally been viewed as a domain with presence or absence of holes. On the contrary, we propose to use a building block model where different segments (primitives) overlap to give shape and volume to the final topology. The topology is completely defined by a set of n nodes and widths representing connectivity between each pair of nodes. The methodology is explained through a topology optimization problem of a cantilever system. The domain dimensions are $120mm \times 40mm$ and a point load of 100 N is applied at the edge as shown in Fig. 1 (b).

2.1 Algorithm Summary

The algorithm can be summarized in the following steps. In the next section, step by step explanation along-with example figures is shown.

Step 1 Define k fixed nodes
Step 2 Interpret variables to obtain positions of m nodes
Step 3 Delaunay triangulation to obtain allowed connectivity between all nodes
Step 4 Allot widths to triangle edges. Remove edges with negative widths.
Step 5 Use CSG tool to obtain the topology after union of all bars
Step 6 Use graph based repair operator to check and correct the topology.
Step 7 Mesh the obtained topology using CSG supporting mesh generator
Step 8 Apply loads and boundary conditions at required nodes
Step 9 Use finite element analysis to solve for deflection, stresses etc.
Step 10 Read required output and return fitness and constraint values

2.2 Description

Having summarized the algorithm, we now give a step by step description of the method.

In a generic topology optimization problem, after the actual problem formulation, the design engineer has a problem domain within which the final topology should be constrained, a set of loads and a set of boundary conditions (like the location of supports). After considering the boundary condition and loads, the user defines k fixed nodes. These nodes represent the spatial locations where material must necessarily be present. The point of application of loads (e.g. point force) or various boundary conditions are usually taken as fixed nodes. As shown in Fig. 2 (a), three fixed nodes (shown by square marker) at (0,0), (120,0) and (0,40) are taken. Thereafter we take m nodes denoting joints, within the domain bounds as shown in Fig. 2 (a) by circular markers. The final mechanism may have any number of joints less than or equal to m. The $n(= m + k)$ points represent the 'joints' in the mechanism.

To find the possible connections between joints, we use triangulation of the nodes as shown in Fig. 2 (b). This gives us the base skeleton. In our simulation we have used *Delaunay triangulation* [3] between the nodes.

In the current work, we have utilized rectangular shaped CSG primitive of various dimensions and union operation as boolean operator. After obtaining the skeleton, some of the edges are replaced by rectangular bars. In figure 3 (a), all the skeleton edges are replaced by bars of different width and CSG union of all such bars is taken to obtain the topology. Edges which are replaced by bars and the width of the bars is controlled by optimization algorithm. It can be noticed that the addition of width to each bar leads to material overlap between bars corresponding to previously non-intersecting edges. Width of one bar may completely overshadow another joint after union operation. We have done CSG calculations using commercial software MATLAB alongwith utilizing its mesh generator.

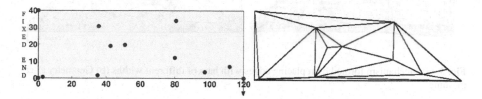

Fig. 2 (a) Fixed and variable nodes within domain (b) Delaunay triangulation of the nodes

Fig. 3 (b) shows the resultant geometry. It can be observed that using only a simple rectangular bar primitive various complex features and holes of different sizes can be obtained. It is possible that a small portion of the geometry may go outside the box formed by the domain. Hence we use the CSG operation of subtraction and any portion outside the domain is trimmed.

Next, meshing of the obtained geometry is done using CSG supporting mesh generator and load and boundary conditions are applied. Post meshing, the 2-D FEA calculation is done and it is found that the tip of the given geometry deflects by 2.086 mm and occupies 62.83% volume. The connectivity between nodes can also be visualized in the form of a connected graph between the numbered nodes which will be utilized for graph repair operator. ABAQUS software has been used for finite element calculations.

In single objective compliance minimization problems, the amount of material available to form the optimum topology is limited. For e.g. a problem may state to find the optimum topology of a cantilever system shown in Fig. 1 (b) with minimum mean compliance using only 50% material. Hence volume occupied by the topology acts as a constraint and any topology using more than 50% volume will be deemed infeasible. Initial trial runs in our study denoted that many geometries tend to be infeasible, hence slowing the convergence of optimization run. Specially when the optimization is near convergence, most of the population members have volume close to 50% boundary and any new member created with slight variation of bar width or node position may slightly increase from the volume limit and would become infeasible. Hence a volume correction operator is proposed to improve GA convergence.

After the formation of final geometry, the volume is calculated. If the volume is above the constraint value (say 60%) the amount of deviation ε from constraint is calculated. If $\varepsilon \leq \delta$, all the width of bars are reduced by a ratio, such that the final topology satisfies the volume constraint. In our simulations, we have taken $\delta = 15$. The volume operator was found to be very effective in improving convergence time. After the volume correction, FEA analysis is done on feasible geometry. To illustrate the operator, we again take the same example as shown in Fig. 3 (b). The topology volume was found to be 62.83% which violated the constraint of 50%. The widths

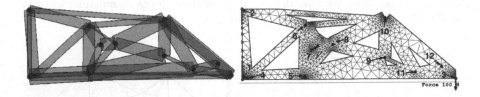

Fig. 3 (a) Mechanism formed by replacing edges with bars of different widths (b) Geometry after meshing

of all bars were reduced such that its volume comes just below 50%. Fig.4 (a) shows the mesh of geometry with reduced widths in original and deflected positions. The new geometry with volume 49.97% satisfies the volume constraint and deflects by 2.972 units.

In the above methodology, the position of the nodes and the widths of the bars are controlled by the optimization algorithm to give different geometries. If the width variable between two nodes is negative or the connection is absent from the base skeleton, then the segment between those nodes is removed. Hence the intersection of remaining segments gives the final geometry. The benefit of removing connections depending on variable values is that some nodes can be completely disconnected and very simple geometries can also be obtained. We now discuss the repair operation to deal with inconsistent geometries.

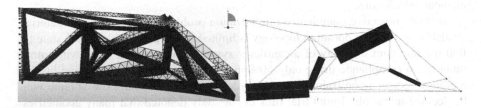

Fig. 4 (a) Final deflected and original geometry after volume correction and FEA analysis (b) Inconsistent geometry to demonstrate repair algorithm

3 Graph based Geometry Repair

3.1 Discussion

The optimization algorithm initializes the variables randomly. It is possible that many bars are absent such that the geometry formed after CSG union is disconnected or broken. A disconnected geometry can be formed in following ways

- The point of application of load is not connected to the geometry
- The geometry is not connected to a fixed nodes corresponding to point of application of boundary conditions
- There are hanging sections in space, that is some section is not connected to any node in main geometry

In any such scenario, the fitness evaluation function must detect inconsistent topology and take corrective actions before FEA analysis. Otherwise the FEA analysis will give error and optimization will prematurely terminate. Rejecting or penalization of such broken geometries is also not desirable as it will reduce the feasible geometries and many function evaluations may be wasted in each generation. Hence we have employed a graph theory approach for anomaly detection and to take corrective action by repairing the geometry.

3.2 Repair algorithm

A brief overview of the algorithm is given below. Here the geometries are viewed as graph of connected nodes. A group of interconnected nodes is termed as a *Set*. To ensure connected geometry between load and boundary conditions as well as no disconnected section, the algorithm aims to form a single set containing all the fixed nodes. In case of absence of such a set, geometry is repaired by adding minimum possible segments to form the connected set. Other disconnected sets are eliminated.

Step 1 Find all *Sets* of interconnected nodes
Step 2 Find the sets containing fixed nodes. These are termed *Fixed Sets* If single such set exists, go to Step 7
Step 3 Using the initial triangulation data, find the connectivity between each pair of sets. A set 'A' is connected to set 'B' if any of its nodes is allowed to be connected to any of the nodes of set 'B' by triangulation
Step 4 Obtain the graph of connectivity between sets. Weight of connection between two nodes is the length of the shortest segment which can connect them
Step 5 Using *BFS* (Breadth first search) from each set in the graph, find the minimum connections required to join all the *Fixed Sets*
Step 6 Add segment between the corresponding nodes between the sets to form a single connected set
Step 7 Ignore all other sets and form the final geometry using the *Fixed Set*

3.3 Repair Example

To illustrate the algorithm, an example broken geometry is derived from the example in Fig. 3 by randomly removing some connections. The broken geometry is shown in Fig. 4 (b). Here all the three fixed nodes are disconnected from each other and two disconnected bars are also present. Using the algorithm, first the graph of nodes is formed and each connected set is identified as shown in Fig. 5. Eight different *Sets* are marked in it of which the first three are the *Fixed Sets*. Fig. 6 (a) shows the connectivity between the eight sets. Using BFS, the shortest path connecting the *Fixed Sets* is calculated and is shown in the figure by bold lines. Once the pair of sets required to be connected is recognized, the nodes connecting them with minimum segment length are identified and final geometry is repaired. Two corrections are required. After applying the corrections, all other sets are ignored, hence all other disconnected bars are eliminated from the final topology. Fig. 6 (b) shows the two bar segments added by the repair operator. The above algorithm ensures that even if a direct connection between fixed sets is not possible, still the shortest path is found through some intermediate set. BFS from each set ensures no bias.

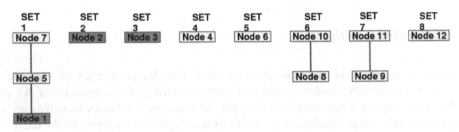

Fig. 5 Connectivity graph of nodes in broken geometry showing connected *Sets*

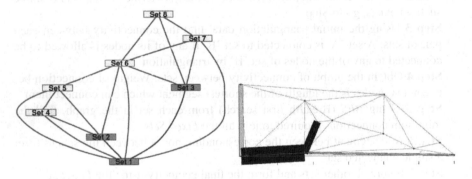

Fig. 6 (a) Graph of *Sets* connected by connections which were previously absent (b) Segment added to repair geometry from graph

4 Single Objective Evolutionary Optimization

We have used Genetic Algorithms [4] to solve the optimization problem. The GA code was integrated with MATLAB CSG tool and ABAQUS FEA solver to convert real variables to geometry, geometry to finite element triangular mesh and finally after carrying the FEA analysis with given loads and boundary conditions, reading the gene fitness from FEA output file for every function evaluation. From the discussions in Section 2, it can be seen that solving any topology optimization problem would initially require the user to decide on number of variable nodes. The fixed nodes depend on problem information. Hence, in a problem with m variable nodes and k fixed nodes, the total number of GA variables required will be $2 \times m + \binom{m+k}{2}$. The $2 \times m$ variables represent the node positions and there variable bounds depend on domain boundary and the $\binom{m+k}{2}$ represent the widths between each pair of numbered nodes.

In the current study we have solved single objective compliance minimization problem using real valued genetic algorithm. The initial population is generated randomly within the variable bounds specified. The next generation of the population is computed using the fitness of the individuals in the current generation. Binary tournament selection, polynomial mutation and SBX crossover are used. The gene representing the m variable nodes is sorted with respect to there x co-ordinates as shown in Fig. 3 (b) and nodes are numbered sequentially.

Since the node numbering is decided on basis of x co-ordinates, the widths are also mapped accordingly. The crossover between the co-ordinates of node i of parent A occurs with corresponding node i in parent B. Similarly the width crossover between parent A and B genes would occur corresponding to bar between nodes (i, j) in both.

4.1 Case Study 1

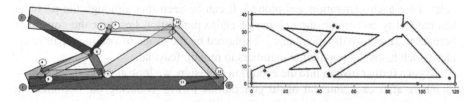

Fig. 7 (a) Patches combination defining the optimized solution by CSG union (b) Optimized solution for 1st benchmark problem. Compliance 178.4 N-mm

The test case considered is benchmark problem taken from [11] to compare our methods with other existing methods in literature. The design domain and bound-

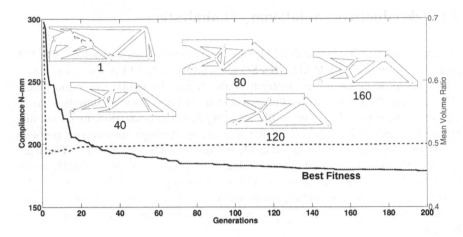

Fig. 8 Variation of fitness and mean volume with generations showing evolution of geometry

aries for the compliance minimization problem was shown in Fig. 1 (b) for a cantilever system. Here the domain is $120mm \times 40mm$ and a point load of 100 N is
applied at the end with material constraint $V/V_0 \leq 0.5$. The volume fraction inequality constraint is $V/V_0 \leq 0.5$ and fitness is reported in N-mm. The material data
for both benchmark test are $E = 10$ GPa and $v = 0.3$

Three fixed nodes have been taken and nine variable nodes. Hence the total number of real variables are 84. A population size of 40 is taken and GA is run for 200
generations. The optimized solution for benchmark test using CG-TOM has been
shown in Fig.7 (b). [11] compares the different methods and filtering schemes for
compliance minimization on this domain and reports 179.1 N-mm mean compliance
as minimum value. The fitness obtained using CG-TOM of 178.4 N-mm is better.
Fig.7 (a) shows the primitive bars that combine to form this optimized solution.

The progress of optimization can be observed by analyzing the improvement of
fitness with each generation. In Fig. 8 the variation of fitness of best population
member is shown with generations. Mean volume of entire population is also calculated for each generation and plotted. It can be seen that although the problem
has inequality constraint, the volume of entire population lies near the constraint
boundary. The sharp dip in volume is explained by the usage of volume repair operator which helps the population members to remain feasible.

To get a further insight into the evolution of design, we look at the best population
members after each interval of 40 generations in Fig. 8 during the optimization
process. It can be visually seen that GA recognizes the optimum geometry shape
quickly and thereafter small shape variations of bar shape lead to the optimized
solution. It can also be seen that the position of nodes gets fixed quickly defining
the basic skeleton.

4.2 Case Study 2

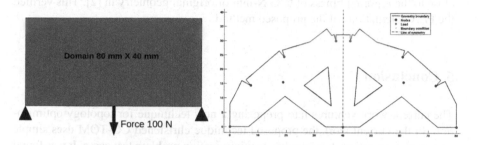

Fig. 9 (a) Domain for 2nd Benchmark problem (b) Optimized symmetrical solution for 2nd benchmark problem

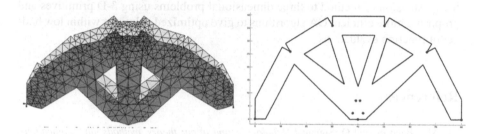

Fig. 10 (a) Original and deflected geometry for optimized solution in ABAQUS software (b) Geometry similar to optimum solution in [2] re-created using CG-TOM method for comparison

The domain of second benchmark problem is shown in Fig. 9(a). For this compliance minimization problem with volume constraint $V/V_0 \leq 0.5$, a symmetric loading of 100 N force is applied on a domain fixed at two corners. The domain dimensions are $80mm \times 40mm$.

The problem has symmetrical loading and boundary conditions, hence the resultant optimized geometry will also be symmetric. CG-TOM is modified to impose symmetry within variable representation itself by taking mirror image of nodes and widths about the line of symmetry. Three fixed nodes have been taken in our method and only four variable nodes. Hence although total 11 nodes will represent the geometry, the position of only 4 nodes needs to be optimized. A population size of 40 is taken and GA is run for 200 generations.

The optimized solution for benchmark test using symmetrical CG-TOM has been shown in Fig.9(b) and Fig.9(b) shows the original and deflected geometry. The fitness obtained by CG-TOM is 8.03 N-mm after 200 generations. To further verify the method, we analyzed the optimum solution reported in [2]. A minimum compliance of 9.83 N-mm is reported for the same problem. Using symmetrical CG-TOM method with four variable nodes, a visually similar solution (exact dimensions were not available) to there optimum solution was generated as shown in Fig.10(b). The fitness calculated using our method for this geometry was 9.82 N-mm which was close to the reported fitness of 9.83 N-mm of original geometry in [2]. This verified the FEA calculations of the proposed method.

5 Conclusion

The current work ventures into proposing a new technique for topology optimization. In the current work,the proposed technique christened CG-TOM uses simple rectangular primitives for compliance minimization problem test cases. It was found to give better results than existing literature in both cases. The current work opens up many possibilities of usage of CSG techniques for topology optimization and inverse finite element problems by integrating CSG modelling technique with optimization. Further research can focus on post-processing to obtain smooth geometries, extension of method to three dimensional problems using 3-D primitives and proposing more efficient GA algorithms to give optimized solutions within low budget of function evaluations.

References

1. M.P. Bendsøe and O. Sigmund, *Topology optimization: theory, methods, and applications*, Springer Verlag, 2003.
2. B. Bochenek and K. Tajs-Zielińska, *Novel local rules of cellular automata applied to topology and size optimization*, Engineering Optimization **44** (2012), no. 1, 23–35.
3. M. De Berg, O. Cheong, and M. Van Kreveld, *Computational geometry: algorithms and applications*, Springer-Verlag New York Inc, 2008.
4. D.E. Goldberg, *Genetic algorithms in search, optimization, and machine learning*, Addison-wesley, 1989.
5. M.J. Jakiela, C. Chapman, J. Duda, A. Adewuya, and K. Saitou, *Continuum structural topology design with genetic algorithms*, Computer Methods in Applied Mechanics and Engineering **186** (2000), no. 2, 339–356.
6. E. Kita and T. Toyoda, *Structural design using cellular automata*, Structural and Multidisciplinary Optimization **19** (2000), no. 1, 64–73.
7. C. Mattheck and S. Burkhardt, *A new method of structural shape optimization based on biological growth*, International Journal of Fatigue **12** (1990), no. 3, 185–190.
8. OM Querin, GP Steven, and YM Xie, *Evolutionary structural optimisation (eso) using a bidirectional algorithm*, Engineering Computations: Int J for Computer-Aided Engineering **15** (1998), no. 8, 1031–1048.
9. G.I.N. Rozvany, *A critical review of established methods of structural topology optimization*, Structural and Multidisciplinary Optimization **37** (2009), no. 3, 217–237.
10. K. Tai, G.Y. Cui, and T. Ray, *Design synthesis of path generating compliant mechanisms by evolutionary optimization of topology and shape*, Journal of Mechanical Design **124** (2002), 492.
11. S. Xu, Y. Cai, and G. Cheng, *Volume preserving nonlinear density filter based on heaviside functions*, Structural and Multidisciplinary Optimization **41** (2010), no. 4, 495–505.

Array P Systems with Hybrid Teams

P. Helen Chandra and S.M. Saroja Theerdus Kalavathy

Abstract Linking the two areas of membrane computing and picture grammars, Array P systems were introduced by Ceterchi et al (2003) which brought out the ability of the bio-inspired model of P system in handling the problem of picture generation. Another well-investigated model of picture generation is the array grammar system wherein prescribed teams of array productions and different derivation modes are introduced by Fernau and Freund (1996), while examining the problem of bounded parallelism for character recognition. Here a kind of array P system with the possibility of different teams having different modes of derivation is considered. We illustrate the model by generating different alphabetic characters. The generative power of such mechanism is also investigated.

Keywords : Membrane Computing, Array P System, Array grammar rules, Array Languages, Teams.

1 Introduction

In the study of generation and description of picture patterns considered as connected digitized finite arrays of symbols, syntactic techniques have played a significant role. Adopting the techniques of formal string language theory, various types of picture or array grammars have been introduced and investigated. The notion of a grammar system and team CD grammar system were introduced and investigated

P. Helen Chandra
Jayaraj Annapackiam College for Women, Periyakulam, Theni District, Tamilnadu, INDIA,
e-mail: chandrajac@yahoo.com

S.M. Saroja Theerdus Kalavathy
Jayaraj Annapackiam College for Women, Periyakulam, Theni District, Tamilnadu, INDIA,
e-mail: kalaoliver@gmail.com

J. C. Bansal et al. (eds.), *Proceedings of Seventh International Conference on Bio-Inspired Computing: Theories and Applications (BIC-TA 2012)*, Advances in Intelligent Systems and Computing 201, DOI: 10.1007/978-81-322-1038-2_21, © Springer India 2013

in [2,5,7] by removing the restriction in the CD grammar system that at each moment only one component is enabled. Fernau [4] and ter Beek [6] studied hybrid (prescribed) team CD grammar system allowing work to be done in teams while at the same time assuming these teams to have different capabilities. In particular in [4], hybrid prescribed team cooperating array grammar systems are considered with the components having context-free or regular array grammar rules.

On the other hand, the area of membrane computing was initiated by Paun [8] introducing a new computability model called as P system, which is a distributed, highly parallel theoretical computing model based on the membrane structure and the behavior of the living cells. Among a variety of applications of this model, the problem of handling array languages using P systems has been considered by Ceterchi et al. by introducing array rewriting P System [1] and thus linking the two areas of membrane computing and picture grammars.

In this paper we introduce a kind of array P system with objects in the regions as arrays and the productions as hybrid prescribed team of CD grammar rules which allow work to be done in team with the possibility of different teams having different modes of derivation. Rewriting is done in parallel in a team. The generative power and the application of the proposed model is examined. Comparison is done with the other existing models.

2 Preliminaries

2.1 Array P systems [1]

We now recall here the basic definition of Array P System.

The array P system (of degree $m \geq 1$) is a construct

$$\Pi = (V, T, \#, , F_1, .., F_m, R_1, .., R_m, i_o),$$

where V is the total alphabet, $T \subseteq V$ is the terminal alphabet, # is the blank symbol, μ is a membrane structure with m membranes labeled in a one-to-one way with $1, 2, ., m$, $F_1, ., F_m$ are finite sets of arrays over V associated with the m regions of $\mu, R_1, .., R_m$ are finite sets of array rewriting rules over V associated with the m regions of μ ; the rules have attached targets *here, out , in* (in general, *here* is omitted), hence they are of the form $\mathcal{A} \rightarrow \mathcal{B}(tar)$; finally, i_o is the label of an elementary membrane of μ (the output membrane).

We emphasize the fact that in an array P system we distinguish terminal and auxiliary symbols in the Lindenmayer sense, that is, no condition is imposed on the symbols appearing in the left hand of rules. The general case, when a set T is distinguished, we speak about an *extended* P system, when $V = T$ we have a *nonextended* system. According to the form of its rules, an array P system can be monotonic, context-free (CF) #-context-free (#CF), or regular (REG). In the extended case, a rule is called regular if it is of one of the following forms:

$$
a\# \to bc\,, \quad \#a \to bc\,, \quad
\begin{matrix} a & b \\ & \to \\ \# & c \end{matrix}\,, \quad
\begin{matrix} \# & b \\ & \to \\ a & c \end{matrix}\,, \quad
a \to b
$$

where all a,b,c are non-blank symbols.(One can also consider #-regular rules, of the form $a \to \#$, but we do not proceed in this way here.) In the non-extended case, we use the notion of a regular rule in the restricted sense; such a rule is of one of the forms :

$$
a\# \to ab\,, \quad \#a \to ba\,, \quad
\begin{matrix} a & a \\ & \to \\ \# & b \end{matrix}\,, \quad
\begin{matrix} \# & b \\ & \to \\ a & a \end{matrix}
$$

where all a,b are non-blank symbols.

A computation in an array P system is defined in the same way as in a string rewriting P system with the successful computations being the halting ones: each array, from each region of the system, which can be rewritten by a rule associated with that region (membrane) should be rewritten; this means that one rule is applied (the rewriting is sequential at the level of arrays); the array obtained by rewriting is placed in the region indicated by the target associated with the used rule (*here* means that the array remains in the same region, *out* means that the array exits the current membrane -thus, if the rewriting was done in the skin membrane, then it can exit the system; arrays leaving the system are "lost" in the environment), and *in* means that the array is immediately sent to one of the directly lower membranes, nondeterministically chosen if several exist (if no internal membrane exists, then a rule with the target indication in cannot be used). A computation is successful only if it stops, a configuration is reached where no rule can be applied to the existing arrays. The result of a halting computation consists of the arrays composed only of symbols from T placed in the membrane with label i_o in the halting configuration.

The set of all such arrays computed (we also say generated) by a system Π is denoted by $AL(\Pi)$. The family of all array languages $AL(\Pi)$ generated by systems Π as above, with at most m membranes, with rules of type $\alpha \in \{REG, CF, \#CF\}$ is denoted by $EAP_m(\alpha)$. If non-extended systems are considered, then we write $AP_m(\alpha)$.

2.1.1 Example [1]

Consider the non extended context free system

$$\Pi_1 = (\{a\}, \{a\}, \#, [_1[_2[_3]_3]_2]_1, \left\{ \begin{array}{c} a \\ a \end{array} \right\}, \emptyset, \emptyset, R_1, R_2, R_3, 3)$$

$$R_1 = \left\{ \begin{array}{ccc} & \# & a \\ & \rightarrow & (in) \\ \# \, a & & \# \, a \end{array} \right\}$$

$$R_2 = \left\{ \begin{array}{cccc} a \, \# & a \, a & a \, \# \, \# & a \, a \, a \\ \rightarrow & (out), & \rightarrow & (in) \\ \# & \# & \# & \# \end{array} \right\},$$

$$R_3 = \emptyset$$

Starting from the unique array present initially in region 1, one grows step by step the two arms of an L-shaped angle, with one pixel up in the skin membrane and with one pixel to the right in membrane 2; at any moment, from membrane 2 we can send the array to membrane 3 (at that step two pixels are added to the horizontal arm) and the computation stops. Thus, $AL(\Pi_1)$ consists of all L-shaped angles with equal arms, each arm being of length at least three.

2.2 Hybrid prescribed team CD grammar system

Cooperating array grammar system [3] was introduced for picture array generation as an extension of the corresponding string model. It was further studied in [6] in the framework of teams. Here we recall the notion of hybrid prescribed team CD grammar system [6].

Definition 1. A hybrid prescribed team CD grammar system [6] is a construct
$\Gamma = (N, T, P_1, ..., P_n, S, (Q_1, f_1), (Q_2, f_2), ..., (Q_m, f_m))$,
where $N, T, , P_1, ..., P_n$ are defined as in the cooperating array grammar system [3]. $Q_1, Q_2, ..., Q_m$ are teams over $N \cup T$, multiset of sets of productions $P_1, ..., P_n$ and $f_1, f_2, ..., f_n$ are modes of derivation.

For a team $Q_i, 1 \leq i \leq m, Q_i = \{P_{ij} | 1 \leq j \leq m_i\}$, and two arrays D_1 and $D_2 \in (N \cup T)^+$ a direct derivation step is defined by $D_1 \vdash_{Q_i} D_2$ if and only if there are array productions $p_j \in P_{ij}, 1 \leq j \leq m_i$, such that in D_1 we can find m_k non-overlapping areas such that the sub-patterns of D_1 located at these areas coincide with the left-hand sides of the array productions p_j and yield D_2 by replacing them by the right-hand sides of the array productions p_j.

An application of the team Q_i to an array D_1 therefore means the following: from each set P_{ij}, one array production p_j is chosen such that $p_1, ..., p_m$, can be applied in a parallel manner to D_1 without disturbing each other. Note that the array productions p_j need not all be different although coming from different sets within the team Q_i. The derivation relations are defined by $\vdash^*_{Q_i}, \vdash^{=k}_{Q_i}, \vdash^{\leq k}_{Q_i}$ and $\vdash^{\geq k}_{Q_i}$ respectively. i.e, derivations with the team Q_i of arbitrary, of exactly k successive steps, of atmost k steps, of atleast k steps, and of as many steps as possible, respectively; this maximal derivation mode t is defined more precisely by: $D_1 \vdash^t_{Q_i} D_2$ if and only if $D_1 \vdash^*_{Q_i} D_2$ and there is atleast one component P_{i,j_0} in the team Q_i such that no array production in P_{i,j_0} can be applied to D_2 anymore. Note that in the t-mode a derivation with a team Q_i can be blocked, although in every P_{i,j_0} we can find an array production which is applicable to the underlying array.

The language generated by Γ is

$$L(\Gamma) = \left\{ X \in T^{**} \ / \ S \underset{Q_1}{\overset{f_1}{\Rightarrow}} X_1 \underset{Q_2}{\overset{f_2}{\Rightarrow}} X_2 \Rightarrow \cdots \underset{Q_m}{\overset{f_m}{\Rightarrow}} X_m = X \right\}$$

3 Array P systems with hybrid teams

Now we introduce Array P systems with hybrid teams which allows work to be done in team with the possibility of different teams having different modes of derivation.

Definition 2. An Array P systems with hybrid teams of degree $m(m > 1)$ is a construct

$$\Pi = (V, T, \#, \mu, F_1, ..., F_m, R_1, ..., R_m, i_o)$$

where V is the total alphabet, $T \subseteq V$ is the terminal alphabet, # is the blank symbol, μ is a membrane structure with m membranes labeled in a one-to-one way with $1, 2, ..., m; F_1, F_2, ..., F_m$ are finite sets of arrays over V initially associated with the m regions of $\mu; R_1, R_2, ..., R_m$ are finite sets of prescribed teams of context - free production rules with the derivation modes associated with the m regions of μ; the rules have attached targets, *here, out, in* (in general, *here* is understood and is omitted); finally, i_o is the label of an elementary membrane of μ (the output membrane).

A computation in the Array P systems with hybrid teams is defined in the same way as in an array rewriting P system with the successful computations being the halting ones; each array, from each region of the system, which can be rewritten by a team of rules associated with that region (membrane), in a specific derivation mode. The array obtained by rewriting is placed in the region indicated by the target associated with the rule used. The term *here* means that the array remains in the same region, *out* means that the array exits the current membrane, thus, if the rewriting was done in the skin membrane, then it can exit the system; arrays leaving the system are *"lost"* in the environment, and *in* means that the array is immediately sent to one of the directly lower membranes, non-deterministically chosen of several exist (if no internal membrane exists, then a rule with the target indication *in* cannot be used). A computation is successful only if it stops and a configuration is reached where no rule can be applied to the existing arrays. The result of a halting computation consists of the arrays composed only of symbols from T placed in the membrane with label i_o in the halting configuration. The set of all such arrays computed (or generated) by a system Π is denoted by $HAL(\Pi)$. The family of all array languages $HAL(\Pi)$ generated by systems Π as above, with at most m membranes with prescribed teams of context - free rules is denoted by $HAP_m(PTCF)$.

3.1 Example

Consider the Array P systems with hybrid teams

$$\Pi_1 = (\{S, U, R\}, \{a\}, \#, [_1[_2[_3]_3]_2]_1, S, \emptyset, \emptyset, R_1, R_2, R_3, 3)$$

where $R_1 = (Q_1, t)_{in}$, $R_2 = \{(Q_2, *)_{here}, (Q_3, t)_{in}\}$, $R_3 = \emptyset$,

$Q_1 = \{P_1\}$, $Q_2 = \{P_2, P_3\}$, $Q_3 = \{P_4, P_5\}$,

$$P_1 = \left\{ \begin{array}{cc} \# & U \\ & \rightarrow \\ S\# & a\ R \end{array} \right\}, P_2 = \left\{ \begin{array}{cc} \# & U \\ & \rightarrow \\ U & a \end{array} \right\},$$

$P_3 = \{R\# \rightarrow aR\}, P_4 = \{U \rightarrow a\}, P_5 = \{R \rightarrow a\}$

The axiom array is initially in the region 1 and the other regions do not have objects. An application of the rules in the team Q_1 with the derivation mode t yields $\begin{array}{cc} U & \\ a & R \end{array}$ and is sent to region 2. In region 2, if the first team Q_2 is applied with the derivation mode *, the array grows equal number of columns to the right of the array and equal number of rows upwards and it remains in region 2 and the process can be repeated. If the second team with the target indication *in* is applied with the derivation mode t, L-shaped angles with equal arms, the length of each arm being at least three are generated and sent to region 3. The picture language generated by Π, consists of rectangular arrays of all right angles (Token L) as depicted in Figure 1.

$$a$$
$$a$$
$$a$$
$$a\ a\ a\ a$$

Figure 1. Array Describing Token L

3.2 Example

Consider the Array P systems with hybrid teams

$$\Pi_2 = (\{S_1, S_2, A_1, A_2, A_3, A_4\}, \{a\}, \#, [_1[_2[_3[_4]_4]_3]_2]_1, S_1, \emptyset, \emptyset, \emptyset, R_1, R_2, R_3, R_4, 4)$$

where $R_1 = (Q_1, t)_{in}$,

$$R_2 = \{(Q_2, *)_{here}, (Q_2, *)_{in}\},$$

$$R_3 = \{(Q_3, t)_{here}, (Q_3, t)_{in}\}, R_4 = \emptyset,$$

$$Q_1 = \{P_1\}, \quad Q_2 = \{P_2\}, \quad Q_3 = \{P_3, P_4, P_5, P_6\}$$

$$P_1 = \left\{ \begin{array}{cc} \# & A_1 \\ S_1 \# \to & a \ S_2 \\ \# & A_2 \end{array} \right\},$$

$$P_2 = \left\{ \begin{array}{cc} \# & A_3 \\ S_2 \# \to a \ S_2 , S2 \to a \ a \\ \# & A_4 \end{array} \right\},$$

$$P_3 = \left\{ \begin{array}{cc} \# & A_1 \\ \to & , A_1 \to a \\ A_1 & a \end{array} \right\},$$

$$P_4 = \left\{ \begin{array}{cc} A_2 & a \\ \to & , A_2 \to a \\ \# & A_2 \end{array} \right\},$$

$$P_5 = \left\{ \begin{array}{cc} \# & A_3 \\ \to & , A_3 \to a \\ A_3 & a \end{array} \right\}$$

$$P_6 = \left\{ \begin{array}{cc} A_4 & a \\ \to & , A_4 \to a \\ \# & A_4 \end{array} \right\}$$

The axiom array is initially in the region 1 and the other regions do not have objects. The array generated an application of the rules in the team Q_1, with the derivation mode t is sent to region 2. In region 2, the rules in the team Q_2 with the derivation mode * is applied and the process is repeated as the target attached to the rule is *here*. The generated array is sent to inner region 3. In region 4, if the first rule in the team Q_3 is applied in the t-mode, one pixel is grown upwards and downwards both in the right side and left side. The process is repeated as the target attached to the rule is *here*. If the second rules in the team Q_3, i.e., P_3, P_4, P_5 and P_6 are applied in parallel, then the array of H shape is obtained and sent to the inner region 4.

If the final array productions $A_i \rightarrow a$ are not applied synchronously taken from $P_{k+i}, 1 \leq i \leq 2$, then the computation in region 3 is blocked without any possibility to yield a terminal array any more.

The picture language generated by Π_2 consists of H shapes with the horizontal line at the middle of the vertical ones as in Figure 2.

$$
\begin{array}{llll}
a & & & a \\
a & & & a \\
a & a & a & a \\
a & & & a \\
a & & & a \\
\end{array}
$$

Figure 2. Array describing pattern H

4 Generative Power

Theorem 1. *The set of all solid rectangles of size $n \times m$ with $n, m \geq 2$ are generated by the Array P systems with hybrid teams of degree 3.*

Proof. Let $\Pi_4 = (\{S, A, A', a\}, \{a\}, \#, [_1 [_2 [_3]_3]_2]_1, S, \emptyset, \emptyset, R_1, R_2, R_3, 3)$

where $R_1 = \{(Q_1, *)_{here}, (Q_1, *)_{in}\}$,

$R_2 = \{(Q_2, \geq k)_{in}, \quad R_3 = \{(Q_3, t)_{here},$

$Q_1 = \{P_1\}, \quad Q_2 = \{P_2, P_3\}, \quad Q_3 = \{P_4\}$

$P_1 = \{S\# \rightarrow AS\}, \quad P_2 = \{S\# \rightarrow AA\},$

$$
P_3 = \left\{ \begin{array}{cc} A & a \\ & \rightarrow \\ \# & A' \end{array} \right\}, \quad P_4 = \{A' \rightarrow a\},
$$

$L(\Pi_3)$ is the set of all solid rectangles of size nxm with $n, m \geq 2$ as depicted in Figure 3.

$$
\begin{array}{l}
a\,a\,a \cdots a \\
a\,a\,a \cdots a \\
\vdots \quad \cdots \\
\vdots \quad \cdots \\
a\,a\,a \cdots a \\
a\,a\,a \cdots a \\
\end{array}
$$

Figure 3. Rectangles of size nxm

Theorem 2. *The classes HAP_3 and PAP_3 [10] have non-empty intersection.*

Proof. Parallel array P system (PAP) has been introduced in [10] and in this system the regions have rectangular array objects and tables of context-free rules. We now compare our hybrid model with PAP.
Consider the parallel array P system

$$\Pi_1 = (V, V, [_1[_2[_3]_3]_2]_1, M_0, \phi, \phi, \Im_1, \Im_2, \phi, 3), \text{ where } V = \{X, .\},$$

$$M_0 = \begin{matrix} X & . \\ X & . \\ X & X \end{matrix} \qquad \Im_1 = \{(R_1, in)\}, \qquad \Im_2 = \{(U, out), (R_2, in)\}$$

$$R_1 = \{X \rightarrow XX, \quad . \rightarrow ..\},$$

$$R_2 = \{X \rightarrow X, \quad . \rightarrow .\} \text{ are right tables.}$$

$$U = \{X \rightarrow \begin{matrix} X \\ X \end{matrix}, \quad . \rightarrow \begin{matrix} . \\ . \end{matrix}\} \text{ is an up table.}$$

The axiom rectangular array M_0 is initially in the region 1. When the rules of the table R_1 are applied to this array it grows one column in the right and the generated array M_1 is sent to region 2. If R_2 is applied, then this array is sent to region 3 where it remains forever and the language collects this array. If U is applied to M_1 in region 2 the array grows upwards and is sent back to region 1. The derivation then continues. The array language generated consists of arrays of the form in Figure 4 where the array represents token L (. is represented as blank) with equal "arms".

$$\begin{matrix} X & . & . & . & . \\ X & . & . & . & . \\ X & . & . & . & . \\ X & . & . & . & . \\ X & X & X & X & X \end{matrix}$$

Figure 4. Array describing token L.

This language also can be generated by HAP_3 as in Example 3.1

Theorem 3. *The family HAP_4 intersects with the family $(R:RIR)SML.$[9]*

Proof. we now compare our model with the model (R:RIR)SML introduced in [9], wherein Siromoney array grammars are studied endowed with the notions of in-dexed nonterminals and indexed production.

Let $G = (G_1, G_2)$ be the $(R : RIR)SMG$ where

$$G_1 = \{\{S, A\}, \{S_1, S_2\}, \{S \rightarrow S_1 A, A \rightarrow S_2, S_1\}, S\}$$

generating strings of intermediates $S_1 S_2^n S_1$ for $n \geq 1$ and $G_2 = (G_{21}, G_{22})$ where

$$G_{21} = \{\{S_1, A_1, A_2, A_3\}, \{x\}, \{g_1, g_2\}, P_{21}, S_1\} \quad \text{with}$$

$$P_{21} = \{S_1 \rightarrow xA_1 g_2, A_1 \rightarrow A_2 g_1, A_1 \rightarrow xA_1 g_1, A_2 g_1 \rightarrow xA_3, A_3 g_1 \rightarrow xA_3, A_3 g_2 \rightarrow x\},$$

$$G_{22} = (\{S_2, B_1, B_2, B_3, B_4\}, \{., x\}, \{f_1, f_2\}, P_{22}, S_2) \quad \text{with}$$

generates the token H of x' s with the horizontal row of x's exactly in the middle which can also be generated by a HAP_4 as in Example 3.2

5 Applications

The Array P systems with Hybrid Teams works as syntactic analysing mechanism for pixel images of patterns like characters. In examples 3.1 and 3.2 the characters like L and H are generated. Now we examine the notion of generating the arrays of token T by a hybrid array P system of degree 3.

Consider the Array P systems with hybrid teams

$$\Pi_3 = (\{S, A, B, a\}, \{a\}, \#, [_1[_2[_3]_3]_2]_1, S, \emptyset, \emptyset, R_1, R_2, R_3, 3)$$

where $R_1 = (Q_1, t)_{in}, \quad R_2 = \{(Q_2, *)_{here}, (Q_2, *)_{in}\}, \quad R_3 = \{(Q_3, t)_{here},$

$Q_1 = \{P_1\}, \quad Q_2 = \{P_2, P_3, P_4\}, \quad Q_3 = \{P_5, P_6\}$

$$P_1 = \left\{ \begin{array}{cc} \# S \# & A\, a\, A \\ & \rightarrow \\ \# & B \end{array} \right\}, \quad P_2 = \{\#A \rightarrow Aa\}, \quad P_3 = \{A\# \rightarrow aA\},$$

$$P_4 = \left\{ \begin{array}{cc} B & a \\ \rightarrow \\ \# & B \end{array} \right\}, \quad P_5 = \{A \rightarrow a\}, \quad P_6 = \{B \rightarrow a\}$$

This Array P systems with hybrid teams Π_3 generates a language consisting of arrays in the shape of token T as shown in Figure 5.

$$a\, a\, a\, a\, a\, a\, a$$
$$a$$
$$a$$
$$a$$

Figure 5. Array Describing Token T

6 Conclusion

In this paper we have proposed a new generative model for picture arrays called Array P systems with hybrid teams. The generative power is compared with other models of picture description. The recognition of specific patterns like characters are generated by the Array P systems with Hybrid Teams. It is worth examining further properties of the system.

References

1. Ceterchi R., Mutyam M.,Paun Gh., Subramanian K.G.: *Array Rewriting P Systems,* Natural Computing **2**,229-249(2003).
2. Csuhaj-Varj E., Dassow J., Kelemen J., Paun Gh.: *Grammar Systems: A grammatical approach to distribution and cooperation.*, Gordon and Breach Science Publishers, Topics in Computer Mathematics 5, Yverdon (1994).
3. Dassow J., Freund R., Paun Gh. : *Cooperating array grammar system*, Int. J. of Pattern Recognition and Artificial Intelligence **9**, 1-25(1995).
4. Fernau H., Freund R.: *Bounded Parallelism in Array Grammars Used for Character Recognition*, In : Perner P., Wang P., Rosenfeld A. (eds.), Advances in Structural and Syntactical Pattern Recognition (Proceedings of the SSPR'96), pp.40-49. Springer,Berlin **1121**(1996).
5. Kari L., Mateescu A. , Paun GH. Salomaa A.: *Teams in cooperating grammar systems*, J.Exper. Th. AI, **7**,347-359(1995).
6. Maurice H. ter Beek: *Teams in grammar systems: hybridity and weak rewriting*, Acta Cybernetica **12**,427-444(1996).
7. Paun Gh., Rozenberg G.: *Prescribed teams of grammars*, Acta Informatica, **31**,525-537(1994).
8. Paun Gh: *Membrane Computing: An introduction* , Springer Verlag, Berlin , Heidelberg (2002)
9. Subramanian K.G., Revathi L., Siromoney R.: *Siromoney Array Grammars and Applications*, Int. Journal of Pattern Recognition and Artificial Intelligence, **3**, 333 - 351(1989).
10. Subramanian K.G., Saravanan R., Geethalakshmi M., Helen Chandra P., Margenstern M.: *P systems with array objects and array rewriting rules*, Progress in Natural Science, **17**, 479 - 485(2007).

An Approach for the Ordering of Evaluation of Objectives in Multiobjective Optimization

Preeti Gupta, Sanghamitra Bandyopadhyay and Ujjwal Maulik

Abstract The computational complexity of the multiobjective optimization (MOO) increases drastically in the presence of the large number of objectives. It is desirable to lower the complexity of the existing MOO algorithms. In this work we present an algorithm which periodically rearranges the objectives in the objective set such that the conflicting objectives are evaluated and compared earlier than non-conflicting ones. Differential Evolution (DE) is used as the underlying search technique. DE is designed especially for the real optimization problems. We have studied the reduction in the number of function computations and timing requirement achieved with the proposed technique. Remarkably, it is found to be much reduced as compared to the traditional approach. The variation of the gain in the number of objective computations vis-a-vis the number of objectives is demonstrated for a large number of benchmark MOO problems. Additionally, the relationship between the frequency of reordering the objectives and the number of objective computations is also established experimentally.

Keywords: Multiobjective Optimization, Manyobjective Optimization, Non-Domination, Differential Evolution.

Preeti Gupta
Machine Intelligence Unit, Indian Statistical Institute, Kolkata-700108 e-mail: preeti.miu@gmail.com

Sanghamitra Bandyopadhyay
Machine Intelligence Unit, Indian Statistical Institute, Kolkata-700108 e-mail: sanghami@isical.ac.in

Ujjwal Maulik
Department of Computer Science and Engineering, Jadavpur University, Kolkata e-mail: umaulik@cse.jdvu.ac.in

J. C. Bansal et al. (eds.), *Proceedings of Seventh International Conference on Bio-Inspired Computing: Theories and Applications (BIC-TA 2012)*, Advances in Intelligent Systems and Computing 201, DOI: 10.1007/978-81-322-1038-2_22, © Springer India 2013

1 Introduction

Evolutionary algorithms are more popular than traditional operations research methods to solve multiobjective optimization problems (MOO) as they have the advantage of finding more than one optimal solution in a single run of the algorithm and they are also not affected by the nature of the functions (such as continuity, convexity, differentiability, etc.). A detailed discussion about MOO can be found in [5, 6]. Several successful MOO techniques have been developed in the past. However, these do not scale well with the number of objectives. In recent times, a significant attention is being paid to many-objective optimization (MaOO) problems [14, 9, 10]. However, MaOO problems are computationally more expensive than MOO problems. In the literature several techniques exists for solving the MaOO problems [4, 8, 11, 12]. It is desirable to devise new strategies which can be incorporated in the existing MOO algorithms so that they can execute faster when the number of objectives is large.

Differential Evolution (DE) is a new approach belonging to the class of evolutionary search algorithms [16]. It has been appreciated for its speed, robustness and ability to search continuous spaces. DE exists in many variants such as DE/rand/1/bin, DE/rand/1/exp, DE/best/1/bin, DE/best/1/exp and many other variants, of which the first one is most popular. This is the variant used in the present work as well. A detailed discussion about DE and its various variants can be found in [13]. In the literature various multiobjective differential evolution algorithm (MODE) also exists [1, 2]. However, the present work is based on the Price and Storn's MODE proposed in [13]. This algorithm is based on the sequential function evaluation approach. The major disadvantage of this method is that the objectives are evaluated in any arbitrary order. Moreover, the ordering of objectives in the objective set remains fixed during the entire search.

Our proposed strategy is based on the observation: If two solutions are non-dominated to each other in a subset of the objectives then they remain non-dominated over the complete set of objectives. Additionally, incomparability relation between the two solutions can be detected earlier if conflicting objectives are computed first and compared. Therefore, the identification of conflicting objectives is important. In the worst case, the determination of non-dominance relation may require computation of the complete objective set. In contrast, the best case requires the comparison with respect to two objectives to establish non-dominance relationship between the two solutions. It is essential to provide an ordering on the evaluation of objectives such that the conflicting objectives appear before the non-conflicting ones.

This paper is organized as follows: In Section 2 the formulation of the multiobjective opimization is provided and the related concepts are described in brief. In Section 3, we have provided the pseudocode of Price and Storn's MODE algorithm and its discussion. In Section 4, the technique COA (correlation based ordering) is proposed and discussed. In the same section multiobjective differential evolution algorithm with correlation based ordering is proposed. It is hereafter, referred to as MODECO. In Section 5, the experimental results are provided. The parameter settings, test problems and the performance measures used are also described. In Section 6 final conclusion is provided.

2 Basic Preliminaries

The formulation of the multiobjective optimization problem is provided below:

$$F(\vec{X}) = \{f_1(\vec{X}), f_2(\vec{X}), \dots, f_m(\vec{X})\} \tag{1}$$

$$g_i(\vec{X}) \leq 0 \tag{2}$$

$$\vec{X} = [x_1, x_2, \dots, x_n]^T \in R^n \tag{3}$$

where, m is the number of objectives, n is the number of decision variables and p represents the number of constraints. Set of vectors satisfying the above equations are known as feasible solutions and the problem is known as constrained multi-objective optimization problem. In the absence of constraints the same problem is referred to as unconstrained multiobjective optimization problem. In mutiobjective optimization optimal solution is not unique but there are several optimal solutions known as Pareto optimal solutions. A decision vector dominates the other if the first improves in at least one objective without getting worse in the other objectives. Two decision vectors are said to be non-dominated if neither of them dominates each other. Pareto optimal solution set in decision space consists of non-dominated solutions and there exists no feasible solution which dominates any Pareto optimal solution. Region described by the Pareto optimal solutions in the objective space is known as Pareto optimal front

3 Price and Storn's Multiobjective Differential Evolution Algorithm

In this section we describe the MODE proposed by Price and Storn in [13]. The pseudo code is formalized in Algorithm 1. This model is based on the sequential function evaluation approach. The parameters of the search and optimization problem are real. A population of solutions of size *population_size* is maintained which is initialized randomly. The *CR* represents the crossover rate while F represents the mutation factor. In each iteration and for each solution in the population, referred to as the parent solution, a trial vector is constructed using other solution vectors existing in the population through various techniques. In Algorithm 1 sol_{r1}, sol_{r2} and sol_{r3} represents the three distinct decision vectors selected to produce the trial solution . The notation $sol_{i,j}$ is used to represent the jth component of the ith solution. Objectives corresponding to the trial vector are evaluated and compared with the parent sequentially. At any stage if the objective of the trial soution becomes worse with respect to the parent's objective, the trial is rejected without computing any of its objectives further. The trial solution is accepted if and only if it dominates the parent solution, in which case all its objectives have been computed. The aforementioned described procedure is performed for the prescribed number of generations, G_Max.

Algorithm 1 Price and Storn's Multiobjective Differential Evolution Algorithm with DE/rand/1/bin Strategy

---

1: Generation = 1.
2: Initialize the population randomly.
3: Evaluate the objective values corresponding to each solution candidate.
4: **for** $(G = 1$ to $G\_Max)$ **do**
5:    **for** $(i = 1$ to $population\_size)$ **do**
6:       Select three distinct integers $r1$, $r2$, $r3$ randomly such that $i \neq r1 \neq r2 \neq r3$ and each $r_i \in [1, population\_size]$.
7:       Generate $j_{rand} \in (1, n)$ randomly, where n is the dimensionality of decision space.
8:       Trial vector is constructed in the following manner:
9:       **if** $(rand(0, 1) \leq CR$ OR $j = j_{rand})$ **then**
10:          $trial_{i,j} = sol_{r3,j} + F * (sol_{r2,j} - sol_{r1,j})$.
11:       **else**
12:          $trial_{i,j} = parent_{i,j}$.
13:       **end if**
14:       **for** $(k = 1$ to $m)$ **do**
15:          Compare parent's and trial's k^{th} objective value.
16:          **if** (trial's k^{th} objective is worse than parent's k^{th} objective) **then**
17:             Reject trial and exit for loop.
18:          **end if**
19:       **end for**
20:       **if** ($(k = m + 1)$ and (trial dominates the parent)) **then**
21:          Accept trial.
22:       **else**
23:          Accept parent.
24:       **end if**
25:    **end for**
26: **end for**
27: From the final population determine non-dominated solutions and output these solutions.

---

4 Proposed Work

First, we formalize the concept that determines the non-dominance as a subset property because of which this can be identified with only a few objective computations. Next, we propose our algorithm for correlation based ordering of objectives (COA) and provide its detailed discussion. Finally, MODECO is proposed which integrates the proposed COA procedure in Price and Storn's MODE

K-Subset Non-dominance Principle: Let F represent the complete objective set of cardinality m and F' represents its subset. If two decision vectors are non-dominated with respect to F' then they remain non-dominated with respect to F; where $2 \leq |F'| \leq m$.

It is to be noted the worst case requires m objective computations and comparisons for the determination of non-domination relationship. However, in the best case non-dominance relationship can be determined with respect to two objectives only. The proposed algorithm, MODECO is designed such that the determination of non-dominance requires fewer computations than its base version, Price and Stron's DE.

4.1 Correlation Coefficient based Ordering Algorithm (COA)

Here, we describe the working procedure of the proposed COA algorithm. The input to COA procedure is a set of k objective vectors. From the set of objectives, any one is picked randomly (say objective i). The correlation coefficient of all objectives with the ith objective is calculated over the k objective vectors. The objectives are ordered in ascending order of their correlation-coefficient with respect to i. It is apparent that objectives which are more conflicting appear first than those which are non-conflicting with respect to the ith objective. This procedure is formalized in Algorithm 2.

Algorithm 2 The COA procedure

INPUT: The matrix F of dimension $k \times m$; where k is the number of objective vectors and m is the dimensionality of the objective space.

Output: A rearrangement of the objectives in objective set stored in the array $A[1..m]$ such $A[i]$ represents i^{th} objective to be computed.

The Steps of the Algorithm are:

1: Select the first objective randomly (say i^{th} objective is chosen)
2: **for** ($j = 1$ to m) **do**
3:    **if** ($j \neq i$) **then**
4:       Calculate the correlation coefficient between the ith and j^{th} objective.
5:    **end if**
6: **end for**
7: Rearrange the objectives in objective set as follows:
8: $A[1] = i$.
9: Sort the objectives in ascending order according to their correlation-coefficient with respect to i and store in the array A.

4.2 Multiobjective Differential Evolution with Correlation Coefficient based Ordering Algorithm (MODECO)

The MODECO is a faster version of Price and Storn's MODE, and is formalized in Algorithm 3. Initially the Price and Storn's MODE procedure is allowed to execute for a user defined fixed g number of generations. The set of objective vectors corresponding to the population of solutions obtained after g generations is provided as input to the COA procedure. The output of COA procedure is a new arrangement of the objectives in the objective set. In the subsequent g number of generations the objectives are evaluated sequentially in accordance with the new arrangement. In MODECO whenever parent and trial vector are incomparable the trial vector may be rejected in a fewer number of objective computations than required in Price and Storn's MODE. However, the decision to reject the trial can be taken after just a single comparison in the best case. That is, if after one comparison between trial and parent, the latter turns out to be better, that trial should be discarded, since it can only be either dominated or be non-dominated with respect to the parent. In MODECO the COA procedure is executed after every g generations. This ensures that existing arrangement of objectives within the objective set remains consistent with the change in conflict between objectives. There exists a trade-off between the frequency of executing COA and the number of objective computations. If COA is executed more frequently, the cost of COA may exceed savings with respect to running time, achieved by reducing the number of objective computations. If COA is executed less frequently then the ordering of the objectives may become inconsistent with their conflict status leading to a larger number of objective computations. In a part of this article we study the relative performance of the proposed algorithm vis-a-vis the frequency of executing COA.

Algorithm 3 MODECO Algorithm

1: Initialize the population randomly.
2: **for** $(G = 1$ to $G\_MAX)$ **do**
3:     **for** $(i = 1$ to $population\_size)$ **do**
4:         Generate trial vector using DE/rand/1/bin strategy.
5:         Compute the objectives corresponding to trial sequentially.
6:         **if** $(g$ is a divisor of $G)$ **then**
7:             Execute COA.
8:         **end if**
9:     **end for**
10: **end for**
11: From the final population determine non-dominated solutions and output these solutions.

---

5 Experimental Studies

First, parameter settings, test problem and performance metrics used are described. Thereafter, the two different studies conducted are enumerated.

5.1 Parameter Settings

The parameter values used are provided below.

Parameter	Value
Population Size	200
Number of Generations	500
Crossover Rate	0.8
Mutation Rate	0.2

5.2 Multiobjective Test Problems

In this work the three test problems namely DTLZ1, DTLZ3, DTLZ5 [7] are used. For all the three test problems we have varied the dimensions as: 3, 5, 7 and 10. The study 1 utilizes all the three test problems while for the study 2 only DTLZ3 is utilized.

5.3 Performance Measures

Various performance measures have been suggested in the literature which evaluate the goodness of any approximated Pareto optimal set achieved with the help of any multiobjective optimizer. There is no such unique metric which alone is the indicator of all the features of a multiobjective optimizer. Purity [3] and spacing [15] are two such measures which are used to establish the similarity of the final output obtained by MODECO and Price and Storn's DE. The purity metric is used to measure the convergence. To use this measure the non-dominated solutions of two or more multiobjective optimizers is combined. This combined set is further used to determine the number of non-dominated solutions in it. Finally, purity of an algorithm is the percentage of non-dominated solutions contributed by it in this set. The spacing metric measures the uniformity of the spread of the solutions along the entire stretch of the approximated Pareto front of the algorithm. This measure is applied individually on the output of each algorithm.

5.4 Study 1: Number of Objective Computations and Timing Comparison

MODECO and MODE, are executed independently for each test problem using different seed values for 100 runs. Results reported are the average values over the hundred runs. The COA procedure is executed after every five generations in MODECO in all the instances for conducting this study. The details can be found in Table 1, Table 2 and Table 3. Additionally, the percentage of the reduced objective computations achieved with MODECO over MODE is investigated with respect to the dimensionality of the objective space. Obtained results are illustrated in figure 1.

Tables 1 to 3 shows the comparative performance of MODE and the proposed MODECO with respect to the running time and the number of function evaluations for DTLZ1, DTLZ3 and DTLZ5 respectively. As can be seen from the tables, the running time of MODECO is in general smaller than the running time of MODE while the number of function evaluations is consistently less for the former. The only exception is for DTLZ3 for three objectives, where the timing requirement is slightly higher as compared with MODE, although the number of function evaluations is still smaller. The reason for this appears to be that the overhead due to COA overcomes the time gain obtained because of reduced number of function evaluations. As can be seen from the figure, as the number of objectives increases, the percentage time gain also increases. This indicates the MODECO is especially useful when the number of objectives is large (more than three). It is to be noted that this time gain is obtained not at the cost of performance. Purity value of all the test functions in the all the cases is exactly one. At the same time the spacing value is also the exactly same. This can be explained as the both multiobjective optimizers, Price and Storn's MODE and MODECO, produce exactly similar outputs in all the cases.

Table 1 Average Number of Function Evaluations and Running Time Corresponding to DTLZ1

Number of Objectives	Price and Storn's MODE		MODECO	
	Running Time (sec)	Function Evaluations	Running Time (sec)	Function Evaluations
	Mean ± SD	Mean ± SD	Mean ± SD	Mean ± SD
3	0.080300±0.00003	153490±180	0.079700±0.000030	131418±071
5	0.091200±0.001126	209564±205	0.080500±0.00005	147104±111
7	0.108100±0.000191	261790±533	0.087300±0.000271	147244±074
10	0.134400±0.000563	342976±134	0.098400±0.000161	157790±152

5.5 Study 2: Frequency of Applying COA

The main objective of this study is to investigate the relationship between the reduction in the number of objective computations vis-a-vis the frequency of executing the COA procedure in MODECO. The COA procedure is executed in each instance

Table 2 Average Number of Function Evaluations and Running Time Corresponding to DTLZ3

Number of Objectives	Price and Storn's MODE		MODECO	
	Running Time (sec)	Function Evaluations	Running Time (sec)	Function Evaluations
	Mean ± SD	Mean ± SD	Mean ± SD	Mean ± SD
3	0.121300±0.000874	144059±142	0.121500±0.000854	136045±008
5	0.158300±0.000171	184971±97	0.140000±0.000101	143141±138
7	0.189300±0.001075	221740±936	0.1483000±0.000171	148460±010
10	0.240600±0.001065	276957±380	0.173600±0.000362	153809±040

Table 3 Average Number of Function Evaluations and Running Time Corresponding to DTLZ5

Number of Objectives	Price and Storn's MODE		MODECO	
	Running Time (sec)	Function Evaluations	Running Time (sec)	Function Evaluations
	Mean ± SD	Mean ± SD	Mean ± SD	Mean ± SD
3	0.068800±0.000121	183465±4	0.068300±0.000171	157025±021
5	0.093600±0.000643	225955± 9	0.079900±0.001015	158131±245
7	0.114000±0.000603	271780±89	0.092600±0.000261	168644±045
10	0.169200±0.000925	342079±307	0.115700±0.000432	170750±013

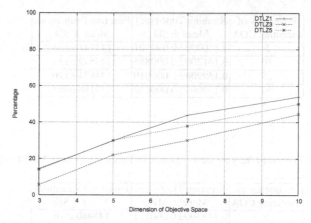

Fig. 1 Plot of percentage reduction in the number of objective computations of MODECO versus the dimensionality of the objective space

of MODECO after 5, 10, 15 and 50 generations. The running time and the number of objective computations are reported in Tables 4, 5, 6 and 7 for DTLZ3 test problem of the dimensions 3, 5, 7 and 10 respectively. Values reported are the average value obtained over 100 runs.

Tables 4 to 7 depict the trade-off between the frequency of executing the COA procedure in MODECO and the number of objective computations. The best values with respect to running time and number of objectives is highlighted using boldface. It can be noted that the objective computations is minimum in all the instances when the rate of executing the COA is maximum. This indicates the conflict information amongst the objectives changes as the search progress and hence, needs to be relearned periodically. However, the running time is minimum in all instances when the COA is executed after every 25 generations. This can be explained as the overhead due to COA exceeds the reduced objective computations.

Table 4 Results Obtained by Varying the Frequency of Executing COA in MODECO for DTLZ3 of Dimension 3

Frequency of executing COA	Running Time (sec) Mean ± SD	Function Evaluations Mean ± SD
5	0.121500±0.000854	**136045**±08
10	0.120500± 0.000955	136231±21
25	**0.119700**± 0.000300	136610±16
50	0.120000± 0.001005	136924±05

Table 5 Results Obtained by Varying the Frequency of Executing COA in MODECO for DTLZ3 of Dimension 5

Frequency of executing COA	Running Time (sec) Mean ± SD	Function Evaluations Mean ± SD
5	0.140000±0.000101	**143141**±138
10	0.141200±0.000884	143549±134
25	**0.139900**±0.000010	144874±110
50	0.139600±0.000040	147188±083

Table 6 Results Obtained by Varying the Frequency of Executing COA in MODECO for DTLZ3 of Dimension 7

Frequency of executing COA	Running Time (sec) Mean ± SD	Function Evaluations Mean ± SD
5	0.148300±0.000171	**148460**± 10
10	0.142800±0.001286	149203±117
25	**0.141800**±.000824	151560±101
50	0.146100±0.000392	155559±29

Table 7 Results Obtained by Varying the Frequency of Executing COA in MODECO for DTLZ3 of Dimension 10

Frequency of executing COA	Running Time (sec) Mean ± SD	Function Evaluations Mean ± SD
5	0.173600±0.000362	**153809**±40
10	0.177100±0.000291	154406±51
25	**0.171700**±0.016251	157812±75
50	0.174000±0.000603	158904±16

6 Conclusions and Future Work

In this paper, we have presented MODECO, a multiobjective DE algorithm integrated with correlation based ordering of objectives (COA). Additionally, we make a simple but interesting observation that if two solutions vectors are non-dominated with each other in a small subset of the objectives, then they remain non-dominated even when all the objectives are compared (Note that this need not be true for domination relationship). The proposed algorithm is able to reduce the number of objective computations significantly. It is to be noted that the gain in the computational time and the function computations is achieved without any loss of performance as the output produced by both the algorithms, Price and Storn's DE and MODECO is identical. It is established experimentally that the percentage reduction in the objective computations increases with the increase in the number of objectives. Moreover, the relationship between the rate of executing COA and the number of computations is established experimentally. The main motivation of the present article is to demonstrate the effectiveness of using COA for reducing the objective computations. In the future new techniques of ordering and incorporation of diversity preservation mechanism in MODECO will be performed.

Acknowledgments

The authors acknowledge support from Department of Science and Technology, Government of India for Indo-Mexico project (DST/INT/MEX/RPO/04/08). The first author also acknowledges Council of Scientific and Industrial Research, Government of India for providing senior research fellowship (File No: 9/93 (0/23)/2010 EMR-I).

References

1. Hussein A. Abbas, Rahul Sarker, and Charles Newton. PDE: A Pareto-frontier Differential Evolution Approach for Multi-objective Optimization Problems. *In Proceedings of the Congress on Evolutionary Computation 2001 (CEC'2001)*, pages 971–978, Piscataway, New Jersey, (2001).
2. B. V. Babu, M. Mathew, and Leenus Jehan. Differential Evolution For Multiobjective Optimisation. *In Proceedings of the 2003 Congress on Evolutionary Computation(CEC'2003)*, pages 2696–2703, Canberra,Australia, (2003).
3. S. Bandyopadhyay, S. K. Pal, and B. Aruna. Multi-objective GAs, quantitative indices and pattern classification. *IEEE Transactions on Systems, Man and Cybernatics-B*, 34(5):2088–2099, (2004).
4. D. Brockhoff and E. Zitzler. Objective Reduction in Evolutionary Multiobjective Optimization: Theory and Applications. *Evolutionary Computation*, 17(2):135–166, (2009).
5. C. Coello, D. V. Veldhuizen, and G. Lamont. *Evolutionary Algorithms for Solving Multi-Objective Problems*. Wiley, (2002).

6. K. Deb. *Multiobjective Optimization Using Evolutionary Algorithms*. Wiley, (2001).
7. Kalyanmoy Deb, Amrit Pratap, Sameer Agarwal, and T. Meyariavan. Scalable Test Problems for Evolutionary Multiobjective Optimization. In *IEEE Congress on Evolutionary Computation*, pages 825–830, Honolulu, May 12-17 (2002).
8. M. Farina and P. Amato. A fuzzy definition of "optimality" for many-criteria optimization problems. *IEEE Transactions on Systems, Man, and Cybernetics Part A—Systems and Humans*, 34(3):315–326, May (2004).
9. P. J. Fleming, R. C. Purhouse, and R. J .Lygoe. Many objective optimization: An engineering design perspective. In *lecture Notes in Computer Science 3410: Evolutionary Multi-Criterion Optimisation-EMO*, pages 14–32. Springer, Berlin, March (2005).
10. E. J. Hughes. Evolutionary many objective optimization: Many once or one many ? In *IEEE Congress on Evolutionary Computation*, pages 222–227, Edinburgh, September 2-5 (2005).
11. Antonio López Jaimes, Hernán Aguirre, Kiyoshi Tanaka, and Carlos A. Coello Coello. Objective Space Partitioning Using Conflict Information for Many-Objective Optimization. In Robert Schaefer, Carlos Cotta, Joanna Kołodziej, and Günter Rudolph, editors, *Parallel Problem Solving from Nature–PPSN XI, 11th International Conference, Proceedings, Part I*, pages 657–666. Springer, Lecture Notes in Computer Science Vol. 6238, Kraków, Poland, September (2010).
12. Julin Molina, Luis V. Santana, Alfredo G. Hernndez-Daz, Carlos A. Coello Coello, and Rafael Caballero. g-dominance: Reference point based dominance for Multi-Objective Metaheuristics. *European Journal of Operational Research*, 197(2):685–692, September (2009).
13. K. Price, R. Storn, and J. Lampien. *Differential Evolution- A Practical Approach to Global Optimization*. Springer, Berlin, (2005).
14. R. C. Purhouse and P. J. Fleming. Evolutionary many objective optimization: An exploratory analysis. In *IEEE Congress on Evolutionary Computation*, pages 2066–2073, Canberra, December 8-12 (2003).
15. J. R. Schott. *Fault tolerant design using single and multi-criteria genetic algorithms*. PhD thesis, (1995).
16. R. Storn and K. Price. Differential Evolution-a Simple and Efficient Heuristic for Global Optimization over Continuous Spaces. *Journal of Global Optimization, Kluwer Academic Publishers*, 11:341–359, (1997).

Extended Forma : Analysis and An Operator Exploiting it

Dharmani Bhaveshkumar C.

Dhirubhai Ambani Institute of Information & Communication Technology (DAIICT),
Gandhinagar, Gujarat, 382007, India
dharmanibc@gmail.com

Abstract. There exists a long discussion over the issue of whether minimal alphabet or maximal alphabet gives maximum schemata. The article generalizes the concept of schemata to dependency relation based 'extended formae'. Further, it proves that theoretical maximum schemata could be achieved through an operator exploiting extended formae and using maximal alphabet. It shows that the previous conclusion of minimal alphabet giving maximum schemata is also true for some operators. It is known that maximum schemata is advantageous to achieve maximum implicit parallelism. The article raises a discussion over the requirement of availing maximum schemata by showing some disadvantages also. As a conclusion, it suggests to use an intermediate level alphabet for representation, balancing maximal alphabet to avail maximum schemata and minimal alphabet to overcome the disadvantages due to maximum schemata.

Keywords: Genetic Algorithm (GA), schema, schemata, forma, formae, implicit parallelism, dependency relation, extended forma, extended formae, p-schemata, diploidy

1 Introduction

Genetic Algorithm (GA) has been fairly successful as a huiristic search and optimization technique, in many practical engineering and design problems. Towards the success of GA, Holland made a powerful observation of progress in search through prorogation (inheritance) of similarities within chromosomal representation of the solution space. To signify the importance of similarities among the chromosomes, he gave mathematical notion to it deriving the concept of Schema and Schema Theorem.

The significance of the schema concept can be understood by the generalizations it has obtained. Towards the efforts to understand real coded GA (RCGA), the schema concept got generalized as Wright's schema by Wright [12], as interval-schema by Eshelman [3] and as virtual alphabet by Goldberg [6]. Schema, as a way to define similarity among the set of chromosomes, got generalizations as forma induced by an equivalence relation by Radcliffe [9, 8], as a predicate by Vose [11] and also through work by Antonisse [1].

Though available many generalizations, the maximum schemata[1] has not achieved the theoretical maximum possible. Towards achieving maximum theoretical pos-

---

[1] Schemata - is a plural of Schema

J. C. Bansal et al. (eds.), *Proceedings of Seventh International Conference on Bio-Inspired Computing: Theories and Applications (BIC-TA 2012)*, Advances in Intelligent Systems and Computing 201, DOI: 10.1007/978-81-322-1038-2_23, © Springer India 2013

sible schemata[2] and a more successful GA, this article contributes in the following ways:

1. It provides generalization of the schemata concept through further generalization of equivalence relation based formae definition to the dependency relation based extended formae definition. This makes it possible to achieve theoretical maximum possible schemata for both string and non-string representations, which would not have been possible through formae or other previous definitions of schemata.

2. It signifies that similarity has to be an operator perspective. To be able to exploit similarity information through extended forma definition -

   (a) it proposes the use of already existing ploidy representation and derives for it a novel ploidy schemata (p-schemata) definition. Thus, it adds the notion of past similarities to define schemata.

   (b) it derives a specific Mendelian Crossover Operator to exploit the similarity information through p-schemata.

3. It proves that not the minimal but the maximal alphabet gives maximum schemata, through extended formae definition. It also proves that extended formae could achieve theoretical maximum schemata.

4. It provides discussions on the relevance of previous concept of minimal alphabet giving maximum schemata [7, 4–6] and the need of maximum schemata for efficient GA. Based on the discussions and inspiration from nature it provides a conclusion.

The next section 2 starts with the preliminary schema definitions and motivation for further generalization of schema concept. Then, the section 3 defines the extended forma induced by an arbitrary dependency relation. It compares extended forma with the existing schema definitions, in terms of the definition and available maximum schemata. Section 4, signifies the role of operator by concluding that the schemata has to be an operators perspective. Towards, achieving an operator exploiting extended formae it proposes a ploidy representation and a Mendelian crossover operator in section 5. Then after, section 6 derives the important theorems and interpretation to achieve maximum schemata. Section 7 provides discussion on the requirement of maximum schemata and then finally the section 8 provides conclusion.

2 Concept of Schemata and the Motivation for Further Generalization

Let ρ be a mapping of the n-point discrete search space S onto the space of chromosomes C.

$$\rho : S \longrightarrow C$$

Let the chromosomes be a k-ary string of length l; set A be the set of all k symbols of k-ary alphabet, $A = \{0, 1, \ldots, k-1\}$, $P(A)$ be the power set of A and $\Omega = P(A) - \{$set of all subsets of $P(A)$ with cardinality 0 and 1$\}$.

There are different ways possible to define the similarity in C.

---

[2] To avoid confusion, this article uses 'schemata' nomenclature as a more generalized form for similarity subsets through any existing (original Holland's schemata, o-schemata, formae, predicates etc.) or futuristic representation and definition; and 'Holland's schemata' as that specifically defined by Holland.

Holland's Schemata Holland [7] defined schema, S as a similarity template describing a subset of k-ary strings with fixed values, say either 0 or 1 for $k = 2$, at certain positions and 'don't care' (#) at the other positions. For example, a Holland's schema 1#00 implies a subset $\{1000, 1100\}$.

Generalization of schemata as an equivalence relation based Formae

Radcliffe [9,8] defined similarity through an arbitrary equivalence relation on $\mathcal{C}$ and identified it as the Forma. That generalized the concept of schemata to be applicable for string and non-string structures.

The similarity among a subset of chromosomes in $\mathcal{C}$ can be represented by a string of symbols ■ and □, indicating 'fixed' or 'matching' values at ■ symbol positions and 'don't care' at the □ symbol positions. There each symbol in the string was called a *component* and the string containing symbols may be named as a *relational string*. More precisely, a relation $\sim \in \Psi$ be defined for $\eta, \zeta \in \mathcal{C}$

$$\eta \sim \zeta \Leftrightarrow (\forall i \in \mathbb{Z}_l; (\sim_i = \blacksquare) : \eta_i = \zeta_i) \tag{1}$$

where, $\mathbb{Z}_l = \{1, 2, \ldots, l\}$; $\sim_i$ indicates ith component of the relational string. The properties of $=$, defines $\sim$ as an equivalence relation, satisfying the properties of symmetry, reflexivity and transitivity. An equivalence relation string may be written for example as ■□□■. A specific fixed value at ■ symbol, in a given equivalence relation string will induce equivalence classes, called formae. The formae, induced by a particular equivalence relation, will divide $\mathcal{C}$ into a disjoint partition.

Theoretical maximum possible schemata and motivation for further generalization of schemata
Each chromosome could be a member of multiple schemata. So, a fitness calculation of a chromosome implicitly imparts information about the mean fitness of the multiple regions. Holland [7] identified this phenomena as an implicit parallelism. Holland, First, inferred that schemata maximization would add to the implicit parallelism and then, he proved that minimal alphabet will provide maximum schemata[3].

The maximum possible correlated subsets among the n-point search region S (NS_{max}) would give an upper bound on the theoretical maximum possible schemata in $\mathcal{C}$ (NC_{max}).

$$NS_{max} = \binom{n}{1} + \binom{n}{2} + \ldots + \binom{n}{n} = 2^n - 1 \tag{2}$$

If ρ is a one-one mapping, $NC_{max} = NS_{max}$. Without loss of generality, this is the case assumed for rest of the article. In case of, an under-specified or over-specified representation $NC_{max} = \alpha NS_{max}$, where $\alpha = |\mathcal{C}|/|\mathcal{S}|$ and $|\cdot|$ indicates cardinality of a set.

It could be observed that the theoretical maximum possible $2^n - 1$ subsets of $\mathcal{C}$ (NC_{max}) are not disjoint. Many of theses subsets share some common elements. This implies that forma definition can not achieve all correlated subsets in S as schemata in $\mathcal{C}$ i.e. $NC_{max} \neq NS_{max}$. By discarding the transitivity condition,

---

[3] Later on in the section 7, there has been provided discussion on the requirement of maximum schemata. But, before that the maximum schemata has been assumed significant.

an equivalence relation gets generalized to dependency relation (finite tolerance relation)[4]. This will allow dependency relation based extended formae to define similarity subsets also for those schemata in $\mathcal{C}$, which are not available through forma definition. As has been proved later in Theorem 3, section 6 that this will achieve for extended formae $NC_{max} = NS_{max}$.

3   Proposed dependency relations based extended formae

Towards the goal of schema definition, the similarity in $\mathcal{C}$ can be obtained defining an arbitrary dependency relation (finite toletrence relation). Let,

$$\Psi = \{\blacksquare, \square\}^l$$

is the space of all dependency relations on $\mathcal{C}$. The related chromosome strings are having 'fixed' or 'matching' values from $\mathcal{A}$, at the $\blacksquare$ symbol positions and are having either of the elements of the specific subset $\square_{i..j}^{(p)}$ at symbol $\square$ positions, where $\square_{i..j}^{(p)} \in \Omega$. So, the $\square$ could be interpreted as 'either or' symbol reminding selection of either of the element of Ω as the specific subset. The specific subset $\square_{i..j}^{(p)}$ could be interpreted as 'p-level either or'; where $p = |\square|, |\cdot|$ indicates cardinality of a set, $1 < p \le k$ and the suffix $i..j$ is the list of all elements in that subset. For ease, $\square^{(k)}$ may be used and the suffix may be avoided to indicate 'k-level either or' or 'don't care'. For the same reason of ease, the cardinality representation as power may be avoided. Given, say $l = 4$, a relation $\sim \in \Psi$, $(\blacksquare, \square, \blacksquare, \square)$ could be represented as the relational string $\blacksquare \square \blacksquare \square$ and $\sim_i$ indicates ith component of the relational string.

More precisely, a relation $\sim \in \Psi$ be defined for $\eta, \zeta \in \mathcal{C}$ and $\mathbb{Z}_l = \{1, 2, \ldots, l\}$

$$\eta \sim \zeta \Leftrightarrow \left(\forall i \in \mathbb{Z}_l; (\sim_i = \blacksquare) : \eta_i = \zeta_i, (\sim_i = \square) : \eta_i, \zeta_i \in \square_{i..j}^{(p)} \, \square_{i..j}^{(p)} \in \Omega \right) \quad (3)$$

As all the members of Ω are not disjoint, based on the definition and properties of $=$ and $\in$ operators, the relation $\sim$ satisfies the properties of symmetry and reflexivity but not transitivity. Also, due to finite chromosome space assumed, it is a dependency relation or a finite tolerance relation.

Fixing the value of $\blacksquare$ and $\square$ at the given positions, induces the similarity subsets (schemata) for each dependency relation in Ψ. As dependency relations are generalization of equivalence relations and schemata induced by equivalence relations are called formae, the schemata induced by dependency relations could be named 'extended formae'.

The comparision of equation 3 with equation 1 and the used relational strings makes clear the difference between formae and extended formae. As the first step, let equivalence relation be defined as $\blacksquare\square^{(3)}\square^{(3)}\blacksquare$ for $k = 3, \mathcal{A} = \{0, 1, 2\}$ alphabet. Fixing the values for $\blacksquare$ symbols, will induce a disjoint subset of a partition. Let, $1\square^{(3)}\square^{(3)}2$ is one of the disjoint subset of a partition. Now, as the second step, let $\square$ be interpreted as 'either or' interpretation taking different possible values from Ω. Extending the example, a relational string $1\square_{12}\square_{01}2$ and $1\square_{01}\square_{12}2$

---
[4] Tolerance relation is a relation which is symmetric and reflexive but not necessarily transitive. Dependency relation is a finite tolerance relation.

are under consideration. It could be observed that this is like dividing the previous disjoint subset of a partition further into correlated subsets. This further division is correlated as having a non-empty intersection. The common string between the considered two correlated subsets would be $\{1112\}$. This makes clear the relation between the formae and extended formae, and also the nomenclature 'extended formae'.

For example: k=2 (a binary alphabet), then $\mathcal{P}(\mathcal{A}) = \{\{\}, \{0\}, \{1\}, \{0,1\}\}$ and $\Omega = \{\{0,1\}\}$. Accordingly, $\square = \{0,1\}$ is the only possibility. An extended forma $1\square00 = \{1000, 1100\}$

Similarly, for k=4, a 4-ary alphabet with $\mathcal{A} = \{0,1,2,3\}$, corresponding $\Omega = \{\{0,1\}, \{1,2\}, \{2,3\}, \{3,0\}, \{0,2\}, \{1,3\}, \{0,1,2\}, \{1,2,3\}, \{2,3,0\}, \{3,0,1\}, \{0,1,2,3\}\}$, Symbolically, $\Omega = \{\square_{01}, \square_{12}, \square_{23}, \square_{30}, \square_{02}, \square_{13}, \square_{012}, \square_{123}, \square_{230}, \square_{301}, \square_{0123}\}$. $\square$ could be any of these 11 sets. An extended formae $20\square_{12}1 = \{2011, 2021\}$, where $\square_{12} = \{1,2\}$.

Antonisse work [1] and comparision It must be acknowledged that the article [1] by Antonisse suggested almost similar generalization of Holland's schemata. The article interpreted the symbol # in the Holland's Schema as any subset from Ω. But, the dependency relation was not used there for schemata definition. The proposed arbitrary dependency relation based extended forma definition is applicable to both string and non-string structures. Also, that article [1] did not tried to answer from where or how these extra schemata will be available to GA. There it was almost neglected that the similarity has to be an operator perspective. Rather, in the concluding remark it hoped to achieve the extra schemata with crossover operator being intact. Compare to this, the present work, identifies the significance of an operator (section 4) and derives an operator (section 5) exploiting the extra schemata available through dependency relation.

4 Schemata and Operators

GA has to find the optimal and generally, is blind i.e. there is an absence of any information about the correlated regions in the fitness landscape. The goal of schema definition is to get subsets in the set of chromosomes, with some similarities (schemata) based on two assumptions:

1. The similarities in the set of chromosomes (representational landscape) gives possible information about the correlations in the actual fitness landscape. To validate this assumption for a wide range of fitness landscapes, the notion of schemata definition has to be based upon the most common search principles of searching through say, locality, periodicity etc. For a specific fitness landscape there could be thought upon specific notions of similarity.

2. There exists a GA operator, using this information about the similarities or dissimilarities in the set of chromosomes to direct the next generation search towards the corresponding most correlated or uncorrelated regions.

The success of GA or the used schemata definition and a corresponding representation depends upon the degree by which both the assumptions are satisfied.

Holland interpreted the used symbol # as a 'wild-card' or a 'don't care'. But, the then existing crossover operators, were interpreting the symbol as '2-level either or', as they select a specific value from either of the two parents. The assumed interpretation of 'don't care' and the actual interpretation of '2-level either or' by a crossover operator - are same only in case of binary. The difference between the interpretations is important as it will create difference in the number of schemata possible. If assumed the conventional single-point, multi-point or uniform crossover then the maximum possible schemata per position (NCP_{max}) for k-ary would be, $NCP_{max} = \binom{k}{1} + \binom{k}{2}$; which is more than $k+1$ schemata for all $k > 2$. An example of an operator, which interprets Holland's schemata as he defined i.e. 'don't care' or a 'wild card' interpretation - is the Random Respectful Recombination (R^3) Operator by Radcliffe [9]. So, for this operator NCP_{max} with k-ary symbol would be, $NCP_{max} = k+1$, as expected by Holland.

As mention earlier, Antoisse expressed schemata in a way similar to extended formae. But, then existing conventional crossover operator is an operation between two parents. So, these operators would not be able to interpret schema position with # as any subset with cardinality $|\#| > 2$ from a power set $\mathcal{P}(\mathcal{A})$. Say, for $k = 4$, a schema $20\#_{013}$ indicates the set of strings with value at not matching positions to be '3-level either or' from the subset $\{0,1,3\}$. But, the available single-point, multi-point or uniform crossover can not operate on this interpretation.

So, this section concludes that for a successful GA, similarity has to be an operator perspective or schemata should be the perception of an operator. Accordingly, there must be designed a GA operator exploiting similarity information through extended formae.

5   GA operator - exploiting extended formae

Achieving 'p-level either or' with $2 < p < k$ interpretation for $\square_{i..j}^{(p)}$ necessitates two things.

1. *Representation Perspective*: There has to be available at least p allele values at a position. This could be achieved either by using multiple parents (p parents) for recombination or using a ploidy (multiplicity of chromosomes) representation with atleat $p/2$ copies of chromosomes for both parents or multiple parents and ploidy representation both together for each parent. But, it is better to use a solution approved by nature.
2. *Operator Perspective*: There has to be an operator implementing 'p-level either or'.

5.1   Inspiration from nature for representation perspective

It is known that in nature the most complex organisms are diploid or polyploid. Humans, most animals and many plants are diploid (pairs of chromosomes). Polyploidy is also existing in few plants and animals. The ploidy structure makes it possible to store and inherit the past genetic history in terms of the unexpressed. It is one of the sources of wide diversity in nature. Also, the ploidy structures, specifically the diploidy are not new to the GA community and have already proved their significance for non-stationary or time-varying applications [10, 5]. The multi-parent recombination operators [2] have also been found useful in some applications, but are not that popular compare to ploidy structures.

Finally, there has been decided to follow nature and to use ploidy representation and corresponding operator than a multi-parent haploidy, for the further development of the topic.

5.2 Deriving p-schemata for ploidy representations

Conventionally, schemata definition is applied to get similarities among the parents or chromosomes from different individuals. In case of ploidy representations, there exists two way similarities. The first one is, within an individual among the multiple copies of chromosomes i.e. intra-parent similarity. The other one is, similarity between (among) the parental ploidy representations - inter-parent similarity. It could be observed that the intra-parent similarities and the inter-parent similarities - both could be obtained by extended forma definition.

For example, let $k = 4$, $\mathcal{A} = \{0, 1, 2, 3\}$ and ploidy structurewith four copies of chromosomes per parent is in use. Let, the parent P_1 be with chromosomes 0123, 1103, 2133, 2103 and the parent P_2 be with copies 3120, 2130, 0120, 0130. The intra-parent extended forma for parent P_1 could be written as $\square_{012}1\square_{023}3$ and that for parent P_2 could be written as $\square_{023}1\square_{23}0$. Then, inter-parent extended forma for parent P_1 and P_2 could be written as $\square_{0123}1\square_{023}\square_{0,3}$.

It is possible to give importance to the multiplicity of alleles and precedence of the chromosomes in terms of the generation in which they were expressed. If the operator needs to use these information, then instead of the used conventional set structure for $\square$; there has to be used ordered multiset [5] structure.

Overall, the application of the extended forma definition to ploidy representation for similarity subsets has been achieved. Let both the intra-parent and inter-parent schemata be given a common name 'p-schemata', where p reminds us the term ploidy. To be more precise, if required, the intra-parent schemata may specifically be named as 'p^1-schemata' and the inter-parent schemata may specifically be named as 'p^2-schemata'.

The introduction to p-schemata, introduces the notion of either past similarities or past inheritance or past successful changes in allele values or past, in general - as the search strategy to define schemata. GA using diploidy representations are existing. That means the past or the unexpressed genetic information was already in use for search in GA. But, the notion of past with the already existing notion of search through locality, periodicity etc. for schemata definition is new and may find useful.

5.3 Inspiration from nature for operator perspective

Towards the goal of deriving an operator exploiting extended formae, the first necessity of availing more than p alleles at a gene position has been achieved through deriving ploidy representation and p-schemata for it. The next step is to derive an operator implementing 'p-level either or'. Organisms with ploidy, follow largely two major inheritance mechanisms. Accordingly, there could be suggested crossover operators matching Mendelian Inheritance or Non-Mendelian Inheritance using p-schemata.

Mendelian Inheritance Though basically defined for diplody, can be made applicable to all ploidy. There are two basic principles of Mendelian inheritance, described here in the way they are understood now.

---

[5] Multiset is a generalization of set to allow repeatation of elements.

1. *Law of Segregation* : It says that during gameteeogenesis[6] through meiosis, in both the parents, the sets of chromosomes is segregated and gametes have only half of the total chromosomes.

   During fertilization, any of the gamete is randomly selected and without any exchange of genetic material just passed on to the embryo. The embryo will have full sets of chromosomes, adding half from each gamete.

   Generalizing, the inheritance does not depend upon the allele at a gene position being expressed or unexpressed. The dominancy decides what is expressed and does not affect the probability of what is inherited.

2. *Law of Independent Assortment or Inheritance Law*: This law states that all genes are passed on independent of each other to the gametes during gameteeogenesis. This says that no two genes are having linkage to get inherited always together. Practically, crossover point could be anywhere in the string of genes.

Non-Mendelian Inheritance Non-Mendelian inheritance is a collective nomenclature for inheritance mechanisms not following laws of Mendelian Inheritance. Say, importance of dominancy to the inheritance, assuming unequal probabilities for crossover points etc. could be considered Non-Mendelian.

5.4 Mendelian Crossover Operator Exploiting Extended Formae

Just to give example of an operator exploiting extended formae there is described a Mendelian Crossover Operator assuming diploidy and following Mendelian inheritance. The operator is three step:

1. During gameteeogenesis through meiosis, there will be a crossover between the two copies of chromosomes in each parent. In humans, there are created 4 haploid gametes through separate crossover for each. For ease of operation and representation, let the simulated operator be using the expressed (i.e. dominant) allele combination and the unexpressed (i.e. dormnat) allele combination as two chromosomes in a parent. Also, two crossover operations would be sufficient - one, modifying the expressed and the other, modifying the unexpressed[7].

2. During the fertilization step a child gets one copy of chromosome from a gamete from each parent randomly.

3. Then, the dominance mechanism decides the alleles which are expressed and unexpressed. In case of no knowledge about dominant alleles, there could be assumed random dominancy mechanism i.e. any allele at a gene position is equally likely to get expressed. This could be implemented using a one more crossover operator application on the available two haploids with the child.

The above operator requires three crossovers (two during gameteeogenesis and one for dependency). It could be assumed that the gene locations affected by the crossover during gameteeogenesis are the same as those affected by the crossover during dominancy decision. This assumption, can bring the same overall result through a single crossover between randomly selected one chromosome from

---

[6] The process of gamete formation, before fertilization

[7] The conventional crossover will keep the most significant gene position unaffected. So, the modified chromosome is the one with the most significant gene position allele intact.

each parent. This has been shown in table 1. Other than computational reductions, use of single crossover will also bring reduction in the population diversity. The table, specifically the last column entries for equivalent single crossover among the selected chromosomes from parents, reminds the outcomes of Mendel's experiments in terms of the probabilities of inheritance of the dominant and dormant alleles.

Table 1. Mendelian Crossover Operator

Parent p_1	Expressed Chromosome (p1e): A1B1; Unexpressed Chromosome (p1u): a1b1			
Parent p_2	Expressed Chromosome (p2e): A2B2; Unexpressed Chromosome (p2u): a2b2			
	Gametogenesis			
Parent p_1	gamete 1 (g11) : A1b1; gamete 2 (g12) : a1B1;			
Parent p_2	gamete 1 (g21) : A2b2; gamete 2 (g22) : a2B2;			
	Possible children after fertilization and dominancy mechanism			
Child	Gametes Selected for fertilization	Expressed Chromosome	Unexpressed Chromosome	Equivalent Parent Chromosomes for single crossover
c_1	g11-g21	A1b2	A2b1	p1e-p2u
c_2	g11-g22	A1B2	a2b1	p1e-p2e
c_3	g12-g21	a1b2	a2B1	p1u-p2u
c_4	g12-g22	a1B2	a2B1	p1u-p2e
c_5	g21-g11	A2b1	A1b2	p2e-p1u
c_6	g21-g22	A2B1	a1b2	p2e-p1e
c_7	g22-g21	a2b1	A1B2	p2u-p1u
c_8	g22-g22	a2B1	a1B2	p2u-p1e

6   Whether small alphabet or large alphabet?

Till now, dependency relation based extended formae and a Mendelian operator exploiting that has been derived. So, it will be interesting to re-look at the issue of whether small alphabet or large alphabet to get maximum schemata, based on this new schemata definition.

Theorem 1. *An alphabet of cardinality k will induce $n_s = (2^k - 1)^{1/\log_2 k}$ schemata or similarity subsets per bit of information, based on extended formae definition.*

Proof. Let s be the set of all k symbols and S_p be the set of all schemata per position, through extended formae definition. Then, the maximum schemata per position,

$$NCP_{max} = |S_p| = |\mathcal{P}(s) - \{\}| = \binom{k}{1} + \binom{k}{2} + \ldots + \binom{k}{k} = 2^k - 1$$

Each position represents $\log_2 k$ bits.
So,

$$n_s = |S_p|^{1/\log_2 k} = (2^k - 1)^{1/\log_2 k} \quad \square$$

Corollary 1. *The set of extended formae as schemata for k-ary alphabet contains all extended formae (schemata) induced by all m-ary alphabets, m < k.*

Proof. Let A_k be the set of all k symbols and S_k be the set of all schemata induced by k-ary alphabet. Let $P(A_k)$ be the power set of A_k. Similarly, let A_m be the set of all m symbols and S_m be the set of all schemata induced by m-ary alphabet. Let $P(A_m)$ be the power set of A_m.
Given $m < k$, $A_m \subset A_k \Rightarrow P(A_m) \subset P(A_k)$
As schema position takes any value from the set $\{P(\cdot) - \{\}\}$, $S_m \subset S_k$.   □

Theorem 2. *For a given information content, strings coded with larger alphabets give more extended formae as schemata per bit of information than that coded with smaller alphabets.*

Proof. From Theorem 1,

$$n_s = (2^k - 1)^{1/\log_2 k}$$

We are interested in the rate of change of n_s with respect to k. Then,

$$\log_2 n_s = \frac{\log_2(2^k - 1)}{\log_2 k}$$

$$\Rightarrow \frac{1}{n_s} \frac{dn_s}{dk} = \frac{1}{\log_2 k} \left[\frac{2^k}{2^k - 1} - \frac{1}{(\ln 2)^2} \frac{\log_2(2^k - 1)}{k} \right] = \frac{1}{\log_2 k} [p(k) - q(k)]$$

where, $p(k) = \frac{2^k}{2^k - 1}$ and $q(k) = \frac{1}{(\ln 2)^2} \frac{\log_2(2^k - 1)}{k}$. As $p(k) > 1$ and $q(k) < 1$, the derivative is always positive, n_s is monotonically increasing function. This proves that for a given information content, larger alphabets give more extended formae as schemata than that coded with smaller alphabet.   □

Theorem 3. *For an n-point search space S, maximum schemata induction requires n-ary alphabet and schemata as extended formae.*

Proof. Without loss of generality, let us assume a one-one mapping $\rho : S \longrightarrow C$. Given n-point search grid, the length of binary coded chromosome string will be $\log_2 n$. Let also assume that a k-ary alphabet gives maximum schemata. As already discussed in section 2 and the calculation of maximum schemata per position (NCP_{max}) in section 4; the maximum possible schemata in C is

$$NC_{max} = NS_{max} = 2^n - 1$$

and neither the Holland's schemata nor the formae can achieve these maximum schemata.
Now, from Theorem 1, total extended formae as schemata per bit of information are $n_s = (2^k - 1)^{1/\log_2 k}$. Then, the total extended formae (schemata) for the n-point search grid using k-ary alphabet (N_k),

$$N_k = \left(\left(2^k - 1 \right)^{(1/\log_2 k)} \right)^{(\log_2 n)}$$

$$N_k = NC_{max}, \text{ iff } k = n$$

□

7 Discussion and interpretations

The old result had stated that the minimal alphabet gives maximum schemata [7, 4, 5]. The current article proves that the maximal alphabet gives maximum schemata same as maximum theoretically possible. Both the coclusions seem to be contradictory. But, both of them are correct with respect to specific set of operators. The R^3 like operators, interpreting $\square$ as 'don't care' will give maximum schemata for minimal alphabet. The uniform crossover like operators interpreting $\square$ as '2-level either or' will give maximum schemata for maximal alphabet, but available schemata will not be same as the theoretical maximum possible. It is the combination of ploidy representation and the derived operator 5.4, interpreting $\square$ as 'p-level either or' with $2 \leq p \leq k$ that will provide theoretical maximum possible schemata. Also, it is important to ask whether successful GA requires maximum schemata? It is true that implicit parallelism increases with the increased schemata. This may suggests to maximize the schemata. The schema generalizations have shown that other than the traditional Holland's schemata, these generalized schemata may also contribute to the implicit parallelism. Radcliffe [8] emphasized that only the schemata with correlated performances (with least fitness variances) are significant and not the total number. In a blind case, it may be a valid assumption that maximization of schemata gives maximization of correlated schemata. But, the previous Theorem 3 motivates us to think of some disadvantages of the way maximum schemata made available. It says that maximum schemata for an n-point search space could be available through n-ary alphabet. That means, single position schema would be required. This is correct intuitively also, as more the chromosome string length - the 'Hamming Cliff' [8] causes more schemata representing correlated nearby search points to be missed. But, at the same time single digit schema implies search through just locality principle. Due to the lack of other allele positions, no exchange of genetic material and so periodicity principle is not in use for search. The total schemata is more but the search is based on the principle of similarity through locality only. In general, it could be observed that larger the alphabet, larger the locality region and smaller the periodicity (number of repeatations). For a successful GA, based on the application and the progress in search, both the principles of searching through locality and searching through periodicity may be given varying significances and accordingly the minimal or maximal or intermediate alphabet be chosen. As a general strategy, there may be selected a representation providing a balance between - maximal alphabet providing maximum schemata and a minimal alphabet providing search regions with smaller locality width and more periodicity.

The nature also supports the latter argument. The genetic information is coded in terms of four nucleobases. The four DNA-bases[9] are cytosine (C), guanine (G), adenine (A) and thymine (T) and four RNA-bases[10] are A, G, C and Uracil (U). Thus, nature balances the small alphabet and large alphabet using 4-ary and diploidy representation.

---

[8] Hamming Cliff is the effect of the used representation causing points nearer in the actual search space to be far in the representation space. Say, 7 and 8 are the given search points nearer to each other. But when represented using binary coding, they are mapped as 0111 and 1000 and are at the farthest hamming distance.

[9] Nucleobses for Deoxyribo-nucleic acid

[10] Nucleobses for Ribo-nucleic acid

8 Conclusion

The article generalizes the concept of schemata to dependency relation based extended formae. The generalization makes it possible to achieve theoretical maximum possible schemata for both the conventional string and non-string representations.

Further, the article signifies that the similarity has to be an operator perspective i.e. schemata should be able to be processed by an operator. It proves that whether minimal alphabet gives maximum schemata or maximal alphabet gives maximum schemata is decided by the operator used. It develops a Mendelian crossover oparator exploiting the extended formae. For that it uses the ploidy representation and derives p-schemata definition. This brings the notion of 'past' information with existing locality, periodicity and other notions for schemata definition.

Also, successful GA requires maximum schemata for intrinsic parallelism, as well the proper use of search through locality and periodicity principles. As a balance between these two contradictory reuirements, the article suggests to use neither small nor large but an intermediate level alphabet. The nature inspires to use 4-ary, diploidy representation.

References

1. Antonisse., J.: A new interpretation of schema notation that overturns the binary coding constraint. In: Proceedings of the Third International Conference on Genetic Algorithms. pp. 86–91. Morgan Kaufmann (San Mateo) (1989)
2. Eiben, A., Raué, P., Ruttkay, Z.: Genetic algorithms with multi-parent recombination. Parallel Problem Solving from NaturePPSN III pp. 78–87 (1994)
3. Eshelman, L.: Real-coded genetic algorithms and interval-schemata. Foundations of genetic algorithms 2, 187–202 (1993)
4. Goldberg, D.E.: Zen and the art of genetic algorithms. In: Proceedings of the third international conference on Genetic algorithms. pp. 80–85. Morgan Kaufmann Publishers Inc., San Francisco, CA, USA (1989)
5. Goldberg, D.: Genetic algorithms in search, optimization, and machine learning. Addison-wesley (1989)
6. Goldberg, D.: Real-coded genetic algorithms, virtual alphabets, and blocking. Complex Systems 5, 139–167 (1991)
7. Holland, J.: Adaptation in natural and artificial systems. No. 53, University of Michigan press (1975)
8. Radcliffe, N.: Equivalence class analysis of genetic algorithms. Complex Systems 5(2), 183–205 (1991)
9. Radcliffe, N.: Forma analysis and random respectful recombination. In: Proceedings of the fourth international conference on genetic algorithms. pp. 222–229. San Marco CA: Morgan Kaufmann (1991)
10. Smith, R.E., Goldberg, D.E.: Diploidy and dominance in artificial genetic search. Complex Systems 6(3), 251–285 (1992)
11. Vose, M.D.: Generalizing the notion of schema in genetic algorithms. Artif. Intell. 50(3), 385–396 (Aug 1991)
12. Wright, A.: Genetic algorithms for real parameter optimization. Foundations of genetic algorithms 1, 205–218 (1991)

Incorporating Great Deluge with Harmony Search for Global Optimization Problems

Mohammed Azmi Al-Betar, Osama Nasif Ahmad, Ahamad Tajudin Khader, and Mohammed A. Awadallah

Abstract Harmony search (HS) algorithm is relatively a recent metaheuristic optimization method inspired by natural phenomenon of musical improvisation process. despite its success, the main drawback of harmony search are contained in its tendency to converge prematurely due to its greedy selection method. This probably leads the harmony search algorithm to get stuck in local optima and unsought solutions owing to the limited exploration of the search space. The great deluge algorithm is a local search-based approach that has an efficient capability of increasing diversity and avoiding the local optima. This capability comes from its flexible method of accepting the new constructed solution. The aim of this research is to propose and evaluate a new variant of HS. To do so, the acceptance method of the great deluge algorithm is incorporated in the harmony search to enhance its convergence properties by maintaining a higher rate of diversification at the initial stage of the search process. The proposed method is called Harmony Search Great Deluge (HS-GD) algorithm. The performance of HS-GD and the classical harmony search algorithm was evaluated using a set of ten benchmark global optimization functions. In addition, five benchmark functions of the former set were employed to compare the results of the proposed method with three previous harmony search variations including the classical harmony search. The results show that HS-GD often outperforms the other comparative approaches.

Keywords: Harmony Search, Great Deluge, Global Optimization, Diversification, Intensification

Mohammed Azmi Al-Betar
Department of Computer Science, Jadara University, PO Box 733, Irbid, Jordan
e-mail: mohbetar@cs.usm.my

Mohammed Azmi Al-Betar · Osama Nasif Ahmad · Ahamad Tajudin Khader · Mohammed A. Awadallah
School of Computer Sciences, Universiti Sains Malaysia, 11800 USM, Penang, Malaysia e-mail: {mohbetar, osnasif, tajudin, mohawad}@cs.usm.my

J. C. Bansal et al. (eds.), *Proceedings of Seventh International Conference on Bio-Inspired Computing: Theories and Applications (BIC-TA 2012)*, Advances in Intelligent Systems and Computing 201, DOI: 10.1007/978-81-322-1038-2_24, © Springer India 2013

1 Introduction

Evolutionary algorithms (EAs) are a class of optimization methods which normally start with a population of random solutions. These solutions are evolutionary evolved using effective operators that drive the search either randomly or structurally based on the objective function of the current population. These operators explore and exploit the problem search space to come up with an optimal solution. However, this class of algorithms easily gets trapped into a chronic premature convergence problem due to identical population in the last stage of search [10].

Harmony Search (HS) algorithm [14] is a recent EA that imitates the behavior of a group of musicians when improvising a musical harmony. It has several impressive characteristics related to its simplicity, flexibility, adaptability, generality, and scalability [4]. As such, HS algorithm has been successfully adapted to a plethora of optimization problems such as Structural Design, Clustering, Bioinformatics, nurse restoring and timetabling [1, 6, 3, 2, 5, 9, 8, 7] and many others.It has been the subject of various researches that have been conducted to further improve its performance [16, 6]. The Theory of the HS has also undergone improvement as shown in [4].

HS is initiated with a population of random solutions stored in Harmony Memory (HM). Iteratively, it generates a new solution using three operators: i) Memory Consideration, which exploits the current population, ii) pitch adjustment, which locally refines some solutions in the current population, and iii) random consideration, which explores new solutions. The new solution is then evaluated to replace the worst solution in HM, if better. This process is repeated until a termination rule is reached.

As aforementioned, HS greedily accepts the new solution to enter the population, if and only if, it is better than the worst solution in HM. This acceptance criteria is similar to Hill climbing optimizer. The main weakness of Hill climbing is related to its simplicity to get stuck in local optima due to the lack of exploration capability. Therefore, several variations of Hill climbing that inject an explorative strategy with hill climbing were proposed. Examples include Simulated Annealing and Great Deluge [12]. The acceptance criteria of SA and GD algorithms substitutes the current solution with the new solution, if better or if it is accepted by certain threshold though it is worst [12]. This acceptance criteria empowers an explorative capability of SA and GD and eventually avoids the trap of local optima.

According to Geem et al., [15], the majority of researches and studies have been conducted in order to enhance the solution accuracy and speed up the convergence rate of HS. The resulting variations have advantages in terms of the implementation time but they still suffer from deficiencies in terms of avoiding the premature convergence. In this paper, Harmony Search- Great Deluge (HS-GD) algorithm is proposed. In HS-GD algorithm, the acceptance method of GD and its main related concepts are incorporated with the greedy acceptance method of classical HS algorithm to empower its explorative properties and eventually avoid a chronic premature convergence problem. Using standard mathematical functions, the experimental results show that the proposed HS-GD algorithm has improve the performance of the classical HS.

2 Harmony search Great Deluge (HS-GD) algorithm

Harmony Search (HS) is an evolutionary algorithm (EA) inspired by the musical improvisation process [14], where a group of musicians improvise the pitches of their musical instruments, *practice after practice*, seeking for a pleasing harmony as determined by an audio-aesthetic standard. Analogously in optimization, a set of decision variables is assigned with values, *iteration by iteration*, seeking for a '*good enough*' solution as evaluated by an objective function.

The great deluge(GD) algorithm has a special acceptance method. It accepts the new generated solution if its quality is better than or equal to a predefined linearly boundary (called "level" or "ceiling") increasing (in maximization) or decreasing (in minimization) that increases or decreases according to a fixed rate. This method is incorporated in HS. The flowchart of the HS-GD algorithm is shown in Figure 1 where the acceptance method of GD is highlighted by the red diamond. The HS-GD algorithm has five main steps illustrated as follows:

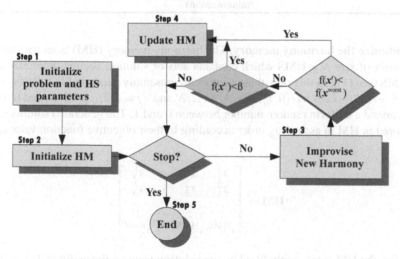

Fig. 1 The flowchart of the HS-GD algorithm

1. **Initialize the problem and HS-GD parameters.** Normally, the optimization problem is initially modeled as: $\min\{f(x)\,|\,x \in \mathbf{X}\}$, where $f(x)$ is the objective function; $x = \{x_i\,|\,i = 1,\dots,N\}$ is the set of decision variables. $\mathbf{X} = \{X_i\,|\,i = 1,\dots,N\}$ is the possible value range for each decision variable, where $X_i \in [LB_i, UB_i]$, where LB_i and UB_i are the lower and upper bounds for the decision

variable x_i respectively and N is the number of decision variables. The parameters of the HS-GD algorithm required to solve the optimization problem are also specified in this step as described in Table 1. Since this work only attends to the minimization problems, it is always referred to the boundary of accepting a worse solution with 'ceiling' instead of 'level'.

Table 1 Parameters of HS-GD algorithm

Parameter	Description
The Harmony Memory Consideration Rate (HMCR)	It is used in the improvisation process to determine whether the value of a decision variable is to be selected from the solutions stored in the Harmony Memory (HM)
The Harmony Memory Size (HMS)	It is similar to the population size in Genetic Algorithm
The Pitch Adjustment Rate (PAR)	It decides whether the decision variables are to be adjusted to a neighbouring value.
Number of Improvisations (NI)	It corresponds to the number of iterations.
Distance Bandwidth (BW)	It determines the distance of adjustment in the pitch adjustment operator.
Ceiling (β)	The ceiling is the boundary of the acceptability of the quality of a candidate solution. It is called "level" in maximization and "ceiling" in minimization problems
Decay rate ($\Delta\beta$)	The decay rate is the rate that the ceiling changes during the search process (increases in maximization and decreases in minimization)

2. **Initialize the harmony memory.** The harmony memory (HM) is an augmented matrix of size $N \times$ HMS which contains sets of solution vectors determined by HMS (see (1)). In this step, these vectors are randomly generated as follows: $x_i^j = LB_i + (UB_i - LB_i) \times U(0,1), \forall i = 1, 2, \ldots, N$ and $\forall j = 1, 2, \ldots,$ HMS, and $U(0,1)$ generate a uniform random number between 0 and 1. The generated solutions are stored in HM in ascending order according to their objective function values.

$$
\mathbf{HM} = \begin{bmatrix} x_1^1 & x_2^1 & \cdots & x_N^1 \\ x_1^2 & x_2^2 & \cdots & x_N^2 \\ \vdots & \vdots & \cdots & \vdots \\ x_1^{HMS} & x_2^{HMS} & \cdots & x_N^{HMS} \end{bmatrix} .
\tag{1}
$$

After the HM is randomly filled by the solution vectors, the ceiling (β) is initialized to be assigned the fitness value of the worst solution vector in HM.

In an analogous manner to the HS-GD algorithm, the decay rate ($\Delta\beta$) of the ceiling is calculated using the following formula.

$$
\Delta\beta = \frac{\beta_0 - f(x^{opt})}{\text{NI}}
\tag{2}
$$

where β_0 is the initial value of the ceiling that is set to the fitness value of the worst solution in the initial HM, $f(x^{opt})$ is the estimated quality of the final solution (the global optimum) and NI is the maximum number of improvisations.

3. **Improvise a new harmony.** In this step, the HS-GD algorithm generating (improvising) a new harmony vector from scratch, $x' = (x'_1, x'_2, \cdots, x'_N)$, based on three operators: (1) memory consideration, (2) random consideration, and (3) pitch adjustment.

   Memory consideration. In memory consideration, the value of the first decision variable x'_1 is randomly *selected* from the historical values, $\{x_1^1, x_1^2, \ldots, x_1^{HMS}\}$, stored in HM vectors. Values of the other decision variables, $(x'_2, x'_3, \ldots, x'_N)$, are sequentially selected in the same manner with probability (w.p.) HMCR where HMCR $\in (0, 1)$.

   Random consideration. Decision variables that are not assigned with values according to memory consideration are randomly assigned according to their possible range by random consideration with a probability of (1-HMCR) as follows:

   $$x'_i \leftarrow \begin{cases} x'_i \in \{x_i^1, x_i^2, \ldots, x_i^{HMS}\} & \text{w.p.} \quad \text{HMCR} \\ x'_i \in X_i & \text{w.p.} \quad 1 - \text{HMCR}. \end{cases}$$

   Pitch adjustment. Each decision variable $x'_i, i \in \{1, 2, \ldots, N\}$ of a new harmony vector, that has been assigned a value by memory considerations is pitch adjusted with the probability of PAR where PAR $\in (0, 1)$ as follows:

   $$\text{Pitch adjusting decision for } x'_i \leftarrow \begin{cases} \text{Yes} & \text{w.p.} \quad \text{PAR} \\ \text{No} & \text{w.p.} \quad \text{1-PAR}. \end{cases}$$

   If the pitch adjustment decision for x'_i is Yes, the value of x'_i is modified to its neighboring value as follows: $x'_i = x'_i + U(-1, 1) \times \text{BW}$

4. **Update the harmony memory.**
   This section describes the main focus of this research. The worst vector in HM is excluded and the new harmony x' is included if its fitness value meets the following two conditions:

   - Better than the fitness value of the worst solution vector in HM (greater in the maximization problems and less in minimization problems).
   - Better than or equal to the current ceiling (β).

   The ceiling (β) is degraded with every improvisation of a new solution vector by subtracting the decay rate $(\Delta\beta)$ value.

5. **Check the stop criterion.** Step 3 and step 4 of HS algorithm are repeated until NI is reached.

The procedure of HS-GD algorithm can be presented as in Algorithm 1:

Algorithm 1 HS-GD algorithm

Set HMCR, PAR, NI, HMS, BW.

x^{opt} = expected optimal solution of the minimization problem.

$x_i^j = LB_i + (UB_i - LB_i) \times U(0,1), \forall i = 1,2,\ldots,N$ and $\forall j = 1,2,\ldots,$HMS {generate HM solutions}

Calculate$(f(x^j)), \forall j = (1,2,\ldots,HMS)$

$\beta_0 = \beta = f(x^{\text{worst}})$ of the initial HM.

$\Delta\beta = (\beta_0 - f(x^{opt}))/$ NI.

$itr = 0$

while ($itr \leq$ NI) **do**

    $x' = \phi$

    for $i = 1, \cdots, N$ **do**

        if $(U(0,1) \leq$ HMCR$)$ **then**

            $x_i' \in \{x_i^1, x_i^2, \ldots, x_i^{\text{HMS}}\}$ {memory consideration}

            if $(U(0,1) \leq$ PAR$)$ **then**

                $x_i' = x_i' + U(-1,1) \times$ FW { pitch adjustment }

            end if

        else

            $x_i' = LB_i + (UB_i - LB_i) \times U(0,1)$ { random consideration }

        end if

    end for

    if $(f(x') < f(x^{\text{worst}}))$ OR $(f(x') < \beta)$ **then**

        Include x' to the **HM**.

        Exclude x^{worst} from **HM**.

    end if

    $\beta = \beta - \Delta\beta$

    $itr = itr + 1$

end while

---

3 Experimental results

3.1 Benchmark Functions

Table 2 overviews a summary for 10 global minimization benchmark functions used to evaluate HS-GD algorithm most of which previously used in [16, 4]. These benchmark functions provide a trad-off between unimodal and multimodal functions. The benchmark functions were implemented with $N=30$, with the exception of Six-Hump Camel-Back function which is two-dimensional.

3.2 Sensitivity Analysis of HS-GD Algorithm

The effect of different parameter settings of the HS-GD parameters (β_0 and $\Delta\beta$) is investigated. The parameters (HMCR, HMS and PAR) are set as recommended in the previous work of HS as follows [4]: HMS=50, HMCR=0.98, PAR=0.3,

Table 2 Benchmark functions used to evaluate HS variations

Function Name	Expression	Search Range	Optimum Value	Category [17]
Sphere function	$f_1(x) = \sum_{i=1}^{N} x_i^2$	$x_i \in [-100, 100]$	$\min(f_1) = f(0,\dots,0) = 0$	unimodal
Schwefel's Problem 2.22 [20]	$f_2(x) = \sum_{i=1}^{N} |x_i| + \prod_{i=1}^{N} |x_i|$	$x_i \in [-10, 10]$	$\min(f_2) = f(0,\dots,0) = 0$	unimodal
Step function	$f_3(x) = \sum_{i=1}^{N} (\lfloor x_i + 0.5 \rfloor)^2$	$x_i \in [-100, 100]$	$\min(f_3) = f(0,\dots,0) = 0$	unimodal & discontinues
Rosenbrock function	$f_4(x) = \sum_{i=1}^{N-1} (100(x_{i+1} - x_i^2)^2 + (x_i - 1)^2)$	$x_i \in [-30, 30]$	$\min(f_4) = f(1,\dots,1) = 0$	multimodal
Rotated hyper-ellipsoid function	$f_5(x) = \sum_{i=1}^{N} \left(\sum_{j=1}^{i} x_j \right)^2$	$x_i \in [-100, 100]$	$\min(f_5) = f(0,\dots,0) = 0$	unimodal
Schwefel's problem 2.26 [20]	$f_6(x) = -\sum_{i=1}^{N} \left(x_i \sin(\sqrt{|x_i|}) \right)$	$x_i \in [-500, 500]$	$\min(f_6) = f(420.9687,\dots,$ $420.9687) = -12569.5$	multimodal
Rastrigin function	$f_7(x) = \sum_{i=1}^{N} (x_i^2 - 10\cos(2\pi x_i) + 10)$	$x_i \in [-5.12, 5.12]$	$\min(f_7) = f(0,\dots,0) = 0$	multimodal
Ackley's function	$f_8(x) = -20\exp\left(-0.2\sqrt{\frac{1}{30} \sum_{i=1}^{N} x_i^2} \right) -$ $\exp\left(\frac{1}{30} \sum_{i=1}^{N} \cos(2\pi x_i) \right) + 20 + e$	$x_i \in [-32, 32]$	$\min(f_8) = f(0,\dots,0) = 0$	multimodal
Griewank function	$f_9(x) = \frac{1}{4000} \sum_{i=1}^{N} x_i^2 - \prod_{i=1}^{N} \cos\left(\frac{x_i}{\sqrt{i}} \right) + 1$	$x_i \in [-600, 600]$	$\min(f_9) = f(0,\dots,0) = 0$	multimodal
Six-Hump Camel-Back function	$f_{10}(x) = 4x_1^2 - 2.1x_1^4 + \frac{1}{3}x_1^6 + x_1 x_2 - 4x_2^2 + 4x_2^4$	$x_i \in [-5, 5]$	$\min(f_{10}) = f(-0.08983,$ $0.7126) = -1.0316285$	multimodal

NI=100,000, BW=0.03. Note that the default value of ceiling β_0 is $f(x^{worst})$ where x^{worst} is the worst vector in the initial HM. Furthermore, $\Delta\beta$ is $\frac{\beta_0 - f(x^{opt})}{NI}$.

3.2.1 The effect of initializing the ceiling (β)

In this section, the performance of the proposed HS-GD using two different ceiling values (β) is investigated. Table 3 summarizes the results of two values for initializing the ceiling parameter (β_0) in HS-GD. Firstly, the ceiling is initialized by assigning it the cost function of the best solution vector of the initial HM. Secondly, the ceiling is initialized by assigning it the cost function of the worst solution vector of the initial HM.

The results show that HS-GD is sensitive to the ceiling value of setting the initial ceiling. The best results are obtained when the ceiling is assigned the cost function of the worst harmony $f(x^{worst})$ of the initial HM. Notably, assigning the best cost function $f(x^{best})$ of the initial HM to (β_0) might hinder accepting the worst moves at

Table 3 The effect of initializing the ceiling in HS-GD for ten benchmark functions

Function	$\beta_0 = f(x^{best})$	$\beta_0 = f(x^{worst})$
f_1: Sphere	7.026E-09	**5.555E-09**
	(3.862E-09)	**(2.602E-09)**
f_2: Rosenbrock	1.020E+00	**8.338E-01**
	(5.344E-01)	**(5.140E-01)**
f_3: Ackley	9.423E-05	**8.722E-05**
	(2.587E-05)	**(2.637E-05)**
f_4: Griewank	5.547E-02	**4.347E-02**
	(3.057E-02)	**(2.755E-02)**
f_5: Rastrigin	8.374E-07	**7.684E-07**
	(4.384E-07)	**(4.986E-07)**
f_6: Schwefel Problem 2.22	**1.245E-04**	1.341E-04
	(3.290E-05)	(3.580E-05)
f_7: Step	7.140E-09	**6.681E-09**
	(4.589E-09)	**(5.605E-09)**
f_8: Rotated hyper-ellipsoid	7.754E+01	**5.664E+01**
	(7.554E+01)	**(4.243E+01)**
f_9: Schwefel Problem 2.26	**-4.190E+03**	**-4.190E+03**
	(7.531E-09)	**(3.571E-09)**
f_{10}: Camel-Back	**-1.032E+00**	**-1.032E+00**
	(0)	**(0)**

the early stage of the search process (especially when the fitness values of both the worst and the best harmonies of the initial HM are relatively distant). This probably allows the greedy replacement to dominate at the initial stage of the run, and thus to speed up the convergence towards, probably, unsought solutions.

3.2.2 Effect of the decay rate ($\Delta\beta$)

The performance of the proposed method using high and low decay rate ($\Delta\beta$) is investigated in this section. The different values of $\Delta\beta$ indicate the speed that the ceiling of accepting the worst moves is degraded.

– The effect of using high decay rates

The results for the ten benchmark functions using varying high ($\Delta\beta$) values (i.e. $\Delta\beta \times 20$, $\Delta\beta \times 2$, $\Delta\beta \times 1.05$ and the standard $\Delta\beta$) are summarized in Table 4.

The fast decay of the ceiling narrows the condition of accepting the worst moves. The value ($\Delta\beta \times 20$) means that the decay rate is proportional to 5% of the entire run. The value ($\Delta\beta \times 2$) means that the decay rate is proportional to 50% of the entire run. The value ($\Delta\beta \times 1.05$) means that the decay rate is proportional to roughly 95% of the entire run.

It is observed that the best results for the majority of the benchmark functions are obtained when the decay rate is proportional to the entire run.

Table 4 The effect of using high decay rates for ten benchmark functions

Function	$\Delta\beta \times 20$	$\Delta\beta \times 2$	$\Delta\beta \times 1.05$	$\Delta\beta \times 1$
f_1: Sphere	6.840E-09	6.084E-09	6.618E-09	**5.555E-09**
	(5.155E-09)	(4.588E-09)	(4.395E-09)	**(2.602E-09)**
f_2: Rosenbrock	9.338E-01	1.127E+00	1.027E+00	**8.338E-01**
	(5.636E-01)	(4.854E-01)	(5.901E-01)	**(5.140E-01)**
f_3: Ackley	**8.639E-05**	8.913E-05	8.765E-05	8.722E-05
	(2.422E-05)	(2.640E-05)	(2.877E-05)	(2.637E-05)
f_4: Griewank	4.547E-02	6.363E-02	5.431E-02	**4.347E-02**
	(2.425E-02)	(3.275E-02)	(3.020E-02)	**(2.755E-02)**
f_5: Rastrigin	9.965E-07	8.325E-07	9.026E-07	**7.684E-07**
	(6.988E-07)	(4.553E-07)	(6.025E-07)	**(4.986E-07)**
f_6: Schwefel Problem 2.22	1.370E-04	**1.276E-04**	1.279E-04	1.341E-04
	(3.544E-05)	**(3.528E-05)**	(3.666E-05)	(3.580E-05)
f_7: Step	7.192E-09	7.212E-09	8.045E-09	**6.681E-09**
	(3.162E-09)	(3.961E-09)	(3.506E-09)	**(5.605E-09)**
f_8: Rotated hyper-ellipsoid	6.822E+01	6.262E+01	6.602E+01	**5.664E+01**
	(5.222E+01)	(5.442E+01)	(7.556E+01)	**(4.243E+01)**
f_9: Schwefel Problem 2.26	**-4.190E+03**	**-4.190E+03**	**-4.190E+03**	**-4.190E+03**
	(5.414E-09)	**(2.011E-09)**	(4.096E-09)	(3.571E-09)
f_{10}: Camel-Back	**-1.032E+00**	**-1.032E+00**	**-1.032E+00**	**-1.032E+00**
	(0)	**(0)**	**(0)**	**(0)**

– The effect of using low decay rates

The results for the same ten benchmark functions using different low $\Delta\beta$ values (i.e. $\Delta\beta \div 20$, $\Delta\beta \div 2$, $\Delta\beta \div 1.05$ and the standard $\Delta\beta$) are summarized in Table 5.

The slow decay of the ceiling expands the condition of accepting the worst moves. The value ($\Delta\beta \div 20$) means that the entire run is proportional to 5% of the entire decay of the ceiling. The value ($\Delta\beta \div 2$) means that the entire run is proportional to 50% of the entire decay. The value ($\Delta\beta \div 1.05$) means that the entire run is proportional to roughly 95% of the entire decay.

It is observed that the best results for the majority of the benchmark functions are obtained at the standard value of $\Delta\beta$ where the entire run is proportional to the entire decay of the ceiling β.

3.3 Comparative Analysis

Numerous variations of harmony search are presented in the literature [13]. To have a fair comparison, two improved variations of HS in addition to the basic HS algorithm are selected taking into consideration that all the selected methods are comparable to the proposed method in terms of the number of the basic operations that each method achieves (i.e. the time complexity). Furthermore, comparable recommended setting of the unified parameters among the selected methods is also considered. The

Table 5 The effect of using low decay rates for ten benchmark functions

Function	$\Delta\beta \div 20$	$\Delta\beta \div 2$	$\Delta\beta \div 1.05$	$\Delta\beta \div 1$
f_1: Sphere	7.264E-09	6.908E-09	6.802E-09	**5.555E-09**
	(4.325E-09)	(4.685E-09)	(4.461E-09)	**(2.602E-09)**
f_2: Rosenbrock	9.319E-01	1.134E+00	1.027E+00	**8.338E-01**
	(5.586E-01)	(4.878E-01)	(5.924E-01)	**(5.140E-01)**
f_3: Ackley	**8.302E-05**	8.658E-05	9.585E-05	8.722E-05
	(1.608E-05)	(2.956E-05)	(2.952E-05)	(2.637E-05)
f_4: Griewank	4.699E-02	6.363E-02	5.430E-02	**4.347E-02**
	(2.714E-02)	(3.274E-02)	(3.018E-02)	**(2.755E-02)**
f_5: Rastrigin	8.801E-07	8.748E-07	9.141E-07	**7.684E-07**
	(4.818E-07)	(4.250E-07)	(6.031E-07)	**(4.986E-07)**
f_6: Schwefel Problem 2.22	**1.257E-04**	1.343E-04	1.279E-04	1.341E-04
	(3.733E-05)	(3.976E-05)	(3.666E-05)	(3.580E-05)
f_7: Step	8.274E-09	9.000E-09	8.131E-09	**6.681E-09**
	(6.788E-09)	(4.578E-09)	(4.004E-09)	**(5.605E-09)**
f_8: Rotated hyper-ellipsoid	6.703E+01	6.708E+01	6.503E+01	**5.664E+01**
	(5.417E+01)	(6.124E+01)	(7.482E+01)	**(4.243E+01)**
f_9: Schwefel Problem 2.26	**-4.190E+03**	**-4.190E+03**	**-4.190E+03**	**-4.190E+03**
	5.902E-09	**(1.965E-09)**	(2.939E-09)	(3.571E-09)
f_{10}: Camel-Back	**-1.032E+00**	**-1.032E+00**	**-1.032E+00**	**-1.032E+00**
	(0)	**(0)**	**(0)**	**(0)**

methods that are used to compare the proposed variant of HS are summarized in Table 6.

Table 6 The methods used in the comparative study.

Method	Denotation	Reference
A new heuristic optimization algorithm: harmony search	HS	Geem et al., [14]
Self-adaptive harmony search algorithm	SaHS	Wang and Huang, [19]
An improved harmony search algorithm with differential mutation operator	DHS	Chakraborty et al., [11]

Table 7 summarizes the results of comparing the proposed method with the selected methods for ten-dimensional objective function (n=10). The results of HS, SaHS and DHS are obtained from [18].

As can be seen from the results, the proposed HS-GD algorithm outperforms the other methods for the majority of the benchmark optimization functions. f_1 is a unimodal problem, and it is straightforward and easy to solve. Thus the HS variant that tends to converge prematurely seems to be a more efficient choice for such problems.

f_2 is a unimodal and can be considered as a multimodal problem. It has a narrow valley from its local optima to its global optimum. f_3 has one narrow global optimum valley and many shallow local optima. f_4 is more difficult when the dimensions of the function decrease. f_5 is considered a more complex multimodal problem with multiple local optima that may lead to an ambush into a local optima

Table 7 Mean and standard deviation of five benchmark function's optimization results (N=10)

Function	HS	SaHS	DHS	HS-GD
Sphere	**3.52E-09**	1.90E-02	8.13E-02	5.56E-09
	-6.75E-09	-1.95E-02	-5.21E-02	-2.60E-09
Rosenbrock	1.05E+00	5.66E+00	6.03E+00	**8.34E-01**
	-4.97E-01	-2.58E+00	-1.65E+00	**-5.14E-01**
Ackley	9.56E-05	5.82E-02	1.47E-01	**8.72E-05**
	-2.68E-05	-4.90E-02	-6.70E-02	**-2.64E-05**
Griewanks	5.91E-02	8.42E-02	1.57E-01	**4.35E-02**
	-3.37E-02	-3.67E-02	-5.00E-02	**-2.76E-02**
Rastrigin	8.89E-07	1.39E-02	3.48E-02	**7.68E-07**
	-6.04E-07	-1.39E-02	-2.31E-02	**-4.99E-07**

easily. Therefore, an algorithm that is more efficient in maintaining a higher rate of diversity may be more capable of obtaining better results for these functions. Apparently, the proposed HS-GD algorithm seems to be the best choice for solving such problems.

4 Conclusion and future work

This paper has proposed a Harmony Search Great Deluge (HS-GD) algorithm which is able to avoid the premature convergence situation by means of employing the acceptance method of GD. In this context, the HS-GD algorithm is able to maintain the right balance between diversification (exploration) and intensification (exploitation) during the search. The HS-GD is evaluated using ten benchmark mathematical functions circulated in the literature. The proposed method is able to perform better than the classical HS. Additionally, comparative evaluation shows that the proposed method can also be considered as an efficient technique for global optimization problems. For future work, the following three directions can be recommended:

1. Incorporating other methods, such as local search techniques, in HS-GD is suggested to reinforce the intensification specifically at the advanced stage of the search process.
2. Modifying HS-GD itself in a way that the decay rate changes dynamically throughout the decrease of the ceiling (in minimization) or the increase of the level (in maximization).
3. The performance of the proposed method can be further investigated by utilizing it in solving the combinatorial problems such as the Travelling Salesman Problem (TSP), the Knapsack Problem (KP) and Timetabling Problems. These problems are constrained and the algorithm that attempts to solve them is more likely to fall into local minima easily. Accordingly, an algorithm that maintains a higher rate of diversity, such as HS-GD, seems to be an efficient choice.

References

[1] M.S. Abual-Rub, M.A. Al-Betar, R. Abdullah, and A.T. Khader. A hybrid harmony search algorithm for ab initio protein tertiary structure prediction. *Network Modeling and Analysis in Health Informatics and Bioinformatics*, pages 1–17.

[2] M. A. Al-Betar, A. T. Khader, and F. Nadi. Selection mechanisms in memory consideration for examination timetabling with harmony search. In *GECCO '10: Proceedings of Genetic and Evolutionary Computation Conference*. ACM, Portland, Oregon, USA, July 7–11 2010.

[3] M. A. Al-Betar, A. T. Khader, and J. J. Thomas. A combination of metaheuristic components based on harmony search for the uncapacitated examination timetabling. In *8th International Conference on the Practice and Theory of Automated Timetabling (PATAT 2010)*, Belfast, Northern Ireland, August 10–13 2010.

[4] M.A. Al-Betar, I.A. Doush, A.T. Khader, and M.A. Awadallah. Novel selection schemes for harmony search. *Applied Mathematics and Computation*, 218(10), 2011.

[5] M.A. Al-Betar and A.T. Khader. A harmony search algorithm for university course timetabling. *Annals of Operations Research*, 194:1–29, 2012.

[6] M.A. Al-Betar, A.T. Khader, and M. Zaman. University course timetabling using a hybrid harmony search metaheuristic algorithm. *Systems, Man, and Cybernetics, Part C: Applications and Reviews, IEEE Transactions on*, (99):1–18.

[7] O. Alia, M. Al-Betar, R. Mandava, and A. Khader. Data clustering using harmony search algorithm. *Swarm, Evolutionary, and Memetic Computing*, pages 79–88, 2011.

[8] M. Awadallah, A. Khader, M. Al-Betar, and A. Bolaji. Nurse rostering using modified harmony search algorithm. *Swarm, Evolutionary, and Memetic Computing*, pages 27–37, 2011.

[9] M.A. Awadallah, A.T. Khader, M.A. Al-Betar, and A.L. Bolaji. Nurse scheduling using harmony search. In *Bio-Inspired Computing: Theories and Applications (BIC-TA), 2011 Sixth International Conference on*, pages 58–63. IEEE, 2011.

[10] Christian Blum and Andrea Roli. Metaheuristics in combinatorial optimization: Overview and conceptual comparison. *ACM Comput. Surv.*, 35(3):268–308, 2003.

[11] P. Chakraborty, G.G. Roy, S. Das, D. Jain, and A. Abraham. An improved harmony search algorithm with differential mutation operator. *Fundamenta Informaticae*, 95(4):401–426, 2009.

[12] G. Dueck. New optimization heuristics. *Journal of computational physics*, 104(1):86–92, 2005.

[13] Z. Geem. State-of-the-art in the structure of harmony search algorithm. *Recent Advances In Harmony Search Algorithm*, pages 1–10, 2010.

[14] Z. W. Geem, J. H. Kim, and G. V. Loganathan. A New Heuristic Optimization Algorithm: Harmony Search. *Simulation*, 76(2):60–68, 2001.

[15] Z.W. Geem, M. Fesanghary, J. Choi, MP Saka, J.C. Williams, M.T. Ayvaz, L. Li, S. Ryu, and A. Vasebi. Recent advances in harmony search. *Advance in evolutionary algorithms, I-Teach Education and Publishing, Vienna, Austria*, pages 127–142, 2008.

[16] M. G. H. Omran and M. Mahdavi. Global-best harmony search. *Applied Mathematics and Computation*, 198(2):643–656, 2008.

[17] Quan-Ke Pan, P.N. Suganthan, M. Fatih Tasgetiren, and J.J. Liang. A self-adaptive global best harmony search for continuous optimization problems. *Applied Mathematics and Computation*, 216(3):830 – 848, 2010.

[18] AK Qin and F. Forbes. Dynamic regional harmony search with opposition and local learning. In *Proceedings of the 13th annual conference companion on Genetic and evolutionary computation*, pages 53–54. ACM, 2011.

[19] C.M. Wang and Y.F. Huang. Self-adaptive harmony search algorithm for optimization. *Expert Systems with Applications*, 37(4):2826–2837, 2010.

[20] Xin Yao, Yong Liu, and Guangming Lin. Evolutionary programming made faster. *IEEE Transactions on Evolutionary Computation*, 3(2):82 –102, 1999.

Boundary Handling Approaches in Particle Swarm Optimization

Nikhil Padhye[1], Kalyanmoy Deb[2] and Pulkit Mittal[3]

[1] Department of Mechanical Engineering
Massachusetts Institute of Technology, Cambridge, MA-02139
npdhye@mit.edu
[2] Department of Mechanical Engineering
Indian Institute of Technology Kanpur, Kanpur-208016, U.P., India
[3] Department of Electrical Engineering
Indian Institute of Technology Kanpur, Kanpur-208016, U.P., India
deb,pulkitm@iitk.ac.in

Abstract. In recent years, Particle Swarm Optimization (PSO) methods have gained popularity in solving single objective and other optimization tasks. In particular, solving constrained optimization problems using swarm methods has been attempted in past but arguably stays as one of the challenging issues. A commonly encountered situation is one in which constraints manifest themselves in form of variable bounds. In such scenarios the issue of constraint-handling is somewhat simplified. This paper attempts to review popular *bound handling* methods, in context to PSO, and proposes new methods which are found to be robust and consistent in terms of performance over several simulation scenarios. The effectiveness of *bound handling* methods is shown PSO; however the methods are general and can be combined with any other optimization procedure.

Keywords: Constrained Optimization, Evolutionary Algorithms, Particle Swarm Optimization

1 Introduction

Optimization problems are wide-spread in several domains of science and engineering. The usual goal is to minimize or maximize some pre-defined objective(s) and specified constraints. Without any loss of generality, the most general form of constrained optimization problem can be written as a nonlinear programming problem (NLP):

$$Minimize \quad f(\bar{x})$$
$$Subject to$$
$$g_j(\bar{x}) \geq 0, \quad j = 1,...,J$$
$$h_k(\bar{x}) = 0, \quad k = 1,...,K$$
$$x_i^l \leq x_i \leq x_i^u, i = 1,...,n \tag{1}$$

J. C. Bansal et al. (eds.), *Proceedings of Seventh International Conference on Bio-Inspired Computing: Theories and Applications (BIC-TA 2012)*, Advances in Intelligent Systems and Computing 201, DOI: 10.1007/978-81-322-1038-2_25, © Springer India 2013

The above NLP problem contains n variables (i.e. $\bar{x}$ is vector of size n), J greater-than-equal-to type inequality constraints and K equality type constraints. The problem variables x_is are bounded within upper and lower limits.

The classical optimization algorithms employ several constraint handling methods such as penalty function, Lagrange multiplier, complex search, cutting plane, reduced gradient, gradient projection, etc. For details see [10, 2].

In context to Particle Swarm Optimization (PSO), several bound handling methods have already been proposed [8, 6, 1, 7]. However, many of these past proposals exploit the information about location of the optimum and fail to perform when location of optimum changes [9]. The goal of this paper is to come up with robust bound handling techniques which never fail to perform. This is achieved by proposing two stochastic and adaptive distributions to bring particles back into the feasible region once they fly out the search space. The existing and proposed bound handling methods are tested on four standard test problems under different scenarios and their performances are compared.

The rest of the paper is organized as follows: Section 2 reviews different bound handling techniques and provides a detailed description two newly proposed adaptive bound handling methods. Section 3 provides a description on the test problems, simulation carried out along with results and discussions. Finally, conclusions are made in Section 4.

2   Bound Handling Mechanisms

When objective function is not defined in the infeasible region only following alternatives are available: (a) creation of feasible-only solutions during the evolutionary search, or (b) an explicit mechanism to repair an infeasible solution i.e. bringing the infeasible solution back into the feasible search space. Generation of feasible-only solutions in case of EAs is not always straight- forward task, if not impossible [3]. Particularly, in context to PSO we need to have an explicit scheme which can bring an infeasible particle back into the feasible search region.

In past, several methods have been proposed to bring PSO particles back into feasible regions. In this study we shall refer to them as *bound handling methods*. The *bound handling methods* can be broadly divided into two groups A and B. Group A techniques carry out feasibility search variable wise, whereas group B techniques carry out feasibility search vectorially. According to group A techniques, for every solution, each variable is tested for its feasibility with respect to its bounds and made feasible if found to violate any bound. Here, only the variables violating their bounds are altered, independently, and other variables remain unchanged until they are tested and found to violate any bounds. In the group B techniques, if a solution (vector location) is found to violate any of the variable bounds, it is brought back into the search space along a vector direction. In such cases, the variables which have not violated any bound are also modified. It can be speculated that for separable problems (where variables are not linked with one-another), techniques belonging to group A are likely to perform well. However, for problems where optimization of the function requires high-degree of correlated variable alterations, group B techniques may become more useful (one such method named *Inverse Parabolic (IP) Distribution* is proposed in this paper and utilizes this fact while bringing solutions back into the feasible regions in a hopefully meaningful way). Next, we discuss some popular *bound handling methods*, using which an infeasible solution (violating variable bounds) can be made feasible.

2.1   Existing Bound Handling Methods

Random   This is one of the simplest strategies and belongs to group A. One-by-one, each variable is checked for bound-violations and modified if necessary. If X_C is the current infeasible variable location with L and U denoting the upper and lower bounds for the corresponding dimension (the same notation is used for rest of the text), then a new feasible location is selected randomly in [L,U].

Periodic   This strategy maps an infeasible location, for each variable violating the bounds, to a feasible location by assuming an infinite search space and was originally proposed in [11]. This is done by placing repeated copies of original search space along the dimension of interest. For example, let X_C denote the current location of a solution along d^{th} dimension, then X_C is mapped to X_C^{new} as follows:

$$X_C \rightarrow X_C^{new} = \begin{cases} U - (L - X_C)\%S_d & \text{IF } X_C < \text{L} \\ L + (X_C - U)\%S_d & \text{IF } X_C > \text{U} \end{cases}$$

The *Periodic* method handles all the variables separately and allows the infeasible solution to re-enter the search space from an end which is opposite to where it left the search space. For problems where optima is at the boundary this approach is rendered ineffective, as majority of solutions approaching the boundary optima shall fall outside the search space and will be brought back into the search space from opposite end. The entire effort carried out in locating the optima may be lost. For unimodal problems with optima at the center of the boundary the *Periodic* approach may be useful. For PSO, two variants of this strategy are studied (depending on the way in which velocity is computed), details of which are provided later.

Set on Boundary   As the name suggests, according to this strategy the individual is reset on the bound of the variable which it exceeds. The strategy belongs of group A. For example, along d^{th} dimension, let X_C denote a current location of a solution, then X_C is set to X_C^{new} as follows:

$$X_C \rightarrow X_C^{new} = \begin{cases} L & \text{IF } X_C < \text{L} \\ U & \text{IF } X_C > \text{U} \end{cases}$$

Clearly this approach biases the infeasible solutions on the search boundaries and can be highly helpful in cases where problem optima lies on the boundary of the variables. For PSO, three variants of this strategy can be done (depending on the way in which velocity is computed). The details on the three variants are provided later.

SHR   Originally *Shrink (SHR.)* method was introduced in context to PSO in [1]. The goal of the *SHR.* method was to re-adjust the particle's velocity such that particles just lands on the closest boundary along its path. To make X_C feasible the solution is dragged back along its line of movement till it reaches the nearest boundary. It should be noted that this mechanism belongs to group B, as the movement is carried out vectorially.

Exponential Distribution: This method is similar to *EXP.*, which was proposed in [1]. According to this approach a particle is brought back inside the search space, dimension-wise, in the region between particle's old position and the violated bound. The new particle positions are sampled in such a manner that higher probability is assigned to regions near the boundary, and the probability of sampling a location decreases exponentially as one moves away from the boundary. We have tried two version for *EXP.* method: (i) the new position is re-sampled between the particle's original location and the bound (lower or upper) that has been violated, and (ii) the new position is re-sampled in the entire region between between the upper and lower bounds of the dimension being violated.

2.2 Proposed Boundary Handling Methods

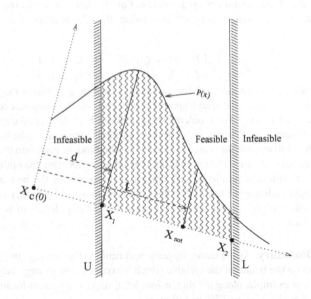

Fig. 1. Vector based *Inverse Parabolic Distribution* strategy for handling bounds.

The distance by which particle exceeds the boundary can also provide useful information. One way to utilize this distance information is to bring particles back into the search space with higher probabilities at the boundary when *falling-out* distance is small. In situations when particles are too far outside the search space, i.e. the *falling-out* distance is large, particles be brought back more uniformly. Since the distribution of particles back into search space varies it is unlikely that search becomes stagnated.

Consider a scenario, Figure 1, in which particle was originally located at a vector point X_{not} and after updation it moves to a new vector location X_C (which is infeasible). The goal is to bring particle back into the feasible region along the vector joining X_{not} and X_C. For the purpose of illustration, we consider a case where

X_C violates the bound (U) along some particular dimension as shown in Figure 1. It should be noted that, in general, more than one bounds may be violated. In such case, the bound intersecting the line joining X_{not} and X_C and lying closest to the old location (X_{not}) is selected. Let the intersection of selected bound and line joining X_{not} and X_C be X_1.

Let the intersection of the line joining X_C and X_{not} with bound on the opposite side be X_2. At this stage we propose two strategies, namely *Inverse Parabolic Spread Distribution* and *Inverse Parabolic Confined Distribution*, to re-sample a location x' in region between X_1 and X_2. Both, these strategies utilize the following probability distribution function:

$$P(x) = \frac{a}{(x-d)^2 + \alpha^2 d^2} \quad s.t. \ \ 0 \le x \le \infty \tag{2}$$

In above equation a is a constant to be determined and α is kept as a user defined parameter (we choose α equal to 1.2 in this study). According to the Figure 1, it can be seen that the proposed probability density function has a peak at location X_1. The peak is heightened if α is lowered. The calculation for the distribution constant a is done by equating the cumulative probability equal to one. The limits are chosen from X_C (taken as origin) till infinity.

$$\int_0^\infty \frac{a}{(x-d)^2 + \alpha^2 d^2} \, dx = 1 \implies a = \frac{\alpha^2 d^2}{\frac{\pi}{2} + \tan^{-1}\frac{1}{\alpha}} \tag{3}$$

1. Inverse Parabolic Spread Distribution: This strategy aims to sample a location between (and inclusive of) X_1 and X_2, thereby maintaining diversity while bringing the particles back into the feasible region. The bringing back is done by redistributing the probability in infeasible region probability into the feasible region as follows:

$$let, \int_d^{|X_2-X_1|+d} \frac{a}{(x-d)^2 + \alpha^2 d^2} \, dx = p_1 \tag{4}$$

Then, the probability distribution function is reconstructed as:

$$P_1(x) = \frac{a}{p_1((x-d)^2 + \alpha^2 d^2)} \quad s.t. \ \ d \le x \le |X_2 - X_1| + d \tag{5}$$

Let X' denote the sampled location, r be a uniformly distributed random number in [0,1] then $|X'|$ can be found as follows:

$$r = \int_d^{|X'|} \frac{a}{p_1((x-d)^2 + \alpha^2 d^2)} \, dx \tag{6}$$

$$\implies |X'| = d + \alpha d \tan(r \tan^{-1} \frac{(|X_2|-d)}{\alpha d}) \tag{7}$$

Once $|X'|$ is calculated, the new vector position X' between X_1 and X_2 can be easily found.

2. Inverse Parabolic Confined Distribution: This is similar to method 1, with the only difference that the re-sampled location in this case (denoted as X'') lies between (and inclusive of) X_1 and X_{not}. As the name suggests, bringing back of particle by this method is confined in the region on the line joining old position and the nearest bound. The probability distribution function and computation of a remain same as before. The redistribution of probability is carried out in the region between X_1 and X_{not}, and new location X'' is calculated as follows :

$$let, \int_d^{|X_{not}-X_1|+d} \frac{a}{(x-d)^2+\alpha^2 d^2}\, dx = p_2 \tag{8}$$

$$P_2(x) = \frac{a}{p_2((x-d)^2+\alpha^2 d^2)} \quad s.t. \quad d \leq x \leq |X_{not}-X_1|+d \tag{9}$$

$$r = \int_d^{|X''|} \frac{a}{p_2((x-d)^2+\alpha^2 d^2)}\, dx \tag{10}$$

$$\implies |X''| = d + \alpha d \tan(r \tan^{-1} \frac{(|X_{not}|-d)}{\alpha d}) \tag{11}$$

3   Simulations and Results

We have considered four standard test problems: Ellipsoidal (F_{elp}), Schwefel (F_{sch}), Ackley (F_{ack}) and Rosenbrock (F_{ros}) with 20 variables. Three different scenarios are considered for each problem such that optimum lies (i) on the boundary, (ii) in the center, and (iii) just inside the boundary of the search space. The test problems are given as follows:

$$F_{elp} = \sum_{i=1}^{n} i x_i^2 \tag{12}$$

$$F_{sch} = \sum_{i=1}^{n} \left(\sum_{j=1}^{i} x_j\right)^2 \tag{13}$$

$$F_{ack} = -20 exp\left(-0.2\sqrt{\frac{1}{n}\sum_{i=1}^{i=n} x_i^2}\right) - exp\left(\frac{1}{n}\sum_{i=1}^{n} cos(2\pi x_i)\right) + 20 + e \tag{14}$$

$$F_{ros} = \sum_{i=1}^{n-1} (100(x_i^2 - x_{i+1})^2 + (x_i - 1)^2) \tag{15}$$

F_{elp}, F_{sch} and F_{elp} have their minimum at $x_i^* = 0$, whereas F_{ros} has its minimum at $x_i^* = 1$. All the functions have minimum value of $F^* = 0$. F_{elp} is the only variable separable problem.

After initializing the population randomly and uniformly in the search domain, we count the number of function evaluations needed for the algorithm to find a solution close to the optimal solution. We choose a function value of 10^{-10} for F_{elp}, F_{sch} and F_{elp} and 10^{-3} for F_{ros} (which is a relatively difficult problem). This is similar to what has been proposed in [5, 4].

Each algorithm is tested on a test problem 50 times (each run starting with a different initial population). A particular run is concluded if termination criterion is met or the number of function evaluations exceed one million. If only a few out of 50 runs are successful then we report the count of successful runs in the bracket. In this case, the best, median and worst number of function evaluations are computed from the successful runs. If none of the runs are successful, we denote this by marking with *(DNC)* (Did Not Converge). In such cases, we report the best, median and worst attained function values of the best solution at the end of each run.

Once the particle is brought back into the search space the velocity can either be left unchanged or re-computed. By re-computed we mean that if X'_{t+1} is the new feasible location corresponding to X_t, then the velocity is re-adjusted as $V_{t+1} = X'_{t+1} - X_t$. For *Inverse Parabolic Spread Distribution, Inverse Parabolic Confined Distribution* and *Exponential Distribution* the velocity is re-computed. For *Periodic, Random* and *SetOnBoundary* velocity is either left unchanged and re-computed. For *SetOnBoundary* additional strategy of velocity reflection is also tried i.e. if a particle is set on the ith boundary, then v_i^{t+1} is changed to $-v_i^{t+1}$. The goal of the velocity reflection is to explicitly allow particles to move back into the search space. For *SHR.* the particle is placed on the boundary as discussed earlier and velocity is set to zero. To this end, a total of 13 different bound handling cases are tested. The simulations results for four test problems in three different settings are provided in Tables 1 to 4.

Following key inferences can be drawn from the tabulated results:

1. The *bound handling methods* show a large variation in the performance depending on the choice of test problem and location of the optimum with respect to the search space.

2. The performance of *bound handling* is comparable when optimum lies in the center. This can be understood intuitively from the fact that tendency of particles to fly out of the search space is little when the optimum is in the center of the search space. For e.g., the *Periodic* methods fail in all the cases but are able to show convergence for all the test problems when optimum is in the center. When optimum is on the boundary or close to the boundary then effect of bound handling method becomes critical.

3. *Inverse Parabolic Spread Distribution* never failed in any of the 12 simulation scenarios. *Inverse Parabolic Confined Distribution, Exponential Confined Distribution* and *Exponential Spread Distribution* are successful in 11, 10 and 8 cases, respectively. It is speculated that proposed IP distributions allow greater chances of creating an offspring close to the boundary then the exponential probability distribution, and hence the performance using IP method is better.

4. *SHR. (Vel. Recomputed and Vel. Set Zero)* methods succeeded in 10 cases. *Set on Boundary: Vel. Recomputed, Vel., Reflected and Vel. SetZero* succeeded 7, 7, 9 times, respectively. Random (Velocity re-computed and Vel. Set Zero) succeed in 5 cases.

5. On lowering α *Inverse Parabolic Distributions* showed an improvement in the performance. For the sake of brevity we exclude those results.

4 Conclusion

In this paper, we have compared existing and newly proposed *bound handling* methods in context to PSO. Four test problems with three different settings of optimum with respect to the search space were tried. The performance of the *bound handling* strategy dependent upon the test problem and location of the optimum. *Inverse Parabolic Spread Distribution* was found to be most the most robust method and never failed. *Inverse Parabolic Confined Distribution* and *Exponential Spread Distribution* were found competitive. *SHR.* methods were successful too but took larger number of function evaluations. Other bound handling strategies were either too deterministic or did not utilize the information of the particle's location properly. It can be concluded that probabilistic way of bringing the solutions back into the search space while utilizing the information of their initial location is an appropriate approach, and guarantees a robust performance. Although, the illustrations made in this paper are in context to PSO, but the bound handling strategies can be applied to any other evolutionary algorithm.

References

1. Alvarez-Benitez, J.E., Everson, R.M., Fieldsend, J.E.: A MOPSO algorithm based exclusively on pareto dominance concepts. In: EMO. vol. 3410, pp. 459–473 (2005)
2. Deb, K.: Optimization for Engineering Design: Algorithms and Examples. Prentice-Hall, New Delhi (1995)
3. Deb, K.: An efficient constraint handling method for genetic algorithms. In: Computer Methods in Applied Mechanics and Engineering. pp. 311–338 (1998)
4. Deb, K., Annand, A., Jhoshi, D.: A computationally efficient evolutionary algorithm for real-parameter optimization. Evol. Comput. 10(4), 371–395 (2002)
5. Deb, K., Padhye, N.: Development of efficient particle swarm optimizers by using concepts from evolutionary algorithms. In: Proceedings of the 12th annual conference on Genetic and evolutionary computation. pp. 55–62 (2010)
6. Helwig, S., Wanka, R.: Particle swarm optimizatio in high-dimensional bounded search spaces. In: Proceedings of the 2007 IEEE Swarm Intelligence Symposium. pp. 198–205
7. Helwig, S., Branke, J., Member, S.M.: Experimental analysis of bound handling techniques in particle swarm optimization. IEEE Transactions on Evolutionary Computation (99) (2012)
8. Padhye, N., Branke, J., Mostaghim, S.: Empirical comparison of mopso methods - guide selection and diversity preservation -. In: Proceedings of CEC. pp. 2516 – 2523. IEEE (2009)
9. Padhye, N.: Development of Efficient Particle Swarm Optimizers and Bound Handling Methods. Master's thesis, IIT Kanpur, India (2010)
10. Reklaitis, G.V., Ravindran, A., Ragsdell, K.M.: Engineering Optimization Methods and Applications. Willey, New York (1983)
11. Zhang, W.J., Xie, X.F., Bi, D.C.: Handling boundary constraints for numericaloptimization by particle swarm flying in periodic search space. In: Proceedings of Congress on Evolutionary Computation. pp. 2307–2311 (2004)

Table 1. Results on F_{elp}

Strategy	Best	Median	Worst
F_{elp} in [0,10]			
IP Spread Dist.	39,900	47,000	67,000
IP Confined Dist.	47,900 (49)	88,600	140,800
Exponential Spread Dist.	3.25e-01	5.02e-01	1.08e+00
Exponential Confined Dist.	4,600	5,900	7,500
Periodic(Vel. Recomputed)	3.94e+02 (DNC)	6.63e+02 (DNC)	1.17e+03(DNC)
Periodic(Vel. Unchanged)	8.91e+02 (DNC)	1.03e+03 (DNC)	1.34e+03 (DNC)
Random(Vel. Recomputed)	1.97e+01 (DNC)	3.37e+01 (DNC)	8.10e+01(DNC)
Random(Vel. Unchanged)	5.48e+02 (DNC)	6.69e+02 (DNC)	9.65e+02 (DNC)
SetOnBoundary(Vel. Recomputed)	900 (44)	1,300	5,100
SetOnBoundary(Vel. Reflected)	242,100	387,100	811,400
SetOnBoundary(Vel. Set Zero)	1,300 (48)	1,900	4,100
SHR.(Vel. Recomputed)	8,200 (49)	10,900	14,300
SHR.(Vel. Set Zero)	33,000	40,700	53,900
F_{elp} in [-10,10]			
IP Spread Dist.	31,600	34,000	37,900
IP Confined Dist.	30,900	33,800	38,500
Exponential Spread Dist.	30,500	34,700	38,300
Exponential Confined Dist.	31,900	35,100	38,200
Periodic(Vel. Recomputed)	32,200	35,100	37,900
Periodic(Vel. Unchanged)	33,800	36,600	41,200
Random(Vel. Recomputed)	31,900	34,800	37,400
Random(Vel. Unchanged)	31,600	34,900	38,100
SetOnBoundary(Vel. Recomputed)	31,900	35,500	40,500
SetOnBoundary(Vel. Reflected)	50,800 (38)	83,200	484,100
SetOnBoundary(Vel. Set Zero)	31,600	35,000	37,200
SHR.(Vel. Recomputed)	32,000	34,400	48,200
SHR.(Vel. Set Zero)	31,400	34,000	37,700
F_{elp} in [-1,10]			
IP Spread Dist.	28,200	31,900	35,300
IP Confined Dist.	28,300	32,900	44,600
Exponential Spread Dist.	28,300	30,700	33,200
Exponential Confined Dist.	29,500	33,000	44,700
Periodic(Vel. Recomputed)	4.86e+01 (DNC)	1.41e+02 (DNC)	4.28e+02 (DNC)
Periodic(Vel. Unchanged)	2.84e+02 (DNC)	5.46e+02 (DNC)	8.28e+02 (DNC)
Random(Vel. Recomputed)	36,900	41,900	45,600
Random(Vel. Unchanged)	1.13e+02 (DNC)	2.26e+02 (DNC)	4.35e+02 (DNC)
SetOnBoundary(Vel. Recomputed)	1.80e+01 (DNC)	7.60e+01 (DNC)	3.00e+02 (DNC)
SetOnBoundary(Vel. Reflected)	2.13e-01 (DNC)	2.17e+01 (DNC)	1.06e+02 (DNC)
SetOnBoundary(Vel. Set Zero)	31,700 (2)	31,700	32,600
SHR.(Vel. Recomputed)	29,500 (6)	36,100	42,300
SHR.(Vel. Set Zero)	28,400 (36)	32,700	65,600

Table 2. Results on F_{sch}

Strategy	Best	Median	Worst
F_{sch} in [0,10]			
IP Spread Dist.	67,200	257,800	970,400
IP Confined Dist.	112,400 (6)	126,500	145,900
Exponential Spread Dist.	3.79e+00 (DNC)	8.37e+00 (DNC)	1.49e+01 (DNC)
Exponential Confined Dist.	4,900	6,100	13,500
Periodic(Vel. Recomputed)	4.85e+03 (DNC)	7.82e+03(DNC)	1.34e+04 (DNC)
Periodic(Vel. Unchanged)	7.69e+03 (DNC)	1.11e+04 (DNC)	1.51e+04 (DNC)
Random(Vel. Recomputed)	2.61e+02 (DNC)	5.44e+02 (DNC)	1.05e+03 (DNC)
Random(Vel. Unchanged)	5.30e+03 (DNC)	7.60e+03 (DNC)	1.22e+04 (DNC)
SetOnBoundary(Vel. Recomputed)	800 (30)	1,100	3,900
SetOnBoundary(Vel. Reflected)	171,500	241,700	434,200
SetOnBoundary(Vel. Set Zero)	1,000 (40)	1,600	5,300
SHR.(Vel. Recomputed)	6,900	9,100	11,600
SHR.(Vel. Set Zero)	17,900	31,900	49,800
F_{sch} in [-10,10]			
IP Spread Dist.	106,700	127,500	144,300
IP Confined Dist.	111,500	130,100	149,900
Exponential Spread Dist.	112,300	131,400	149,000
Exponential Confined Dist.	116,400	131,300	148,200
Periodic(Vel. Recomputed)	113,400	130,900	150,600
Periodic(Vel. Unchanged)	121,200	137,800	159,100
Random(Vel. Recomputed)	112,900	129,800	151,100
Random(Vel. Unchanged)	117,000	130,600	148,100
SetOnBoundary(Vel. Recomputed)	118,500 (49)	132,300	161,100
SetOnBoundary(Vel. Reflected)	3.30e-06 (DNC)	8.32e+01(DNC)	2.95e+02 (DNC)
SetOnBoundary(Vel. Set Zero)	111,900	132,200	149,700
SHR.(Vel. Recomputed)	111,800 (49)	131,800	183,500
SHR.(Vel. Set Zero)	108,400	125,100	143,600
F_{sch} in [-1,10]			
IP Spread Dist.	107,200	130,400	272,400
IP Confined Dist.	120,100 (44)	171,200	301,200
Exponential Spread Dist.	92,800	109,200	126,400
Exponential Confined Dist.	110,200	127,400	256,100
Periodic(Vel. Recomputed)	8.09e+02 (DNC)	2.01e+03 (DNC)	5.53e+03(DNC)
Periodic(Vel. Unchanged)	2.16e+03 (DNC)	4.36e+03 (DNC)	6.87e+03 (DNC)
Random(Vel. Recomputed)	123,300	165,600	280,000
Random(Vel. Unchanged)	8.17e+02 (DNC)	1.96e+03 (DNC)	2.68e+03 (DNC)
SetOnBoundary(Vel. Recomputed)	2.50e+00 (DNC)	1.25e+01 (DNC)	5.75e+02 (DNC)
SetOnBoundary(Vel. Reflected)	1.86e+00 (DNC)	7.76e+00 (DNC)	5.18e+01 (DNC)
SetOnBoundary(Vel. Set Zero)	1.00e+00 (DNC)	5.00e+00 (DNC)	4.21e+02 (DNC)
SHR.(Vel. Recomputed)	5.00e-01 (DNC)	3.00e+00 (DNC)	1.60e+01 (DNC)
SHR.(Vel. Set Zero)	108,300 (8)	130,300	143,000

Table 3. Results on F_{ack}

Strategy	Best	Median	Worst
F_{ack} in [0,10]			
IP Spread Dist.	150,600 (49)	220,900	328,000
IP Confined Dist.	*4.17e+00 (DNC)*	*6.53e+00 (DNC)*	*8.79e+00 (DNC)*
Exponential Spread Dist.	*2.76e-01 (DNC)*	*9.62e-01 (DNC)*	*2.50e+00 (DNC)*
Exponential Confined Dist.	7,800	9,600	11,100
Periodic(Vel. Recomputed)	*6.17e+00 (DNC)*	*6.89e+00 (DNC)*	*9.22e+00 (DNC)*
Periodic(Vel. Unchanged)	*8.23e+00 (DNC)*	*9.10e+00 (DNC)*	*9.68e+00 (DNC)*
Random(Vel. Recomputed)	*3.29e+00 (DNC)*	*3.40e+00 (DNC)*	*4.19e+00 (DNC)*
Random(Vel. Unchanged)	*6.70e+00 (DNC)*	*7.46e+00 (DNC)*	*8.57e+00(DNC)*
SetOnBoundary(Vel. Recomputed)	800	1,100	2,100
SetOnBoundary(Vel. Reflected)	420,600	598,600	917,400
SetOnBoundary(Vel. Set Zero)	1,100	1,800	3,100
SHR.(Vel. Recomputed)	33,800 (5)	263,100	690,400
SHR.(Vel. Set Zero)	*3.65e+00 (DNC)*	*6.28e+00 (DNC)*	*8.35e+00 (DNC)*
F_{ack} in [-10,10]			
IP Spread Dist.	53,900 (46)	58,600	66,500
IP Confined Dist.	54,800 (49)	59,200	64,700
Exponential Spread Dist.	55,100	59,300	63,600
Exponential Confined Dist.	56,800	59,600	65,000
Periodic(Vel. Recomputed)	55,700 (48)	59,900	64,700
Periodic(Vel. Unchanged)	57,900 (49)	62,100	66,700
Random(Vel. Recomputed)	55,100 (47)	59,400	65,100
Random(Vel. Unchanged)	56,300	59,700	65,500
SetOnBoundary(Vel. Recomputed)	55,100 (49)	58,900	65,400
SetOnBoundary(Vel. Reflected)	86,900 (4)	136,400	927,600
SetOnBoundary(Vel. Set Zero)	53,900 (49)	59,600	67,700
SHR.(Vel. Recomputed)	55,800 (47)	58,700	65,800
SHR.(Vel. Set Zero)	55,700 (49)	58,900	62,000
F_{ack} in [-1,10]			
IP Spread Dist.	54,600 (5)	55,100	56,600
IP Confined Dist.	63,200 (1)	63,200	63,200
Exponential Spread Dist.	51,300	55,200	58,600
Exponential Confined Dist.	*1.42e+00(DNC)*	*2.17e+00 (DNC)*	*2.92e+00 (DNC)*
Periodic(Vel. Recomputed)	*2.88e+00 (DNC)*	*4.03e+00 (DNC)*	*5.40e+00 (DNC)*
Periodic(Vel. Unchanged)	*6.61e+00 (DNC)*	*7.46e+00 (DNC)*	*8.37e+00 (DNC)*
Random(Vel. Recomputed)	60,300 (45)	66,200	72,200
Random(Vel. Unchanged)	*4.21e+00 (DNC)*	*4.93e+00 (DNC)*	*6.11e+00 (DNC)*
SetOnBoundary(Vel. Recomputed)	*2.74e+00 (DNC)*	*3.16e+00 (DNC)*	*3.36e+00 (DNC)*
SetOnBoundary(Vel. Reflected)	824,700 (1)	824,700	824,700
SetOnBoundary(Vel. Set Zero)	*1.70e+00 (DNC)*	*2.63e+00 (DNC)*	*3.26e+00 (DNC)*
SHR.(Vel. Recomputed)	*1.45e+00 (DNC)*	*2.34e+00 (DNC)*	*2.73e+00 (DNC)*
SHR.(Vel. Set Zero)	*2.01e+00 (DNC)*	*3.96e+00 (DNC)*	*6.76e+00 (DNC)*

Table 4. Results on F_{ros}

Strategy	Best	Median	Worst
F_{ros} in [1,10]			
IP Spread Dist.	89,800	195,900	243,300
IP Confined Dist.	23,800	164,300	209,300
Exponential Spread Dist.	*9.55e-01 (DNC)*	*2.58e+00 (DNC)*	*7.64e+00 (DNC)*
Exponential Confined Dist.	3,700	128,100	344,400
Periodic(Vel. Recomputed)	*1.24e+04 (DNC)*	*2.35e+04 (DNC)*	*4.24e+04 (DNC)*
Periodic(Vel. Unchanged)	*6.99e+04 (DNC)*	*1.01e+05 (DNC)*	*1.45e+05 (DNC)*
Random(Vel. Recomputed)	*6.00e+01 (DNC)*	*1.37e+02 (DNC)*	*4.42e+02 (DNC)*
Random(Vel. Unchanged)	*2.32e+04 (DNC)*	*3.90e+04 (DNC)*	*8.22e+04 (DNC)*
SetOnBoundary(Vel. Recomputed)	900(45)	1,600	89,800
SetOnBoundary(Vel. Reflected)	*2.14e-03 (DNC)*	*6.01e+02 (DNC)*	*5.10e+04 (DNC)*
SetOnBoundary(Vel. Set Zero)	1,400(48)	3,000	303,700
SHR.(Vel. Recomputed)	3,900(44)	5,100	406,000
SHR.(Vel. Set Zero)	15,500	136,200	193400
F_{ros} in [-8,10]			
IP Spread Dist.	302,300(28)	774,900	995,000
IP Confined Dist.	296,600(32)	729,000	955,000
Exponential Spread Dist.	208,800(24)	754,700	985,200
Exponential Confined Dist.	301,100(33)	801,400	961,800
Periodic(Vel. Recomputed)	26,200(27)	705,100	986,200
Periodic(Vel. Unchanged)	247,300(32)	776,800	994,900
Random(Vel. Recomputed)	311,200(30)	809,300	990,800
Random(Vel. Unchanged)	380,100(29)	793,300	968,300
SetOnBoundary(Vel. Recomputed)	248,700(35)	795,600	973,900
SetOnBoundary(Vel. Reflected)	661,900(1)	661,900	661,900
SetOnBoundary(Vel. Set Zero)	117,400(25)	858,400	995,400
SHR.(Vel. Recomputed)	347,900(33)	790,500	996,300
SHR.(Vel. Set Zero)	353,300(26)	788,700	986,800
F_{ros} in [1,10]			
Strategy	Best	Median	Worst
IP Spread Dist.	184,600(47)	442,200	767,500
IP Confined Dist.	229,900(40)	457,600	899,200
Exponential Spread Dist.	19,400(47)	378,200	537,300
Exponential Confined Dist.	*6.79e-03 (DNC)*	*4.23e+00 (DNC)*	*6.73e+01 (DNC)*
Periodic(Vel. Recomputed)	*1.51e-02 (DNC)*	*3.73e+00 (DNC)*	*5.17e+02 (DNC)*
Periodic(Vel. Unchanged)	*1.92e+04 (DNC)*	*2.86e+04 (DNC)*	*6.71e+04(DNC)*
Random(Vel. Recomputed)	103,800	432,200	527,200
Random(Vel. Unchanged)	*2.33e+02 (DNC)*	*1.47e+03 (DNC)*	*4.23e+03 (DNC)*
SetOnBoundary(Vel. Recomputed)	*1.71e+01 (DNC)*	*1.87e+01 (DNC)*	*3.13e+02 (DNC)*
SetOnBoundary(Vel. Reflected)	*6.88e+00 (DNC)*	*5.52e+02 (DNC)*	*2.14e+04 (DNC)*
SetOnBoundary(Vel. Set Zero)	*6.23e+00 (DNC)*	*1.80e+01 (DNC)*	*3.12e+02 (DNC)*
SHR.(Vel. Recomputed)	350,300(3)	350,900	458,400
SHR.(Vel. Set Zero)	163,700(26)	418,000	531,900

Diversity Measures in Artificial Bee Colony

Harish Sharma, Jagdish Chand Bansal, and K V Arya

Abstract Artificial Bee Colony (ABC) is a recent swarm intelligence based approach to solve nonlinear and complex optimization problems. Exploration and exploitation are the two important characteristics of the swarm based optimization algorithms. Exploration capability of an algorithm is the capability of exploring the solution space to find the possible solution while exploitation capability of an algorithm is the capability of exploiting a particular region of the search space for a better solution. Usually, exploration and exploitation capabilities are contradictory in nature, i.e., a better exploration capability results a worse exploitation capability and vice versa. An economic and efficient algorithm can explore the complete solution space and shows a convergent behavior after a finite number of trials. Exploration and exploitation capabilities, are quantified using various *diversity measures*. In this paper, an analytical study has been carried out for various diversity measures for ABC process.

Keywords: Diversity measures, Swarm intelligence, Exploration-Exploitation, Artificial Bee Colony

Harish Sharma
ABV-Indian Institute of Information Technology and Management, Gwalior, India
e-mail: harish.sharma0107@gmail.com

Jagdish Chand Bansal
South Asian University, New Delhi, India
e-mail: jcbansal@gmail.com

K V Arya
ABV-Indian Institute of Information Technology and Management, Gwalior, India
e-mail: kvarya@gmail.com

J. C. Bansal et al. (eds.), *Proceedings of Seventh International Conference on Bio-Inspired Computing: Theories and Applications (BIC-TA 2012)*, Advances in Intelligent Systems and Computing 201, DOI: 10.1007/978-81-322-1038-2_26, © Springer India 2013

1 Introduction

Swarm Intelligence has become an emerging and interesting area in the field of nature inspired techniques that is used to solve optimization problems during the past decade. It is based on the collective behavior of social creatures. Swarm based optimization algorithms find solution by collaborative trial and error. Social creatures utilizes their ability of social learning to solve complex tasks. Peer to peer learning behavior of social colonies is the main driving force behind the development of many efficient swarm based optimization algorithms. Researchers have analyzed such behaviors and designed algorithms that can be used to solve nonlinear, nonconvex or discrete optimization problems. Previous research [1, 2, 3, 4] have shown that algorithms based on swarm intelligence have great potential to find solutions of real world optimization problems. The algorithms that have emerged in recent years include ant colony optimization (ACO) [1], particle swarm optimization (PSO) [2], bacterial foraging optimization (BFO) [5], artificial bee colony optimization (ABC) [6] etc.

Artificial bee colony (ABC) optimization algorithm introduced by D.Karaboga [6] is a recent addition in this category. This algorithm is inspired by the behavior of honey bees when seeking a quality food source. Like any other population based optimization algorithm, ABC consists of a population of potential solutions. The potential solutions are food sources of honey bees. The fitness is determined in terms of the quality (nectar amount) of the food source. ABC is relatively a simple, fast and population based stochastic search technique in the field of nature inspired algorithms.

There are two fundamental processes which drive the swarm to update in ABC: the variation process, which enables exploring different areas of the search space, and the selection process, which ensures the exploitation of the previous experience. Diversity has a significant effect on the performance of an algorithm [7]. It shows the behavior of the algorithm during the solution search process. A large value of diversity implies exploration of the search space i.e. the algorithm is discovering a true solution in whole search space. A low value of diversity implies exploitation i.e. the algorithm is exploiting a selected search space found during the search process. It is expected that an optimization algorithm retains high diversity value in early stage of the search process and proportionally decreases the value of diversity as search progresses. Study of diversity quantification is important because with this it is possible to rank two or more algorithms in their performance. There are many diversity measures in the literature [7, 8, 9, 10, 11, 12]. A good study on diversity measures for Particle Swarm Optimization process is given in [13]. In this paper, seven important diversity measures have been considered to quantify the diversity of *ABC*. The considered diversity measures have been tested on the five well known benchmark problems to quantify the dispersion of individuals in the swarm of ABC algorithm. Further, effect of outliers has been analyzed over the diversity measures.

Rest of the paper is organized as follows: Section 2 describes brief overview of the basic ABC. Various diversity measures are described in Section 3. In section 4, importance and behavior of diversity measures are discussed. Experimental results are shown in section 5. At last, in section 6, paper is concluded.

2 Artificial Bee Colony(ABC) algorithm

The ABC algorithm is relatively recent swarm intelligence based algorithm. The algorithm is inspired by the intelligent food foraging behavior of honey bees. In ABC, each solution of the problem is called food source of honey bees. The fitness is determined in terms of the quality of the food source. In ABC, honey bees are classified into three groups namely employed bees, onlooker bees and scout bees. The number of employed bees are equal to the onlooker bees. The employed bees are the bees which searches the food source and gather the information about the quality of the food source. Onlooker bees which stay in the hive and search the food sources on the basis of the information gathered by the employed bees. The scout bee, searches new food sources randomly in places of the abandoned foods sources. Similar to the other population-based algorithms, ABC solution search process is an iterative process. After, initialization of the ABC parameters and swarm, it requires the repetitive iterations of the three phases namely employed bee phase, onlooker bee phase and scout bee phase. Each of the phase is described as follows:

2.1 Initialization of the swarm

The parameters for the ABC are the number of food sources, the number trials after which a food source is considered to be abandoned and the termination criteria. In the basic ABC, the number of food sources are equal to the employed bees or onlooker bees. Initially, a uniformly distributed initial swarm of SN food sources where each food source $x_i (i = 1, 2, ..., SN)$ is a D-dimensional vector, generated. Here D is the number of variables in the optimization problem and x_i represent the ith food source in the swarm. Each food source is generated as follows:.

$$x_{ij} = x_{minj} + rand[0, 1](x_{maxj} - x_{minj}) \tag{1}$$

where x_{minj} and x_{maxj} are bounds of x_i in j^{th} direction and $rand[0, 1]$ is a uniformly distributed random number in the range $[0, 1]$

2.2 Employed bee phase

In employed bee phase, employed bees modify the current solution (food source) based on the information of individual experience and the fitness value of the new solution. If the fitness value of the new solution is higher than that of the old solution, the bee updates her position with the new one and discards the old one. The position update equation for i^{th} candidate in this phase is

$$v_{ij} = x_{ij} + \overbrace{\phi_{ij}(x_{ij} - x_{kj})}^{\text{Step size}} \tag{2}$$

where $k \in \{1, 2, ..., SN\}$ and $j \in \{1, 2, ..., D\}$ are randomly chosen indices, k must be different from i. ϕ_{ij} is a random number between [-1, 1] and component $\phi_{ij}(x_{ij} - x_{kj})$ is the step size of the ith food solution.

2.3 Onlooker bees phase

After completion of the employed bees phase, the onlooker bees phase starts. In onlooker bees phase, all the employed bees share the new fitness information (nectar) of the new solutions (food sources) and their position information with the onlooker bees in the hive. Onlooker bees analyze the available information and select a solution with a probability, $prob_i$, related to its fitness. The probability $prob_i$ may be calculated using following expression (there may be some other but must be a function of fitness):

$$prob_i = \frac{fitness_i}{\sum_{i=1}^{SN} fitness_i} \tag{3}$$

where $fitness_i$ is the fitness value of the solution i. As in the case of the employed bee, it produces a modification on the position in its memory and checks the fitness of the candidate source. If the fitness is higher than that of the previous one, the bee memorizes the new position and forgets the old one.

2.4 Scout bees phase

If the position of a food source is not updated up to predetermined number of cycles, then the food source is assumed to be abandoned and scout bees phase starts. In this phase the bee associated with the abandoned food source becomes scout bee and the food source is replaced by a randomly chosen food source within the search space. In ABC, predetermined number of cycles is a crucial control parameter which is called *limit* for abandonment.

Assume that the abandoned source is x_i. The scout bee replaces this food source by a randomly chosen food source which is generated as follows

$$x_{ij} = x_{minj} + rand[0, 1](x_{maxj} - x_{minj}), \text{ for } j \in \{1, 2, ..., D\} \tag{4}$$

where x_{minj} and x_{maxj} are bounds of x_i in j^{th} direction.

2.5 Main steps of the ABC algorithm

Based on the above explanation, it is clear that there are three control parameters in ABC search process: The number of food sources SN (equal to number of onlooker or employed bees), the value of *limit* and the maximum number of iterations. The pseudo-code of the ABC is shown in Algorithm 1 [14]:

Algorithm 1 Artificial Bee Colony Algorithm:

Initialize the parameters;
while Termination criteria is not satisfied **do**
    Step 1: Employed bee phase for generating new food sources.
    Step 2: Onlooker bees phase for updating the food sources depending on their nectar amounts.
    Step 3: Scout bee phase for discovering the new food sources in place of abandoned food sources.
    Step 4: Memorize the best food source found so far.
end while
Output the best solution found so far.

---

3 Diversity Measures

There are many strategies available in the literature for measuring the diversity of swarm. Basically, all the measures are based on the distance metric of individuals. Generally, the diversity measures differ in terms of distance metric or normalization of parameters. Further, the measures are differed based on the choice of swarm center which may be global best solution found so far or may be spatial. In this section, seven different diversity measures, which are based on the Euclidean distance metric, are described. Further, global best swarm center is used in this paper wherever required opposed to a spatial swarm center. Generally, spatial swarm center and global best swarm center can be considered equivalent where the global best is not necessarily centered position in the swarm. Further, for normalization of parameters, the swarm diameter is used, opposed to the radius of swarm.

1. **Swarm Diameter:** The swarm diameter is defined as the distance between two farthest individuals, along any axis [15], of the swarm as shown in Fig. 1. The diameter D is calculated using equation (5):

$$D = \max_{(i \neq j) \in [1, N_p]} \left(\sqrt{\sum_{k=1}^{I} (x_{ik} - x_{jk})^2} \right) \tag{5}$$

where N_p is the swarm size, I is the dimensionality of the problem and x_{ik} is the kth dimension of the ith individual position.

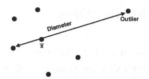

Fig. 1: Swarm Diameter

In Figure 1, an outlier individual is also shown. In a swarm, a significantly deviated individual from the remaining individuals is often termed as *outlier*. From Figure 1, it can be seen that the presence of an outlier can significantly affect the diameter of a swarm.

2. **Swarm Radius:** The radius of a swarm is defined as the distance between the swarm center and the individual in the swarm which is farthest away from it [15], as shown in Fig. 2. The swarm radius is calculated using equation (6):

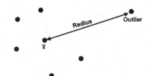

Fig. 2: Swarm Radius

$$R = \max_{i \in [1, N_p]} \left(\sqrt{\sum_{k=1}^{I} (x_{ik} - \bar{x}_k)^2} \right) \tag{6}$$

where the parameters have same meaning as for the swarm diameter. $\bar{x}$ is the position of swarm center and $\bar{x}_k$ represents the kth dimension of $\bar{x}$.

Now, it is evident that the swarm diameter (D) and radius (R) are two important diversity measures. A large value of D or R exhibits exploration of the search region while low value results exploitation. However, both are badly affected with outliers.

3. **Average Distance around Swarm Center** The average distance from the swarm center D_A, can be defined as the average of distances of all individuals from the swarm center. This measure is given in [10] and defined in equation (7)

$$D_A = \frac{1}{N_p} \sum_{i=1}^{N_p} \left(\sqrt{\sum_{k=1}^{I} (x_{ik} - \bar{x}_k)^2} \right) \tag{7}$$

where the notations have their usual meaning. A low value of this measure shows swarm convergence around the swarm center while a high value shows large dispersion of individuals from the swarm center.

4. **Geometric Average Distance around the Swarm Center:** Geometric average is not significantly affected by outliers in the swarm on the high end. The geometric average distance around the swarm center is defined in equation (8).

$$D_{GM} = \left(\prod_{i=1}^{N_p} \sqrt{\sum_{k=1}^{I} (x_{ik} - \bar{x}_k)^2} \right)^{\frac{1}{N_p}} \tag{8}$$

5. **Normalized Average Distance around the Swarm Center:** This diversity measure is almost same as the average distance of all individuals from the swarm center. The only difference is that, the average distance is normalized using the swarm diameter. This normalization can also be done by the radius of the swarm. This diversity measure is given in [15] and described by equation (9):

$$D_N = \frac{1}{N_p \times D} \sum_{i=1}^{N_p} \left(\sqrt{\sum_{k=1}^{I} (x_{ik} - \bar{x}_k)^2} \right) \tag{9}$$

6. **Average of the Average Distance around all Particles in the Swarm:** In this measure, first the average distances, considering each individual as a swarm center are calculated and then an average is taken of all these averaged distances. It is described by equation (10).

$$D_{all} = \frac{1}{N_p} \sum_{i=1}^{N_p} \left(\frac{1}{N_p} \sum_{j=1}^{N_p} \sqrt{\sum_{k=1}^{I} (x_{ik} - x_{jk})^2} \right) \tag{10}$$

This diversity measure shows average dispersion of every individual in the swarm from every other individual in the swarm.

7. **Swarm Coherence:** This diversity measure is given in [9] and described by equation (11):

$$S = \frac{s_c}{\bar{s}} \tag{11}$$

where s_c represents the step size of swarm center which is defined in equation (12):

$$s_c = \frac{1}{N_p} \left\| \sum_{i=1}^{N_p} \tilde{s}_i \right\|_2 \tag{12}$$

where $\tilde{s}_i$ is the vector of step size for i^{th} individual as indicated in equation (2) and $\bar{s}$ shows the average individual step size in the swarm and is defined by equation (13). $\|.\|_p$ is the Euclidean p-norm.

$$\bar{s} = \frac{1}{N_p} \sum_{i=1}^{N_p} \|\tilde{s}_i\|_2 \tag{13}$$

This diversity measure, is calculated by averaging the step sizes of all the individuals in a swarm with respect to swarm center.

The dispersion of the individuals in ABC could be quantified, at some extent, using the various measures of diversity described in this section. The diversity measures shows a trend of exploration or exploitation of the swarm and helps to analyze the behavior of the swarm based algorithms.

4 Discussion

Exploration and Exploitation are two important properties of swarm based algorithms. Most of the time, a better exploration capability contradicts a better exploitation capability and vice-versa. In initial iterations, exploration requires to explore the search region and later exploitation is used to thoroughly search the selected search area. Hight value of diversity measure shows the exploration whereas low value exhibits exploitation. A decreasing diversity measures through iterations represents the transition of exploration to exploitation. On the basis of these characteristics, following conclusion have been drawn:

- The swarm diameter presents a required decrease by iterations, as $(x_{ik} - x_{jk})^2$ (refer equation (5)) tends to zero for all the individuals as the swarm converges to a solution. The same behavior is shown by the swarm radius as the distance between each individual to swarm center decreases as the swarm converges with iterations. Further, it is clear from the equations (5) and (6) that the swarm diameter and swarm radius both are very sensitive to the outlier individual. Considering

Fig. 3: Swarm Diameter and outlier

Figure 1 and Figure 3, it can be shown that the diverse behavior of the current swarm shown in Figure 1 and Figure 3 is same, if swarm radius or swarm diameter is the diversity measure, while in Figure 1 the individuals are more diversified than in Figure 3.

- The diversity measure D_A, which is shown in equation (7) is robust measure as compared to the swarm diameter and swarm radius because it is based on the average distance of all individuals in the swarm from the swarm center. Hence, this diversity measure is considered less affected due to the outliers as compared to the swarm diameter and swarm radius. But an extreme farthest outlier may skew the individual's dispersion significantly in the swarm. Further, $(x_{ik} - \bar{x}_k)^2 \to 0$

(refer equation (7)) for all individuals in the swarm as swarm converges. The same behavior is shown by the diversity measure D_{all} given by equation (10) because for all individuals in the swarm, the component $(x_{ik} - x_{jk})^2$ also approaches to zero as swarm converges.

- The diversity measure D_{GM} shown in equation (8) is again a robust measure for measuring diversity. In statistics, geometric average is relatively less affected from outliers.

- The diversity measure D_N shown in equation (9) is the ratio of the average distance D_A and the swarm diameter D. Here, diameter is considered as a normalization parameter used to normalize the average distance around the swarm center. In this measure of dispersion, as swarm converges, D_N and D, both approaches to zero with iterations. Further, in this dispersion, as the normalization is done by the swarm diameter or the swarm radius, it is significantly influenced by the outlier individuals. Therefore, D_A and D_{all} still may be considered as a better choice for measuring diversity of the swarm.

- The diversity measure S, which is shown in equation (11), is the ratio of the absolute step size of the swarm center to the average step size of all individuals in the swarm. A high value of the swarm center step size implies that all the individuals in the swarm are moving in the same direction. Further, a low value implies that a most of the individuals are moving to opposite directions. A high value of average individual's step size in swarm implies the the solutions are significantly changing the positions which implies exploration of the search space while, a low value shows the convergence in the swarm i.e. exploiting the solution search space found so far. Thus, S could be used to analyse the diversity behavior of the algorithm.

Fig. 4 shows a large swarm diversity and small swarm coherence situation i.e. the individuals are dispersed in the solution search space whereas the step size of swarm center is low relatively.

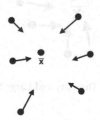

Fig. 4: High Particle Diversity and Small Swarm Coherence

Fig. 5 shows a large swarm diversity with high swarm coherence value i.e. the individuals are dispersed in the solution search space whereas the step size of swarm center is also high.

Further, by analyzing Fig. 6 and Fig. 7, it is clear that the value of S does not depends completely over the diversity of swarm. Therefore, it could be concluded

Fig. 5: High Particle Diversity and Large Swarm Coherence

that the swarm coherence is not proportional to swarm diversity of individuals in the ABC and is not a true measure of diversity.

Fig. 6: Low Particle Diversity and Small Swarm Coherence

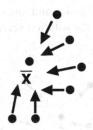

Fig. 7: Low Particle Diversity and Large Swarm Coherence

The outcome of above discussion is that the effect of outlier is significant on most of the diversity measures and it biases the measure of dispersion. However, the effect of outliers could be minimized, it can not be ignored completely.

5 Experimental Results

To analyze the various diversity measures for ABC process, experiments have been carried out on five well known benchmark test problems listed in Table 1. For these experiments following experimental setting is adopted:

- Colony size $NP = 50$ [16, 17],
- $\phi_{ij} = rand[-1, 1]$,
- Number of food sources $SN = NP/2$,
- $limit = 1500$ [18, 19],
- The stopping criteria is either maximum number of function evaluations (which is set to be 200000) is reached or the acceptable error (mentioned in Table 1) has been achieved,
- The number of runs =100 and graphs are plotted using the mean of each run.
- Scaling, which is used to constrict the graph outputs to the interval $[0, 1]$, is shown below:

$$\bar{y} = \frac{\bar{y} - \min(\bar{y})}{\max(\bar{y}) - \min(\bar{y})}.$$

This is done to make comparisons of all measures in the same range.

Table 1: Test problems

Test Problem	Objective function	Search Range	I	Acceptable Error
Sphere	$f_1(x) = \sum_{i=1}^{I} x_i^2$	[-5.12 5.12]	30	$1.0E - 15$
Griewank	$f_3(x) = 1 + \frac{1}{4000} \sum_{i=1}^{I} x_i^2 - \prod_{i=1}^{I} \cos(\frac{x_i}{\sqrt{i}})$	[-600 600]	30	$1.0E - 15$
Rosenbrock	$f_4(x) = \sum_{i=1}^{I} (100(x_{i+1} - x_i^2)^2 + (x_i - 1)^2)$	[-30 30]	30	$1.0E - 15$
Rastrigin	$f_5(x) = 10I + \sum_{i=1}^{I} [x_i^2 - 10\cos(2\pi x_i)]$	[-5.12 5.12]	30	$1.0E - 15$
Ackley	$f_6(x) = -20 + e + exp(-\frac{0.2}{I}\sqrt{\sum_{i=1}^{I} x_i^3})$ $-exp(\frac{1}{I}\sum_{i=1}^{I} \cos(2\pi x_i)x_i)$	[-1 1]	30	$1.0E - 15$

Figures 8-12 show the swarm diversity, as returned by diversity measure D, R, D_A, D_{GM}, D_N and D_{all} with respect to the error of the considered benchmark problems. Figures 13-17 illustrate swarm diversity, as returned by the swarm coherence measure.

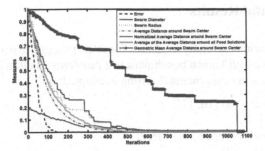

Fig. 8: Diversity Measures with respect to Error on the Spherical Function

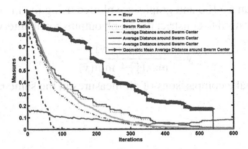

Fig. 9: Diversity Measures with respect to Error on the Griewank Function

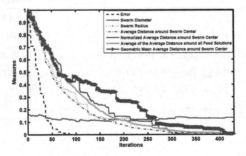

Fig. 10: Diversity Measures with respect to Error on the Rosenbrock Function

6 Conclusion

In this paper, various diversity measures are studied and analyzed to measure the dispersion in the swarm of Artificial Bee Colony algorithm (ABC). In swarm based algorithms, diversity measures are used to investigate the exploration and exploita-

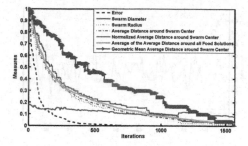

Fig. 11: Diversity Measures with respect to Error on the Rastrigin Function

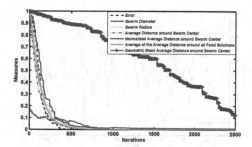

Fig. 12: Diversity Measures with respect to Error on the Ackley Function

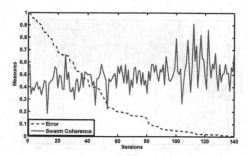

Fig. 13: Swarm Coherence with respect to Error on the Spherical Function

tion characteristics of the algorithms. Further, the diversity measures are analyzed on five well known benchmark problems. The outcome of the experiments shows that the value of diversity measures proportionally decreases with the iterations of the ABC algorithm. A high value of diversity measure shows dispersion in the swarm where as the low value shows convergence of the individuals in the swarm to a solution point. Further, it is found that the diversity measures are more or less effected

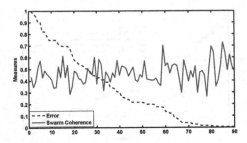

Fig. 14: Swarm Coherence with respect to Error on the Griewank Function

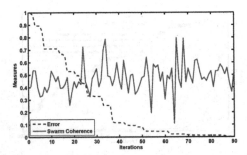

Fig. 15: Swarm Coherence with respect to Error on the Rosenbrock Function

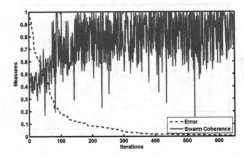

Fig. 16: Swarm Coherence with respect to Error on the Rastrigin Function

by the outliers in the swarm. Diversity measures like the average distance around swarm center, the Geometric Mean average distance around the Swarm Center and the average of average distance around swarm center are less affected by the outliers and could be used for analyzing the exploration and exploitation of the solution search space in the swarm.

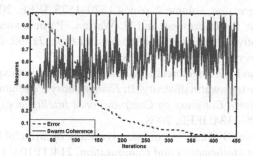

Fig. 17: Swarm Coherence with respect to Error on the Ackley Function

References

[1] M. Dorigo and G. Di Caro. Ant colony optimization: a new meta-heuristic. In *Evolutionary Computation, 1999. CEC 99. Proceedings of the 1999 Congress on*, volume 2. IEEE, 1999.

[2] J. Kennedy and R. Eberhart. Particle swarm optimization. In *Neural Networks, 1995. Proceedings., IEEE International Conference on*, volume 4, pages 1942–1948. IEEE, 1995.

[3] K.V. Price, R.M. Storn, and J.A. Lampinen. *Differential evolution: a practical approach to global optimization*. Springer Verlag, 2005.

[4] J. Vesterstrom and R. Thomsen. A comparative study of differential evolution, particle swarm optimization, and evolutionary algorithms on numerical benchmark problems. In *Evolutionary Computation, 2004. CEC2004. Congress on*, volume 2, pages 1980–1987. IEEE, 2004.

[5] K.M. Passino. Biomimicry of bacterial foraging for distributed optimization and control. *Control Systems Magazine, IEEE*, 22(3):52–67, 2002.

[6] D. Karaboga. An idea based on honey bee swarm for numerical optimization. *Techn. Rep. TR06, Erciyes Univ. Press, Erciyes*, 2005.

[7] A.P. Engelbrecht. Fundamentals of computational swarm intelligence. *Recherche*, 67:02, 2005.

[8] TM Blackwell. Particle swarms and population diversity i: Analysis. In *GECCO*, pages 103–107, 2003.

[9] T. Hendtlass and M. Randall. A survey of ant colony and particle swarm meta-heuristics and their application to discrete optimization problems. In *Proceedings of the Inaugural Workshop on Artificial Life*, pages 15–25, 2001.

[10] T. Krink, J.S. VesterstrOm, and J. Riget. Particle swarm optimisation with spatial particle extension. In *Evolutionary Computation, 2002. CEC'02. Proceedings of the 2002 Congress on*, volume 2, pages 1474–1479. IEEE, 2002.

[11] J.S. Vesterstrom, J. Riget, and T. Krink. Division of labor in particle swarm optimisation. In *Evolutionary Computation, 2002. CEC'02. Proceedings of*

the 2002 Congress on, volume 2, pages 1570–1575. IEEE, 2002.

[12] A. Ratnaweera, S. Halgamuge, and H. Watson. Particle swarm optimization with self-adaptive acceleration coefficients. In *Proc. 1st Int. Conf. Fuzzy Syst. Knowl. Discovery*, pages 264–268, 2003.

[13] O. Olorunda and AP Engelbrecht. Measuring exploration/exploitation in particle swarms using swarm diversity. In *Evolutionary Computation, 2008. CEC 2008.(IEEE World Congress on Computational Intelligence). IEEE Congress on*, pages 1128–1134. IEEE, 2008.

[14] D. Karaboga and B. Akay. A comparative study of artificial bee colony algorithm. *Applied Mathematics and Computation*, 214(1):108–132, 2009.

[15] J. Riget and J.S. Vesterstrøm. A diversity-guided particle swarm optimizer-the arpso. *Dept. Comput. Sci., Univ. of Aarhus, Aarhus, Denmark, Tech. Rep*, 2:2002, 2002.

[16] K. Diwold, A. Aderhold, A. Scheidler, and M. Middendorf. Performance evaluation of artificial bee colony optimization and new selection schemes. *Memetic Computing*, pages 1–14, 2011.

[17] M. El-Abd. Performance assessment of foraging algorithms vs. evolutionary algorithms. *Information Sciences*, 2011.

[18] D. Karaboga and B. Akay. A modified artificial bee colony (abc) algorithm for constrained optimization problems. *Applied Soft Computing*, 2010.

[19] B. Akay and D. Karaboga. A modified artificial bee colony algorithm for real-parameter optimization. *Information Sciences*, 2010.

Digital Video Watermarking using Scene Detection

Dinesh Tiwari, K. V. Arya, Mukesh Saraswat

Abstract : Video is one of the most commonly used multimedia data used in daily life. This increases the occurrences of abuse and copyright infringement which happen to video data content. To overcome from this problem video security and copyright protection techniques are required. Video watermarking is one of the most popular techniques for the providing security and copyright protection. Researchers have introduced many methods for video watermarking but still there are limitations of frame dropping, averaging problems that are to be dealt properly. Therefore, in this paper, a robust video watermarking method is proposed which is based on the scene detection method and it effectively deals with frame dropping, averaging and statistical analysis.

Keywords: Watermark Detection, Image Processing, Scene Detection

1 Introduction

Presently the Internet and multimedia is highly used in academics and industry for transferring data, text and multimedia documents across the internet. This increases the risk of copyright infringement and therefore, some tools are required to protect the same. There are two basic techniques; encryption and watermarking; to provide

Dinesh Tiwari
Feroze Gandhi Institute of Engineering & Technology, Raebareli, India.
e-mail: dk_tiwari_1975@yahoo.com

K. V. Arya
ABV- Indian Institute of Information Technology & Management, Gwalior, India.
e-mail: kvarya@iiitm.ac.in

Mukesh Saraswat
ABV- Indian Institute of Information Technology & Management, Gwalior, India.
e-mail: saraswatmukesh@gmail.com

J. C. Bansal et al. (eds.), *Proceedings of Seventh International Conference on Bio-Inspired Computing: Theories and Applications (BIC-TA 2012)*, Advances in Intelligent Systems and Computing 201, DOI: 10.1007/978-81-322-1038-2_27, © Springer India 2013

copy and copyright protection. But encryption techniques can only provide protection during transmission of data from the sender to the receiver and after receiving protection finished. To overcome this problem, watermarking techniques can provide the solution by embedding a secret imperceptible signal known as watermark into data in such a way that it always remains present. Therefore, watermarking algorithm is one of the techniques used extensively for this purpose as it deals with embedding owner's information into the multimedia data transparently or undetectable. Out of other multimedia data, video data is most commonly used across the internet and requires robust methodology to minimize the abuse and copyright infringement. That's why, in recent years, an emphasis on the video watermarking have been increased among the researchers.

There are different challenges for video watermarking which increase the complexity of embedding the watermarking in the video. Video data are susceptible to increased attacks than any other media. The quality of video data is subjective in nature and watermarking may degrade the quality. Further, video compression techniques reduces the space for watermarking computation. Frame dropping, frame averaging, frame swapping, statistical analysis etc. are common problems in video data transmission [1, 2]. This leads to embed the watermark so that it can be recoverable in case of frame dropping. Watermarking should be robust enough against these phenomenons. However, the current techniques have some limitations for solving these problems effectively. Therefore in this paper, we have proposed a robust video watermarking technique which embeds the different watermarks on different parts of the video using scene detection and scrambled watermarks. The result shows that the proposed method is robust against frame dropping, frame averaging and statistical analysis.

Rest of the paper is organized as follows: Section 2 describes the techniques presented in the literature in the past decade for video watermarking. Proposed methodology is described in Section 3. Experimental results are discussed in Section 4 and Section 5 concludes the paper.

2 Related Work

Video watermarking can be performed for both compressed video and uncompressed video. The watermarking can be performed in spatial domain as well as in frequency domain. In the spatial domain, the pixel values are directly effected in watermarking whereas in frequency domain the watermarking can be performed after transforming the image into one of the transforms: discrete fourier transform (DFT), discrete cosine transform (DCT) or discrete wavelet transform (DWT). The different watermarking schemes includes least significant bit (LSB) based watermarking [3], threshold-based correlation watermarking [4], direct sequence watermark using m-frame [5], DFT with template matching [6], DWT based watermarking [7], DCT based watermarking [8], and spread spectrum [9] watermarking scheme.

In case of uncompressed video, spread spectrum technique can be used as described by Hartung and Girod [10]. Whereas, gain adjustment, drift adjustment and bit-rate control for containing data rate and artifacts are the approaches which may be used in compressed domain video. In transmission of video, compressed domain watermarking is more acceptable as compare to uncompressed domain processing. In case of lossy compression, added noise, and geometric transformation, frequency domain watermarking schemes perform better than the spatial domain methods. In spite of the development of various techniques for video watermarking, most of the techniques are based on the digital image watermarking methods which laid some limitations on video watermarking such as the developed algorithms are not robust to frame dropping, averaging and statistical analysis in case of video [1], [2]. The watermarks which are embedded does not consists of whole information of the message and bit rate of the watermark is kept low. Therefore, to overcome from these cases, we have introduced a robust watermarking scheme which will effectively deal with these problems.

3 Proposed Method

The proposed digital video watermarking technique is a three step process as depicted in Figure 1. The input video is given to scene detection phase as described by Patrick [11] which find out the number of scene in the video using entropy of frame difference. The watermark message which is used in this method is represented as an image. This image is preprocessed before embedding in the video. Then the message is embedded into the video in subsequent phase. The watermark detection also requires the scene detection followed by watermark detection. The whole process is described in following subsections.

3.1 Scene Change Detection

The introduced method is based on uncompressed video content. To detect the scene change in the image sequence, the frames from the input video are extracted. Let there are n frames each represented as f_i for $i = 1, 2, ..., n$ in the video. To find the scene change, each frame image is converted into HSI model and intensity component is taken for scene detection. The bit wise difference between two consecutive frames f_i and f_{i-1} is calculated which provide $n - 1$ difference images as output. Now, the entropy of nth difference image is calculated as given in Eq. 1.

$$E_n = -\sum p_i log_{256} p_i \tag{1}$$

where p_i is the probability of ith intensity value in the difference image. This entropy is used to calculate the scene change by comparing it with a threshold value

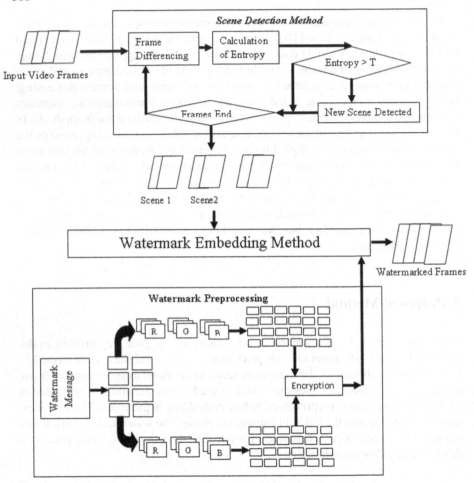

Fig. 1 Detailed flowchart of proposed method.

T. If the entropy of difference image is greater than T then there is a change in the scene. For performing the difference at least three frames are required to process. The information regarding scene change, number of frames per scene are stored for further processing.

3.2 Watermark Preprocess

In this paper, a RGB watermark image is used for embedding into the frames of the video. For the same, the watermark image is first divided into subparts and these subparts are embedded in the frames of different scene. First, the watermark image is resized to, as given by Pik et al. [12], $16.2^p \times 16.2^q$ where p and q are defined as shown in Eq. 2 and 3.

$$2^n \leq m; \quad n > 0 \tag{2}$$

$$p + q = n; \quad p, q > 0 \tag{3}$$

Here m is the number of scene change and n is an integer. From this image, 2^n small images of size 16×16 have been cropped. Further, each plane of small image is decomposed into 8 bit-planes which are arranged one by one to form a large image of size 64×96 as shown in Figure 1. Thus, the pixels in the resultant image consists of only 0's and 1's. This provides 2^n independent watermarks which are to be embedded into the video frames. Before embedding the watermark, it goes through a process of encryption.

3.3 Watermark Embedding

To embed the modified watermark images into the video frames, we have used least-significant bit modification method. In this method, LSB plane of the image is replaced with the given watermark. As the size of the video frames is much larger than the generated watermark image, it can be embedded easily into the frames. The watermark image is embedded into the red component of each frame. One small watermark image is embedded into all frames of one scene. This minimizes the effect of the attacks that are carried out at some designated frames.

3.4 Watermark Detection

The watermark detection goes through the similar process of scene detection followed by extraction of the part of watermark through image averaging from different frames of each scene. All the extracted 8 bit RGB planes are combined to reform the 16 x 16 size image which is the part of the original watermark. From these parts of watermark image, the original large watermark image can be obtained.

4 Experimental Results

To implement the proposed watermarking scheme, ten different types of videos were captured from fuji films digital cam (fxs2000HD). Each video contains number of frames of size 512 x 256. Figure 2 represents one frame from four different considered videos. To measure the similarity of the extracted and the input watermarks, normalized cross-correlation has been used which is given in Eq. 4.

$$cc = \frac{\sum [f(x,y) - \overline{f_{u,v}}][t(x-u,y-v) - \overline{t}]}{\{\sum [f(x,y) - \overline{f_{u,v}}]^2 \sum [t(x-u,y-v) - \overline{t}]^2\}^{0.5}} \qquad (4)$$

here f, $\overline{t}$, and $\overline{f_{u,v}}$ are the image, mean of the watermark image and mean of input watermark image respectively. To simulate the method, MatLab tool is used. The accuracy of scene change detection is given by Eq. 7.

(a)   (b)

(c)   (d)

Fig. 2 Representative frames from four different videos captured from HS2000HD digital camera.

$$Recall = \frac{N_c}{N_c + N_m} \qquad (5)$$

$$Precision = \frac{N_c}{N_c + N_f} \qquad (6)$$

$$Accuracy = \frac{2 * Recall * Precision}{Recall + Precision} \qquad (7)$$

where N_c, N_m and N_f stands for the number of correctly detected scenes, number of missed ones and number of false detections respectively. The results of scene

Table 1 Accuracy results of scene detection methodology.

Video	Recall	Precision	Accuracy
1	91.6	69.2	78.8
2	93.3	75.1	83.2
3	81.3	67.5	73.8
4	97.1	76.2	85.4
5	92.5	79.4	85.5
6	81.3	66	72.9
7	86.6	71.6	78.4
8	73.2	87.3	79.6
9	75.9	68.6	72.1
10	85.5	81.2	83.3

detection are compared with the manual counting of the scenes. The results for all considered videos are shown in Table 1.

For embedding the watermark, it is subdivided into smaller parts numbered less than or equals to number of scene detected in the video. The generated binary watermarks were embedded into the red component of each frames of the video. Each frame from one scene contains similar watermark part. The original watermark image is recovered by applying the watermark detection algorithm as described in section 3.4. The original large watermark image can be reconstructed by finding enough scenes and all parts of the watermark image. Figure 3 depicts one of the original frame, the watermarked frame, subparts of preprocess watermark, and the embedded watermark from the video.

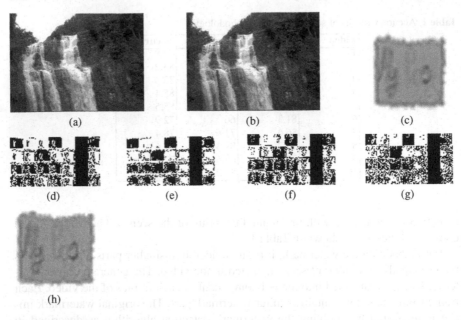

Fig. 3 (a) Original frame (b) Watermarked frame (c) Embedded watermark (d-g) Subparts of pre-process watermark (h) Extracted watermark.

5 Conclusion

This paper presents a robust video watermarking method based on scene detection and LSB modification. The whole process is divided into three step: scene change detection, watermark embedding and watermark detection. The scene change detection is calculated using the entropy of frame difference. The watermark image is preprocessed before embedding. For this, RGB watermark image is taken which is converted into different parts of binary image. The parts of the watermark are embedded into the frames using LSB modification technique. Experiments are conducted on a set of video database. The average accuracy of 79.3% for the scene change detection is observed. The proposed algorithm works well against attacks by frame dropping, frame averaging, and statistical analysis. Future works includes testing the method on a large set of video database and improving the accuracy of the system.

References

1. S.-J. Lee and S.-H. Jung, "A survey of watermarking techniques applied to multimedia," in *Proc. of IEEE International Symposium on Industrial Electronics*, 2001.

2. T. Jayamalar and V. Radha, "Survey on digital video watermarking techniques and attacks on watermarks," *International Journal of Engineering Science and Technology*, vol. 2, pp. 6963–6967, 2010.
3. A. Bamatraf, R. Ibrahim, and M. Salleh, "Digital watermarking algorithm using lsb," in *Proc. of International Conference on Computer Applications and Industrial Electronics*, 2010.
4. G. Langelaar, I. Setyawan, and R. Lagendijk, "Watermarking digital image and video data," *IEEE Signal Processing Magazine*, vol. 17, p. 2046, 2000.
5. B. Mobasseri, "Direct sequence watermarking of digital video using m-frames," in *Proc. of International Conference on Image Processing*, 1998.
6. S. Pereira and T. Pun, "Robust template matching for affine resistant image watermarks," *IEEE Transactions on Image Processing*, vol. 9, pp. 1123–1129, 2000.
7. M. Ejima and A. Miyazaki, "A wavelet-based watermarking for digital images and video," in *Proc. of International Conference on Image Processing*, 2000.
8. L. Jianfeng, Y. Zhenhua, Y. Fan, and L. Li, "A mpeg2 video watermarking algorithm based on dct domain," in *Workshop on Digital Media and Digital Content Management*, 2011, pp. 194–197.
9. M. George, J. V. Chouinard, and N. Georganas, "Digital watermarking of images and video using direct sequence spread spectrum techniques," in *Proc. of IEEE Canadian Conference on Electrical and Computer Engineering*, 1999.
10. F. Hartung and B. Girod, "Watermarking of uncompressed and compressed video," *Signal Processing*, vol. 66, pp. 283–301, 1998.
11. P. Seeling, "Scene change detection for uncompressed video," in *Proc. of the Int. Joint Conferences on Computer, Information, and Systems Sciences and Engineering*, 2008.
12. P. W. Chan, M. Lyu, and R. Chin, "A novel scheme for hybrid digital video watermarking: approach, evaluation and experimentation," *IEEE Transactions on Circuits and Systems for Video Technology*, vol. 15, pp. 1638–1649, 2005.

Self Adaptive Acceleration Factor in Particle Swarm Optimization

Shimpi Singh Jadon, Harish Sharma, Jagdish chand Bansal, and Ritu Tiwari

Abstract Particle swarm optimization (PSO), which is one of the leading swarm intelligence algorithms, dominates other optimization algorithms in some fields but, it also has the drawbacks like it easily falls into local optima and suffers from slow convergence in the later stages. This paper proposes improved version of PSO called Self Adaptive Acceleration Factors in PSO (SAAFPSO) to balance between exploration and exploitation. We converted the constant acceleration factors used in standard PSO into function of particle's fitness. If a particle is more fit then it gives more importance to global best particle and less to itself to avoid local convergence. In later stages, particles will be more fitter so all will move towards global best particle, thus achieved the convergence speed. Experiment is performed and compared with standard PSO and Artificial bee colony (ABC) on 14 unbiased benchmark opti-

Shimpi Singh Jadon
ABV-Indian Institute of Information Technology and Management, Gwalior
Tel.: +09926262325,
Fax: + 0751-2449813
e-mail: shimpisingh2k6@gmail.com

Harish Sharma
ABV-Indian Institute of Information Technology and Management, Gwalior
Tel.: +09479810157,
Fax: + 0751-2449813
e-mail: harish.sharma0107@gmail.com

Jagdish Chand Bansal
ABV-Indian Institute of Information Technology and Management, Gwalior
Tel.: +0751-2449819,
Fax: + 0751-2449813
e-mail: jcbansal@gmail.com

Ritu Tiwari
ABV-Indian Institute of Information Technology and Management, Gwalior
Tel.: +09479810157,
Fax: + 0751-2449813
e-mail: tiwariritu2@gmail.com

J. C. Bansal et al. (eds.), *Proceedings of Seventh International Conference on Bio-Inspired Computing: Theories and Applications (BIC-TA 2012)*, Advances in Intelligent Systems and Computing 201, DOI: 10.1007/978-81-322-1038-2_28, © Springer India 2013

mization functions and one real world engineering optimization problem (known as pressure vessel design) and results show that proposed algorithm SAAFPSO dominates others.

Key words: Particle Swarm Optimization, Artificial Bee Colony, Swarm intelligence, Acceleration factor, Optimization

1 Introduction

After being inspired from social behavior of fish schooling and birds flocking while searching the food, Kennedy and Eberhart [15], [8] developed a swarm intelligence based optimization technique called Particle swarm optimization (PSO) in 1995. PSO is a population based, easy to understand and implement, robust meta heuristic optimization algorithm. PSO can be a better choice for multi model, non convex, non linear and complex optimization problems but like any other evolutionary algorithm, it also has drawbacks like trapping into local best solutions[19], inefficient computationally as measured by the number of function evaluations required [5]. These points restrict PSO to less applicability [18]. Researchers are continuously working to achieve these goals i.e., increasing convergence speed and ignoring the local optima to explore PSO applicability. As a result, a huge variants of PSO algorithm have been proposed [22],[19],[31],[32],[13], [5] to get rid of these weaknesses. However, achieving both goals simultaneously is difficult like liang et al. proposed the comprehensive-learning PSO (CLPSO) [19] which aims at ignoring the local optima, but results show that it also suffers from slow convergence. Ratnaweera et al. [22] proposed time varying acceleration factors to balance between cognitive and social component in initial and later stages. Zhan et al. [31] also tried to adapt the acceleration factors increasing or decreasing depending on different exploring or exploring search space stages. Zhang et al. [32] studied effect of these factors on position expectation and variance and suggested that setting the cognitive acceleration factor as 1.85 and the social acceleration factor as 2 works good for improving system stability. Gai-yun et al. [13] also worked for self adaption of cognitive and social factors. We proposed self adaptive acceleration factors in PSO algorithm after analyzing the drawbacks of PSO.

Rest of the paper is organized as follows: Standard PSO is explained in section 2. In section 3, self adaptive particle swarm optimization is proposed. In Section 4, performance of the proposed strategy is analyzed. Finally, in section 5, paper is concluded.

2 Standard Particle Swarm Optimization Algorithm

PSO is an optimization technique which simulates the birds flocking behavior. PSO is a dynamic population of active, interactive agents with very little in the way of inherent intelligence. In PSO, whole group is called *swarm* and each individual is called *particle* which represents possible candidate's solution. The Swarm finds food for its self through social learning by observing the behavior of nearby birds who appeared to be near the food source. Initially each particle is initialized within the search space randomly and keeps the information about its personal best position known as *pbest*, swarm best position known as *gbest* and current velocity V with which it is moving, in her memory. Based on these three values, each particle updates its position. In this manner, whole swarm moves in better direction while following collaborative trail and error method and converges to single best known solution.

For an D dimensional search space, the ith particle of the swarm at time step t is represented by a D- dimensional vector, $X_i^t = (x_{i1}, x_{i2}, ..., x_{iD})$. The velocity of this particle is represented by another D-dimensional vector $V_i^t = (v_{i1}, v_{i2}, ..., v_{iD})$. The previously best visited position of the ith particle is denoted as $P_i^t = (p_{i1}, p_{i2}, ..., p_{iD})$. g is the index of the best particle in the swarm. PSO swarm uses two equations for movement called *velocity update equation* and *position update equation*. The velocity of the ith particle is updated using the velocity update equation given by (1) and the position is updated using (2).

$$v_{id}^{t+1} = v_{id}^t + c_1 r_1 (p_{id}^t - x_{id}^t) + c_2 r_2 (p_{gd}^t - x_{id}^t) \tag{1}$$

$$x_{id}^{t+1} = x_{id}^t + v_{id}^{t+1} \tag{2}$$

where $d = 1, 2, ..., D$ represents the dimension and $i = 1, 2, ..., S$ represents the particle index. S is the size of the swarm and c_1 and c_2 are constants (usually $c_1 = c_2$), called cognitive and social scaling parameters respectively or simply acceleration coefficients. r_1 and r_2 are random numbers in the range $[0, 1]$ drawn from a uniform distribution.

The right hand side of velocity update equation (1) consists of three terms, the first term v_{id}^t is the memory of the previous direction of movement which can be thought of as a momentum term and prevents the particle from drastically changing direction. The second term $c_1 r_1 (p_{id}^t - x_{id}^t)$ is called cognitive component or persistence which draws particle back to their previous best situation and enables the local search in swarm. The last term $c_2 r_2 (p_{gd}^t - x_{id}^t)$ is known as social component which allows individuals to compare themselves to others in it's group and is responsible for global search. The Pseudo-code for Particle Swarm Optimization, is described as follows :

Based on the neighborhood size, initially two versions of PSO algorithm were presented in literature, namely, global version of PSO which is the original PSO (PSO-G) and the local version of PSO (PSO-L)[23] The only difference between PSO-G and PSO-L is that the term p_g in social component in velocity update equa-

Algorithm 1 Particle Swarm Optimization Algorithm:

Initialize the parameters, w, c_1 and c_2;
Initialize the particle positions and their velocities in the search space;
Evaluate fitness of individual particles;
Store gbest and pbest;
while stopping condition(s) not true **do**
    for each individual, X_i **do**
        for each dimension d, x_{id} **do**
            (i) Evaluate the velocity v_{id} using (1);
            (ii) Evaluate the position x_{id} using (2);
        end for
    end for
    Evaluate fitness of updated particles;
    Update gbest and pbest;
end while
Return the individual with the best fitness as the solution;

---

tion (1). For PSO-G, it refers the best particle of whole swarm while for PSO-L it represents the best particle of the individual's neighborhood. The social network employed by the PSO-G reflects the star topology which offers a faster convergence but it is very likely to converge prematurely. While PSO-L uses a ring social network topology where smaller neighborhoods are defined for each particle. It can be easily observed that due to the less particle inter connectivity in PSO-L, it is less susceptible to be trapped in local minima but at the cost of slow convergence. In general, PSO-G performs better for unimodal problems and PSO-L for multimodal problems.

Velocity update equation in PSO determines the balance between exploration and exploitation capability of PSO. In Basic PSO, no bounds were defined for velocity, due to which in early iterations the particles far from gbest, will take large step size and are very much intended to leave the search space. Thus to control velocity so that particle update step size is balanced, velocity clamping concept was introduced. In velocity clamping, whenever velocity exceeds from its bounds, it is set at its bounds. To avoid the use of velocity clamping and to make balance between exploration and exploitation, two new parameters called inertia weight [25] and constriction factor [6] were introduced in velocity update equation as:

$$v_{id}^{t+1} = w * v_{id}^t + c_1 r_1 (p_{id}^t - x_{id}^t) + c_2 r_2 (p_{gd}^t - x_{id}^t) \tag{3}$$

$$v_{id}^{t+1} = \chi * (v_{id}^t + c_1 r_1 (p_{id}^t - x_{id}^t) + c_2 r_2 (p_{gd}^t - x_{id}^t)) \tag{4}$$

where inertia weight and constriction factor are denoted by w and χ respectively. In the velocity update equation(1) of PSO, if we remove the first part then the flying particle's velocity is only determined by its current position and its best position in history and velocity itself is memory less. Therefore, it can be imagined that the search process for PSO without the first part is a process where the search space statistically shrinks through the generations. PSO starts to behave like a local search algorithm. Only when the global optimum is within the initial search space, then there is a chance for PSO to find the solution. On the other hand, by adding the first part, the particles have a tendency to expand the search space, that is, they have the ability to explore the new area. Considering of this, both terms w and χ were introduced to balance the global search and local search but for constriction factor, velocity clamping was completely ignored. As far as the optimal values to set these parameters are concerned, researchers are continuously trying to fine tune these and suggested that alternative values in[25, 12, 20, 30, 10, 3, 17] can be used, depending on the nature of problem, as for different problems, there should be different balances between the local search ability and global search ability. Eberhart and Shi [9] concluded that as a rule of thumb, the best approach to use with particle swarm optimization is to utilize the constriction factor approach while limiting V_{max} to X_{max} or utilize the inertia weight approach while selecting w, c_1, and c_2 according to equation:

$$\chi = \frac{2}{|2 - \phi - \sqrt{\phi(\phi - 4)}|}, \text{where } \phi = c_1 r_1 + c_2 r_2, \phi > 4. \quad (5)$$

3 Self adaptive acceleration factor in Particle Swarm Optimization

However the standard PSO has the capability to get a good solution at a significantly faster rate but, when it is compared to other optimization techniques, it is weak to refine the optimum solution, mainly due to less diversity in later search [2]. On the different side, problem-based tuning of parameters is also important in PSO, to get optimum solution accurately and efficiently[26]. In standard PSO's velocity update equation (1) contains three terms. The first term has the global search capability, the second and third terms are the particles cognitive and social information sharing capability respectively. More cognitive capability force particle to move towards personal best position fast and more social information force particle to move towards global best position fast. It can be seen from (1), the movement of swarm towards optimum solution is guided by the acceleration factor c_1 and c_2. Therefore, acceleration coefficient c_1 and c_2 should be tuned carefully to get the desired solution.

Kennedy and Eberhart [15] explained that more value of the cognitive component compared to the social component, results in excessive wandering of individuals through the search space while on the other hand more value of the social component may results that particles will converge prematurely toward a local optimum. Moreover, they advised to set both coefficient to 2 so that mean of both factors in velocity update equation is unity in order to move particles over only half the time of feasible region. Since then, most of the researchers are extensively using this phenomena. Suganthan [27] tested time varying linearly decreasing both factors and noticed that the fixed value 2 of these factors produce better solutions. However, he also advised that the acceleration factors should not be equal to 2 all the time, based on empirical studies. Generally, in population-based optimization methods, in early stages individuals should be diverged through the entire feasible region and should not gathered at one point while in later stages, the solutions should be fine tuned to concentrate towards the global optima.

After analyzing these concerns, we propose fitness based acceleration factors as a new parameter strategy for PSO, simply to explore search space in early stages and to fine tune solutions in final stages. The proposed modification can be mathematically seen as follows:

$$c_1 = 1.2 - prob_i \tag{6}$$

$$c_2 = 0.5 + prob_i \tag{7}$$

where c_1 and c_2 are cognitive component's coefficient and social component's coefficient respectively and $prob_i$ is the relative fitness of ith particle in population at time t which is calculated as:

$$prob_i = \frac{fit_i}{maxfit} \tag{8}$$

where fit_i is the fitness of ith particle and $maxfit$ is the fitness of best particle in the current population so that it will be in range $[0, 1]$. And the constants $1.2, 0.5$ in equations (6),(7) respectively are used based on empirical studies. In this way, after each iteration, probability of each particle is calculated and used to decide acceleration factor's value. Initially each particle has low fitness so coefficient value will be more, therefore particles will be more able to explore search space and in later stages all particle will have higher fitness so coefficient will be low, therefore particles will be able to search in small neighborhood of existing solutions i.e., fine tuning of existing solutions.

4 Experiments and Results

4.1 Test problems under consideration

To validate the discussion, we compared the proposed strategy with standard PSO over 14 unbiased benchmark optimization functions which are given in Table 1. Further, to see the robustness of the proposed strategy, a well known engineering optimization problem (f_{15} namely, pressure vessel (confinement method) [28] is also solved. The considered engineering optimization problem is described as follows:

Pressure Vessel design: The pressure vessel design is an optimization problem in which the total cost of the material, forming, and welding of a cylindrical vessel is minimized [28]. There are four design variables involved: x_1, (T_s, shell thickness),

x_2 (T_h, spherical head thickness), x_3 (R, radius of cylindrical shell), and x_4 (L, shell length). The mathematical formulation of this typical constrained optimization problem is as follows:

$$f_{15}(\mathbf{x}) = 0.6224x_1x_3x_4 + 1.7781x_2x_3^2 + 3.1611x_1^2x_4 + 19.84x_1^2x_3$$

subject to

$$g_1(\mathbf{x}) = 0.0193x_3 - x_1$$
$$g_2(\mathbf{x}) = 0.00954x_3 - x_2$$
$$g_3(\mathbf{x}) = 750 \times 1728 - \pi x_3^2(x_4 + \frac{4}{3}x_3)$$

The search boundaries for the variables are

$$1.125 \leq x_1 \leq 12.5,$$

$$0.625 \leq x_2 \leq 12.5,$$

$$1.0 \times 10^{-8} \leq x_3 \leq 240$$

and

$$1.0 \times 10^{-8} \leq x_4 \leq 240.$$

The best known global optimum solution is $f(1.125, 0.625, 55.8592, 57.7315) = 7197.729$ [28]. For a successful run, the minimum error criteria is fixed to be $1.0E - 05$ i.e. an algorithm is considered successful if it finds the error less than acceptable error in a specified maximum function evaluations.

Table 1: Test problems

Test Problem	Objective function	Search Range	Optimum Value	D	Acceptable Error
Beale function	$f_1(x) = [1.5 - x_1(1-x_2)]^2 + [2.25 - x_1(1-x_2^2)]^2 + [2.625 - x_1(1-x_2^3)]^2$	[-4.5,4.5]	$f(3,0.5) = 0$	2	1.0E-05
Colville function	$f_2(x) = 100[x_2 - x_1^2]^2 + (1-x_1)^2 + 90(x_4 - x_3^2)^2 + (1-x_3)^2 + 10.1[(x_2-1)^2 + (x_4-1)^2] + 19.8(x_2-1)(x_4-1)$	[-10,10]	$f(1) = 0$	4	1.0E-05
Branins's function	$f_3(x) = a(x_2 - bx_1^2 + cx_1 - d)^2 + e(1-f)\cos x_1 + e$	$-5 \le x_1 \le 10, 0 \le x_2 \le 15$	$f(-\pi, 12.275) = 0.3979$	2	1.0E-05
Kowalik function	$f_4(x) = \sum_{i=1}^{11} [a_i - \frac{x_1(b_i^2 + b_i x_2)}{b_i^2 + b_i x_3 + x_4}]^2$	[-5,5]	$f(0.1928, 0.1908, 0.1231, 0.1357) = 3.07E-04$	4	1.0E-05
2D Tripod	$f_5(x) = p(x_2)(1 + p(x_1)) + |(x_1 + 50p(x_2)(1-2p(x_1)))| + |(x_2 + 50(1 - 2p(x_2)))|$	[-100,100]	$f(0,-50) = 0$	2	1.0E-04
Shifted Rosenbrock	$f_6(x) = \sum_{i=1}^{D-1}(100(z_i^2 - z_{i+1})^2 + (z_i - 1)^2) + f_{bias}, z = x - o + 1, x = [x_1, x_2, ...x_D], o = [o_1, o_2, ...o_D]$	[-100, 100]	$f(o) = f_{bias} = 390$	10	1.0E-01
Shifted Sphere	$f_7(x) = \sum_{i=1}^{D} z_i^2 + f_{bias}, z = x - o, x = [x_1, x_2, ...x_D], o = [o_1, o_2, ...o_D]$	[-100,100]	$f(o) = f_{bias} = -450$	10	1.0E-05
Shifted Rastrigin	$f_8(x) = \sum_{i=1}^{D}(z_i^2 - 10\cos(2\pi z_i) + 10) + f_{bias} z=(x-o), x=(x_1,x_2,.......x_D), o=(o_1,o_2,.......o_D)$	[-5,5]	$f(o) = f_{bias} = -330$	10	1.0E-02
Shifted Schwefel	$f_9(x) = \sum_{i=1}^{D}(\sum_{j=1}^{i} z_j)^2 + f_{bias}, z = x - o ,x = [x_1, x_2, ...x_D], o = [o_1, o_2, ...o_D]$	[-100,100]	$f(o) = f_{bias} = -450$	10	1.0E-05
Shifted Griewank	$f_{10}(x) = \sum_{i=1}^{D} \frac{z_i^2}{4000} - \prod_{i=1}^{D} \cos(\frac{z_i}{\sqrt{i}}) + 1 + f_{bias}, z = (x - o), x = [x_1, x_2, ...x_D], o = [o_1, o_2, ...o_D]$	[-600,600]	$f(o) = f_{bias} = -180$	10	1.0E-05
Goldstein-Price	$f_{11}(x) = (1 + (x_1 + x_2 + 1)^2 \cdot (19 - 14x_1 + 3x_1^2 - 14x_2 + 6x_1 x_2 + 3x_2^2)) \cdot (30 + (2x_1 - 3x_2)^2 \cdot (18 - 32x_1 + 12x_1^2 + 48x_2 - 36x_1 x_2 + 27x_2^2))$	[-2,2]	$f(0, -1) = 3$	2	1.0E-14
Easom's function	$f_{12}(x) = -\cos x_1 \cos x_2 e^{(-(x_1-\pi)^2 - (x_2-\pi)^2)}$	[-10,10]	$f(\pi, \pi) = -1$	2	1.0E-13
Shubert	$f_{13}(x) = -\sum_{i=1}^{5} i\cos((i+1)x_1 + 1) \sum_{i=1}^{5} i\cos((i+1)x_2 + 1)$	[-10, 10]	$f(7.0835, 4.8580) = -186.7309$	= 2	1.0E-05
Sinusoidal	$f_{14}(x) = -[A\prod_{i=1}^{n} \sin(x_i - z) + \prod_{i=1}^{n} \sin(B(x_i - z))], A = 2.5, B = 5, z = 30$	[0, 180]	$f(90 + z) = -(A+1)$	10	1.00E-02

4.2 Experimental setting

To prove the efficiency of *SAAFPSO*, it is compared with *PSO* and *ABC*. To test *SAAFPSO*, *PSO* and *ABC* over considered problems, following experimental setting is adopted:

Parameter setting for PSO and SAAFPSO:

- Swarm size $NP = 50$,
- Inertia weight $w = 0.8$,
- Acceleration coefficients $c_1 = c_2 = 0.5 + log2$ (for PSO)[16] and for SAAFPSO as per equations (6) and (7),
- The number of simulations/run =100.

Parameter setting for ABC:

- Colony size $NP = 50$ [7, 11],
- $\phi_{ij} = rand[-1,1]$,
- Number of food sources $SN = NP/2$,
- $limit = D \times SN$ [14, 1],
- The number of simulations/run =100.

4.3 Results Comparison

Numerical results with experimental setting of subsection 4.2 are given in Table 2. In Table 2, standard deviation (*SD*), mean error (*ME*), average function evaluations (*AFE*), and success rate (*SR*) are reported. Table 2 shows that most of the time *SAAFPSO* outperforms in terms of reliability, efficiency and accuracy as compare to the *PSO* and *ABC*. Some more intensive analyses based on acceleration rate (AR) [21], performance indices and boxplots have been carried out for results of *SAAFPSO*, *PSO* and *ABC*.

Table 2: Comparison of the results of test problems

Test Function	Algorithm	SD	ME	AFE	SR
f_1	PSO	2.78E-06	5.01E-06	2717	100
	SAAFPSO	2.71E-06	5.22E-06	2269.5	100
	ABC	1.66E-06	8.64E-06	16520.09	100
f_2	PSO	1.98E-04	8.40E-04	52701	99
	SAAFPSO	1.98E-04	8.41E-04	22996	100
	ABC	1.03E-01	1.67E-01	199254.48	1
f_3	PSO	3.94E-06	5.98E-06	30590	86
	SAAFPSO	3.43E-06	5.78E-06	22826	90
	ABC	6.83E-06	6.05E-06	1925.52	100

to be cont'd on next page

Table 2: Comparison of the results of test problems (Cont.)

Test Function	Algorithm	SD	ME	AFE	SR
f_4	PSO	8.28E-05	9.90E-05	39473.5	99
	SAAFPSO	1.07E-05	9.16E-05	21835.5	100
	ABC	7.33E-05	1.76E-04	180578.91	18
f_5	PSO	3.54E-01	1.20E-01	33546	89
	SAAFPSO	2.86E-01	9.01E-02	29985.5	91
	ABC	2.69E-05	6.42E-05	8771.65	100
f_6	PSO	7.97E+00	2.53E+00	188678	52
	SAAFPSO	8.35E+00	1.55E+00	86424.5	82
	ABC	1.05E+00	6.36E-01	176098.02	23
f_7	PSO	1.35E-06	8.35E-06	15862	100
	SAAFPSO	1.56E-06	7.88E-06	8652.5	100
	ABC	2.42E-06	7.16E-06	9013.5	100
f_8	PSO	5.35E+00	3.61E+01	200050	0
	SAAFPSO	5.23E+00	3.95E+01	200050	0
	ABC	1.21E+01	8.91E+01	200011.71	0
f_9	PSO	3.22E+03	1.40E+03	200050	0
	SAAFPSO	3.39E+03	3.68E+03	200050	0
	ABC	3.54E+03	1.11E+04	200029.02	0
f_{10}	PSO	3.01E-02	4.63E-02	197240.5	2
	SAAFPSO	5.54E-03	3.29E-03	121521.5	62
	ABC	2.21E-03	6.95E-04	61650.9	90
f_{11}	PSO	2.85E-15	5.70E-15	9722.5	100
	SAAFPSO	2.82E-15	5.54E-15	7688	100
	ABC	5.16E-06	1.04E-06	109879.46	62
f_{12}	PSO	2.90E-14	5.19E-14	9758.5	100
	SAAFPSO	3.18E-14	5.12E-14	8953.5	100
	ABC	4.44E-05	1.60E-05	181447.91	17
f_{13}	PSO	6.25E-04	1.96E-04	89428	68
	SAAFPSO	8.86E-05	2.48E-05	34221.5	93
	ABC	5.34E-06	4.86E-06	4752.21	100
f_{14}	PSO	2.91E-01	4.09E-01	178643	21
	SAAFPSO	3.67E-01	4.03E-01	163543.5	35
	ABC	1.83E-03	7.77E-03	54159.26	99
f_{15}	PSO	3.17E-05	2.92E-05	91222	65
	SAAFPSO	9.36E-02	1.56E-02	87339	76
	ABC	1.07E+01	1.69E+01	200024.49	0

SAAFPSO, *PSO* and *ABC* are compared through *SR*, *ME* and *AFE* in Table 2. First *SR* is compared for all these algorithms and if it is not possible to distinguish the algorithms based on *SR* then comparison is made on the basis of *AFE*. *ME* is used for comparison if it is not possible on the basis of *SR* and *AFE* both. Outcome of this comparison is summarized in Table 3. In Table 3, '+' indicates that the *SAAFPSO* is better than the considered algorithms and '-' indicates that the algorithm is not better or the difference is very small. The last row of Table 3, establishes the superiority of *SAAFPSO* over *PSO* and *ABC*.

Further, we compare the convergence speed of the considered algorithms by measuring the average function evaluations (AFEs). A smaller AFEs means higher convergence speed. In order to minimize the effect of the stochastic nature of the algo-

Table 3: Summary of Table 2 outcome

Function	SAAFPSO Vs PSO	SAAFPSO Vs ABC
f_1	+	+
f_2	+	+
f_3	+	-
f_4	+	+
f_5	+	-
f_6	+	+
f_7	+	+
f_8	-	+
f_9	-	-
f_{10}	+	-
f_{11}	+	+
f_{12}	+	+
f_{13}	+	-
f_{14}	+	-
f_{15}	+	+
Total number of + sign	13	9

Table 4: Acceleration Rate (AR) of $SAAFPSO$ compare to PSO and ABC

Test Problems	PSO	ABC
f_1	1.197179996	7.27917603
f_2	2.291746391	8.664745173
f_3	1.340138439	0.084356436
f_4	1.807767168	8.269969087
f_5	1.118740725	0.292529723
f_6	2.183154082	2.037593738
f_7	1.833227391	1.041722046
f_8	1	0.999808598
f_9	1	0.999895126
f_{10}	1.623091387	0.507325041
f_{11}	1.264633195	14.29233351
f_{12}	1.089908974	20.26558441
f_{13}	2.613210993	0.13886621
f_{14}	1.092327118	0.33116119
f_{15}	1.044458947	2.290208154

rithms, the reported function evaluations for each test problem is the average over 100 runs. In order to compare convergence speeds, we use the acceleration rate (AR) [21] which is defined as follows, based on the AFEs for the two algorithms $ALGO$ and $SAAFPSO$:

$$AR = \frac{AFE_{ALGO}}{AFE_{SAAFPSO}}, \qquad (9)$$

where, $ALGO \in \{PSO, ABC\}$ and $AR > 1$ means $SAAFPSO$ is faster. In order to investigate the AR of the proposed algorithm, compare to the PSO and ABC, results of Table 2 are analyzed and the value of *AR* is calculated using equation (9). Table 4 shows a clear comparison between *SAAFPSO* and *PSO*, *SAAFPSO* and *ABC* in terms of AR. It is clear from the Table 4 that convergence speed of *SAAFPSO* is faster among all the considered algorithms.

For the purpose of comparison in terms of consolidated performance, boxplot analyses have been carried out for all the considered algorithms. The empirical distribution of data is efficiently represented graphically by the boxplot analysis tool [29]. The boxplots for *PSO*, *SAAFPSO* and *ABC* are shown in Figure 1. It is clear from this figure that *SAAFPSO* is better than the considered algorithms as interquartile range and median are comparatively low.

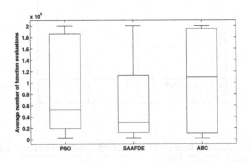

Fig. 1: Boxplots graphs for average function evaluation

Further, to compare the considered algorithms, by giving weighted importance to the success rate, the standard deviation and the average number of function evaluations, performance indices (*PI*) are calculated [4]. The values of *PI* for the *PSO*, *SAAFPSO* and *ABC* are calculated by using following equations:

$$PI = \frac{1}{N_p} \sum_{i=1}^{N_p} (k_1 \alpha_1^i + k_2 \alpha_2^i + k_3 \alpha_3^i)$$

Where $\alpha_1^i = \frac{Sr^i}{Tr^i}$; $\alpha_2^i = \begin{cases} \frac{Mf^i}{Af^i}, & \text{if } Sr^i > 0. \\ 0, & \text{if } Sr^i = 0. \end{cases}$; and $\alpha_3^i = \frac{Mo^i}{Ao^i}$

$i = 1, 2, ..., N_p$

- Sr^i = Successful simulations/runs of ith problem.
- Tr^i = Total simulations of ith problem.
- Mf^i = Minimum of average number of function evaluations used for obtaining the required solution of ith problem.
- Af^i = Average number of function evaluations used for obtaining the required solution of ith problem.
- Mo^i = Minimum of standard deviation obtained for the ith problem.

- Ao^i = Standard deviation obtained by an algorithm for the ith problem.
- N_p = Total number of optimization problems evaluated.

The weights assigned to the success rate, the average number of function evaluations and the standard deviation are represented by k_1, k_2 and k_3 respectively where $k_1 + k_2 + k_3 = 1$ and $0 \leq k_1, k_2, k_3 \leq 1$. To calculate the *PIs*, equal weights are assigned to two variables while weight of the remaining variable vary from 0 to 1 as given in [24]. Following are the resultant cases:

1. $k_1 = W, k_2 = k_3 = \frac{1-W}{2}, 0 \leq W \leq 1$;
2. $k_2 = W, k_1 = k_3 = \frac{1-W}{2}, 0 \leq W \leq 1$;
3. $k_3 = W, k_1 = k_2 = \frac{1-W}{2}, 0 \leq W \leq 1$

The graphs corresponding to each of the cases (1), (2) and (3) for *PSO, SAAFPSO* and *ABC* are shown in Figures 2(a), 2(b), and 2(c) respectively. In these figures the weights k_1, k_2 and k_3 are represented by horizontal axis while the *PI* is represented by the vertical axis.

In case (1), average number of function evaluations and the standard deviation are given equal weights. *PIs* of the considered algorithms are superimposed in Figure 2(a) for comparison of the performance. It is observed that *PI* of *SAAFPSO* are higher than the considered algorithms. In case (2), equal weights are assigned to the success rate and standard deviation and in case (3), equal weights are assigned to the success rate and average function evaluations. It is clear from Figure 2(b) and Figure 2(c) that the algorithms perform same as in case (1).

5 Conclusion

In this paper, a self adaptive strategy is presented for the acceleration factors to balance exploration and exploitation in the search space. In the proposed strategy, acceleration factors are adaptively modified by a probability which is a function of fitness. Therefore, in this strategy, there are different acceleration factors for every individual based on its fitness. It is showed by the acceleration factor that the proposed strategy speedup the convergence ability of PSO. With the help of experiments over 14 unbiased benchmark test functions and one real world engineering optimization problem, it is proved that the proposed strategy outperforms over PSO and basic ABC in terms of reliability, efficiency and accuracy.

References

[1] B. Akay and D. Karaboga. A modified artificial bee colony algorithm for real-parameter optimization. *Information Sciences*, 2010.

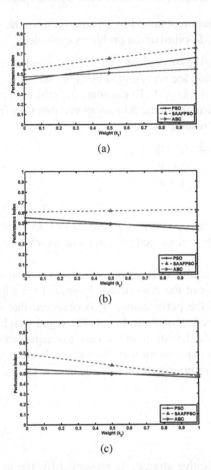

Fig. 2: Performance index for test problems; (a) for case (1), (b) for case (2) and (c) for case (3).

[2] P. Angeline. Evolutionary optimization versus particle swarm optimization: Philosophy and performance differences. In *Evolutionary Programming VII*, pages 601–610. Springer, 1998.

[3] M.S. Arumugam and MVC Rao. On the performance of the particle swarm optimization algorithm with various inertia weight variants for computing optimal control of a class of hybrid systems. *Discrete Dynamics in Nature and Society*, 2006, 2006.

[4] J.C. Bansal and H. Sharma. Cognitive learning in differential evolution and its application to model order reduction problem for single-input single-output systems. *Memetic Computing*, pages 1–21, 2012.

[5] G. Ciuprina, D. Ioan, and I. Munteanu. Use of intelligent-particle swarm optimization in electromagnetics. *Magnetics, IEEE Transactions on*, 38(2):1037–1040, 2002.

[6] M. Clerc. The swarm and the queen: towards a deterministic and adaptive particle swarm optimization. In *Evolutionary Computation, 1999. CEC 99. Proceedings of the 1999 Congress on*, volume 3. IEEE, 1999.

[7] K. Diwold, A. Aderhold, A. Scheidler, and M. Middendorf. Performance evaluation of artificial bee colony optimization and new selection schemes. *Memetic Computing*, pages 1–14, 2011.

[8] R. Eberhart and J. Kennedy. A new optimizer using particle swarm theory. In *Micro Machine and Human Science, 1995. MHS'95., Proceedings of the Sixth International Symposium on*, pages 39–43. Ieee, 1995.

[9] R.C. Eberhart and Y. Shi. Comparing inertia weights and constriction factors in particle swarm optimization. In *Evolutionary Computation, 2000. Proceedings of the 2000 Congress on*, volume 1, pages 84–88. IEEE, 2000.

[10] R.C. Eberhart and Y. Shi. Tracking and optimizing dynamic systems with particle swarms. In *Evolutionary Computation, 2001. Proceedings of the 2001 Congress on*, volume 1, pages 94–100. IEEE, 2001.

[11] M. El-Abd. Performance assessment of foraging algorithms vs. evolutionary algorithms. *Information Sciences*, 2011.

[12] J. Ememipour, M.M.S. Nejad, M.M. Ebadzadeh, and J. Rezanejad. Introduce a new inertia weight for particle swarm optimization. In *Computer Sciences and Convergence Information Technology, 2009. ICCIT'09. Fourth International Conference on*, pages 1650–1653. IEEE, 2009.

[13] W. Gai-yun and H. Dong-xue. Particle swarm optimization based on self-adaptive acceleration factors. In *Genetic and Evolutionary Computing, 2009. WGEC'09. 3rd International Conference on*, pages 637–640. IEEE, 2009.

[14] D. Karaboga and B. Basturk. Artificial bee colony (abc) optimization algorithm for solving constrained optimization problems. *Foundations of Fuzzy Logic and Soft Computing*, pages 789–798, 2007.

[15] J. Kennedy and R. Eberhart. Particle swarm optimization. In *Neural Networks, 1995. Proceedings., IEEE International Conference on*, volume 4, pages 1942–1948. IEEE, 1995.

[16] J.J. Kim, S.Y. Park, and J.J. Lee. Experience repository based particle swarm optimization for evolutionary robotics. In *ICCAS-SICE, 2009*, pages 2540–2544. IEEE, 2009.

[17] H.R. Li and Y.L. Gao. Particle Swarm Optimization Algorithm with Exponent Decreasing Inertia Weight and Stochastic Mutation. In — *2009 Second International Conference on Information and Computing Science*, pages 66–69. IEEE, 2009.

[18] X. D. Li and A. P. Engelbrecht. Particle swarm optimization: An introduction and its recent developments. *Genetic Evol. Comput. Conf.*, pages 3391–3414, 2007.

[19] J.J. Liang, AK Qin, P.N. Suganthan, and S. Baskar. Comprehensive learning particle swarm optimizer for global optimization of multimodal functions. *Evolutionary Computation, IEEE Transactions on*, 10(3):281–295, 2006.

[20] R.F. Malik, T.A. Rahman, S.Z.M. Hashim, and R. Ngah. New particle swarm optimizer with sigmoid increasing inertia weight. *International Journal of Computer Science and Security (IJCSS)*, 1(2):pages 35, 2007.

[21] S. Rahnamayan, H.R. Tizhoosh, and M.M.A. Salama. Opposition-based differential evolution. *Evolutionary Computation, IEEE Transactions on*, 12(1):64–79, 2008.

[22] A. Ratnaweera, S.K. Halgamuge, and H.C. Watson. Self-organizing hierarchical particle swarm optimizer with time-varying acceleration coefficients. *Evolutionary Computation, IEEE Transactions on*, 8(3):240–255, 2004.

[23] J. Kennedy R.C. Eberhart. A new optimizer using particle swarm theory, proceedings of 6th symp. micro machine and human science. In *Proceedings of 6th symp. Micro Machine and Human Science*, pages 39–43. Nagoya, Japan, 1995.

[24] H. Sharma, J. Bansal, and K. Arya. Dynamic scaling factor based differential evolution algorithm. In *Proceedings of the International Conference on Soft Computing for Problem Solving (SocProS 2011) December 20-22, 2011*, pages 73–85. Springer, 2012.

[25] Y. Shi and R. Eberhart. A modified particle swarm optimizer. In *Evolutionary Computation Proceedings, 1998. IEEE World Congress on Computational Intelligence., The 1998 IEEE International Conference on*, pages 69–73. IEEE, 1998.

[26] Y. Shi and R. Eberhart. Parameter selection in particle swarm optimization. In *Evolutionary Programming VII*, pages 591–600. Springer, 1998.

[27] P.N. Suganthan. Particle swarm optimiser with neighbourhood operator. In *Evolutionary Computation, 1999. CEC 99. Proceedings of the 1999 Congress on*, volume 3. IEEE, 1999.

[28] X. Wang, XZ Gao, and SJ Ovaska. A simulated annealing-based immune optimization method. In *Proceedings of the International and Interdisciplinary Conference on Adaptive Knowledge Representation and Reasoning, Porvoo, Finland*, pages 41–47, 2008.

[29] D.F. Williamson, R.A. Parker, and J.S. Kendrick. The box plot: a simple visual method to interpret data. *Annals of internal medicine*, 110(11):916, 1989.

[30] J. Xin, G. Chen, and Y. Hai. A Particle Swarm Optimizer with Multi-stage Linearly-Decreasing Inertia Weight. In *Computational Sciences and Optimization, 2009. CSO 2009. International Joint Conference on*, volume 1, pages 505–508. IEEE, 2009.

[31] Z.H. Zhan, J. Zhang, Y. Li, and H.S.H. Chung. Adaptive particle swarm optimization. *Systems, Man, and Cybernetics, Part B: Cybernetics, IEEE Transactions on*, 39(6):1362–1381, 2009.

[32] W. Zhang, H. Li, Z. Zhang, and H. Wang. The selection of acceleration factors for improving stability of particle swarm optimization. In *Natural Computation, 2008. ICNC'08. Fourth International Conference on*, volume 1, pages 376–380. IEEE, 2008.

Applying Case Based Reasoning in Cuckoo Search for the expedition of Groundwater Exploration

Daya Gupta[1], Bidisha Das[2], V.K. Panchal[3]

[1] Department of Computer Engineering,
Delhi Technological University, Delhi, India
[2] Department of Computer Engineering,
Delhi Technological University, Delhi, India
[3] Defence Terrain Research Laboratory,
Defence Research & Development Organization, Delhi, India

{ dgupta@dce.ac.in[1];das.bidisha86@gmail.com[2];vkpans@gmail.com[3]}

Abstract. A new metaheuristic algorithm named Cuckoo search came up in the recent years. Though lots of metaheuristic algorithms exist but the main advantage of this algorithm is that its search space is extensive in nature. Due to its new arrival it hardly has any footprint in any application. Thus we have adapted this new nature inspired algorithm in our application. The main objective of our application is to find out the potentiality of groundwater in any area as queried by the user. To detect the presence of groundwater, we not only applied Cuckoo search but also case based reasoning. Thus our paper tries to integrate the above mentioned techniques. Our expert provided us with different geographical attributes such as landform, soil, lineament, geology, landuse, and slope. Depending on these attribute values, our proposed algorithm finds out the intensity of groundwater in such areas. Basically the intensity values are narrowed down to high, moderate and low. Thus, once a problem case is given by the user, the case based reasoning uses the K nearest neighbor algorithm to find out the best possible match. After the best possible match is obtained, we apply some propositional logic conditions. The need of propositional logic arises because we have observed a lot of varieties in our case base. Thus to maintain consistency in our output propositional logic is required. Our algorithm achieved 99% efficiency. Thus we can use our proposed work in any real life problem where groundwater detection is necessary. In our application, cases are basically nests.

Keywords: Groundwater detection, nature inspired, meta-heuristic, cuckoo search, K-nearest neighbor.

J. C. Bansal et al. (eds.), *Proceedings of Seventh International Conference on Bio-Inspired Computing: Theories and Applications (BIC-TA 2012)*, Advances in Intelligent Systems and Computing 201, DOI: 10.1007/978-81-322-1038-2_29, © Springer India 2013

1 Introduction

"If the wars of the twentieth century were fought over oil the wars of this century will be fought over water" [The World Bank]. One of the most essential requirements of life is water. A survey on groundwater showed that groundwater constitutes only 0.6% of all the water on this earth planet, 97.4% accounts for seawater and 2% for snow and ice on the poles [1]. Water is a great commodity for human beings. The accessibility of groundwater depends on various geographical features. It had already shown its purpose in domestic, industrial and agricultural use. The study of groundwater existence had been increased due to increase in population.

Though already huge numbers of metaheuristic algorithms are devised but in the year 2010, Yang and Deb [2] formulated another metaheuristic algorithm called Cuckoo search. The nature-inspired algorithms have been used in many optimization problems including NP-hard problems [3]. The main aim of all nature inspired algorithm is that it can intimate the best feature of nature. Thus many nature-inspired metaheuristic algorithms already came into existence like PSO, GA but cuckoo search is different from them. The significant differences can be formulated down [3]:

- CS though a population based algorithm but its selection is similar to harmony search.
- The randomization is more efficient as the step length is heavy tailed and thus large step is possible.
- No of parameters in CS are less than GA and PSO.

Even CS follows two important characteristics of modern metaheuristic algorithm: intensification and diversification. In case of intensification the problem initially searches for current best solution and then finally for global solution while diversification means the algorithm explores the search space efficiently.

The field of case based reasoning arouse due to cognitive science [4]. The classic definition of CBR was coined by Riesbeck and Schank [5]:

"A Case based reasoned solves problems by using or adapting solutions to old problems"

The main aim of CBR is to acquire new skills based on our past experience. An example of CBR is medical diagnosis. When a doctor examines a new patient then he analyzes his current symptoms and compares with those patients those who were having similar symptoms. The treatments of those similar patients are used and modified. Thus CBR system maintains a structured memory of cases (called case base) which represents the experience and a means for specifying the similarity between cases [6]. Thus the main advantages of CBR are basically knowledge acquisition, high solution efficiency and easy knowledge accumulation [7].

The main aim of this paper is to detect the presence of groundwater. Our experts provide us a database that contains various geographical attribute values along with the potentiality of groundwater. We use this database as our case base. The user inputs query to our software and our proposed algorithm generates the potentiality of groundwater. The problem case i.e. query of user is matched with similar cases in the case base. The most relevant case is found out by CBR and then we use propositional logic to detect the presence of groundwater. Usage of propositional logic [8] is necessary because our case base contain cases which are varied in nature.

The organization of this paper is as follows: - Section 2 represents the theoretical aspects of Cuckoo search and CBR. Section 3 represents the methodology adapted to generate our proposed work. The sub-sections are organized as – introduction, algorithm, detailed description of our algorithm. Section 4 represents experiments and simulation – the dataset, the user interface of our system and the result along with its graphical representation. Section 5 represents conclusion and future scope of our proposed work.

2 The Omnipresence of CBR & CR

2.1 Case Based Reasoning

CBR is a methodology to solve new problems based on strategy of learning by past experience. It was first proposed by Roger Scank and his students at Yale University [9]. The main idea of CBR is to gain experience while solving new problems. Basically CBR has two main portions: case retriever and case reasoner. It has a repository called Case Base which stores past cases. The case retriever fetches most relevant case and case reasoned find the most appropriate solution to a given problem. CBR life cycle has the following 4 activities [10]:

- Repartition: a case base to form satisfactory one.
- Retrieve: similar cases depending upon the problem case.
- Reuse: a solution by similar case
- Revise: solution that fit best.
- Retain: the new solution once it has been confirmed.

Figure 1 [11] depicts the entire CBR life cycle.

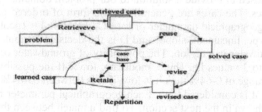

Fig. 1. CBR life cycle

2.2 *Cuckoo Search*

CS originates from the behavior of certain species of cuckoo which lay eggs in other birds nest in parasitic manner [12]. If cuckoo's egg adapts the behavior of host nests it will exist otherwise the host bird will discard cuckoo's egg. The process of evolving to best lay parasitic egg is the main aim of CS [13]. The entire process of CS is described in figure 2. The algorithm is described below:

i)   A comparison is done between cuckoo's egg and set of host nests.

ii)   Lévy flight is used to introduce randomness to choose host nests.

iii)   Comparison produces two sets of solution; one is quality solution and the other discarded solution.

iv)   Depending on the ranking function best solution is found out from quality solution and worst nests are removed with probability p_a.

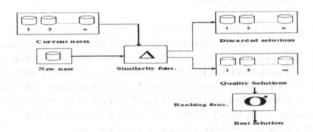

Fig. 2. The procedure of CS

3   Methodology

3.1 *Introduction*

The aim of this section is to describe our problem area and along with it an algorithm is proposed to provide a solution to our problem domain. Experts give the database of cases. The cases are generally in the form of ordered pair C= {F, D}, where F is the geographical parameter of case C, therefore F= {geology, landform, land use, soil type, lineament, slope} and D is the decision attribute i.e. the potentiality of groundwater in a region. The possibility of groundwater or the solution is scaled down into 3 values i.e. {high, moderate, low}. If the possibility of groundwater is in the range of 0-64% then it is low, if it is between 65-84% then it is moderate otherwise it is considered high. Each geographical parameter has certain values which are shown in the next section. To find a match between the problem case and the set of cases in the case base we have used K-nearest neighbor algorithm (KNN) [14]. Our experts provided us a case base of 136 cases. But for our validation test case, we have eliminated 21 cases from the case base. Thus we have worked with a total of 115 cases.

3.2. *Integration with CBR*

In our application CBR has immense importance for retrieval of similar cases from the stored database. In this CBR system, geographical parameters and their corresponding solutions i.e. the possibility of groundwater (High, Moderate and Low) are stored as cases in the case base. By the attribute importance, six attributes are selected as inputs. The possibility of groundwater is selected as the output decision attribute of the case. The cases are collected from the expert's knowledge. The case is represented by the ordered pair C= <F,D> where F is the geographical parameter of case C i.e. F= {Geology, Landform, Soil, Lineament, Slope, Land type} and D is the decision attribute i.e. the possibility of groundwater in the given region.

The case searching and matching is a key step in case retrieving and it directly influences the retrieval efficiency and accuracy. The case retrieval in essence is to find the most similar case in the case base to target case. The nearest neighbor is widely used for its advantage of clear physical concept and simple calculation for case searching. A sum of similarities is calculated according to the similarity between each feature in problem case and the cases in case base. Evaluate the fitness of each nest for case retrieval according to the following equation:

$$Similarity \ (F_i, F_j) = \sum_{i=1}^{n} f \ (F_i, F_j) \ X \ w_i$$

$$if \ (F_i == F_j)$$
$$f(F_i, F_j) = 1;$$
$$else$$
$$f(F_i, F_j) = 0.$$

where Similarity (F_i, F_j) represents the similarity degree of problems case (P) and case (C) stored in case base, i = attribute of case, n = number of attributes, F_i = problem case, F_j= cases stored in case base and w_i = weight of attribute i, the more important attributes should be assigned larger weights than less important ones. The larger the weighted sum, the more similar the two cases will be. The view of our case base is shown in the following figure 3. It contains 7 columns. The first 6 are for attributes and the last column tells about the potentiality of groundwater in such region.

GEOLOGY	LAND_FORM	SOIL	LAND_USE	SLOPE	LINAMENT	SOLUTION
SEDIMENTARY	FLOODPLAIN	SANDYLOAM	AGRICULTURALLAND	GENTLE	ABSENT	HIGH
SEDIMENTARY	INTERMONTAINVALLEY	SANDYGRAVEL	FOREST	STEEP	PRESENT	HIGH
SEDIMENTARY	PEDIMENT	COARSESAND	CULTIVATED	GENTLE	PRESENT	MODERATE
SEDIMENTARY	FLOODPLAIN	SANDYLOAM	AGRICULTURALLAND	GENTLE	PRESENT	HIGH
YOUNGERALLUVIUM	FLOODPLAIN	SANDYLOAM	AGRICULTURALLAND	GENTLE	PRESENT	HIGH

Fig. 3. Cases from the case base

3.3   Algorithm

Input: User query
Output: Potentiality of groundwater
1. Generate objective function of each host nest i.e. the objective function f(x), $x = \{x_1, x_2, \ldots\ldots x_d\}$. /*Host nests are basically stored cases in the case base. Expert has provided us 136 host nests. Thus our application d=136 */
2. Get a cuckoo egg (say i) /* Query input from user*/
3. Initialize (count_no_of_host_nests) = 1.
4. Initialize (max_generation) = 115. /* We have taken a total of 115 for our database.*/
5. While (count_no_of_host_nests <= max_generation)
  [a] Consider the current host nest (Say j) /*Each case
       from the case base*/
  [b] Evaluate Similarity $(F_i, F_j) = \sum_{i=1}^{n} f(Fi, Fj) \times w_i$
       /*Applying K-Nearest Neighbour (KNN) */
  [c] Store the similarity value of F_j in another
       database named (Temp_nest_quality_solution)
  [d] count_ no_of_host_nests = count_ no_of_host_nests + 1

  End While.
6. Nests having maximum objective function value are stored from (Temp_nest_quality_solution) to (Final_nest_quality_solution)
7. Keep all these maximum value nests (quality solutions)
8. Worse_nests = (max_generation – Final_nest_quality_solution)
9. Abandon worst nests.
10. Rank the quality solutions from (Final_nest_quality_solution)
  [a] Find a correlation between a cuckoo egg and
       the host nests in (Final_nest_quality_solution)
  [b] The maximum value of correlation function is
       ranked as the best solution.

11. The nest with highest correlation value is considered.
12. Post process the results
  [a] Apply all possible propositional logic condition.
  [b] Depending on the condition the outcome can be
       {High, Moderate or Low}.

3.4   Detailed Description of Algorithm

Step 1: In this step basically we have to find out number of host nests needed to be considered. Though objective function has to be found out for each host nest (case) in the case base but the real evaluation is done in step 5(ii). Thus we have considered a total of 115 nests (cases) from our case base. The left out cases are needed for the validation process.

Step 2: In this step, the user input is taken. User input is to simply get a new cuckoo egg. Thus user has to enter the values of 6 attributes. A popup menu appears for each attribute and in that menu all the attribute values are displayed. (Say i)

Step 3: A counter is required for keeping track of host nests (case) in the case base. Thus we initialize a counter name count_ no_of_host_nests to 1. This counter basically points to the current host nest (case) in the case base.

Step 4: In this step we consider the total host nests needed for our processing purpose. Thus we initialize a counter named max_generation to 115.

Step 5: A while condition is used. This loop is terminated until all the host nests in the case base are not encountered.

Step 5.1: Point the counter count_no_of_host_nests to the first host nest in the database (say j).

Step 5.2: Evaluate the objective function $Sim(F_i, F_j) = \sum_{i=1}^{n} f(Fi, Fj) X w_i$. This step applies KNN formula.

Where F_i is the cuckoo egg (problem case), F_j is the host nest. N= number of attributes, in our application it is 6. The function f returns the 1 if $F_i = = F_j$ otherwise 0 is returned and w_i is the weight of the attribute. Figure 4 shows the weight of each attribute.

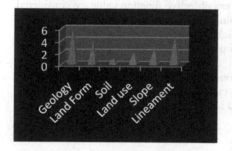 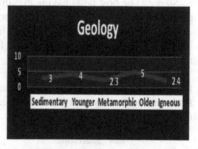

Fig. 4. Prioritizing geographical parameter with geology as the highest and soil as the lowest parameter.

Fig. 5. Distribution of geology parameter values

Y axis represents the weight whereas X-axis represents the attributes. It shows geology has a weight of 6 whereas soil has weight 1. It means geology has immense significance in groundwater possibility whereas soil has least importance.

In the following figures shows the distribution of attribute values. Figure 5 shows the geology values. X-axis represents the attribute values and Y-axis gives their numerical values. Figure 5 shows $F_{geology}$. The attribute geology is scaled in the range of 5 to 2.3. And other attributes like landform, landuse and soil is scale in the range of 5 to 0. If attribute value is 5 it means it has highest importance in that attribute and 0 means least importance. The following figure 6 shows the attribute value for lineament. It means a sort of ridge in the rock. Thus if lineament is present high possibility of water otherwise low. It depicts $F_{lineament}$.

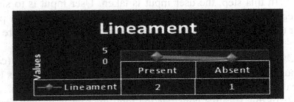

Fig. 6. Highlighting 2 aspects of lineament parameter

Step 5.3: As each host nest of the case base is encountered, their similarity function is evaluated and their function values is stored in a smaller database. This database is named as Temp_nest_quality_solution. It contains all the function values when compared between the cuckoo egg and the host nest.

Step 5.4: Increment count_ no_of_host_nests by 1.

Step 6: The database named as Temp_nest_quality_solution contains all host nests having similarity values. Now we need to find out all those host nests which are most suitable for a cuckoo egg. All the maximum valued host nests are stored in final case base named Final_nest_quality_solution. Among these final host nests cuckoo has to find out the nest which matches it requirements.

Step 7: The host nests having maximum objective function value are considered as quality solutions. All these quality solutions are stored.

Step 8: Find out those host nests which are not considered as quality solutions. Those ignored host nests are found out by max_generation – Final_nest_quality_solution.

Step 9: The ignored host nests are basically worst nests. They are abandoned for later uses. They are ignored for the current cuckoo egg. If some other cuckoo finds out nests they will be considered again. Here we find a fraction of worsts nests

$$p_a = \frac{worst\_nest}{max\_generation}$$

Step 10: Rank the best solution. In the following steps we will find out the nest that mostly suits with the current cuckoo egg. From the Final_nest_quality_solution we need to find out the nest that matches perfectly with the cuckoo egg. It is considered as current best. To find out the current best following steps are performed

Step 10.1: Take each host nest stored in Final_nest_quality_solution. The nest from Final_nest_quality_solution correlates maximum with the current cuckoo egg is considered.

Step 10.2: The maximum correlated host nests is chosen as best solution.

Step 11: The maximum correlated host nest undergoes some post processing.

Step 12: This is the final step of our proposed algorithm. Post processing is performed in this step wherein the best nest undergoes through some propositional logic condition. Based on the outcome of these conditions, the groundwater potentiality is detected. Quite a few number of propositional logic is framed but only 2 of them is given below:

i)   P= "geology is igneous"
        Q = "low groundwater"

$$P \rightarrow Q$$

ii)   P= "slope is gentle"
        Q= "lineament is absent"
        R= "high groundwater"

$$(P \wedge Q) \leftrightarrow R$$

Step 12.1: Depending on the case base some propositional logic formula is derived. The logic which derives the outcome is formalized and all the conditions are discussed later.

Step 12.2: If the condition is met, the following outcome is possible.

i)   High
ii)   Moderate
iii)   Medium.

4 Experiment & Simulation

4.1 *Dataset*

The dataset that is provided by domain expert (courtesy of DTRL, DRDO) to predict groundwater possibility is given in figure 7.

Attributes	Values
Geology	Sedimentary, Younger alluvium, Older alluvium, Igneous,Metamorphic
Landform	Floodplain,Intermontanevalley,Pediment,Alluvialfans,Bajada,Pediplain, Buriedpediment,AlluvialPlain,DeltaicPlain,Wadi,Riverterraces,Oldmeander etc.
Soil	Sandyloam,Sandygravel,Coarsesand,Clayloam,Alluvialsand,Gravelsand, GravelSandPebbles,Sand,Rocky etc.
Land use	Agriculturalland,Forest,Cultivatedland,Fallowland,Waterbody,Wasteland, Swampy land, Buildup, Urban ,Grass, Shrubs ,mixed vegetation etc.
Slope	Gentle, Steep
Lineament	Absent, Present

Fig. 7. The dataset provided by experts

4.2 *Graphical User Interface*

This section shows the user interface that has been developed. Figure 8 shows the graphical user interface for our system.

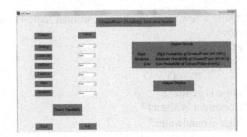

Fig. 8. Screenshot of our GUI

A query is given by the user as follows:
- Geology : igneous
- Landform: denundation hill
- Soil type: gravel sand pebble
- Land use: trees
- Slope: steep
- Lineament: absent

Once these values are entered the output should be low as suggested by our expert. Figure 9 shows the validation of this query.

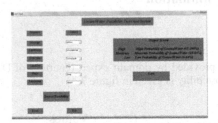

Fig. 9. Depiction of query as given by user along with the generated result i.e. "Low"

4.3 Graphical *Representation of our result*

When the query is provided by user then the first objective function is evaluated and all the intial nests values are depicted. The X axis shows all the possible nests that cuckoo will search i.e. host nests and Y-axis represent their objective function value. They are plotted on the graph according to their values. It is illustrated in figure 10.

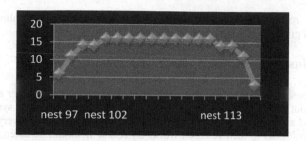

Fig. 10. Plot of quality solutions from nest 97 to nest 115, remaining nests had objective function value 0, thus not depicted in the graph.

Now among the all the host nests the quality functions are retrieved. This is done by scanning all the nests and those nest with highest objective function. From these quality function it has been observed that for our given case function with value 16 has the highest possibility. In many cases there may exists many nests with same objective function value. In those cases the best solutions are retrived by comparing all the best objective funtion valued nests with the problem case.

In our case 10 nests with highest objective funtion. Thus our graph will be interpreted and drawn accordingly. Figure 11 shows the best solutions from the objective function. Fom the objective funtion the best solution is interpreted by correlating the problem case with the objective funtions.

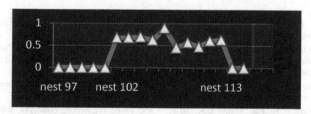

Fig. 11. Depicting best solution among quality solutions. It shows Nest 106 has highest objective function value.

It shows that nest number 106 is the best solution. Thus taking the most relevant one we have applied the propositional logic stored in our database. It follows the logic,

P= "geology is igneous"
Q= "lineament is absent"
R= "slope is steep"
S= "groundwater is low"

$$(P \wedge Q \wedge R) \rightarrow S \qquad\qquad (1)$$

Thus the result is indeed low. Thus our detector has proved almost 99% accurate and gives most efficient output in all cases.

5 Conclusion & Future Scope

Our proposed work is able to determine the existence of groundwater at various depths without even drilling wells [15]. It produces a cost effective system and hence it plays a crucial role in the economic sector. Our system can be very productive to all geologists in their research work. Of course it should be taken into consideration that only 6 mentioned attributes are queried by geologist. If they would like to vary the attributes then some minor changes are required in our software.

Our algorithm has shown improvement when compared with other existing algorithm for detecting potentiality of groundwater [6] [15]. In both cases i.e. Particle Swarm Optimization (PSO) and Biogeography Based Optimization (BBO) the searching is done only in one step but in our method we have applied searching twice, one for finding the quality solution and the other for finding the best solution. Thus our method is highly optimized in nature and moreover we have used propositional logic due to dissimilarities in our case base.

Though our work had proved 99% efficiency but there still remain some problems. Though we have obtained a reasonable amount of cases from our experts but still huge case base is desirable to achieve 100% accuracy. Consequently the efficiency may be improved and our work can be adapted in further research.

References

[1] I. Pufford, H. Flowers, "Environmental chemistry at a glance," Great Britain: Blackwell, pp 46-47, 2006.

[2] Xin-She Yang, Suash Deb, "Cuckoo search via Lévy flights," In Proceedings of World Congress on Nature & Biologically Inspired Computing, (NaBIC), IEEE Publications USA, pp 210-214, Dec. 2009, India

[3] E. Bonabeau, M. Dorigo, G. Theraulaz, "Swarm Intelligence From Nature to Artificial Systems," Oxford University Press, 1999.

[4] Sankar K. Pal, Simon C.K. Shiu, "Case based reasoning: concepts, features & soft computing," Artificial Intelligence 21, pp 233-238, 2004.

[5] C.K. Reisbeck, R.C. Schank, "Inside Case based reasoning", L. Erlbaum Associates Inc. Hillsdale, New Jersey, 1989.

[6] V.K. Panchal, Harish Kundra, Navpreet Kaur, "A Novel Approach to Integration of waves of swarms with case based reasoning to detect groundwater potential," 8th Annual Asian Conference & Exhibition of Geospatial information technology & application, Map Asia, Singapore, 2009

[7] Chunhua Yang, Hongqui Zhu, Weihua Gui, "Permeability prediction model for imperial smelting furnance based on improved case based reasoning," IEEE Proceedings of the 7th World Congress on Intelligent Control & Automation, June 25-27, Chongqing, China, 2008

[8] Li feng Li, "Consistency of finite theory in three types of many-valued propositional logic systems," 2009 WRI World Computer science & Information engineering, vol. 4, pp 651, Los Angeles, March 31, 2009.

[9]   K.J.Mammond, "Case based reasoning planning: viewing planning as a memory task", In JL Kolodner (ed.) Proceedings of the DARPA Case based reasoning workshop, Morgan Kaufmann, CA, USA, 1988.

[10] I. Watson :Case based reasoning is a methodology not a technology. International Journal of Elsevier, Knowledge based system 12, 303-308 (1999).

[11] Zhao Yong, Lujian, "Case based reasoning for emergency disposal of the traffic accidents in regional highway network," 2010 IEEE International Conference on Emerging management & managerial science, (ICEMMS), pp 258-260, Beijing, 2010.

[12] Ereck R. Speed, "Evolving a Mario agent using cuckoo search and softmax heuristics," 2010 2nd International IEEE Consumer Electronics society's games innovations conference, ICE-GIC, pp 1-7, Hong Kong.

[13] Xin-She Yang, Suash Deb :Engineering Optimization by Cuckoo search. International Journal Mathematical Modelling and Numerical Optimization, vol. 1, no.4, 330-343 (2010)

[14] David W. Aha :The Omnipresence of case based reasoning in science & applications. International Journal of Elsevier, Knowledge Based systems 11, 261-273 (1998)

[15] V.K.Panchal, Harish Kundra, Amanpreet Kaur :Biogeography based groundwater exploration. 2010 International Journal of Computer Application, IJCA, vol. 1, no. 8, 0975-8887 (2010).

Reversible OR Logic gate design using DNA

Pradipta Roy[1], Debarati Dey[1], Swati Sinha[2], Debashis De[1]

[1] Department of Computer Science & Engg. West Bengal University of Technology
BF-142, Sector 1, Salt Lake City. Kolkata – 700 064. West Bengal, India.

[2] Department of Zoology, Ananda Mohan College. University of Calcutta.
102/1, Raja Rammohan Sarani, Kolkata - 700 009. West Bengal, India.

{dr.debashis.de@gmail.com; pradiptoroy@gmail.com}

Abstract. In today's world DNA technology is promoted as an alternative approach for advancement over silicon technology. The DNA technology is also used to detect diseases. It needs some molecular computation for which the development of basic circuit unit is required. Basic circuit comprises the AND, OR and NOT gate. In this paper we proposed the design of two-input OR gate with E6 deoxyribozyme whose internal loop is not fixed. OR logic helps to express Boolean expression in Sum of Product (SOP) form and sometimes it uses to minimize the Boolean function. The DNA technology can be used as a substitute method not only for lower time complexity and low power consumption but also this technology is reversible in nature.

Keywords: DNA logic gate, Oligonucleotide, Deoxyribozyme, Reversible logic, Venn diagram.

1 Introduction

Nucleic acid is one of the essential materials in the field of engineering artificial biochemical circuits. It is used due to its several enviable properties. Distinct signals can be encoded and can be accurately restricted via nucleic acid sequences. Some supplementary restriction enzymes are used to design biochemical circuits using DNA [1, 2]. To perform molecular computation some structural properties such as hairpins has been used [3-5]. Some inventive molecular circuits or devices have been developed with this idea [6, 7]. To construct large-scale, modular circuits of some simpler designs have been projected. The strand displacement property of DNA is used to construct circuits to process information where the computation is performed by chemical reaction. Various digital logic circuits have been implemented with this property of DNA. The catalytic feature of DNA acts as proficient molecular devices [8, 9]. The strand displacement property of DNA is useful to produce uncomplicated but robust circuit which computes in high speed [9].

J. C. Bansal et al. (eds.), *Proceedings of Seventh International Conference on Bio-Inspired Computing: Theories and Applications (BIC-TA 2012)*, Advances in Intelligent Systems and Computing 201, DOI: 10.1007/978-81-322-1038-2_30, © Springer India 2013

A range of modeling approaches has also been developed for DNA computation [10, 11]. Such operations can effectively model after the Adleman's experiment [12], where DNA was used to figure out the Hamiltonian path in a given graph. This experiment has shown a new prospect and advantages over silicon technology. In the last decade, there are huge progresses in DNA computing. In nano scale several logic gates has already been designed [13-15]. The aspect that helped the advancement of molecular computing than silicon computing as a substitute technology was the awareness gained by study and experiment of the DNA over the decades. Several logic gates, such as, AND, NOT, Ex-OR gate and some combinational logic, such as, half-adder has already designed [16]. Molecular scale computation can be done using DNA. Thus biologists are involved to find out the connection between DNA and its relation with data processing. Nowadays, researchers take the benefit of the knowledge in the area of DNA computing and deal with the chemical reaction which the biologists are familiar with. The advancement in biology will result in advancement in molecular computing. DNA computing take considerably smaller time to solve complicated problems than conventional silicon based computing.

Nowadays, the DNA computing becomes more popular to solve huge, intractable problems than traditional computing. The main advantages of DNA computing are considerably lower time complexity, possibility of execution of parallel processing, the low power utilization of the circuit and the high speed of molecular incident. We can find a different subject of study if we would be capable to achieve all these advantages. The practical use of DNA computing is to construct DNA based circuits for detecting diseases. The first step to construct DNA circuit is the evolution of fundamental binary devices such as logic gates.

2   Structure of DNA

DNA has natural characteristics to store information and inherits them from parental generation of the same species. From the structural point of view DNA can be compare with the stair-case with handrail. It is actually a double stranded molecule in double helix structure. Each strands of DNA is a molecule that can be linked to other identical molecule to form a polymer. These are called nucleotides. A nucleotide is made of with 5-carbon sugar, a nitrogen consisting base attached to the sugar and a phosphate group. Four different bases are there hence four different types of nucleotides. They are Adenine (A), Thymine (T), Cytosine (C) and Guanine (G). The 5-carbon sugar (deoxyribose) is numbered like $1'$, $2'$, $3'$, $4'$ and $5'$. The DNA backbone is formed by the link between the phosphate group and the hydroxyl group on the $5'$-end and $3'$-end. The hydrogen bond between nucleotides with complementary bases are tied together to form polynucleotide chain. The one of the bases are connected in one strand and its complementary base is connected in opposite side strand. Among the four bases, Adenine (A) settle connection with Thymine (T) and Cytosine (C) settle connection with Guanine (G) [17].

A short chain of nucleic acid is the oligonucleotide. It is actually a set of nucleotides that are connected in a single strand. Oligonucleotides are chosen as inputs and output of the binary logic gate due to various reasons. Some of the reasons are associated with the objective of this paper and we can have the power to fabricate computational molecular system through DNA computing for detecting diseases. We can choose particular series of oligonucleotides as inputs which is the complementary series of specific part of deoxyribozyme (DNA based catalysts) that generate desired impact. We obtained the output from the cleaved outcome of oligonucleotide, which we can achieve from the splitting of initial substrate.

Normal DNA structure is not suitable to instigate in chemical reaction without enzyme. The enzymes behave like a catalyst which takes part in the chemical reaction but do not modify itself. It is useful to develop catalytic function within DNA, if it is configured correctly. The catalytic DNA is actually known as deoxyribozyme. The deoxyribozyme catalyze the chemical reaction when it came into contact with substrate if they are configured with special structure of oligonucleotides. The deoxyribozymes are not available in the physical world. Based on the model of ribozyme it created in the laboratory [18]. A ribozyme is Ribo Nucleic Acid (RNA) that acts as an enzyme.

In this paper, mainly two types of deoxyribozyme are used to construct logic gate, they are E6 deoxyribozyme [18] and 8-17 deoxyribozyme [19, 20]. The E6 deoxyribozyme consist a catalytic core and an internal loop. The internal loops are supposed to change by a desired sequence, that is, they are not fixed. E6 deoxyribozyme is shown in Figure 1 (a). On the contrary, the 8-17 deoxyribozyme consists of a catalytic core and a fixed internal loop, which cannot be changed. Figure 1 (b) shows 8-17 deoxyribozyme.

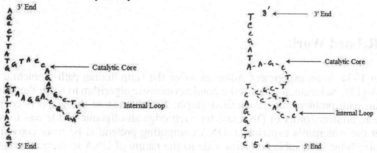

Fig. 1. (a) E6 Deoxyribozyme.   (b) 8-17 Deoxyribozyme.

Stem loop can be added in the branch or extension of the oligonucleotides. These are stimulates with specific sequence of nucleotides. In the inactive form of oligonucleotide, the stem loops remain tied up when the complementary sequence of nucleotides are not mixed in the solution with deoxyribozyme which is shown in Figure 2.

The oligonucleotides become active when the complementary sequences are mixed in the chemical solution and as a result of that the loops are infolded.

To build the logic gates the deoxyribozymes are used as a catalyst, but they are not precisely so. A catalyst always remains constant before and after the reaction.

If the input is "0" then the catalytic oligonucleotides do not change but they become active if the input is "1".

By breaking the chemical bond the molecules are divided into simpler molecules. This method is called cleaving. In logic gate, a substrate can be splitted by the deoxyribozyme. A substrate is a sequence of bases that ended by fluorescein donor (F) in 5' -end and tetremethylrhodamine acceptor (R) in 3'-end. Hence the substrate cleaved into two oligonucleotides: a fluorescein donor (F) and tetremethylrhodamine acceptor (R). We take fluorescein donor (F) in the 5'-end part output in the experiment.

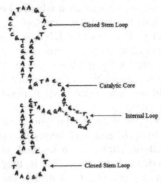

Fig. 2. Stem Loops are attached with 3' end and 5' end of E6 deoxyribozyme.

3   Related Work

In 1994, in an experiment Adleman solve the Hamiltonian path problem using DNA [12]. Adleman also devised a nondeterministic algorithm to solve the Hamiltonian path problem for the directed graph. Each vertex of the graph is coupled with an arbitrary order of DNA and for each edge an oligonucleotide was formed. After the Adleman's experiment, DNA computing pretend to be more convenient to solve huge intractable problems. Due to the nature of DNA to store information naturally and inherits information from previous generation of the same species, which is beneficial to solve problems that are hard to deal with. The biologists Breaker and Joyce reveal the concept of using deoxyribozyme in DNA application [18]. The design of deoxyribozyme base logic gates were first proposed by Stojanovic et. al [21]. They have proposed the structural design of AND, NOT and Ex-OR gate. AND gate was devised by 8-17 deoxyribozyme and E6 deoxyribozyme was used to make NOT and Ex-OR gate [21]. They described the construction of deoxyribozyme-based logic gates which is competent to produce any Boolean function.

Methoxybenzodeazaadenine (^{MD}A), an artificial nucleobase was developed by Okamoto et al in 2004 in an experiment over DNA logic gate [22]. This nucleobase was used to develop DNA logic gate through efficient hole transport method. Multiple series of ^{MD}A bases produces AND logic gate strand. Use of OR logic is mandatory for Sum of Product (SOP) form of any Boolean expression. The hole transport efficiency method demonstrate the OR logic strand [22].

4 Reversible 2-input OR Gate Design using DNA

To build combinational circuits it is necessary to design simple basic logic gates. With the help of basic logic gates we can construct any Boolean circuit. The OR logic is very important for SOP form of Boolean expression. In this paper, we proposed the design of two-input OR logic gate based on E6 deoxyribozyme.

4.1 Two-input OR gate

We have chosen E6 deoxyribozyme to develop a two-input OR gate. The well known truth table of two-input OR gate is shown in Table 1. The OR logic gives respond only when at least one of its input is "1", otherwise it produce output "0". Two stem loops are connected at the end of each arm of the oligonucleotides as shown in Figure 2. The sequences of nucleotides are different in the two loops hence they react with different inputs.

Table 1. Well known truth table of two-input OR logic gate.

Input		Output
I_B	I_A	O_F
0	0	0
0	1	1
1	0	1
1	1	1

Case - I. For input combination $I_B = 0$ and $I_A = 0$

When no inputs are added to the solution with the gate and the substrate the stem loops remain closed hence the substrate will not cleaved. So, output is "0" as shown in Figure 3.

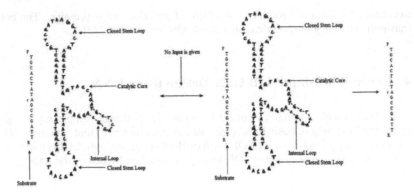

Fig. 3. Behaviour of two-input OR gate for input combination of logic 00.

Case - II. For input combination $I_B = 0$ and $I_A = 1$

If first input (I_A) is given to the solution which is the complementary sequence of the associated loop, after the reaction with the input and the gate the loop will open. The other loop will remain closed. Due to the reaction the substrate will cleaved and it will generate the output oligonucleotide, which is treated as "1" as shown in Figure 4.

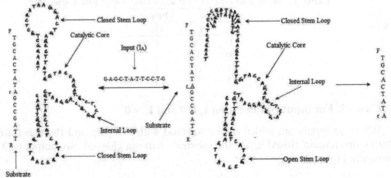

Fig. 4. Behaviour of two-input OR gate for input combination of logic 01.

Case - III. For input combination $I_B = 1$ and $I_A = 0$

Same event happen when only the second input (I_B) is added to the solution. The corresponding loop will react with the solution and will because of the complementary sequence of input. Now, the first loop remains closed. Due to the open stem loop the substrate cleavage is done and we get the output oligonucleotide. The output is treated as "1" as shown is Figure 5.

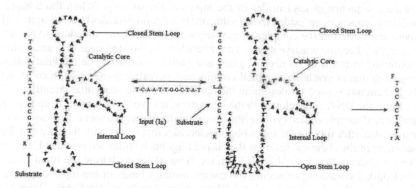

Fig. 5. Behaviour of two-input OR gate for input combination of logic 10.

Case - IV. For input combination IB = 1 and IA = 1

If both the inputs are mixed in the solution, both the input sequence will hybridize with the stem loops because of they are complementary sequence. Due to the hybridization both stem loops will open hence the substrate will cleaved. Output oligonucleotide is produced from the cleaved substrate. The output is treated as "1" as shown in Figure 6.

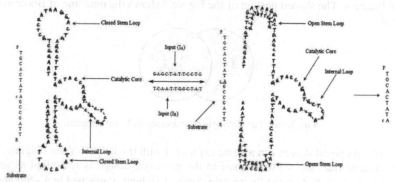

Fig. 6. Behaviour of two-input OR gate for input combination of logic 11.

4.2 Reversible logic in OR gate

Nowadays, reversible logic has shown its potential in computational prototype having its applications in low power computing, quantum computing, nanotechnology, optical computing and DNA computing. The classical set of gates such as AND, OR, and Ex-OR are not reversible. The reversible logic can be incorporated in DNA computing. We can remove the input by washing or adding the comple-

mentary sequence of nucleotides of the inputs to the solution. When the complementary sequences are added to the solution the oligonucleotides become active. If
these are washed or the complementary sequence of the inputs are added to the solution they become inactive again. The reversible circuits designed here are highly
optimized in terms of number of gates and garbage outputs. The modularization
approach that is synthesizing small circuits and thereafter using them to construct
bigger circuits is used for designing the optimal reversible sequential circuits.

Since the DNA is developed in the laboratory, it is very convenient to configure
the desired sample of deoxyribozyme according to our requirement. It is discussed
earlier that after mixing the input oligonucleotides that reacts with the gate and the
substrate in the chemical reaction the output oligonucleotides are produced. Hence
we get some output from the DNA circuit. Now, if we withdraw the input oligonucleotide by washing or adding the complementary bases of the input oligonucleotide then the chemical reaction will revert back to its original state. Thus we
can construct reversible circuit.

5 Results and Discussion

In this paper, we have proposed the design of OR logic gate using DNA. The
Boolean OR logic can be described with the help of Venn diagram [23] as shown
in Figure 7. The shaded portion of the Figure 7 shows the outcome of Boolean OR
logic.

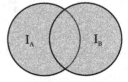

Fig. 7. Boolean OR logic is expressed in Venn diagram.

Our proposed design can also be expressed with the help of Venn diagram. Let
us assume that, the substrate used in the proposed design is identified as set S,
where S_F, $S_R \in S$. S_F is the fluorescein donor (F) which is attached in 5′-end and S_R
is the tetremethylrhodamine acceptor (R) which is attached in 3′-end. The E6
deoxyribozyme is identified as set D and Input oligonucleotide is expressed as set
I, where two inputs I_A, $I_B \in I$. I_A and I_B represents the two inputs of the two-input
OR gate. The set of double stranded DNA molecule is a set (D) consisting of D =
([A/T], [G/C], [C/G], [T/A]). The set consisting of only (A, T, C, G) is suitable for
single stranded DNA [24].

Here Case – II is considered for example. In step 1, the E6 deoxyribozyme solution is mixed with substrate solution. This scenario can be expressed as shown in
Figure 8.

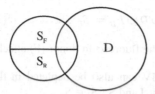

Fig. 8. The solution mixture of E6 deoxyribozyme and substrate is shown with the help of Venn diagram.

In step 2, the input oligonucleotide is mixed with the solution found in step 1. The mixture of the solution will take part in the chemical reaction and the output oligonucleotide is found from the cleavage of the substrate. The S_F is the output of the reaction which is represented by fluorescein donor (F) attached in 5'-end of the single ribonucleotide (rA). The process is shown in Figure 9 with the help of Venn diagram.

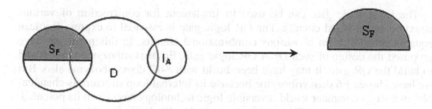

Fig. 9. The reaction process and the outcome is shown with the help of Venn diagram for input combination 01 in case of two-input OR gate.

The set equation of the above mention process is

$$O = S \cup D \cup I_A = S_F \qquad (1)$$

Figure 9 clearly shows that the deoxyribozyme (D) is mixed with substrate (S) and input (I_A). After the reaction the fluorescein donor (F) attached in 5'-end (S_F) will come out as output.

The other cases (Case – I, Case – III and Case - IV) of two-input OR gate can also be represented by Venn diagram. The equation for Case – I is shown in Eqn. 2. In this case, no input is given and as a result of that cleavage of the substrate is not done. So, no oligonucleotide will appear as output.

$$O = S \cup D = \varnothing \qquad (2)$$

In Case – III, the input oligonucleotide I_B is given into the mixture. After the chemical reaction the substrate cleavage is done. The equation for this case is shown in Eqn. 3, where $S_F, S_R \in S$.

$$O = S \cup D \cup I_B = S_F \qquad (3)$$

After the chemical reaction, the fluorescein donor (F) attached in $5'$-end (S_F) will come out as output.

The equation for Case – IV can also be depicted in the same procedure as shown in Eqn. 4, where I_A, $I_B \in I$ and S_F, $S_R \in S$.

$$O = S \cup D \cup I = S_F \qquad (4)$$

Here, both the inputs I_A and I_B are mixed in the solution and the fluorescein donor (F) attached in $5'$-end (S_F) will come out as output.

6 Conclusion

The DNA logic gate can be used to implement for construction of various deoxyribozyme based circuits. The OR logic gate is essential to express Boolean equation in SOP form of various combinational circuits. In this paper we have proposed the design of such DNA OR logic gate. E6 deoxyribozyme is proposed to build the OR gate. It may have been build with 8-17 deoxyribozyme also. But we have chosen E6 deoxyribozyme because its internal loop structure is changeable. In today's computer world reversible logic technology has shown its potential. According to traditional silicon technology AND, OR, NOT, Ex-OR gates are not reversible. As a result of that, formation of reversible logic circuit in silicon technology is not possible. Output to input mapping can be done in reversible logic design if the input is unknown. Nowadays various computation is performed on reversible logic hence the design of reversible basic logic gates are very crucial. Any Boolean circuit can be implemented with the help of basic logic gate. Thus with the advent of reversible DNA logic gate any large circuit can be designed using reversible technology.

7 Acknowledgement

Authors are grateful to University Grants Commission (UGC) for sanctioning a research project having File no.: 41-631/2012(SR) under which this paper has been completed.

References

[1] Yaakov Benenson, Tamar Paz-Elizur, Rivka Adar, Ehud Keinan, Zvi Livneh and Ehud Shapiro.: Programmable and autonomous computing machine made of biomolecules. Nature. 414, 430–434 (2001).

[2] Yaakov Benenson, Rivka Adar, Tamar Paz-Elizur, Zvi Livneh, and Ehud Shapiro.: DNA molecule provides a computing machine with both data and fuel. Proc. National Acad. Sci. USA. 100 (5), 2191–2196 (2003).

[3] Sakamoto, K., Gouzu, H., Komiya, K., Kiga, D., Yokoyama, S., Yokomori, T. and Hagiya, M.: Molecular computation by DNA hairpin formation Science. 288, 1223–1226 (2000).

[4] Yaakov Benenson, Binyamin Gil, Uri Ben-Dor, Rivka Adar and Ehud Shapiro.: An autonomous molecular computer for logical control of gene expression. Nature. 429, 423–429 (2004).

[5] Yin, P., Choi, H. M. T., Calvert, C. R. and Pierce, N. A.: Programming biomolecular self-assembly pathways. Nature. 451, 318–322 (2008).

[6] Bernard Yurke, Andrew J. Turberfield, Allen P. Mills Jr, Friedrich C. Simmel and Jennifer L. Neumann.: A DNA-fuelled molecular machine made of DNA. Nature. 406, 605–608 (2000).

[7] Venkataraman S., Dirks R. M., Rothemund P. W. K., Winfree E. and Pierce N. A.: An autonomous polymerization motor powered by DNA hybridization. Nature Nano technology. 2, 490–494 (2007).

[8] Seelig, G., Soloveichik, D., Zhang, D. Y. and Winfree E.: Enzyme-free nucleic acid logic circuits. Science. 314, 1585–1588 (2006).

[9] Zhang, D. Y., Turberfield, A. J., Yurke, B. and Winfree E.: Engineering entropy-driven reactions and networks catalyzed by DNA. Science. 318, 1121–1125 (2007).

[10] Kari, L., Paun, G., Rozenberg, G., Salomaa, A. and Yu, S.: DNA computing, sticker systems, and universality, Acta Informatica. 35, 401–420 (1998).

[11] Paun, G. and Rozenberg, G.: Sticker systems. Theoritical Computer Science. 204, 183–203 (1998).

[12] Leonard M. Adleman.: Molecular computation of solutions to combinatorial problem. Science, New Series. 266 (5187), 1021–1024 (1994).

[13] Kunal Das and Debashis De: A study on Diverse Nanostructure for implementing Logic Gate design for QCA. Int. Journal of Nanoscience. 10 (01n02), 1–7 (2011).

[14] Kunal Das and Debashis De: Novel Approach to design A Testable Conservative Logic Gate for QCA Implementation. Proc. IEEE 2nd International Advance Computing Conference. 82–87 (2010).

[15] Kunal Das and Debashis De: A Novel Approach of And-Or-Invert (AOI) Gate design for QCA. Proc. Int. Conference on Computers and Devices for Communication. (2009).

[16] Stojanovic, M. N. and Stefanovic, D.: Deoxyribozyme-Based Half-Adder. Journal of American Chemical Society. 125, 6673–6676 (2003).

[17] Watson, J. D. and Crick, F. H. C.: The Structure of DNA. Cold Spring Harbor Symposia Quantitative Biology. 123–131 (1953).

[18] Breaker, R. R. and Joyce, G. F.: A DNA enzyme with Mg2+-dependent RNA phosphoesterase activity. Chemistry & Biology. 2, 655–660 (1995).

[19] Jing Li, Wenchao Zheng, Angela H. Kwon, and Yi Lu.: In vitro selection and characterization of a highly efficient Zn(II)-dependent RNA-cleaving deoxyribozyme. Nucleic Acids Res. 28 (2), 481–488 (2000).

[20] Santoro Stephen W. and Joyce, Gerald F.: A general purpose RNA-cleaving DNA•enzyme. Proc. National Acad. Sci. USA. 94 (9), 4262–4266 (1997).

[21] Stojanovic M. N., Mitchell T. E. and Stefanovic, D.: Deoxyribozyme-Based Logic Gates. Journal of American Chemical Society. 124, 3555–3561 (2002).

[22] Okamoto A., Tanaka K. and Saito, I.: DNA Logic Gates. Journal of American Chemical Society. 126, 9458–9463 (2004).

[23] Natalia S. Akopyants, Robin S. Matlib, Elena N. Bukanova, Matthew R. Smeds, Bernard H. Brownstein, Gary D. Stormo, Stephen M. Beverley.: Expression profiling using random genomic DNA microarrays identifies differentially expressed genes associated with three major developmental stages of the protozoan parasite Leishmania major. Molecular & Biochemical Parasitology Elsevier. 136, 71–86 (2004).

[24] Tim Head.: Formal Language Theory and DNA: An analysis of the generative capacity of specific recombinant behaviors. Bulletin of Mathematical Biology. 49 (6), 737–759 (1987).

Performance Enhanced Hybrid Artificial Neural Network for Abnormal retinal image classification

D.Jude Hemanth[1] and J.Anitha[1]

[1]Department of ECE, Karunya University, Coimbatore, India
{jude_hemanth@rediffmail.com;rajivee1@rediffmail.com}

Abstract. Artificial Neural Networks (ANN) is becoming increasingly important in the medical field for diagnostic applications. The popularity of ANN is mainly due to the high accuracy and the nominal convergence rate. But, the major drawback is that these characteristic features are not simultaneously available in the same network. While supervised neural networks are highly accurate, the requirement for convergence time is high. On the other hand, unsupervised neural networks are sufficiently faster but less accurate. This problem is tackled in this work by proposing a Modified Neural Network (MNN) which possesses the features of both the supervised neural network and the unsupervised neural network. The applicability of this network is explored in the context of abnormal retinal image classification. Images from four abnormal categories such as Non-Proliferative Diabetic Retinopathy (NPDR), Choroidal Neo-Vascularization Membrane (CNVM), Central Serous Retinopathy (CSR) and Central Retinal Vein Occlusion (CRVO) are used in this work. Suitable features are extracted from these images and further used for the training and testing process of the proposed ANN. Experimental results are analyzed in terms of classification accuracy and convergence rate. The experimental results are also compared with the results of conventional networks such as Back Propagation Network (BPN) and the Kohonen Network (KN). The results of the proposed modified ANN are promising in terms of the performance measures.

Keywords: Modified Neural Network, Back Propagation Network, Kohonen Network and Retinal images.

J. C. Bansal et al. (eds.), *Proceedings of Seventh International Conference on Bio-Inspired Computing: Theories and Applications (BIC-TA 2012)*, Advances in Intelligent Systems and Computing 201, DOI: 10.1007/978-81-322-1038-2_31, © Springer India 2013

1 Introduction

Retinal diseases are mostly gradual in nature and hence early detection of the type of abnormality is extremely important to avoid any vision loss. The accuracy of such detection process is extremely important since the treatment planning is based on the detection process. Conventional disease identification methodologies involve human intervention which is highly prone to error. Hence, several automated techniques have been developed to assist the ophthalmologists in accurately diagnosing the retinal abnormality.

One of the significant automated techniques is the Artificial Neural Network (ANN) which falls under the broad category of Artificial Intelligence. Literature survey reveals the usage of ANN for retinal image classification applications. Multi layer perceptron is used for ophthalmologic disease classification in [1]. But the success rate of this methodology is based on input feature set. Different stages of Diabetic Retinopathy (DR) retinal image classification is reported in [2]. Back Propagation Neural Network (BPN) is used as the classifier in this work. The application of neural networks to detect the keratoconous abnormal retinal images is explored in [3]. Experimental analysis is performed in this work based on sensitivity and specificity. Neural techniques for exudates detection in diabetic retinal images are also available in the literature [4]. This methodology is implemented on color retinal images. The comparative analysis of the different image processing techniques for DR detection in retinal images is presented in [5].

Fuzzy ART neural network has been successfully used for DR identification in retinal images [6]. Lack of quantitative analysis & lack of multi-level classification is the major limitations of this system. Different stages of DR images are differentiated using the BPN network [7]. The number of features used in this work is very less which accounts for the inferior classification accuracy. Perceptron is also used for disease identification in retinal images [8]. The proposed system is a bi-level classification system used for differentiating the normal and DR images. Color retinal image processing for disease identification is proposed in [9]. Associative neural networks are used for image classification. Support Vector Machine based disease diagnosis is implemented in [10].

Though several techniques are available, the performance of these methodologies is superior only in terms of either convergence rate or classification accuracy. Both these advantages are not found simultaneously in these approaches which accounts for the practical limitations of such systems. In this work, a novel approach is proposed for image classification which involves the concept of both supervised and unsupervised methodology in the same network. The process of determining the weights occurs without any iteration which accounts for the low convergence rate. This process takes place without compromising the accuracy. Experimental results show encouraging results for the proposed approach.

2 Materials and Methods

The framework of the proposed automated system is shown in Fig. 1.

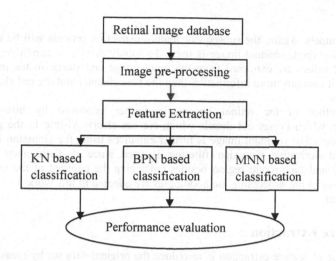

Fig. 1. Proposed Methodology

The abnormal retinal images from four different categories are collected from ophthalmologists and used in this work for disease identification system. The raw images are initially processed to enhance the contrast. An extensive set of features are extracted from these processed images which ultimately aid in enhancing the accuracy of the automated image classification system. Three classifiers are used in this work (BPN, KN and MNN) and a comparative analysis is performed among them in terms of the performance measures.

3 Retinal Image Database & Image Pre-Processing

The image dataset consists of 420 digital retinal images obtained using the imaging camera. All the images are collected from Lotus Eye Care Hospital, Coimbatore, India. The images are stored as colour TIFF images and are 1504×1000 pixels in size for all the objects. The intensity value of all the retinal images ranges from 0 to 255 (for each R, G & B planes). The real time images are collected from four abnormal categories namely Non-Proliferative Diabetic Retinopathy (NPDR), Central Retinal Vein Occlusion (CRVO), Choroidal Neo-Vascularisation Membrane (CNVM) and Central Serous Retinopathy (CSR).

The pre-processing step is a mandatory task in automated image classification system. Pre-processing algorithms are implemented to enhance the original image so that it can increase the chances for success of subsequent processes. In this work, the objective of the pre-processing technique is twofold: (a) Contrast enhancement of the raw image and (b) Converting the original 3-channel image (RGB) to single channel image (G). Literature survey shows that the anatomical structures/background contrast effect is high in the green channel than in the blue and red channels. Again, the subsequent feature extraction process will be made simpler if the single channel image is used. To satisfy these two conditions, the green pixel values are extracted from the input image and stored in the matrix form since it contains more information than the blue channel and the red channel of the image.

The contrast of the retinal images is further improved by histogram equalization which brings out details which are not clearly visible in the green channel image. The resultant image is further enhanced using the Gaussian filter. The second derivative Gaussian filtering is used since it distinguishes the background and foreground region besides enhancing the contrast of the image. Thus, a series of pre-processing methodologies are adopted in this work for image enhancement.

4 Feature Extraction

The purpose of feature extraction is to reduce the original data set by measuring certain properties, or features, that distinguish one input pattern from another pattern. The extracted feature should provide the characteristics of the input type to the classifier by considering the description of the relevant properties of the image into a feature space. Seven textural features are used in this work. The formulas for calculating these features are given below:

Angular Second Moment (ASM):

$$f_1 = \sum_i \sum_j \{p(i,j)\}^2 \tag{1}$$

where $p(i,j)$ represents the input image.
Contrast:

$$f_2 = \sum_{n=0}^{N_g-1} n^2 \left\{ \sum_{i=1}^{N_g} \sum_{j=1}^{N_g} p(i,j) \right\} \tag{2}$$

where N_g is the number of gray levels in the original image.
Correlation:

$$f_3 = \frac{\sum_i \sum_j (ij)p(i,j) - \mu_x \mu_y}{\sigma_x \sigma_y} \tag{3}$$

where $\mu_x, \mu_y, \sigma_x, \sigma_y$ are the means and standard deviations of the input image in the row-wise and column-wise order.
Variance:

$$f_4 = \sum_i \sum_j (i - \mu)^2 p(i,j). \tag{4}$$

where μ is the mean of the whole image.

Inverse Difference Moment (IDM):

$$f_5 = \sum_i \sum_j \frac{1}{1+(i-j)^2} p(i,j). \tag{5}$$

Entropy:

$$f_6 = -\sum_i \sum_j p(i,j) \log(p(i,j)). \tag{6}$$

Skewness:

$$f_7 = \frac{1}{\sigma^3} \sum_i \sum_j (p(i,j) - \mu)^3 \tag{7}$$

Kurtosis:

$$f_8 = \frac{1}{\sigma^4} \sum_i \sum_j (p(i,j) - \mu)^4 - 3 \tag{8}$$

These features are commonly used for medical image analysis [11]. These features are specifically found to be suitable for image classification applications.

5 Classifiers

In this work, three classifiers are used. Representatives of supervised networks (BPN) and unsupervised networks (KN) are implemented along with the proposed approach (MNN). Since the focus of this work on the proposed approach, MNN is dealt in detail followed by the brief explanation of the BPN and the KN.

5.1 MNN Based Retinal Image Classification

In this work, a modified version of the conventional neural network is proposed. This network comprises the supervised and unsupervised methodology and hence it enjoys the benefits of both the methodologies. This network is similar to Counter Propagation Neural (CPN) network in terms of architecture but the training process takes place without iterations. The novelty of this network lies in the training methodology of the network.

5.1.1 Architecture of MNN

The proposed MNN is a 2-layered network with input layer, hidden layer and the output layer. The number of neurons in the input layer is 8 which are equal to the number of input features. The number of output layer neurons is 4 which are equal to the number of pre-defined classes. The number of neurons in the hidden layer is set to 12 which is selected based on trial and error method. The architecture of MNN is shown in Fig. 2.

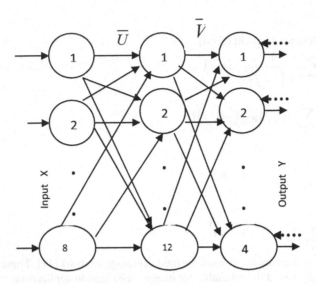

Fig. 2. Topology of MNN

In this network, the weight adjustment between the input and the hidden layer follows the unsupervised mode and the weight adjustment between the hidden and the output layer employs the supervised methodology. The target is supplied to the output layer which is shown as dotted lines in Fig. 2.

5.1.2 Training Algorithm of MNN

The training algorithm of MNN is divided into two phases. The first phase deals with the weight adjustment procedure of input layer and hidden layer. The second phase deals with the weight adjustment procedure of hidden layer and output layer. But, the significant aspect is that the weights are estimated without any iteration which eventually minimizes the convergence time requirement.

Phase 1:

The first part of the network is based on the unsupervised methodology. Hence, it is closely related with the "winner take-all algorithm". The procedure is illustrated with the following mathematical steps.

Step 1: The relation between the inputs and weights using the "winner take-all algorithm" is given by

$$D(j) = \sum_i \left(u_{ij} - x_i\right)^2 \tag{9}$$

where 'j' denotes the number of neurons in the hidden layer, 'u' denotes the weight values, 'x' denotes the inputs supplied to the input layer and 'D' denotes the distance measure.

Step 2: In the conventional algorithm, the above mentioned equation is calculated for several iterations with an objective to make the distance measure to be minimum. The minimum distance measure will lead to highly stabilized weights. This concept is used in this work in a different angle.

Step 3: The distance measure is preset to be a minimum value of 0.01. The distance measure is set to be the same value for all the hidden layer neurons ensuring more accuracy. It may be noted that the minimum value of zero is not possible practically. Thus, from Eqn. (9), the weight values can be easily calculated mathematically using the following formula:

$$u_{ij} = x_i + 0.01 \tag{10}$$

Thus, by normalizing the input values, the weight matrices can be easily calculated without the necessity for iterations. The distance measure can be any value greater than zero but 0.01 is found to be an optimum value for ensuring accuracy. Whenever the input changes, the weight value also changes correspondingly which results in stabilized weight matrices. Thus, this approach has also avoided the 'low accuracy' problem of the conventional unsupervised network.

Phase 2:

The second part of the proposed MNN is based on supervised methodology and hence the weight adjustment is based on error values. The training process is performed in the reverse direction unlike the conventional supervised approaches. The mathematical steps of this phase are given below:

Step 4: The error value is calculated using the formula:

$$E = T - O \tag{11}$$

where 'O' denote the output of the output layer and 'T' denote the target values which is supplied by the user. This calculation is repeated for several iterations in the conventional approach with an objective to minimize the error value. To avoid these iterations, the error value is preset to be a minimum value of 0.01 since an error value of zero is practically not feasible.

Step 5: Thus the output values of the output layer can be estimated using Eqn. (11) since the target vectors and the error values are known.

Step 6: Using the output values of the output layer, the NET value of the output layer can be estimated using the sigmoid activation function.

$$O = \frac{1}{1 + e^{-NET}} \tag{12}$$

$$NET = \ln\left[\frac{O}{1 - O}\right] \tag{13}$$

Eqn. (12) corresponds to the standard activation function and Eqn. (13) is derived from Eqn. (12)

Step 7: Further, the weight matrices of the output layer can be estimated using the following equation

$$NET = \sum v_{ij} D_j \tag{14}$$

where 'v' denote the weight values of the output layer and 'D' denote the input for the output layer (which is equivalent to the output of the hidden layer). Since 'D' is already known, the weight matrices of the output layer are easily calculated using mathematical equations.

Thus, MNN has eliminated the 'low accuracy' problem of unsupervised neural networks by involving the standard convergence condition (D=preset value, E=preset value) in the training approach. The 'high convergence time requirement' problem of the supervised neural networks is also eliminated by estimating the weights without any iteration. Thus, the proposed approach enjoys the benefits of both training methodologies simultaneously which is further proved quantitatively in the following section.

5.2 BPN Based Image Classification

BPN networks belong to the supervised category which requires the assistance of target vectors. They are found to be accurate but the major drawback is the huge convergence time requirement of such techniques. The high accuracy is guaranteed only at the cost of high convergence time. Thus, it is evident that both the parameters are not simultaneously available within the same network. A detailed explanation of the architecture and the training algorithm is illustrated in [12]. Experiments are also performed in this work using these networks for comparative analysis.

5.3 KN Based Image Classification

Kohonen neural networks are usually single layered networks which follows the unsupervised training methodology of weight adjustments. Though these networks are quicker than supervised networks, the accuracy of these networks is not convincing which is evident from the earlier research works. Thus, it is very clear that the conventional Kohonen neural networks have failed to yield high accuracy and high convergence rate simultaneously. A detailed explanation of the architecture and the training algorithm is illustrated in [13]. Experiments are also performed in this work using these networks for comparative analysis.

6 Experimental Results and Discussions

The experiments are carried out on an IBM PC Pentium with processor speed 1.66 GHz and 1GB RAM. The software used for the implementation is MATLAB (version 7.0) (Mathworks 2002), developed by Math works Laboratory [14]. The dataset used for the implementation is shown in Table 1.

Table 1. Dataset for retinal image classification

Class	Training data	Testing data	Number of images
CNVM	35	64	99
CRVO	35	60	95
CSR	35	73	108
NPDR	35	83	118
Total images			420

The performance of these classifiers is analyzed based on classification accuracy, sensitivity and specificity. These performance measures are defined as follows:

Sensitivity=TP/(TP+FN)   (15)

Specificity=TN/(TN+FP)   (16)

Accuracy=(TP+TN)/(TP+FP+FN+TN)   (17)

In the above expressions, the parameters are defined with an example as follows: TP=True Positive (an image of CNVM type is categorized correctly to the same type), TN=True Negative (an image of Non-CNVM type is categorized as Non-CNVM type), FP= False Positive (an image of Non-CNVM type is categorized wrongly as CNVM type) and FN=False Negative (an image of CNVM type is categorized wrongly as Non-CNVM type). Initially, the results of image pre-processing are analyzed followed by the analysis of performance measures of the classifiers. Sample pre-processed images are shown in Fig. 3.

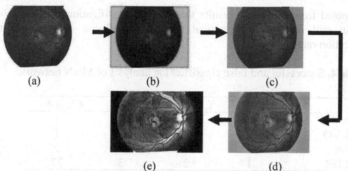

(a)   (b)   (c)

(e)   (d)

Fig.3. Sample output images. (a) RGB image (b) Green band processed image (c) Gray Scale image (d) Adaptive Histogram image (f) Gaussian Filtered image

Thus, these results have shown the necessity for pre-processing techniques. Fig. 3(e) is more suitable for further processing than Fig. 3(a). An extensive feature set is then extracted from these images and given as input to the classifiers. Initially the successful and false classification rates of the classifiers are analyzed individually followed by the performance measure analysis. The successful and false classification rates are illustrated in the form of confusion matrix. This analysis is performed on the testing images. Table 2 illustrates the misclassification rates of the Kohonen neural network.

Table 2. Successful and false classification analysis of Kohonen neural network

	Class1	Class 2	Class 3	Class 4
CNVM	49	4	6	5
CRVO	3	47	4	6
CSR	6	5	57	5
NPDR	4	7	5	67

In the above method, 49 images have been successfully classified in CNVM, 47 images have been successfully classified in CRVO, 57 images have been successfully classified in CSR and 67 images have been successfully classified in NPDR. The incorrect weight values due to the irregular convergence conditions lead to the inferior results. Table 3 depicts the misclassification rates of BPN.

Table 3. Successful and false classification analysis of BPN network

	Class 1	Class 2	Class 3	Class 4
CNVM	54	3	4	3
CRVO	1	53	2	4
CSR	4	2	64	3
NPDR	3	4	3	73

It may be noted from the above results that the misclassification rate is slightly reduced but still inferior for practical applications. Table 4 illustrates the misclassification rates of MNN.

Table 4. Successful and false classification analysis of MNN network

	Class 1	Class 2	Class 3	Class 4
CNVM	60	1	2	1
CRVO	0	57	2	1
CSR	1	2	68	2
NPDR	1	2	3	77

The level of misclassification is highly reduced in comparison to BPN and KN. The performance measure of the classifiers is further analyzed by estimating the TN, TP, FP and FN from the confusion matrices. Table 5 shows the performance analysis of the conventional Kohonen neural network. This analysis is initially performed individually on the different classes of images.

Table 5. Performance measure analysis of conventional Kohonen network

	TP	TN	FP	FN	Sensitivity	Specificity	Accuracy (%)
CNVM	49	171	13	15	0.76	0.92	88.7
CRVO	47	173	16	13	0.78	0.91	88.3
CSR	57	163	15	16	0.76	0.91	87.6
NPDR	67	153	16	16	0.77	0.90	87.3
			Average value		0.77	0.91	88

Table 5 shows the inferior nature of the conventional Kohonen neural network in terms of sensitivity. These measures directly impact the TP value which must be significantly high for accurate classification. But, the low values of these measures show the low efficiency of the system. The classification accuracy of this system is also low which limits the practical application of the conventional Kohonen neural network. The performance measures of BPN are shown in Table 6.

Table 6. Performance measures of the BPN network

	TP	TN	FP	FN	Sensitivity	Specificity	Accuracy (%)
CNVM	54	205	8	10	0.84	0.96	93.5
CRVO	53	211	9	7	0.88	0.96	94
CSR	64	198	9	9	0.87	0.95	93.5
NPDR	73	187	10	10	0.90	0.95	92.8
Overall value					0.87	0.96	94

The performance measures of BPN are improved over the unsupervised neural network. But, the dependency of iterations does have impact on the performance measures. These drawbacks are eliminated in the proposed system which is analyzed in Table 7.

Table 7. Performance measures of the MNN network

	TP	TN	FP	FN	Sensitivity	Specificity	Accuracy (%)
CNVM	60	202	2	4	0.94	0.99	97.7
CRVO	57	205	5	3	0.95	0.98	97
CSR	68	194	7	5	0.93	0.97	95.6
NPDR	77	185	4	6	0.93	0.98	96.3
Overall value					0.94	0.98	97

The performance measures are sufficiently higher than the conventional system which proves the efficiency of the system. A significant increase in classification accuracy is also achieved by the MNN over the conventional system. In terms of convergence rate, the time requirement for supervised neural network is 1650 CPU seconds, the time requirement for the Kohonen neural network is 990 CPU seconds and the time requirement for the proposed approach is only 10 seconds. The huge improvement in the convergence rate is evident from the experimental results. Thus, the efficiency of the MNN network is verified through the experimental results.

7 Conclusion

A modified neural network is proposed in this work for retinal image classification. Significant improvement in the classification accuracy is achieved over the supervised and unsupervised neural networks. The convergence rate is also considerably increased in comparison to the conventional neural networks. Hence, the proposed network is capable of yielding accurate results at quick convergence time. Thus, the proposed approach possesses the characteristic features of the supervised and unsupervised methodologies which are highly essential for practical applications.

Acknowledgment

The authors thank Dr. A. Indumathy, Lotus Eye Care Hospital, Coimabtore, India for her help regarding database validation. The authors also wish to thank Council of Scientific and Industrial Research (CSIR), New Delhi, India for the financial assistance towards this research (Scheme No: 22(0592)/12/EMR-II).

References

[1] Povilas, T., and Vydunas, S.: Neural network as an ophthalmologic disease classifier. Information Technology and Control. 36(4), 365-371 (2007).
[2] Wong, L.U., Rajendra, A. U., Venkatesh, Y.V., Caroline, C., Lim, C. M., Ng, E.Y.K.: Identification of different stages of diabetic retinopathy using retinal optical images. Information Sciences. 178, 106-121 (2008).
[3] Agostino, A. P., Stefano P.: Neural network based system for early keratoconus detection from corneal topography. Journal of Biomedical Informatics. 35, 151-159 (2003).
[4] Alireza, O., Mirmehdi, M., Thomas, B., Markham, R.: Automated identification of diabetic retinal exudates in digital colour images. British Journal of Ophthalmology. 87, 1220-1223 (2003).
[5] Saurabh, G., Jayanthi, S., Gopal, D. J.: Automatic drusen detection from color retinal image. Proceedings of Indian Conference on Medical Informatics and Telemedicine. 84-89 (2006).
[6] Jayakumari, C., Santhanam, T.: Detection of hard exudates for diabetic retinopathy using contextual clustering and fuzzy ART neural network. Asian Journal of Information Technology. 6(8), 842-846 (2007).
[7] Yun, W. etal.: Identification of different stages of diabetic retinopathy using retinal optical images. Information Sciences. 178, 106-121 (2008).
[8] Garcia, M. et al.: Neural network based detection of hard exudates in retinal images. Computer methods and Programs in Biomedicine, 93, 9-19 (2009).
[9] Jayanthi, D., Devi, N., Swarna Parvathi, S.: Automatic diagnosis of retinal diseases from color retinal images. International Journal of Computer Science and Information Security. 7 (1), 234-238 (2010)
[10] Osareh, A., Mirmehdi, M., Thomas, B., Markham R.: Comparative exudates classification using support vector machines and neural networks. Proceedings of the 5th International Conference on Medical Image Computing and Computer-Assisted Intervention (MICCAI)-part II. 413-420 (2002).
[11] Haralick, R.M.: Statistical and structural approaches to texture. IEEE Transactions on System, Man and Cybernatics. 67, 86-804 (1979).
[12] Freeman, J.A., Skapura, D. M.: Neural Networks, Algorithms, Applications and Programming Techniques. Pearson education, (2004).
[13] Willem, M., Ron w., Buydens, L.: Supervised Kohonen networks for classification problems. Chemometrics and Intelligent Laboratory Systems. 83, 99-113 (2006).
[14] MATLAB, User's Guide, The Math Works, Inc., Natick, MA 01760, (1994-2002).

Algorithmic Tile Self-assembly Model for the Minimum Dominating Set problem

Zhen Cheng[1], Jianhua Xiao[2], Yufang Huang[3]

[1]College of Computer Science and Technology, Zhejiang University of Technology,
Hangzhou 310023, China

[2]The Research Center of Logistics Nankai University, Tianjin 300071, China

[3]College of Mathematics, Southwest Jiaotong University, Chengdu 610031, China

{chengzhen0716@163.com; jhxiao@nankai.edu.cn; huangyufang@home.swjtu.edu.cn}

Abstract. Self-assembly is the process in which simple components can sponta-neously form complex complexes. This field has produced a formal model of self-assembly known as the tile self-assembly model. Recently, computation by this model is proved to be a promising technique in nanotechnology. In this paper, aiming to the minimum dominating set problem which is NP-complete, how the tile self-assembly model is used to implement this problem is shown including nondeterministic guess operation, assigning operation and logic OR operation. This method can be successfully performed this problem in $\Theta(n^2)$ steps. Here n is the number of vertices of the given graph in the minimum dominating set problem.
Keywords: Minimum dominating set problem; Algorithmic; Tile; Self-assembly.

1 Introduction

Since 1994, Adleman demonstrated that the DNA recombination techniques can be used to solve the Hamiltonian path problem [1], the field of DNA-based com-puting was established. From then on, many researchers constructed different DNA computing models which made this filed develop very fast [2-4]. This work focuses on computing systems based on DNA molecules, especially the self-assembly of DNA tiles.

In recent years, many researchers began to study the tile self-assembly model, and it was a promising technique for computation [5], by which a simple configu-ration can spontaneously form intricate complexes. In 1998, Seeman proposed DNA nanotechnology to make self-assembled nanostructures from DNA mole-cules [6-7]. Barish et al. [8] experimentally put up with a tile system that can copy an input and count in binary to create crystals with patterns of binary counters.

J. C. Bansal et al. (eds.), *Proceedings of Seventh International Conference on Bio-Inspired Computing: Theories and Applications (BIC-TA 2012)*, Advances in Intelligent Systems and Computing 201, DOI: 10.1007/978-81-322-1038-2_32, © Springer India 2013

Similarly, Rothermund et al. [9] realized a nanostructure-Sierpinski triangle using a DNA implementation of a tile system which can compute the XOR function. In addition, the 3D nanostructures [10-11] can also be created. Considering the special structure of tile, Cook [12] used the tile assembly model to implement arbitrary circuits. In 2006, Rothemund [13] first used the technology of folding DNA to create nanoscale shapes and patterns.

One of the most significant achievements of algorithmic tile self-assembly model is the work of Winfree [14], he proved that 2D tile self-assembly model was capable of Turing-universal and computationally efficient. Mao et al. experimentally implemented cumulative XOR computation based on the algorithmic tile self-assembly model [15], which showed that this model hold basic computational power. Brun proposed the applications of DNA tile self-assembly in arithmetic including addition and multiplication [16], then he used this computational model to solve NP-complete problems such as factoring numbers [17], deciding a path finding problem [18]. Barua et al. showed how the tile self-assembly process can be used for computing finite field multiplication and addition [19]. Furthermore, Lagoudakis et al. [20] presented a tile system that solved satisfiability problem using 2D DNA self-assembly model. Recently, the complexity of self-assembly nanostructure is also studied by some researchers [21-22].

The minimum dominating set problem is a classical NP-complete optimization problem, which has attracted much attention and extensive studies of many researchers [23-24]. Recently, the algorithms which can be used to solve this problem are proposed [25-26]. Guo [27] presented a fast parallel molecular solution to the dominating set problem on massively parallel bio-computing. Liu [28] put up with a biomolecular computing model in vivo for minimum dominating set problem. On this basis, here an important model of DNA computing which is algorithmic tile self-assembly model is proposed to implement the minimum dominating set problem in $\Theta(n^2)$ steps, here n is number of the vertices of the given graph.

The rest of this paper is organized as follows: Section 2 will describe the tile self-assembly model. The process of implementing the minimum dominating set problem based on the tile self-assembly model is shown in Section 3. Section 4 is the conclusion of this work including the complexity analysis of this algorithm.

2   Algorithmic Tile Self-assembly Model

The abstract Tile Assembly Model [29] is a formal model of crystal growth which is designed to model self-assembly of molecules. Winfree [30] defined this model which extended the theory of Wang tilings, and it showed how tiles can be designed to simulate the operation of any Turing Machine. Murata [31] proposed a precise simulation model for DNA tile self-assembly.

For the algorithmic tile self-assembly model, the basic unit is the tile. Each tile can be represented by the binding domains set Σ which is a 4-tuple $\langle\sigma_N, \sigma_E, \sigma_S, \sigma_W\rangle\in\Sigma^4$, here N, E, S, W is labeled as the direction of north, east, south and west of the tile respectively. The set of directions is a set of four functions from positions to positions which is denoted as $D=\{N, E, S, W\}$, i.e. $\mathbb{Z}^2$ to $\mathbb{Z}^2$ such that all positions (x, y), $N(x, y)=(x, y+1)$, $E(x, y)=(x+1, y)$, $S(x, y)=(x, y-1)$, $W(x, y)=(x-1, y)$. For a tile t, for $d\in D$, $bd_d(t)$ is used to represented as the binding domain of tile t on d's side. A special tile $empty=\{null, null, null, null\}$ is referred to the absence of all other tiles. For this computational model, the assembly complexes take place by starting with the seed tile denoted as the basic tile type set, which can be produced the seed configuration S. Here S is a function $S:\mathbb{Z}^2\to\Gamma$, and Γ is a set of tiles which can be used to design the configuration of the tile self-assembly model. A tile model system $\mathbb{S}=\langle T, g, \tau\rangle$ which is designed preparedly is composed of three parameters, here the parameter T is a finite set of tiles containing empty tile. g is a strength function $\Sigma\times\Sigma\to\mathbb{R}$, which denotes the strength of the binding domains, and $\tau\in N$ is the temperature. When the growth of process terminates, these assembled complexes can produce a unique final configuration $\mathbb{S}$ based on the seed configuration S.

Here, as the triple crossover molecule, TAO tile is considered as the basic tile unit and the molecular structure is illustrated in Fig.1. This tile assembly model is used to implement minimum dominating set problem. The binding domains sets including $\sigma_N, \sigma_E, \sigma_S$ and σ_W are on its four sides. The tile has sticky ends that match the complementary sticky ends of other DNA tiles in the mechanism of Watson-Crick, facilitating further assembly into final complexes.

Fig.1. The structure of basic tile unit.

3   Self-assembly Model for Minimum Dominating Set Problem

In this section, the definition and the mathematical model of the minimum dominating set problem are introduced, then this problem can be converted into the solution of the Boolean formula of the given graph which can be solved by algorithmic tile self-assembly model.

3.1 The Minimum Dominating Set Problem

The minimum dominating set problem is a classic NP-complete problem with many real world applications. Mathematically, let $G = (V, E)$ denote an undirected graph. A set $D \subseteq V$ is called a dominating set for G if every vertex of G is either in D, or adjacent to some vertex in D. A dominating set is minimal if all its proper subsets are not dominating. The minimum cardinality of the dominating sets of G is called the domination number of G. The minimum dominating set problem asks to find a dominating set of minimum cardinality according to the following formula:

$$\varphi = \prod_{i=1}^{n} (v_i + \sum_{u \in N(v_i),\, v_i \in V} u)$$

Here, $n = |V|$. The computations "+" and "$\prod$" denote logic OR operation and logic AND operation respectively. $N(v_i)$ represents the vertices which are adjacent to the vertex v_i. $(v_i + \sum_{u \in N(v_i),\, v_i \in V} u)$ denotes the vertices including v_i and the vertices which are adjacent to the vertex v_i. It can be seen that φ is composed of Boolean formula, and the Boolean variable in each clause is v_i (i=1, 2, ... , n) with the value 0 or 1. The Boolean formula can determine an assignment of all the variables v_i (i=1, 2, ... , n) that satisfies $\varphi = 1$. v_i=1 means the vertex v_i is in the dominating set and v_i=0 otherwise.

Take the following undirected graph $G=(V, E)$ with six vertices and five edges as an example, there are six clauses to be composed of the Boolean formula which can be described as $\varphi=(v_1+v_2)\cdot(v_2+v_1+v_3)\cdot(v_3+v_2+v_4)\cdot(v_4+v_3+v_5+v_6)\cdot(v_5+v_4)\cdot(v_6+v_4)$.

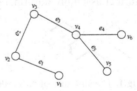

Fig.2. Given a undirected graph $G = (V, E)$, $V=\{v_1, v_2, v_3, v_4, v_5, v_6\}$, $E=\{e_1, e_2, e_3, e_4, e_5\}$.

3.2 The Model for Solving the Minimum Dominating Set Problem

Given an undirected graph $G = (V, E)$, suppose the number of the vertices and edges of this graph be n and m respectively. The Boolean formula is composed of n clauses. In order to check whether the value of the vertices are satisfied with each clause and ensure the Boolean formula $\varphi = 1$, the value of the vertices should be nondeterministically guessed which is 0 or 1. The binding domains set of this nondeterministic guess operation is $\sum_{guess} = \{*, v_1, v_1 1, v_1 0, v_i, v_i 1,$ $v_i 0, i = 1, 2, ..., n\}$ and the basic tile set T_1 of this operation is as follows in Fig.3.

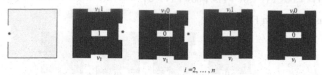

$$i = 2, ..., n$$

Fig.3. The tile set can randomly generate the value of the vertices which is 0 or 1.

The nondeterministic guess operation begins with the label "*" as the first tile in Fig.3. The second and third tiles in Fig.3 can be generated the value of the first vertex v_1 which is 0 or 1. And this value is passed to the upper from the bottom of this tile with the label "$v_1 0$" or "$v_1 1$". The last two tiles can be randomly guessed the value of the vertices from v_2 to v_n. As the same representation, the value is passed to the upper from the bottom of this tile with the label "$v_i 0$" or "$v_i 1$".

Once the value of the vertices is obtained by the nondeterministic guess operation, the assigning operation can get the value of the variables in each clause. The binding domains set of the assigning operation is $\sum_{assigning} = \{v_i, v_j, v_i 11, v_i 00,$ $v_i 1, v_i 0\}$ and the basic tile set T_2 of this operation is as follows in Fig.4.

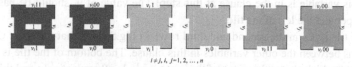

$$i \neq j, i, j = 1, 2, ..., n$$

Fig.4. The tile set of the assigning operation.

The first two tiles in Fig.4 are used to assign the value of the variables. When the right as one input is v_i and the bottom of the tile is "$v_i 1$" or "$v_i 0$", the vertex v_i in this clause is assigned the value 1 or 0 respectively, at the same time, the vertex v_i is passed to the left and the value of this vertex should be passed to the upper of the tile "$v_i 11$" or "$v_i 00$".

The tiles with green in Fig.4 show the vertex v_i is not in the clause and the vertex v_j is in the clause at this step. So this tile passes the value of the vertex v_i to the upper and the vertex v_j is passed to the left of the tile. For the last two tile types in Fig.4, the vertex v_i and v_j are both in the clause at this step, and the vertex v_i is assigned the value 1 or 0 at the previous step. Then this tile pass the value of the vertex v_i to the upper and the vertex v_j is passed to the left of the tile.

The logic OR operation is used to check whether the values of the variables are ensured that the value of each clause is equal to 1. If each clause can be demonstrated to be satisfied, it shows that the vertices with the value 1 are in the dominating set, and the vertices in each clause are adjacent. The logic OR operation is carried out between the variables in each clause. Once there is a variable with the value 1 in this clause, the value of this clause is 1 no matter what the value of the other variables are 0 or 1. The binding domains set of this operation is $\sum_{OR} = \{\#,$ 1, 0, $v_1 11$, $v_1 00$, $v_1 1$, $v_1 0$, $v_i 11$, $v_i 00$, $v_i 1$, $v_i 0\}$ and the basic tile set T_3 of this operation is as follows in Fig.5.

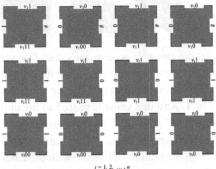

$i = 1, 2, ..., n$

Fig.5. The basic tile set of the logic OR operation.

This logic OR system begins with the label " $\#$ ", and the value of the first variable v_1 can be checked by the tiles in the first row in Fig.5. The bottom of the tile is "$v_1 11$" or "$v_1 00$" which means the value of the vertex v_1 is 1 or 0 respectively and this variable is in the clause at this step. The value of v_1 should be passed to the upper and the left of the tile.

The tile types in the second and third rows in Fig.5 show that the logic OR operations between the other variables in this clause. The bottom of the tile is "$v_i 11$" or "$v_i 00$" which shows the value of the vertex v_i is 1or 0 respectively and this variable is in the clause at this step. The bottom of the tile is "$v_i 1$" or "$v_i 0$" which represents the value of the vertex v_i is 1or 0 respectively and this variable is not in the clause at this step. The right of the tile is 1 or 0 which means the result of the logic OR operations from the right at the previous step is 1or 0 respectively. When the right of the tile is 1 and the bottom of the tile is "$v_i 11$" or "$v_i\, 00$", the result of the logic OR operation at this step is 1. When the inputs with 0 at the right and "$v_i 11$" or "$v_i\, 00$" at the bottom of the tile, the logic OR operation result at this step is 1 or 0 respectively.

The boundary tiles include input and output tiles, and computation boundary tiles which can be shown as following in Fig.6.

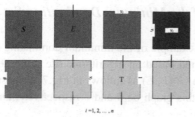

Fig.6. The basic tile set of boundary tiles.

Theorem 1. Suppose the tile system be denoted as $S_s = <T_s, g_s, \tau_s>$, let $g_s = 1$ and $\tau_s = 2$. Let T_s be the set of tiles, here $T_s = T_1 \cup T_2 \cup T_3 \cup T_4$, and let the binding domains set of this tile system be denoted as $\sum_s$, which is composed of four operations introduced above, here $\sum_s = \sum_{guess} \cup \sum_{assigning} \cup \sum_{OR} \cup \sum_{Boundary}$, where T_i (i=1, 2, 3, 4) and $\sum_{guess}, \sum_{assigning}, \sum_{OR}, \sum_{Boundary}$ are defined in Fig.2, Fig.3, Fig.4, Fig.6. Then the tile system S_s can produce a unique final configuration and nondeterministically decide the minimum dominating set problem.

Proof. Considering a dominating set problem, for the given undirected graph $G=(V, E)$ with n vertices, the method of solving this minimum dominating set problem is mainly based on the reduction of this problem into the solution of Boolean formula for this graph. On this basis, the seed configuration is constructed, then the nondeterministic guess operation can generate the value of the vertices in the given graph by the binding domains set $\sum_{guess}$, and the tile set T_1. It is also clear that there is only a single position where a tile $t \in T_2$ can attach to the complexes by these assembly steps, thus the value of the vertices in each clause can be obtained by the assigning operation. And after that tile attaches there will only be a single position where a tile may attach. By induction, because $t \in T_s$, the binding domain $<bd_S(t), bd_E(t)>$ of the tile t is unique, and because $\tau_s = 2$, it follows that on this basis, the complexes can be assembled step by step to generate the final configuration by the tile sets T_2, T_3, T_4 and the binding domains sets $\sum_{guess}, \sum_{assigning}, \sum_{OR}, \sum_{Boundary}$, therefore this tile system S_s produces a unique final configuration. The final assembled complexes contain the identifier tile with "T" in its middle position which means the value of the vertices is the feasible solution of the minimum dominating set problem. Thus the tile system $S_s = <T_s, g_s, \tau_s>$ can nondeterministically decide this problem.

3.3 Examples of the Minimum Dominating Set Problem

Now we take an example of the minimum dominating set problem to show the process of the method based on algorithmic tile self-assembly as follows. Given an undirected graph $G=(V, E)$ as follows in Fig.7, suppose the number of the vertices and edges of this graph be 5 and 6 respectively, and here $V=\{v_1, v_2, v_3, v_4, v_5\}$.

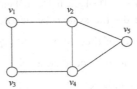

Fig.7. Given an undirected graph $G= (V, E)$, $V=\{v_1, v_2, v_3, v_4, v_5\}$.

Firstly, according to this graph, the Boolean formula $\varphi=\prod_{i=1}^{n}(v_i+\sum_{u\in N(v_i), v_i\in V} u)$ is de-

scribed as $\varphi=(v_1+v_2+v_3)\cdot(v_2+v_1+v_4)\cdot(v_3+v_1+v_4)\cdot(v_4+v_3+v_2+v_5)\cdot(v_5+v_2+v_4)$. Here, the Boolean variable in each clause is v_i with the value 0 or 1. The Boolean formula can determine an assignment of all the variables that satisfies with $\varphi=1$. $v_i=1$ means the vertex v_i is in the dominating set and $v_i=0$ otherwise. The vector $D=(v_1, v_2, v_3, v_4, v_5)$ is used to denote a dominating set.

Secondly, the initial seed configuration is constructed. In the process of growth of the self-assembly complexes, the tile assembly can be executed massively parallel computations at the molecular level. Consequently, the tile assembly system can simultaneously produce various dominating sets in the solution pool, and the final stage can be seen in Fig.8.

The process of the three operations is performed as follows. For this dominating set problem, there are five vertices, so there are five clauses to be composed of the Boolean formula φ. The nondeterministic guess operation can generate the random value of the vertices with some constraints, here the dominating set vector is represented as $D=(1,0,0,0,1)$. The value of the vertices is labeled as $v_i \in \{0,1\}$ $(i=1,2,3,4,5)$. The first clause is composed of the vertices v_1, v_2 and v_3. The value of these three variables can be obtained by the assigning operation which is 1, 0, 0 respectively. Then the logic OR operation can be made between these three variables. And the result of logic OR operation is 1 which makes the value of this clause true with the identifier tile "T". Considering the second clause with the vertices v_1, v_2 and v_4, the assigning operation gives the value of these three variables which is 1, 0, 0 respectively. Then the logic OR operation can be made with these three variables. And the logic OR operation result is 1 which makes the value of this clause true with the identifier tile "T". Thus the other three clauses also be checked whether the values of the vertices are satisfied with them by the same process using the assigning operation and logic OR operation. It can be demonstrated that the vector $D = (1,0,0,0,1)$ can satisfy the other three clauses.

Finally, to output the computation result in the solution pool, some biological operations are carried out including a combination of PCR and gel electrophoresis, and extracting the strands of different lengths which represent the output tiles in the result strand. Through the operations, it can be seen that the strands running fastest are the desired answers to the minimum dominating set problem. In this example, $D = (1,0,0,0,1)$ is the solution of this minimum dominating set problem.

For the nondeterministic algorithm, the same dominating set problem is given to show the failure in attaching tiles which can be shown in Fig.9. Firstly, the nondeterministic assigning system can generate the random value of the vertices which is represented as $D = (0,0,0,1,1)$. The process of the growth for the assembly complexes is the same as the former example. For the first clause, it is composed of the vertices v_1, v_2 and v_3 and the result of the logic OR operation is 0. The result is "No tile can match", thus the value of the vertices which can't be satisfied with the first clause and is not the feasible solution of this problem.

4 Conclusions

Considering the given minimum dominating set problem, for an undirected graph $G = (V, E)$, suppose the number of the vertices of this graph be n. The complexity for designing the minimum dominating set problem using tile self-assembly model contains two aspects: the computation time and number of distinct tiles required. Here, the computation time is the number of the assembled steps which are needed to complete the growth for the complexes, and it is equal to $(n+1)+n(n+1)+1=n^2+2n+1= \Theta(n^2)$.

The basic tiles needed in the process of performing the minimum dominating set problem contain the computation tiles and boundary tiles. For the nondeterministic guess operation which generates the value of the vertices in Fig.3, and the value of each vertex is 1 or 0, so there needs $(2n+1)$ tile types. For the tile types which are used to assign the value of the variables in each clause in Fig.4, there needs $(4n^2-2n)$ tile types. For the logic OR operation in Fig.5, there are $(8n+4)$ tile types. The basic tile types of boundary tiles can be seen in Fig.6, so the number of the boundary tiles is $(3n+5)$. The total number of tiles is $(2n+1)+(4n^2-2n)+(8n+4)+(3n+5)=4n^2+11n+10=\Theta(n^2)$. Therefore the minimum dominating set problem can be carried out with $\Theta(n^2)$ tiles in $\Theta(n^2)$ steps.

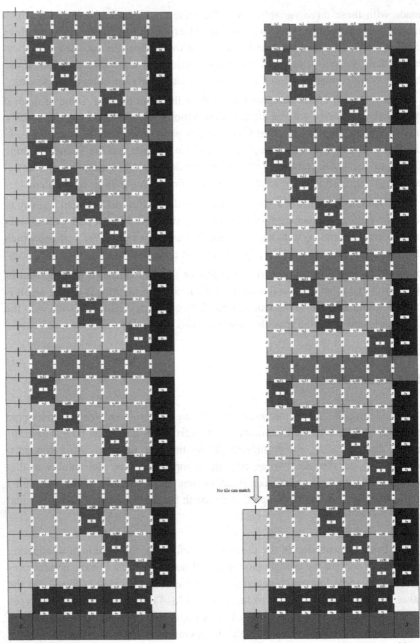

Fig.8. The final stage of the successful example. **Fig.9.** A failure example for this problem.

The algorithm for implementing the minimum dominating set problem using tile self-assembly model is proposed in our work. This method is mainly based on the conversion of the problem into the solution of Boolean formula for the given graph. First, the nondeterministic guess operation can generate the value of the vertices which can be obtained by the assigning operation. Then, the logic OR operation is carried out between these vertices in each clause. The value of the vertices are the solution of the dominating set problem which can satisfy the Boolean formula $\varphi = 1$. In summary, this model can be successfully performed the minimum dominating set in $\Theta(n^2)$ steps, once the initial strands are constructed. Here n is the number of vertices of the given graph in the minimum dominating set problem.

Acknowledgments

This work was supported by the Research Project of Department of Education of Zhejiang Province (Y201120124), and the National Natural Science Foundation of China (Grant Nos. 61202204, 61202011, 60903105, 61170054).

References

[1] Adleman, L.M.: Molecular computation of solutions to combinatorial problems. Science 266, 1021-1024(1994)

[2] Pan, L., Martin-Vide, C.: Solving multidimensional 0-1 knapsack problem by P systems with input and active membranes. J. Parallel Distr. Com. 65, 1578-1584(2005)

[3] Pan, L., Pérez-Jiménez, M.J.: Computational complexity of tissue-like P systems. J. Complexity. 26, 296-315(2010)

[4] Pan, L.Q., Liu, G.W., Xu, J.: Solid phase based DNA solution of the coloring problem. Prog. Nat. Sci. 14, 104-107(2004)

[5] Carbone, A., Seeman N.C.: Molecular Tiling and DNA Self-assembly. Lect. Notes Comput. Sci. 2950, 61-83(2004)

[6] Seeman, N.C.: DNA nanotechnology: novel DNA constructions. Annu. Rev. Biophy. Biomol. Struct. 27, 225-248(1998)

[7] Mao, C., Sun, W., Seeman, N.C.: Designed two dimensional DNA Holliday junction arrays visualized by atomic force microscopy. J. Am. Chem. Soc. 121, 5437-5443(1999)

[8] Barish, R., Rothemund, P.W., Winfree, E.: Two computational primitives for algorithmic self-assembly: copying and counting. Nano Lett. 12, 2586-2592(2005)

[9] Rothemund, P., Papadakis, N., Winfree, E.: Algorithmic self-assembly of DNA Sierpinski triangles. PLoS Biol. 12, 2041-2053(2004)

[10] Zheng, J., Birktoft, J.J., Chen, Y.: From molecular to macroscopic via the rational design of a self-assembled 3D DNA crystal. Nature 461, 74-77(2009)

[11] Douglas, S., Dietz, H., Liedl, T.: Self-assembly of DNA into nanoscale three-dimensional shapes. Nature 459, 414-418(2009)

[12] Cook, M., Rothemund, P.W., Winfree, E.: Self-assembled circuit patterns. In Proceedings of the 9th International Meeting on DNA Based Computers, Madison, WI, USA, 91-107, June 2003.

[13] Rothemund, P.W.: Folding DNA to create nanoscale shapes and patterns. Nature 440, 297-302(2006)

[14] Winfree, E.: On the computational power of DNA annealing and ligation. DNA Based Computers II: DIMACS Workshop, American Mathematical Society Publishers, Rhode Island (1996)

[15] Mao, C., LaBean, T.H., Reif, J. H.: Logical computation using algorithmic self-assembly of DNA triple-crossover molecules. Nature 407, 493-496(2000)

[16] Brun, Y.: Arithmetic computation in the tile assembly model: addition and multiplication, Theor. Comput. Sci. 378, 17-31(2006)

[17] Brun, Y.: Nondeterministic polynomial time factoring in the tile assembly model. Theor. Comput. Sci. 395, 3-23 (2008)

[18] Brun, Y., Reishus, D.: Path finding in the tile assembly model. Theor. Comput. Sci. 410, 1461-1472(2009)

[19] Barua, R., Das, S.: Finite field arithmetic using self-assembly of DNA tilings. In Proceedings of congress on evolutionary computation, 2529-36(2003)

[20] Lagoudakis, M.G., LaBean, T.H.: 2D DNA self-assembly for satisfiability. DIMACS Series in Discrete Mathematics and Theoretical Computer Science 54, 141-154(1999)

[21] Jonoska, N., McColm, G.L.: Complexity classes for self-assembling flexible tiles. Theor. Comput. Sci. 410, 332-346(2009)

[22] Reif, J.H., Sahu, S., Yin, P.: Complexity of graph self-assembly in accretive systems and self-destructible systems. Theor. Comput. Sci. 412: 1592-1605(2011)

[23] Wu, J., Cardei, M., Dai, F., Yang, S.: Extended Dominating Set and Its Applications in Ad Hoc Networks Using Cooperative Communication. IEEE T PARALL DISTRI 17, 851-864(2006)

[24] Bian, Y., Yu, H., Zeng, P.: Construction of a fault tolerance connected dominating set in wireless sensor network. International Conference on Measuring Technology and Mechatronics Automation 1, 610-614(2009)

[25] Basavanagoud, B., Teredhahalli, I.M.: On minimal and vertex minimal dominating graph. Journal of Informatics and Mathematical Sciences 1, 139-146(2009)

[26] Ho, C. K., Singh, Y. P., Ewe, H.T.: An enhanced ant colony optimization metaheuristic for the minimum dominating set problem. Applied Artificial Intelligence 20, 881-903, (2006)

[27] Guo, M.Y., Michael, H., Chang, W.L.: Fast parallel molecular solution to the dominating-set problem on massively parallel bio-computing. Parallel Computing 30, 1109-1125(2004)

[28] Liu, X.R., Wang, S.D., Xi, F.: A biomolecular computing model in vivo for minimum dominating set problem. Chinese Journal of Computer 32, 2325-2331(2009)

[29] Wang, H.: Proving theorems by pattern recognition. I. Bell System Technical Journal 40, 1-42(1961)

[30] Winfree, E.: Algorithmic self-assembly of DNA. Ph.D. Thesis, Caltech, Pasadena, CA, June (1998)

[31] Fujibayashi, K., Murata, S.: Precise simulation model for DNA tile self-assembly. IEEE T. Nanotechnol. 8, 361-368(2009)

Semantic Sub-tree Crossover Operator for Postfix Genetic Programming

Vipul K. Dabhi[1], Sanjay Chaudhary[2]

[1] I.T. Department, Dharmsinh Desai University, Nadiad, Gujarat, INDIA

[2] DA-IICT, Gandhinagar, Gujarat, INDIA

{vipul.k.dabhi@gmail.com;sanjay_chaudhary@daiict.ac.in}

Abstract. Design of crossover operator plays a crucial role in Genetic Programming (GP). The most studied issues related to crossover operator in GP are: (i) ensuring that crossover operator always produces syntactically valid individuals (ii) improving search efficiency of crossover operator. These issues become crucial when the individuals are represented using linear string representation. This paper aims to introduce postfix GP approach to symbolic regression for solving empirical modeling problems. The main contribution includes (i) a linear string (postfix notation) based genome representation method and stack based evaluation to reduce space-time complexity of GP algorithm (ii) ensuring that sub-tree crossover operator always produces syntactically valid genomes in linear string representation (iii) using semantic information of sub-trees, to be swapped, while designing crossover operator for linear genome representation to provide additional search guidance. The proposed method is tested on two real valued symbolic regression problems. Two different constant creation techniques for Postfix GP, one that explicitly use list of constants and another without use of the list, are presented to evolve useful numeric constants for symbolic regression problems. The results on tested problems show that postfix GP comprised of semantic sub-tree crossover offers a new possibility for efficiently solving empirical modeling problems.

Keywords: Postfix Genetic Programming, Symbolic Regression, Empirical Modeling, Semantic Sub-tree Crossover Operator.

1 Introduction

Many industrial applications require quick and efficient conversion of multivariate data sets into a model (mathematical equation) that can provide an insight into and knowledge about the process or system. Developing mathematical model of a process or system from experimental data is known as empirical modeling. It is

J. C. Bansal et al. (eds.), *Proceedings of Seventh International Conference on Bio-Inspired Computing: Theories and Applications (BIC-TA 2012)*, Advances in Intelligent Systems and Computing 201, DOI: 10.1007/978-81-322-1038-2_33, © Springer India 2013

difficult to solve empirical modeling problem by using traditional mathematical techniques since it is hard to define the structure of a model manually and a priori. Discovering both the structure and appropriate numeric coefficients of a model simultaneously is a challenging task for which no efficient mathematical procedure exists.

In this paper, we use our earlier proposed postfix GP framework [3] for empirical modeling problems that adopts a new method of genome (also called chromosome) representation (in postfix notation) and evaluation (stack based). The genotype-phenotype mapping mechanism follows the convention of postfix notation. The proposed method does not require constructing and traversing of expression trees (as in traditional GP), but uses stacks to decode and evaluate chromosome. Representation of genomes in postfix (linear) notation and use of stack are helpful in reducing time-space complexity of GP algorithm.

Different crossover operators in standard GP focus on producing syntactically valid individuals. However, these individuals must not only be syntactically correct but also be semantically correct. As discussed in [20], the semantic of an individual is determined by the behavior of that individual with respect to a set of input values. Incorporation of semantic information in design of crossover operator produces semantically diverse population and helps in improving search efficiency of GP. We propose semantic sub-tree crossover operator for postfix GP. The semantic sub-tree crossover operator uses approximation of semantics of selected sub-trees, to be swapped, while performing crossover operation.

The next section of this paper presents evolutionary algorithms, focusing on Genetic Algorithm (GA) [9], Genetic Programming (GP) [11] and Gene Expression Programming (GEP) [6], and related work on empirical modeling using symbolic regression via GP. Section 3 describes postfix based genetic programming framework. Section 4 discusses semantic sub-tree crossover operator implemented for postfix GP framework. The experiments and discussion of results are covered in section 5. Section 6 presents conclusions and ideas for future work.

2 Evolutionary Algorithms

Evolutionary Algorithms (EAs) employ theory of natural selection to a population of individuals (genomes) in order to generate better individuals. In EAs, genotype refers to the shape and content of an individual whereas phenotype models behavior of the individual. Two significant classes of EAs are: (i) GA (ii) GP. The basic GA [9] uses fixed length binary string representation to code potential solutions of a problem. The solution representation scheme of GA is not suitable for empirical modeling problems, since it does not permit the model structure to vary during evolution. Moreover, it would not be possible to use only a particular region of the chromosome as a solution.

Standard GP uses tree structured, variable length representation. GP can be applied to solve different types of problems compared to GA, because it makes fewer assumptions about the structure of possible solutions. So, GP approach is suitable to develop nonlinear models from input-output dataset. The chromosome representation in GP does not allow multiple genes per chromosome, where each gene codes for a small sub-expression. Gene Expression Programming (GEP) [6], [4] is a GP variant, in which chromosome is usually composed of more than one gene of equal length. GEP genes are composed of a head and a tail. The head contains elements from both function and terminal sets, whereas the tail contains elements from terminal set. Genetic modification occurs in linear structure (genotype) of chromosome which only later will rise into an expression tree (phenotype). Chromosome representation scheme used by GA, GP and GEP is useful to differentiate between them.

2.1 Related Work

Instead of conventional pointer (tree) based GP, linear (prefix ordering) representation of individual is proposed in [18] to improve performance of GP. Linear representation of individuals reduces both the evaluation (computational) time required for an individual and memory usage to represent an individual. String representation methodology in symbolic regression domain is applied by Torres et. al. [19]. They emphasized on generation of syntactically valid genetic programs after applying genetic operators. Prefix-GEP [13], a variant of GEP, uses prefix (fixed linear string) notation for genotype representation to preserve subcomponents of fittest individuals. However, phenotype of P-GEP is an expression tree, same as in GEP.

Conventional sub-tree and GA like one point crossover operators are proposed for linear representation by Tokui and Iba [18]. They calculated correlation coefficient between fitness values of parents and those of their children to measure efficiency of crossover operator. They conclude that sub-tree crossover is more efficient than GA like one point crossover for symbolic regression problems. Semantic analysis of individuals (represented as Boolean expression trees) is used to improve performance of crossover operation in GP on Boolean test problems in [2]. Their proposed Semantic Driven Crossover (SDC) [2] enhances semantic diversity of population by checking semantic equivalence of offspring produced as a result of crossover operator with their parents. The semantic equivalence of two individuals is checked by transforming these individuals into reduced ordered binary decision diagrams (ROBDDs) [2].

Use of semantic to improve search property of crossover operator in real-valued symbolic regression problems is proposed in [16]. A set of points, selected randomly and uniformly from the specified set of fitness cases, is used to measure semantic of trees/sub-trees. Two trees/sub-trees are said semantically equivalent if the outputs of these trees/sub-trees on the set are close enough. The closeness is measured using a parameter called semantic sensitivity. The resultant operator, called Semantic Aware Crossover (SAC) [16], [20] operator, improves semantic diversity of population by preventing swapping of semantically equivalent subtrees. Semantic similarity rather than semantic equivalence is checked in Semantic

Similarity-based Crossover (SSC) [20] operator. The idea behind the implementation of SSC is that exchange of sub-trees is beneficial if the two sub-trees are semantically different but not wildly different.

We have used postfix (linear string) notation for genotype representation. Mapping mechanism between genotype and phenotype obey the rules of postfix notation expression. We have incorporated semantic information of sub-trees, to be swapped, into design of crossover operator for postfix GP.

3 Postfix Genetic Programming Framework

A number of tools [5], [17], [14], [7] are available in field of GP. Many of these tools are open source and others are available commercially. These tools demand actual source code modification in order to produce required environment. Moreover, solutions produced by these tools are difficult to interpret. Because of this, we got motivated to develop our own framework of GP [3] that considers postfix notation for chromosome representation.

Traditional GP uses tree structured, variable length representation to code solution (chromosome) of a problem. Tree structure with pointer based representation is not efficient because recursive evaluation of tree is time consuming and pointer based representation is not memory efficient due to non-contiguous usage of memory locations. GP individual representation without any pointer reference allows evaluation time to be shortened and memory size to be reduced. In Postfix GP [3], individuals are represented with linear structure in postfix ordering.

As the framework uses postfix notation to represents chromosome, stack can be used for evaluating the fitness value of chromosome, without first converting chromosome to tree and then traversing the tree for evaluation. The algorithm that generates chromosomes in valid postfix notation using given operands, unary, and binary operators uses the idea as follows [3]: when a random operand from terminal set is generated, push it on stack, and add generated operand to arraylist. When a random operator is generated from function set, numbers of arguments (operands) corresponding to that operator are popped from stack, and the generated operator is added to arraylist. In postfix GP framework [3], every chromosome is of fixed length and has three length attributes: MaxLength, MinLength and ValidLength. The MinLength and MaxLength are user defined parameters. These parameters specify the range of length of syntactically valid chromosome.

MinLength: Minimum length of syntactically correct chromosome forming valid postfix equation.

MaxLength: Maximum length of syntactically correct chromosome forming valid postfix equation.

ValidLength: Index of last element of chromosome forming valid postfix equation.

To determine the ValidLength of linear chromosome, we have used the concept of "Stack-Count", introduced by Keith and Martin [10] and used in [18] for implementing Linear Genetic Programming system in C. StackCount for a node equals the number of arguments it pushes on the stack minus the number of arguments it pops off the stack. Thus, StackCount value for an operand is 1, whereas for binary operator it is -1. The sum of StackCount must be one at ValidLength position of chromosome.

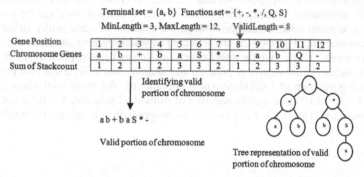

Figure 1 Chromosome in postfix notation and decoding of its syntactically valid portion

As chromosome is represented using linear string in postfix notation, we have set a limit on maximum length of chromosome and not on its depth. Computing the depth of a chromosome requires conversion from linear representation to tree representation and this conversion increases an overhead. In spite of fixed length, each chromosome has potential to form postfix equations of different sizes and shapes, the smallest containing only one element, (i.e. when all elements of a chromosome are terminal) and the biggest containing as many elements as the MaxLength of the chromosome (i.e. when all the elements of the chromosome are forming valid postfix equation) [3]. As shown in Fig. 1, it is possible that a chromosome may have some genes which are not useful for forming syntactically valid postfix equation. These unused genes do not contribute to the fitness of a chromosome. ValidLength attribute refers to index of last element of the chromosome forming valid postfix equation.

Thus, what varies is not the length of chromosome (which is MaxLength), but the ValidLength of the chromosome. Indeed, the ValidLength of a chromosome may be equal to or less than the MaxLength of the chromosome, but must be greater than or equal to MinLength of the chromosome. A chromosome is considered as invalid if its ValidLength is less than MinLength or more than MaxLength. Fig.1 shows chromosome in postfix notation, syntactically valid portion of chromosome and its tree representation. In postfix GP, the transformation of syntactically valid portion of chromosome to tree form and traversal of constructed tree is not required for evaluation of chromosome. The tree representation is shown only for better comprehension of reader. As shown in Fig.1, ValidLength (syntactically valid termination point) of chromosome is greater than MinLength of the chromosome, but less than MaxLength.

In Postfix GP [3], initial population is generated at random with chromosome's ValidLength in between MinLength and MaxLength. In our implementation, while creating initial population, diversity among the chromosomes is forced by generating different individuals in phenotype space to better explore the search space in successive generations. The Postfix GP [3] framework utilizes archive to maintain "so far best" found chromosomes [12]. New population is generated by applying genetic operations (crossover and mutation) between members of archive and current population. Mutations are protected (i.e. a terminal node can be mutated into a terminal node, a function node into a function of same arity). In current implementation, two functions are protected (division by zero and square root of negative number). The roulette wheel selection strategy [8] is applied to select a subset of solutions (from current population and archive) that will be used for producing new generation of individuals (new population). We have used Mean of Absolute Error (MAE) as a quality measure of individual solution. We have used .NET framework [15] as platform and C# as programming language for implementation.

4 Crossover Operators

A common problem in genetic programming is that genetic operators may produce syntactically invalid structures [19]. Due to the mentioned problem the search space in GP remains unexplored and performance of genetic programming gets affected. Following requirements need to be satisfied while designing crossover operator: (i) the operator should produce offspring with minimum bloat. (ii) the operator should explore solutions from a range of different sizes (lengths). First requirement would be fulfilled by fixed-length representation; whereas gratification of second requirement demands variable-length representation. To fulfill these requirements and to explore a range of possible solutions (different lengths chromosomes), we have proposed use of MinLength, MaxLength and Valid-Length attributes of chromosome in the postfix GP framework. However, too large size chromosomes take long time in evaluation and consume precious computational resources. To prevent this, we have set an upper limit on length (Max-Length attribute) of chromosome. In next sub-section we discuss random one-point and semantic sub-tree crossover operators relevant to postfix GP.

4.1 Random One-Point Crossover

This crossover operator is an extension of one-point crossover operator used in fixed length GA. There are three steps involved in implementation of random one-point crossover: (i) alignment of the parents and finding minimum of two Valid-Lengths of selected parents (ii) selection of a common crossover point (iii) swapping of the right-end sides of parents to obtain two offspring.

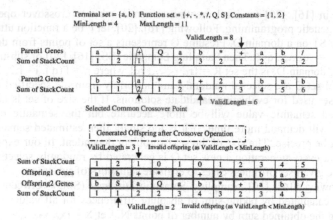

Figure 2 Random One-Point Crossover

To select a common crossover point, the minimum of ValidLengths of two parents is determined. The common crossover point is randomly chosen between values 1 and determined minimum ValidLengh. Applying this crossover operator to valid postfix equations may produce invalid equations. For ex., as shown in Fig. 2, two syntactically valid parents with ValidLength of parent1 equal to value 8 and ValidLength of parent2 equal to value 6. The minimum of these two values is 6. If common crossover point is selected randomly at position 3, the resulting offspring, generated after crossover, are syntactically invalid because the ValidLength of both offspring is less than predefined MinLength.

4.2 Semantic Sub-tree Crossover

Concept of semantic has been applied to different fields of computer science in recent years. Generally, semantic is defined in contrast to syntax: The syntax is what the grammar allows, semantics is what it means. The semantic of a real valued mathematical expression is the result it produces with respect to a set of input values. Two expressions that are same in syntax must have same semantic but the converse is not true. For ex., consider following four expressions:

$$z = 5; \qquad (1) \qquad\qquad z = x + x + x; \qquad (3)$$
$$z = 5; \qquad (2) \qquad\qquad z = x * 3; \qquad (4)$$

Expressions (1) and (2) are same in syntax and have same semantic. But the expressions (3) and (4) have different syntax, but they compute the same result and hence have same semantic.

Semantic information of two sub-trees, to be swapped, can be useful while implementing crossover operator. Semantic of sub-tree can be estimated by evaluating it on a set of given points. We have used the notation of Sampling Semantic (SS), Sampling Semantic Distance (SSD) and Semantic Equivalence (SE) described in [16], [20] for implementing semantic sub-tree crossover operator for postfix genetic programming. Following [16], [20], let F be a function uttered by a sub-tree ST on a domain D. Presume Q represents a set of points from domain D, $Q = \{q_1, q_2, q_3, \ldots, q_N\}$. Then Sampling Semantic (SS) [16], [20] of sub-tree ST on set Q of domain D, is the set $R = \{r_1, r_2, \ldots, r_N\}$ where $r_i = F(q_i)$, i= 1,2,...,N. The points of set Q are selected randomly and uniformly from the specified set of fitness cases used for evaluating candidate solutions. If the size of set is large, then estimated semantic value will be more accurate but the semantic evaluation process will demand more time. If it is too small, then estimated semantic value will not be precise. The size of the set is problem dependent. In our experimental set up, at every generation a new set Q is generated by randomly selecting points from the given set of fitness cases to evaluate semantics of sub-trees.

Following [16], [20], Sampling Semantic Distance (SSD) between two sub-trees is calculated by taking sum of absolute differences for all values of SS and dividing the obtained sum by number of points N. Let $X = (x_1, x_2, \ldots, x_N\}$ and $Y = \{y_1, y_2, \ldots, y_N\}$ be the SS of subtree1 (subtr1) and subtree2 (subtr2) on the same set of sample points. Sampling Semantic Distance (SSD) between subtr1 and subtr2 is:

$$SSD\ (subtr1, subtr2) = |x_1\text{-}y_1| + |x_2\text{-}y_2| + \ldots + |x_N\text{-}y_N| \ / \ N$$

Sampling Semantic Distance is used to evaluate Semantic Equivalence (SE) of two sub-trees [16], [20]. Two sub-trees (subtr1, subtr2) are semantically equivalent if their sampling semantic distance is less than a specified value (ϵ), named semantic sensitivity. In all our experiments we have set value of semantic sensitivity equals to 0.1.

```
SemanticSubtreeCrossover (Parent parent1, Parent parent2)
Begin
    Count = 0;
    Do
            cross_point1 = choose a random crossover point (in between values 1 and ValidLength of parent1) in parent1;
            cross_point2 = choose a random crossover point (in between values 1 and ValidLength of parent2) in parent2;

            Subtree1 Endpos = EndPosition (parent1, cross_point1); // Find end position of sub-tree starting at cross_point1;
            Subtree2 Endpos = EndPosition (parent2, cross_point2); // Find end position of sub-tree starting at cross_point2;

            Subtree1 Length = cross_point1 – Subtree1 Endpos; // Calculate Length of sub-tree
            Subtree2 Length = cross_point2 – Subtree2 Endpos;

            Subtree1 = Extract substring (Subtree1 Endpos, Subtree1 Length) from parent1;
            Subtree2 = Extract substring (Subtree2 Endpos, Subtree2 Length) from parent2;

            Measure SSD between Subtree1 and Subtree2 on P; // Calculate Sampling Semantic Distance

            Count = Count + 1;

            if SE (Subtree1, Subtree2) // Check Semantic Equivalence of two sub-trees
                Continue;
            else
                Swap Subtree1 and Subtree2;
                Offspring1.ValidLength = ValidLength of parent1 –Subtree1 Length + Subtree2 Length;
                Offspring2.ValidLength = ValidLength of parent2 - Subtree2 Length + Subtree1 Length;

    while ((Offspring1.ValidLength < MinLength || Offspring2.ValidLength < MinLength || Offspring1.ValidLength > MaxLength
                        || Offspring2.ValidLength > MaxLength) && Count < MaxTrial);

End
```

Figure 3 Semantic Sub-tree Crossover Algorithm for Postfix GP

In our proposed semantic sub-tree crossover scheme, mating between parents having same fitness values is not allowed. Parents selected for crossover operation are semantically different. Crossover points for both parents are chosen randomly in between values 1 and ValidLength of parent. After choosing crossover point for each parent, the next step is to extract the sub-trees rooted at the chosen crossover points in order to swap them. Following rule is applied to extract the sub-tree from parent: total sum of Stackcount must be one at the end of a sub-tree. After extracting the sub-trees from the parents, their semantic equivalence is evaluated using a set of data points. If the two sub-trees are semantically equivalent, then new crossover points are chosen and the process is repeated until the sub-trees are semantically different or maximum trials has completed. Semantic sub-tree crossover algorithm is presented in Fig. 3.

5   Experiments and Results

This section presents results of solving different empirical modeling problems using symbolic regression via postfix GP framework. The figures show best result obtained out of 10 runs for all symbolic regression problems. The obtained solution in infix notation with its mean absolute fitness (current error) value is shown for each problem. GP parameters setting are shown in figures.

5.1 Symbolic Regression Problem with Constants

The polynomial function identification problem, represented by equation (5), is referred in [13]. The problem is complex, as it involves discovery of different coefficients. For this problem, we have not used explicit list of constants and let the postfix GP algorithm to evolve required constants.

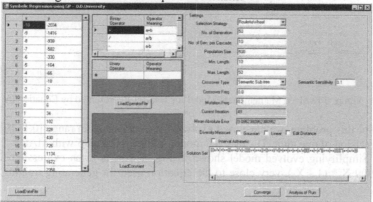

Figure 4 Evolved Mathematical Model for Dataset Represented by Equation (5)

$$y = 3*(x+1)^3 + 2*(x+1)^2 + (x+1) \tag{5}$$

For this problem, a total of twenty-one fitness cases equally spaced along x axis in the range [-10, 10] were chosen. The evolved model is shown in Fig. 4 with Mean Absolute Error (MAE) = 0.0952. Simplifying the evolved model shown in Fig. 4 results into a mathematical equation $3x^3+11x^2+14x+6$, which is an expanded version of mathematical equation shown in (5).

5.2 Time Series Problem : Logistic Map

Logistic map is one dimensional map used to illustrate chaos. For logistic map, value of next data point x_{j+1} is dependent on current data point x_j and control parameter. The logistic map is defined by

$$X_{n+1} = r\, X_n\, (\, 1 - X_n) \tag{6}$$

where r is control parameter. The problem is referred and solved using GP in [1]. We have set value of r to 3.891 and X_0 to 0.1 to generate chaotic time series (with values of embedding dimension $d = 1$ and time delay $\tau = 1$). We have bypassed initial 2000 transient points and taken next 500 data points as training dataset.

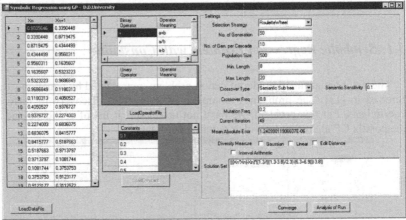

Figure 5 Evolved Mathematical Model for Dataset Represented by Equation (6)

For this problem, we have used explicit list of constants in range [0.1, 10] with a precision of one digit. The evolved model is shown in Fig. 5 with MAE = 1.24E-06. Simplifying evolved model shown in Fig. 5, results in an expression $X_{n+1} = 3.8909\, X_n* (\, 1 - X_n)$, very close to the equation used for generating training dataset.

6 Conclusions

This paper describes a GP framework that uses a linear string based chromosome representation and stack based evaluation to reduce time-space complexity of GP algorithm. The paper proposes use of semantic sub-tree crossover operator to improve search efficiency of the postfix GP algorithm. MinLength, MaxLength and ValidLength attributes of chromosome are used to ensure that crossover operator always produces syntactically valid chromosomes. Two symbolic regression problems are solved using proposed methodology. Two different constant creation techniques for Postfix GP, one that use explicit list of constants and another without use of the list, are successfully applied to solve issue of constant creation in symbolic regression problems. The experimental results indicate that symbolic regression via postfix GP comprised semantic sub-tree crossover can be used for efficiently solving empirical modeling problems. Further research aims at comparing search efficiency of standard sub-tree crossover with semantic sub-tree crossover for developed postfix GP.

References

[1] Dilip P. Ahalpara and Jitendra C. Parikh. Modeling time series of real systems using genetic programming. ArXiv Nonlinear Sciences e-prints, 14 July 2006.

[2] Lawrence Beadle and Colin G. John'son. Semantically driven crossover in genetic programming. In *Proceedings of the IEEE World Congress on Computational Intelligence*, pages 111–116, Piscataway, NJ, USA, 2008. IEEE Press.

[3] V.K. Dabhi and S.K. Vij. Empirical modeling using symbolic regression via postfix genetic programming. In *ICIIP11*, pages 1–6, 2011.

[4] Candida Ferreira. Gene Expression Programming in Problem Solving. In *Soft Computing and Industry Recent Applications*, pages 635–654. Springer, 2001.

[5] Candida Ferreira. *Gene Expression Programming: Mathematical Modeling by an Artificial Intelligence*. Springer, 2nd edition, May 2006.

[6] Cândida Ferreira. Gene expression programming: a new adaptive algorithm for solving problems. *CoRR*, cs.AI/0102027, 2001.

[7] Christian Gagné and Marc Parizeau. Open BEAGLE: A new C++ evolutionary computation framework. In *GECCO 2002: Proceedings of the Genetic and Evolutionary Computation Conference*, page 888, New York, 9-13 July 2002. Morgan Kaufmann Publishers.

[8] David E. Goldberg. *Genetic Algorithms in Search, Optimization and Machine Learning*. Addison-Wesley Longman Publishing Co., Inc., Boston, MA, USA, 1st edition, 1989.

[9] John H. Holland. *Adaptation in Natural and Artificial Systems: An Introductory Analysis with Applications to Biology, Control and Artificial Intelligence*. MIT Press, Cambridge, USA, 1992.

[10] Mike J. Keith and Martin C. Martin. Genetic programming in c++: Implementation issues. *Advances in Genetic Programming*, 1994.

[11] John R. Koza. *Genetic Programming: On the Programming of Computers by Means of Natural Selection.* MIT Press, Cambridge, MA, USA, 1992.

[12] Marco Laumanns, Lothar Thiele, Eckart Zitzler, and Kalyanmoy Deb. Archiving with guaranteed convergence and diversity in multi-objective optimization. In *Proceedings of the Genetic and Evolutionary Computation Conference*, GECCO '02, pages 439–447, San Francisco, CA, USA, 2002.

[13] Xin Li, Chi Zhou, Weimin Xiao, and Peter C. Nelson. Prefix gene expression programming. In Franz Rothlauf, editor, *Late breaking paper at Genetic and Evolutionary Computation Conference (GECCO'2005)*, Washington, D.C., USA, 2005.

[14] Sean Luke, Liviu Panait, Gabriel Balan, and Et. ECJ 16: A Java-based Evolutionary Computation Research System, 2007.

[15] Microsoft. Microsoft .net framework software development kit, 2007.

[16] Quang Uy Nguyen, Xuan Hoai Nguyen, and Michael O'Neill. Semantic aware crossover for genetic programming: The case for real-valued function regression. In *Proceedings of the 12th European Conference on Genetic Programming*, EuroGP '09, pages 292–302, Berlin, Heidelberg, 2009. Springer-Verlag.

[17] Sara Silva and Complex Systems Group. Gplab a genetic programming toolbox for matlab. *October*, (April), 2007.

[18] Nao Tokui and Hitoshl Iba. Empirical and statistical analysis of genetic programming with linear genome. 2007.

[19] Socrates Torres, Monica Larre, and Josi Torres. A string representation methodology to generate syntactically valid genetic programs. In *WSEAS IMCCAS-ISA-SOSM and MEM-MCP*, Mexico, 12-16 May 2002.

[20] Nguyen Quang Uy, Nguyen Xuan Hoai, Michael O'Neill, R. I. Mckay, and Edgar Galván-López. Semantically-based crossover in genetic programming: application to real-valued symbolic regression. *Genetic Programming and Evolvable Machines*, 12:91–119, June 2011.

Exploration Enhanced Particle Swarm Optimization using Guided Re-Initialization

Karan Kumar Budhraja[1], Ashutosh Singh[2], Gaurav Dubey[3], Arun Khosla[4]

National Institute of Technology, Jalandhar, India

{ [1]karank.budhraja, [2]ashutoshs89, [3]gauravdubey.31@gmail.com, [4]khoslaak@nitj.ac.in }

Abstract. Particle Swarm Optimization (PSO) is a stochastic computation technique aimed at finding the optimal solution to a problem. It is a population based technique inspired by the behavior of a flock of birds or school of fish, developed by Dr. Eberhart and Dr. Kennedy in 1995. The original algorithm suffers from drawbacks like premature convergence at local optimum solution (optima), and high computational cost with little robustness in case of multi-modal problems (problems involving multiple optima). This paper introduces a concept aimed at increasing the diversity (exploration of the search space) portrayed by these particles. The algorithm implements a form of teleportation by which particles are randomly re-initialized in the search space once their behavior becomes predictable. Two approaches to the implementation of this idea shall be described and discussed here. The predictability is modeled using a hyper-sphere of variable radius, centered at the best known solution.

Keywords: Particle Swarm Optimization, Evolutionary Computing, Artificial Intelligence, Guided Re-initialization

1. Introduction

Conventional computational algorithms are not suited for solving real world problems [1], [2]. This is due to the complexity of problems faced in multi-dimensional analysis. PSO is one of the various swarm techniques (based on the intelligent collective behavior of several unsophisticated entities) that provide a solution that is computationally cheaper and more robust as compared to other methods. The benefit of this approach is that there are only a few parameters [1], [3], [4] which need to be adjusted. A single version of PSO along with slightly modified variations can cover a large range of applications.

This paper illustrates a method using guided re-initialization of the particles on their approaching the known best value. The highlight of this research lies in the ability to provide a basic level of detection of predictability of particle motion

J. C. Bansal et al. (eds.), *Proceedings of Seventh International Conference on Bio-Inspired Computing: Theories and Applications (BIC-TA 2012)*, Advances in Intelligent Systems and Computing 201, DOI: 10.1007/978-81-322-1038-2_34, © Springer India 2013

present in the locality of the best solution, and its use to remove this redundancy. This is also useful since we can close in on an initial approximate solution and then narrow it down later to a fine one. This theory is combined with the original model of PSO to generate a new hybrid.

The paper is organized as follows. Section 2 briefly introduces the concept of PSO and discusses limitations in the algorithm. Section 3 describes the proposed solution to this drawback and two implementation methods. Experimental details and a comparison between the original and proposed algorithm are included in section 4.

2.   Particle Swarm Optimization

A particle in PSO can be considered as an agent searching through a hyperspace (a multidimensional space comprising of all possible solutions represented as points) with its position at any given time representing a candidate solution to the problem at hand. A new solution is found when the agent moves to a new location. The movement is loosely dependent on the social (specific to the swarm as a whole) and cognitive (specific to the individual) attractions [1].

A. Algorithm

An entity known as a "boid" (artificial bird) is used to simulate the behavior of an individual particle in the swarm. At any given time, a boid is described by two properties, namely its position (x) and velocity (v). As the boids move in the hyperspace, they also keep track of the position in their path for which the value of the fitness function was optimal (on an individual basis). The boids maintain this "personal best" (p or $pBest$) on an individual basis. The knowledge of these values is shared in the swarm, thereby allowing a comparison of these values. Thus emerges a global optima or "global best" (g or $gBest$) for any particular instant in time.

For a boid $[i]$ from the swarm, these values can be represented as x_i, v_i and p_i. For each iteration, the velocity for this particle is updated as follows:

$$v_i = wv_i + c_1 r_1 (p_i - x_i) + c_2 r_2 (g - x_i) \tag{1}$$

Fig. 1 shows a vector representation of the same. In this equation, w is the iner-tial weight representing the momentum factor: c_1 and c_2 are constants representing learning factors. They are named as cognitive parameter and social parameter re-spectively. These parameters are tuned to control the behavior of the algorithm. Parameters r_1 and r_2 are two random numbers between 0 and 1. After the calcula-tion of velocity, the position of the particle is updated as:

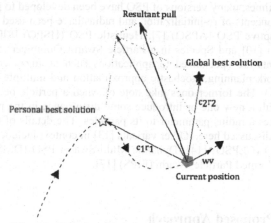

Figure 1. Velocity calculation for a particle in PSO

$$x_i = x_i + v_i \qquad (2)$$

The termination criteria for optimization are generally related to the number of iterations (based upon the computation power available) or the required degree of precision. Once the criteria are met, the simulation is halted, and the best collective performance of the swarm is reported as the optimum answer.

Before proceeding, it should be noted that the terms particle, agent and boid (artificial bird) may be used interchangeably in this context.

B. Limitations

The original version of PSO shows good exploration abilities, but weak exploitation of local optima [5]. Also, due to extensive information flow, PSO results in quick and sometimes premature convergence (particles approaching the same solution), which may lead to a suboptimal solution. It is therefore important to obtain a balanced mix to suit the optimization problem that needs to be solved. Moreover, excessive focus on exploration may lead to only a small fraction of computational efforts being used in the right places.

C. Some PSO Variants

A few factors which have been used in variants include constant inertia weight, linear reduction of inertia weight, dynamic inertia and maximum velocity reduction, tracking of time dependent minima and discrete optimization [5].

In recent times, many versions of PSO have been developed to promote diversity and the concepts of re-initialization and radius have been used by many [6] including Adaptive PSO (APSO) [7], Heuristic PSO (HSPO) [8][9], Perturbation PSO (PPSO) [10] and Species in a Particle Swarm Optimizer (SPSO) [11][12]. These variants are specialized to applications such as fuzzy controller design, power network planning, stochastic approximation and multiple objects tracking (respectively). The former ones take note of when a particle becomes static and replace it with a new one or introduce some disturbance in the particle. The last one introduces a radius parameter to its particles. The details of these algorithms shall not be discussed here. Other variants [13] in context include Comprehensive learning PSO (CLPSO) [14], Dynamic Multi-Swarm PSO (DMS-PSO) [15][16] and Fully Informed Particle Swarm (FIPS) [17].

3.   The Proposed Approach

This concept is inspired by the concept of teleportation or re-spawning, along with the knowledge of particle motion around the best solution. It is realized by placing a region termed as a portal at the location of the global optimum solution, having a certain Radius of Effect (RoE).

A. Algorithm

Particles are given an additional property of being able to stay either within or outside of this region. RoE is variable and its increase or decrease is determined by the motion of particles in and/or out of this region. This is done in two separate phases of the process. The extent to which one phase is exercised depends on the accuracy versus exploration tradeoff for the problem at hand. It must be noted that the pull on the particles towards *gBest* and *pBest* solutions has not been altered. Changes have only been made to enhance the model. This may be implemented using a single portal (single RoE shared between two phases) or a combination of two portals (both phases manipulate independent RoE values).

When particles approach the global optima, it is almost certain that they will only continue to move towards it. Hence, there is little new information that these particles can provide. Knowing this, it is arguable to look for some other use for them. Since they will only come near to each other and the optima, we decide to move them to a random location in the search space and re-initialize their velocity. The particles now start from here.

This ensures continuous investigation of other territories within the hyperspace. In practice, this is implemented by particles which have been given a property to stay outside of this portal. Any such particle found inside of the portal is teleported to a random location outside of the portal on its next update (movement). To compensate for many particles approaching the portal, that too sometimes recurrently, a mechanism is provided to increase the RoE. Whenever a teleport occurs, this value is increased. This amplifies the searching nature of the algorithm. Since this growth may be uncontrolled, it is possible that the portal may expand to cover more than a single candidate solution, which may not be desirable since it could cause overshadowing of that solution. This is harmful if the portal is centered at a sub-optimal solution and may cause overlooking of a better solution. In such situations, a check must be placed on the RoE before it is increased further. This is implemented by checking fitness on multiple points on a random line segment connecting two points on periphery of the portal and passing through *gbest* (center of portal) i.e. along a diameter of this hypersphere. Multiple points on the diameter may be checked for stricter growth of RoE. In our model, we check four such points. Since the terrains mapped on the axes are like mountains (in both positive and negative directions), we can attempt to limit the portal to a single mountain. This is checked by comparing the fitness value at different points for decreasing nature as we move away from the centre (peak of the mountain). More the number of points used, the more accurate the check will be (as shown in Fig. 2). The particles contributing to this comprise the first phase of the process. This phase is responsible for generation of an approximate location of the solution to the mathematical problem.

The second phase of the process involves thinning our broad approximation down to a more accurate and precise solution. It is formed by the particles which have been given the property to remain inside of the RoE. If such particles are found positioned outside of the RoE in an iteration of PSO, they are re-initialized and placed at a location within the portal. Their role can be described as another phase of PSO performed inside of this new search space (the section enclosed by

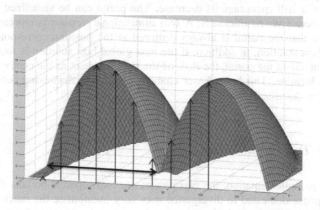

Figure 2. Diagram explaining *RoE* limit. The horizontal arrow marks the maximum span of the *portal* which is to be maintained. Expanding beyond this limit will result in covering of adjacent *mountain*, thereby degrading fitness value (considering that global optimum is a minimum in the case above). Vertical arrows show fitness values.

the portal). As the particles move, decreasing the RoE will only help in achieving higher perfection of fitness value. Thus, we implement a mechanism by which each teleportation involving such a particle causes the RoE to shrink. No restriction is placed on this decrease since it will only improve the results.

At the beginning of our process, all of the particles are assigned with the property to stay outside of the portal. This directs the particles to perform more of exploration activities and maintain high diversity. As time progresses, particles are chosen at random and their property is altered such that they shall now stay inside the portal. This has been implemented by means of a counter (teleport type change interval) for the number of executed iterations of the algorithm. When this counter reaches a certain value, one particle with property to stay outside is selected and this property is changed. This gives a rise to convergence of our particles and the process (in terms of majority of particles) slowly transforms into the second phase. Thus, towards the end of the process, most or all of the particles shall be contributing towards locating a finer solution. This behavior is an enhanced form of the convergence and divergence portrayed by the original model of PSO. In addition, to lower the probability of shadowing better solutions nearby, the portal is re-initialized if the new global best solution is found to be located outside of the current portal.

B. Single Portal Model

As the name suggests, this implementation of the algorithm uses only one portal, which is manipulated by particles inside as well as outside of it.

The particles outside the portal encourage increase of the RoE, whereas those inside of it will encourage its decrease. The portal can be visualized as a hollow hypersphere with no thickness. A two dimensional version is represented in Fig 3a. Note that the portal is always centered at the position corresponding to the global best solution. In addition, if the new global best solution is found to be located outside of the portal, the RoE of the portal is reset as the centre is shifted to its new location. This helps lower chances of the overshadowing better solutions nearby, as discussed above.

C. Two Portal Model

In this implementation, the two phases use two separate portals. The particles of the phase dedicated to divergence may only manipulate the outer RoE. Those belonging to the phase dedicated towards convergence may only manipulate the inner RoE. The terms "outer" and "inner" here are just used to denote that one portal will be located inside of the other one.

Though initially equal, the outer RoE gradually increases while the inner RoE gradually decreases. If, however, the new global best solution is found to be inside of the outer RoE or outside of the inner RoE, then the respective RoE is reset. Due to the nature of initialization of particles, they will not be found in the region lying between the two portals (the region 'B' in Fig. 3b). As the algorithm proceeds, this region increases. Future work may, for some reason, require exploration of this region. In that case, some particles may then be directed to search specifically in that area. It may be observed that the two portals have a common centre, which is the current *gBest*. The pseudo code for a single execution of the single portal based algorithm is provided below. The pseudo code for the two portal model is much like that of the single portal model, except that there are now two different ROE values to be manipulated.

```
Initialize swarm with random particle data
Initialize portal data

For iterationNumber = 1 to numberOfIterations
   Read index of gBest particle
   If (gBest is outside portal) then
      Initialize ROE value
   Endif
   Move portal to location of gBest
   If (teleport type change period completed) then
      Select a particle at random for which teleportType
                                 = STAY_OUTSIDE_PORTAL
      Change teleportType to stay inside portal
   Endif

   For particleNumber = 1 to numberOfParticles
      If (teleportType = STAY_OUTSIDE_PORTAL) then
         If (particle is inside portal) then
            If (roe maximum limit not reached) then
               Increase ROE value
            Endif
            Initialize particle data
            Teleport to location outside of portal
         Endif
      Else
         If (particle is inside portal) then
            Decrease ROE value
            Initialize particle data
            Teleport to location inside of portal
         Endif
         Calculate new velocity and position values
   Endfor
Endfor
```

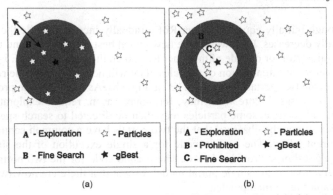

Figure 3. (a) Single *portal* model of the algorithm (b) Two *portal* model of the algorithm

D. Single Portal vs. Two Portal Model

The two models of the algorithm differ only in the portal RoE and its manipulation.

In case of the single portal model, there is a continuous conflict between the two phases of the PSO in terms of manipulation of the RoE. Since particles are initially kept outside of the portal, the RoE is expected to initially increase. It will later decrease when there are sufficient particles inside of the portal to outweigh those outside. No such conflict exists for the two portal model. The outer RoE only increases and the inner RoE only decreases. This allows for better precision since the inner RoE is capable of attaining lower values than the RoE in the single portal model.

Reset of RoE is controlled solely by the particles outside of the portal in the single portal model. On the other hand, only the particles contributing to the particular phase of the algorithm will have an influence on the respective portal. The latter version is better in concept as particles working towards exploration do not interfere in the accuracy of the solution. Similarly, particles working towards making the solution more precise do not interfere in exploration.

Particles in the single portal model are free to explore the entire hyperspace, making it more exhaustive. This is not the case for the particles in the two portal implementation. The region that lies in between the two portals remains unexplored. The single portal model may thus be projected as a simpler realization of the algorithm.

4.   Experimental Studies

A series of experiments were done to study the general behavior of the proposed models. Each of the algorithms was executed multiple times. An average fitness value was then calculated based on the fitness value generated after the completion of each execution. Variation of input and output parameters with time was focused on. These parameters were tuned to obtain their optimum values for use in the proposed models (explained below). The fitness values obtained by the two systems were then compared against those obtained for the original PSO model.

A. Results

To compare the results, some standard optimization problems were solved, and the performance was compared for each of the variations against the original implementation of PSO. The best performance improvement was observed for highly multimodal functions like Rastrigin F1. Among the two variations, two portal variation performed better than single portal model. To emphasize the improvement for the Rastrigin F1 function, we may observe the fitness value as it changes over the course of iterations of the algorithm (Fig. 4).

Tuning of the parameters was done in order to provide a better performance for optimization problems (classified as unimodal or multimodal) in general. This was executed using automated testing by means of a basic version of PSO [3] on top of the proposed approach. The fitness function was calculated as the sum of errors produced by the discussed model, for the fitness functions considered. Testing was performed separately for unimodal and multimodal functions. The following are the tuned parameters for the implementation, specified for unimodal and multimodal optimization functions [18], [19]. It should be noted here that testing was carried out for 1000 iterations, which has been kept as the maximum limit for the respective parameter. It may be noted here that a higher value of teleport type change interval is obtained for multimodal cases. It implies that the algorithm will be slow in movement of particles from the outer to inner portal (change of role of particle). This will ensure that the particles do not converge too quickly. These values were then utilized to optimize some standard optimization problems, the results of which are shown.

Table 1. Tuned values for unimodal and multimodal functions (one portal model)

Parameter	Range	Unimodal	Multimodal
Initial RoE	[0, 1]	0.86135	0.0162373
RoE increase factor	[1, 10]	3.34975	5.11948
RoE decrease factor	[1, 10]	7.84616	6.33022
Teleport type change interval	[1, 1000]	52	775

Table 2. Tuned values for unimodal and multimodal functions (two portal model)

Parameter	Range	Unimodal	Multimodal
Initial inner RoE	[0, 1]	0.048579	0.0461647
Initial outer RoE	[0, 1]	0.999	0.999
Outer RoE increase factor	[1, 10]	9.999	1.001
Inner RoE decrease factor	[1, 10]	9.999	9.84579
Teleport type change interval	[1, 1000]	1	999

Table 3. A comparison of fitness values for different optimization functions

Function	Original Model	Single Portal Model	Two Portal Model
DeJong	0.0000	0.0000	0.9604
Griewank	0.9995	0.9995	0.9995
Rastrigin F1	6.9647	2.9879	0.1392
Rosenbrock	146.7320	123.6780	109.0720
Schaffer F6	0.0372	0.0262	0.0235

B. Discussion

In case of the Dejong (unimodal) function [18], [19], the single portal model provides the same performance as the original model. This may be credited to the quick convergence in the original algorithm. The deviation in fitness value for the two portal model may be attributed to the nature of distribution of particles between the two portals. There is an improvement in case of the Rosenbrock [18], [19] function. This may be credited to the quick convergence in the original algorithm. The fitness values for the single portal model lack precision relative to the two portal model. The probable cause is the conflict between particles inside and outside of the portal, which attempt to decrease and increase the RoE respectively. Since the two portal implementation does not have this problem, it is able to approach fitness values comparable to those of the original model.

The Rastrigin F1 [18], [19] function is a highly multimodal function with sharp peaks (high gradient). From the results, we may argue that the original implementation becomes trapped in a local optimum, thereby producing an approximate result. As can be seen from the graph (Fig.4), this is due to premature convergence (the fitness stagnates shortly after 100 iterations). The other two models manage to continue exploration (even after 900 iterations in case of the two portal model) and thereby provide better values of fitness. Again, the use of two separately manipulated RoEs is able to provide a more precise result. Similar improvement is also noted for the Schaffer F6 function. In case of the Griewank function, all models provide the same result. This common value may be due to the fact that all models

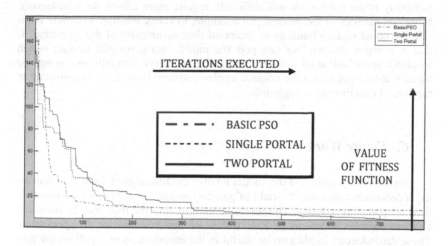

Figure 4. Variation of fitness value for Rastrigin F1 function with iterations of algorithm

share the same base implementation of the algorithm. It may be noted that the original model does not get trapped in optima in this case, and is able to provide well defined results. The performance improvement in case of multimodal functions may be attributed to the decrease in wastage of resources in searching the wrong areas in space (where the absence of global optima is almost certain).

It is observed that in case of the two portal model, the introduction of a forbidden region does not degrade the performance of the algorithm. This behavior is due to the forbidden region being located on the slanted area of the mountain away from the peak.

This model which has been developed on top of the original PSO implementation provides a novel approach to maintaining diversity among particles in the search space. Many parameters such as the factors involved in increase or decrease in RoE, the initial fraction of particles which shall stay out of the portal, and the rate at which particles are changed from being in phase one to phase two, are identified as tunable. They can be customized depending on the desired accuracy of the solution. We can even take advantage of a quick approximation, where a relatively large value of RoE may provide satisfactory results, if that is all we need. Other parameters include general PSO tunings such as swarm size, iterations to be run (which is the stopping criteria used in this implementation) and weights involved in the update of particles. It should be noted that there is a tradeoff between accuracy and diversity. Thus, we can specialize the application by dividing mathematical fitness functions into categories such as multi-peaked (boost exploration) or single peaked functions (boost accuracy), or something that may fall in between.

While the classical PSO model has a little number of tunable constants, this version opens the scope of customizing the behavior of many parameters (discussed above) in order to adapt better for specific applications. However, more flexibility in the parameters will definitely require more efforts for adjustments. More comparisons in the process (for example, to check whether a particle is inside the portal region) translate to increased time complexity of the algorithm. If this is a major concern, we can port the model into a parallel format, which wouldn't be difficult at all since PSO is parallel by nature. Nevertheless, it may be noted that the relative (to the original implementation) time delay experienced in the current experiments is negligible.

C. Future Work

Some possible variants of this model involve modification of the RoE considering a dependency on time, the ratio of particles inside and outside of the portal or their teleportation property. The functions used to alter the RoE may be modified to more complex functions of RoE which are better suited for an application. These dependencies might also be useful in the introduction of a pull on the particles based on their distance from the center of the portal, directed towards the center of the portal. Such a force will help in convergence. Finally, we only convert particle property from staying outside to staying inside. A reverse process may be added in suitable situations.

Furthermore, extensive comparisons with methods (such as CLPSO, DMS-PSO and FIPS) mentioned in section 2C may be conducted in future for a more comparative analysis of performance. CEC 2005 problems may also be incorporated for improved benchmarking.

5. CONCLUSION

In this paper, as an initial attempt, we have discussed PSO and its drawbacks and proposed an improvement taking advantage of the redundancy of information communicated among particles in PSO (the original model described by R. Eberhart and Y. Shi). The model suggested has been more effective at optimizing problems involving multi-peaked functions. The parameters involved may be tuned based on the specific requirements. The proposed method is able to avoid the premature convergence of particles at some sub-optimal point and hence provides more robust optimization algorithm for a varied set of objective functions.

References

[1] Eberhart, R. C., Shi, Y.: Particle Swarm Optimization: Developments, Applications and Resources. In: Proc. of the IEEE Cong. on Evol. Comp., vol. 1, pp. 81–86 (2001)

[2] Poli, R.: An Analysis of Publications on Particle Swarm Optimization Applications. Technical Report CSM-469, Department of Computer Science, University of Essex, Colchester, Essex, UK (2007)

[3] Shi, Y., Eberhart, R. C.: Empirical Study of Particle Swarm Optimization. In: Proc. of the IEEE Cong. on Evol. Comp. (CEC 1999), vol. 3, pp. 1945-1950 (1999)

[4] Hu, X.: Particle Swarm Optimization. http://www.swarmintelligence.org/index.php (2006)

[5] Schutte, J. F.: The Particle Swarm Optimization Algorithm. EGM 6365 - Structural Optimization, Fall 2005, Department of Mechanical and Aerospace Engineering, University of Florida, Gainesville, FL (2005)

[6] Sedighizadeh, D., Masehian, E.: Particle Swarm Optimization: Methods, Taxonomy and Applications. In: International Journal of Computer Theory and Engineering, vol. 1, pp. 1793-8201 (2009)

[7] Xie, X.-F., Zhang, W.-J., Yang, Z. L.: Adaptive Particle Swarm Optimization on Individual Level. In: In Proc. of the 6th Int. Conf. on Signal Processing. vol. 2, pp. 1215-1218 (2002)

[8] Lam, H. T., Nikolaevna, P. N., Quan, N. T. M.: A Heuristic Particle Swarm Optimization. In: Proc. of the 9th Annual Conf. on Genetic and evol. comp. (2007)

[9] Shen, X., Li, Y., Yang, J., Yu, L.: A Heuristic Particle Swarm Optimization for Cutting Stock Problem Based on Cutting Pattern. In: Lecture Notes in Computer Science, vol. 4490, pp. 1175-1178 (2007)

[10] Xinchao, Z.: A Perturbed Particle Swarm Algorithm for Numerical Optimization. In: Applied Soft Computing, vol. 10, pp. 119-124 (2010)

[11] Zhang, X., Hu, W., Li, W., Qu, W., Maybank, S.: Multi-Object Tracking via Species Based Particle Swarm Optimization. In: IEEE 12th Int. Conf. on Computer Vision Workshops, ICCV Workshops, pp. 1105-1112 (2009)

[12] Li, X.: Adaptively Choosing Neighbourhood Bests Using Species in a Particle Swarm Optimizer for Multimodal Function Optimization. In: Proc. Genetic Evol. Comput. Conf., pp. 105–116 (2004)

[13] Sugathan, P. N.: Particle Swarm Optimization & Differential Evolution. http://ewh.ieee.org/cmte/cis/mtsc/ieeecis/tutorial2007/CEC2007/P_N_Suganthan.pdf (2007)

[14] Liang, J. J., Qin, A. K., Suganthan, P. N., Baskar, S.: Comprehensive Learning Particle Swarm Optimizer for Global Optimization of Multimodal Functions. In: IEEE Trans. Evol. Comput., vol. 10, pp. 281–295 (2006)

[15] Zhao, S. Z., Liang, J. J., Suganthan, P. N., Tasgetiren, M. F.: Dynamic Multi-Swarm Particle Swarm Optimizer with Local Search for Large Scale Global Optimization. In: Proc. of IEEE Cong. on Evol. Comp., pp.3845-3852 (2008)

[16] Zhao, S. Z., Suganthana, P. N., Pan, Q.-K., Tasgetiren, M. F.: Dynamic Multi -Swarm Particle Swarm Optimizer with Harmony Search. In: Expert Systems with Application, vol. 38, pp. 3735-3742 (2011)

[17] Mendes, R., Kennedy, J., Neves, J.: The Fully Informed Particle Swarm: Simpler, Maybe Better. In: IEEE Trans. Evol. Comput., vol. 8, pp. 204–210 (2004)

[18] Molga, M., Smutnicki, C.: Test Functions for Optimization Needs. http://www.zsd.ict.pwr.wroc.pl/files/docs/functions.pdf (2005)

[19] Katebi, S.D.: Function Optimization Using GA, ES and EP. http://pasargad.cse.shirazu.ac.ir/~mhaji/ec2/EC_OPT/Project1.htm (2005)

Using Firefly Algorithm to Solve Resource Constrained Project Scheduling Problem

Pejman Sanaei[1], Reza Akbari[2], Vahid Zeighami[3]and Sheida Shams[4]

[1,3]Department of Mathematics, Shiraz University, Shiraz, Iran

[2]Department of Computer Engineering and Information Technology, Shiraz University of Technology, Shiraz, Iran

[4]Department of Management, Shiraz University, Shiraz, Iran

{pejman.sanaei@gmail.com[1], akbari@sutech.ac.ir[2], vahid.zeighami@gmail.com[3], sheida.shams@gmail.com[4]}

Abstract. The Firefly Algorithm (FA) is among the most recently introduced meta-heuristics. This work aims at study the application of FA algorithm to solve the Resource Constrained Project Scheduling Problem (RCPSP). The algorithm starts by generating a set of random schedules. After that, the initial schedules are improved iteratively using the flying approach proposed by the FA. By termination of algorithm, the best schedule found by the method is returned as the final result. The results of the state-of-art algorithms are used in this work in order to evaluate the performance of the proposed method. The comparison study shows the efficiency of the proposed method in solving RCPSP. The proposed method has competitive performance compared to the other RCPSP solvers.

Keywords: Firefly Algorithm, Resource Constrained Project Scheduling problem.

1 Introduction

The RCPSP is among the difficult problems to solve. The difficulty of RCPSP arises from the limitation of resources and precedence constraints. In a single-mode RCPSP, a set of activities A and K renewable resource type R are given as a project. Each activity A_j has duration d_j and requires units of resource R_k during each period of its duration. The first and last activities are known as dummies with zero duration and consume no resource type. In such situation, the objective of a RCPSP solver is to minimize the makespan of a schedule S by considering the constraints.

Due to NP-hardness of the RCPSP problems, the exact methods have difficulty in solving RCPSP problems of large size. Hence other alternative are required. As an alternative, the heuristic methods [1]-[4] can be used. The heuristics methods

J. C. Bansal et al. (eds.), *Proceedings of Seventh International Conference on Bio-Inspired Computing: Theories and Applications (BIC-TA 2012)*, Advances in Intelligent Systems and Computing 201, DOI: 10.1007/978-81-322-1038-2_35, © Springer India 2013

have the ability to produce near-optimal solution even for the large-size RCPSP problems. Meta-heuristic methods can be used as another way to solve the RCPSP problems. These methods can be divided in two classes. The first class contains methods such as tabu search [5] and simulated annealing [6] that maintain only one solution during each iteration. The second class contains methods such as genetic algorithm [7]-[10], Ant Colony Optimization [11], particle swarm optimization [12]-[14], and bee algorithms[15]-[16] that maintain a set of solution during each iteration. The efficiency of the mentioned methods have been combined by each others and a set of hybrid methods have been proposed. ANGEL [17], ACOSS [18], Neurogenetic [19], and Hybrid-GA [20] are among hybrid methods presented in literature. It seems that better performance may be obtained by hybrid methods.

Firefly Algorithm is among recently introduces meta-heuristic method. The algorithm is inspired from the intelligent behaviors of fireflies. The FA is a meta-heuristic method which was first introduced by Yang to solve the numerical problems [21]. Due to the efficiency of the FA on other problems, it seems that the FA may be useful to solve RCPSP problems. It seems that few methods based on FA have been proposed over the RCPSP problem. This idea has been considered by the authors and a new method based on the standard FA is proposed to solve RCPSP problems. The proposed method starts by initializing a population of fireflies. After initialization, the arrangement of the activities in each solution is updated iteration by iteration based on the behavior of fireflies until the termination condition is met. After termination, the best result found by the method is returned as the final result.

The remaining of this paper is organized as follows: Section 2 presents the details of the proposed method for solving RCPSP problems. The experimental results are given in Section 3. Finally, Section 4 concludes this work.

2   Firefly Algorithm for RCPSP

This section presents the proposed FA algorithm for solving RCPSP problem in details. The FA has been designed based on the flashing lights of a swarm of fireflies. The flashing light is vital for a firefly. Using the flashing light, a firefly can find mate, protect himself/herself from predators, and attracts the potential preys. To inspire an algorithm from the behaviors of fireflies it seems that the flashing light intensity should be used. A firefly controls its movement by considering the flashing light intensity of targets. A swarm of fireflies attracts to the brighter and more interesting locations by the flashing light intensity that associated with that location. In the implementation of the FA, it is assumed that the brighter locations represent better solutions. Hence, the algorithm tries to help the fireflies to find such locations in the search space. In general, the brightness of the flashing light is considered as the objective function. The attraction of a firefly depends on the flashing brightness of the target location. The brightness decreases as the distance between a firefly and the target location increases.

In our implementation each firefly represents a schedule for the problem at the hand. If the problem has N activities, the fireflies will fly in the search space with N dimensions. A position is represented as a priority list $\vec{P}(p_1, p_2, ..., p_n)$ where each element of this list fixedly represents an activity and its corresponding value shows the priority of that activity. Based on this representation, the position vector $\vec{x_i} = (x_{i1}, x_{i2}, ..., x_{in})$ of Firefly i represents the priority values of N activities. The lower bound and upper bound of each priority value are set at 0 and 1 respectively. The priority values smaller than 0 are set at 0 and the priority values larger than 1 are set at 1.

Input:

n: number of fireflies in the swarm
G_{max}: maximum number of generation
CS: RCPSP problem

Initialization:

Objective function f(x), $x = (x_1, ..., x_d)$
Generate initial swarm of fireflies xi $(i = 1, 2, ..., n)$
Light intensity I_i at x_i is determined by $f(x_i)$
Define light absorption coefficient γ

Update:

While $t < G_{max}$
    For i = 1 $:n$ all n fireflies
    Evaluate the flashing light intensity of firefly i
        End for
For i = 1 $:n$ all n fireflies
For j = 1 $:i$ all n fireflies
If $(Ij > Ii$), **Move** firefly i towards j in d-dimension
Attractiveness varies with distance r_{ij} via $e^{-\gamma r_{ij}}$
Evaluate new solutions and update light intensity
End if
End for j
End for i
Rank the fireflies and find the current best
End while

Termination:

    Select the best-so-far solution and return it as the final result

---

Fig. 1. Pseudocode of the FA for RCPSP problem.

The pseudo code of FA applied to solve the RCPSP is shown in Fig. 1. The FA for the RCPSP has three main phases. The first phase starts the algorithm. This phase which is called initialization, places the fireflies on the search space randomly. The second phase called "update", receives the initial solutions as input and iteratively update these solution until the termination condition is met. the third phase is called "termination" ends the algorithm and returns the best solution found by the swarm. The details of these phases are given in the next sub-sections.

2.1 Initialization

The "Initialization" is the first of the proposed FA algorithm. After receiving the input parameters such as the size of population (n), the maximum number of generations (G_{max}), and the case study, the FA method starts with the n Fireflies being placed randomly in the solution space. After initialization, the method needs to evaluate the fitness of the solutions proposed by each of the fireflies in the swarm. To evaluate the fitness, the flashing light intensity of the solution should be determined. For this purpose, the method needs to generate the schedule from the priority list. Also, in this phase, the light absorption coefficient γ is determined. The coefficient γ is an absorption coefficient which controls the decrease of the light intensity.

2.2 PositionUpdating

After initializing the swarm of fireflies, the FA method tries to improve the initial solutions based on the behavior of fireflies. The update phase of the proposed method has two main steps: in the first step, the flashing light intensities of the fireflies are calculated and in the second steps the positions of the fireflies are updated using the movement pattern of the standard FA and a permutation method.

Firefly evaluation) To update the positions, we need to compute the fitness of the solutions which are found by the fireflies. For this purpose, the flashing light intensity of the fireflies is computed. For the RCPSP problem, the flashing light intensity is considered as the makspan of the schedule which is presented by a firefly. Hence, we need to use a schedule generation scheme (SGS) such as Serial-SGS or Parallel-SGS [22]. In this work both of the SGS methods are considered. For this purpose, a random number r_{SGS} in range of [0,1] is generated in order to select the SGS which is used by the algorithm to construct the schedule. as shown in equation (5), if r_{SGS} is larger than a predefined threshold th, the Parallel-SGS is select else the Serial-SGS is selected. In our implementation, the threshold th is set at 0.5.

$$SGS = \begin{cases} \text{Serial} - \text{SGS} & \text{if } r_{SGS} \leq \text{th} \\ \text{Parallel} - \text{SGS} & \text{if } r_{SGS} > \text{th} \end{cases} \qquad (1)$$

Using the Serial-Parallel SGS provide the ability for the algorithm to use the potentiality of both scheme. Moreover, the forward-backward improvement (FBI) is used in the proposed FA algorithm. Using serial (parallel) SGS, and FBI, the makespan of each solution presented by a firefly is calculated as its flashing light intensity. The flashing light intensity of the firefly j is calculated using following equation:

$$I_j = fit(\vec{x}_j) = \frac{1}{Make\_Span_j} \qquad (2)$$

where $Make\_Span_j$ is the value of the makespan proposed by the Firefly j.

Update position) After computing the flashing light intensity of all the fireflies in the swarm, the algorithm proceeds to update the position of fireflies. For this a loop of pairwise comparison of flashing light intensity is used. In the comparison of two fireflies, each of the fireflies which has a lower light intensity is attracted toward the other firefly with the higher light intensity. The new position of the firefly is computed using the following equation:

$$x_i = x_i + \beta_0 \times e^{-\gamma r_{ij}^2} \times (x_i - x_j) + \alpha(rand - 0.5) \qquad (3)$$

where x_i and x_j represent the locations of the firefly fireflies i and j respectively. $rand$ is a random number which is drawn from the rang of [0,1]. α is a randomization parameter which can be selected from range [0,1]. The third part of the equation ($\alpha(rand - 0.5)$) controls the random movement of the firefly. The second part controls the movement of the firefly i toward firefly j. β_0 is the initial attractiveness at $r = 0$, and γ is an absorption coefficient which controls the decrease of the light intensity. The value of β_0 can be determined based on the problem. Usually it is set at 1. The attractiveness function of a firefly can be computed as:

$$\beta_{(r)} = \beta_0 \times \exp(-\gamma r^m), with m \geq 1 \qquad (4)$$

In equation (3), r_{ij} represents the distance between two fireflies i and j at x_i and x_j locations in the search space respectively. It can be defined as a Cartesian distance:

$$r_{ij} = \|x_i - x_j\| = \sqrt{\sum_{k=1}^{d}(x_{i,k} - x_{j,k})^2} \qquad (5)$$

where $x_{i,k}$ is the kth component of the spatial coordinate x_i of the firefly i and d represents the number of dimensions of the search space.

The proposed method uses the permutation-based representation which was first introduced in [23]. After calculating the new position of the firefly i using equation (3), the permutation process is used by the algorithm to permute this new solution. To provide the update step in a more clear way, we use an example which is given in Fig. 2. As can be seen in this figure, each of the fireflies i and j represent feasible activity lists and their corresponding priority lists. The priority list of the firefly i is updated using equation (3) and a new priority list will be obtained. After that a random list is generated. Then, each priority value of the firefly i is compared with its corresponding value in the random vector. If the random value is smaller than its corresponding priority value, then the corresponding activity will be swapped with its neighbor. After swapping the priority values and their corresponding activities, the obtained activity list is examined against the constraints and if the constraints satisfaction is violated, the infeasible activity list is resolve to the feasible one.

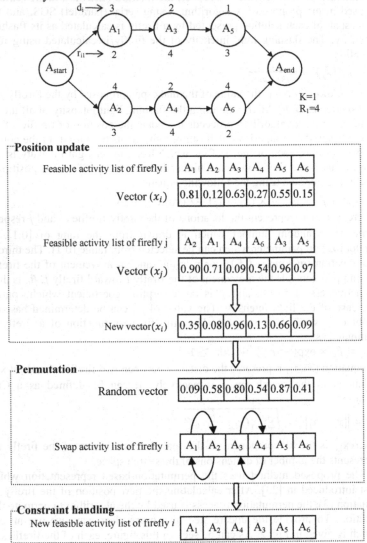

Fig. 2.Updating of the firefly-represented permutation and a precedence graph.

After moving to the new position, the flashing light intensity of the new firefly is evaluated. If the better flashing light intensity obtained, the position of the firefly is updated.

This comparison is done for each couple of the fireflies. After the comparison, the fireflies are ranked based on their flashing light intensity and the best solution found by the swarm is updated.

2.3 Termination

In our implementation the proposed FA algorithm is terminated after predefined number of schedule generations. As can be seen from Fig. 1, G_{max} shows the termination condition. G_{max} is computed as *maximum number of schedules*. After termination, the FA algorithm returns the schedule with minimum makespan as the ultimate output.

3 Experiments

This section presents the performance of the proposed FA method on the Single Mode Data Sets cases in PSPLIB library [24] in terms of success rate and the deviation from the optimal solution. These experiments are conducted under the following setting: The population size (n) is set at 30. Different number of schedules are used here as the termination condition of the FA method. The numbers of schedules are set at 1000, 5000, and 50000 to evaluate the success rate and 5000, and 50000 for the comparative study. The success rates of the FA method are obtained over the j30, j60, j90, and j120 case studies from the PSPLIB. For the comparative study, the j30, j60, and j120 are used. For both of the experiments, the average results of 10 independent runs are reported. The main parameters including β_0, γ and α set to 1, 0.8 and 0.2 respectively.

3.1 Success Rate

The success rate shows the number of instances in a case study which are successfully solved by the FA method. For an instance, the FA method is called successful if it can find the optimal schedule which has the minimum makespan. Table 1 shows the success rates of the proposed algorithm on PSPLIB case studies. From the results, it can be seen that the proposed method has good performance over the PSPLIB case studies. As the number of schedules increases the success rate of the FA method increases too. The complexity of the case studies increases as the number of activities in its instances increase. Hence, it is obvious that the proposed method has more difficulty in solving case studies with more activities. For example the success rate of the proposed method drops from 98.33% over the j30 case study to 35.00% over the j120 case study.

Table 1. Average success rate for PSPLIB case studies.

Case study	#schedules		
	1000	5000	50000
j30	89.58%	93.75%	98.33%
J60	72.91%	74.37%	78.75%
J90	72.30%	75.00%	77.08%
J120	20.33%	32.16%	35.00%

3.2 Comparative Study

The success rates showed the FA method has efficiency in solving RCPSP problem. Although, success rate shows the ability of a method to reach the optimal results, the performance of a method can be considered in term of average deviation from the optimal solution.

Table 2. Average deviation for j30 case study.

Algorithm	Reference	5000	50,000
ACOSS	Chen et al.[18]	0.06	0.01
Scatter Search—FBI	Debels et al. [25]	0.11	0.01
GA—hybrid, FBI	Valls et al. [20]	0.06	0.02
FA – FBI	*This Study*	*0.12*	*0.02*
GA – forw.–backw., FBI	Alcaraz et al. [26]	0.06	0.03
JPSO	Ruey-Maw Chen [27]	0.14	0.04
Sampling—LFT, FBI	Tormos and Lova [28]	0.13	0.05
TS—activity list	Nonobe and Ibaraki [29]	0.16	0.05
Sampling – LFT, FBI	Tormos and Lova[28]	0.16	0.07
GA—self-adapting	Hartmann [8]	0.22	0.08
GA—activity list	Hartmann [30]	0.25	0.08
ANGEL	Tseng and Chen [17]	0.09	-
Sampling – LFT, FBI	Tormos and Lova [28]	0.17	0.09
Neurogenetic	Agrawal et al. [19]	0.10	-
Sampling – random, FBI	Valls et al. [20]	0.28	0.11
GA – late join	Coelho and Tavares [31]	0.33	0.16
SA – activity list	Bouleimen and Lecocq[6]	0.23	-
PSO-Priorities of activities	Zhang [23]	0.42	-
Sampling—adaptative	Schirmer and Riesenberg[32]	0.44	-
TS—schedule scheme	Baar et al.[5]	0.44	-
Sampling—adaptative	Kolisch and Drexl [33]	0.53	-
GA – random key	Hartmann[30]	0.56	0.23
Sampling – LFT	Kolisch[22]	0.53	0.27
Sampling – global	Coelho and Tavares [31]	0.54	0.28
PSO-Permutation	Zhang[23]	0.61	-
GA—priority rule	Hartmann [30]	1.12	0.88
Sampling—WCS	Kolisch [22]	1.28	-
Sampling—LFT	Kolisch [22]	1.29	1.13
Sampling—random	Kolisch [34]	1.48	1.22
GA—problem space	Leon and Ramamoorthy [35]	1.59	-

Average deviation has been considered as an important measure to compare the performance of the investigated methods against each others. This measure shows the ability of an algorithm in converging towards the optimal solution. Here, the average deviation is used to investigate the performance of the proposed method in comparison with the other state-of-art methods. The performance of the state-of-art methods reported here are directly obtained from their original papers. The average deviations of the proposed method in comparison with the other state-of-art methods over the j30 case study are given in Table 2. The results show that the proposed method obtains the fourth rank. Only three hybrid methods called ACOSS, Scatter Search-FBI and GA-hybrid, FBI have better performance than the proposed FA method.

Table 3. Average deviation for j60 case study.

Algorithm	Reference	5000	50,000
ACOSS	Chen et al.[18]	10.98	10.67
Scatter search—FBI	Debels et al.[25]	11.10	10.71
GA—hybrid, FBI	Valls et al.[20]	11.10	10.73
FA - FBI	*This Study*	*11.20*	*10.80*
GA – forw.–backw.FBI	Alcaraz et al.[26]	11.19	10.84
JPSO	Ruey-Maw Chen[27]	11.43	11.00
GA—self-adapting	Hartmann[8]	11.70	11.21
GA—activity list	Hartmann[30]	11.89	11.23
ANGEL	Tseng and Chen [17]	11.27	-
Neurogenetic	Agrawal et al. [19]	11.29	-
Sampling – LFT, FBI	Tormos and Lova [28]	11.62	11.36
Sampling – LFT, FBI	Tormos and Lova [28]	11.82	11.47
GA – forw.–backw	Alcaraz and Maroto[36]	11.86	-
Sampling – LFT, FBI	Tormos and Lova [28]	11.87	11.54
SA – activity list	Bouleimen and Lecocq[6]	11.90	-
TS – activity list	Nonobe and Ibaraki [29]	12.18	11.58
Sampling–random,FBI	Valls et al.[20]	12.35	11.94
Sampling – adaptive	Schirmer [32]	12.58	-
GA – late join	Coelho and Tavares[31]	12.63	11.94
GA – random key	Hartmann[30]	13.32	12.25
GA – priority rule	Hartmann[30]	12.74	12.26
Sampling – adaptive	Kolisch and Drexl[33]	13.06	-
Sampling – WCS	Kolisch[33]	13.21	-
Sampling – global	Coelho and Tavares[31]	13.31	12.83
Sampling – LFT	Kolisch [22]	13.23	12.85
TS – schedule scheme	Baar et al.[5]	13.48	-
GA – problem space	Leon and Ramamoorthy[35]	13.49	-
Sampling – LFT	Kolisch[22]	13.53	12.97
Sampling – random	Kolisch[34]	14.30	13.66
Sampling – random	Kolisch [34]	15.17	14.22

The performance of the proposed FA method over the j60 case study after 5000 and 50000 schedule generation is given in Table 3. Similar to the j30 case study, the fourth rank is obtained by the proposed FA method. The hybrid methods ACOSS Scatter search-FBI and GA-hybrid, FBI obtained the first and second and third ranks respectively. Table 4 shows the results obtained over the j120 case study. The proposed FA algorithm is among the four best algorithms. The results show that the FA algorithm has competitive performance compared to the other methods such as ACOSS,GA-hybrid-FBI, GA-forw-backw FBI, and Scatter Search -FBI.

Table 4. Average deviation forj120case study.

Algorithm	Reference	5000	50,000
ACOSS	Chen et al.[18]	32.48	30.56
GA—hybrid, FBI	Valls et al.[20]	32.54	31.24
GA – forw.–backw.FBI	Alcaraz et al.[26]	33.91	31.49
Scatter Search – FBI	Debels et al.[25]	33.10	31.57
FA - FBI	*This Study*	*34.07*	*32.85*
JPSO	Ruey-Maw Chen [27]	33.88	32.89
GA – self-adapting	Hartmann[8]	35.39	33.21
Sampling – LFT, FBI	Tormos and Lova [28]	34.41	33.71
Neurogenetic	Agrawal et al.[19]	34.15	-
ANGEL	Tseng and Chen[17]	34.49	-
Ant system	Merkle et al.[11]	35.43	-
GA—activity list	Hartmann[8]	36.74	34.03
Sampling – LFT, FBI	Tormos and Lova [28]	35.56	34.77
Sampling – LFT, FBI	Tormos and Lova [28]	35.81	35.01
GA – forw.–backw	Alcaraz and Maroto[36]	36.57	-
TS – activity list	Nonobe and Ibaraki [29]	37.88	35.85
GA – late join	Coelho and Tavares[31]	38.41	36.44
Sampling–random,FBI	Valls et al.[20]	37.47	36.46
SA – activity list	Bouleimen and Lecocq[6]	37.68	-
GA – priority rule	Hartmann[8]	38.49	36.51
Sampling – adaptive	Schirmer [32]	38.70	-
Sampling – LFT	Kolisch [33]	38.75	37.74
Sampling – WCS	Kolisch[22]	38.77	-
GA – random key	Hartmann [8]	42.25	38.83
Sampling – adaptive	Kolisch and Drexl[33]	40.45	-
Sampling – global	Coelho and Tavares [31]	40.46	39.41
GA – problem space	Leon and Ramamoorthy [35]	40.69	-
Sampling – LFT	Kolisch[22]	41.84	40.63
Sampling – random	Kolisch[34]	43.05	41.44
Sampling – random	Kolisch[34]	47.61	45.60

5 Conclusions

A new method for solving resource-constrained project scheduling problem was proposed in this work. This method is based on the Firefly Algorithm which is originally proposed by Yang for optimizing numerical functions. The proposed method gives a set of initial schedules and tries to improve them using the search behavior of the fireflies. For this purpose, the original FA method is adapted to obtain an arrangement of the activities which results the best schedule. the comparative study of the well-known PSPLIB benchmarks showed that the proposed FA algorithm has the ability to produce competitive results compared to the other state-of-art methods. Although, the proposed FA method has efficiency in solving RCPSP problems, more efficiency may be obtained by incorporating local search methods. Although, hybridization of this method with the other meta-heuristics seems interesting.

References

[1] S. Hartmann, R. Kolisch, Experimental evaluation of state-of-the-art heuristics for the resource-constrained project scheduling problem, European Journal of Operational Research 127 (2000) 394–407.

[2] R. Kolisch, S. Hartmann, Experimental investigation of heuristics for resource-constrained project scheduling: an update, European Journal of Operational Research 174 (2006) 23–37.

[3] R. Kolisch, S. Hartmann, Heuristic algorithms for solving the resource-constrained project scheduling problem: classification and computational analysis, in: J. Weglarz (Ed.), Project Scheduling: Recent Models, algorithms and Applications, Kluwer Academic Publishers, Berlin, 1999, pp. 147–178.

[4] R. Kolisch, R. Padman, An integrated survey of deterministic project scheduling, Omega 29 (2001) 249–272.

[5] Baar T, Brucker P, Knust S. Tabu-search algorithms and lower bounds for the resource-constrained project scheduling problem. In: Voss S, Martello S, Osman I, Roucairol C, editors. Meta-heurisitics: advances and trends in local search paradigms for optimization. Dordrecht: Kluwer; 1998. p. 1–8.

[6] K. Bouleimen and H. Lecocq. "A new efficient simulated annealing algorithm for the resource– constrained project scheduling problem and its multiple modes version". European Journal of Operational Research, 149, (2003) 268–281.

[7] S. Hartmann, A competitive genetic algorithm for resource-constrained project scheduling, Naval Research Logistics 45 (1998) 733–750.

[8] S. Hartmann, A self-adapting genetic algorithm for project scheduling under resource constraints, Naval Research Logistics 49 (2002) 433–448.

[9] J.J.M. Mendes, J.F. Goncalves, M.G.C. Resende, A random key based genetic algorithm for the resource constrained project scheduling problem, Computers & Operations Research 36 (2009) 92–109.

[10] M. Ranjbar, F. Kianfar, S. Shadrokh, Solving the resource availability cost problem in project scheduling by path relinking and genetic algorithm, Applied Mathematics and Computation 196 (2008) 879–888.

[11] D. Merkle, M. Middendorf, H. Schmeck, Ant colony optimization for resource-constrained project scheduling, IEEE Transactions on Evolutionary Computation 6 (2002) 333–346.

[12] B. Jarboui, N. Damak, P. Siarry, A. Rebai, A combinatorial particle swarm optimization for solving multi-mode resource-constrained project schedulingproblems, Applied Mathematics and Computation 195 (2008) 299–308.

[13] Luo, X., Wang, D., Tang, J., & Tu, Y. (2006). An improved PSO algorithm for resourceconstrainedproject scheduling problem, intelligent control and automation,2006. In The sixth world congress on WCICA 2006 (Vol. 1, pp. 3514–3518).

[14] Zhang, C., Sun, J., Zhu, X., & Yang, Q. (2008). An improved particle swarmoptimization algorithm for flowshop scheduling problem. InformationProcessing Letters, 108(4), 204–209.

[15] K. Ziarati, R. Akbari, and V. Zeighami, "On the Performance of Bee Algorithms for Resource Constrained Project Scheduling Problem", Journal of Applied Soft Computing, Elsevier, Vol.11, No. 4, pp. 3720-3733, 2011.

[16] Akbari R., Zeighami V., and Ziarati K., Artificial Bee colony for resource constrained project scheduling problem, International Journal of Industrial Engineering Computations, DOI: 10.5267/j.ijiec.2010.04.004.

[17] Tseng, L.Y., & Chen, S. C.(2006). A hybrid metaheuristic for the resource-constrained project scheduling problem, European Journal of Operational Research, 175, 707–721.

[18] Chen, W., Shi, Y.J., Teng, H.F., Lan, X. P., Hu, L. C.(2010). An efficient hybrid algorithm for resource-constrained project scheduling. Information Sciences, 180, 1031–1039.

[19] Agarwal, A., Colak, S., &Erenguc, S.(2010). A Neurogenetic approach for the resource-constrained project scheduling problem. Computers & Operations Research doi:10.1016/j.cor.2010.01.007.

[20] Valls V., Ballestın F., & Quintanilla, S.(2008). A hybrid genetic algorithm for the resource-constrained project scheduling problem. European Journal of Operational Research, 185, 495–508.

[21] X.-S. Yang, "Firefly Algorithms for multimodal optimization", Lecture Notes in Computer Science, vol. 5792, pp. 169-178, 2009.

[22] Kolisch R. Serial and parallel resource-constrained project scheduling methods revisite: theory and computation. European Journal of Operational Research 1996;90:320–33.

[23] Hong Zhang, Xiaodong Li, Heng Li, Fulai Huang, "Particle swarm optimization-based schemes for resource-constrained project scheduling", in *Journal of Automation in Construction*, Vol. 14 (2005) 393– 404.

[24] Project Scheduling Problem Library – PSPLIB: <http://www.129.187.106.231/psplib/>.

[25] D. Debels, B. De Reyck, R. Leus, and M. Vanhoucke. "A hybrid scatter search / Electromagnetism meta–heuristic for project scheduling". European Journal of Operational Research, 2004. To appear.

[26] J. Alcaraz, C. Maroto, and R. Ruiz. Improving the performance of genetic algorithms for the RCPS problem . Proceedings of the Ninth International Workshop on Project Management and Scheduling, (2004) pages 40–43, Nancy,.

[27] Ruey-Maw Chen "Particle swarm optimization with justification and designed mechanisms for resource-constrained project scheduling problem" Expert Systems with Applications, Volume 38, Issue 6, June 2011, Pages 7102-7111

[28] Tormos P, Lova A. Integrating heuristics for resource constrained project scheduling: one step forward. Technical Report, Department ofStatistics and Operations Research, Universidad Politecnica de Valencia; 2003.

[29] Nonobe K, Ibaraki T. Formulation and tabu search algorithm for the resource constrained project scheduling problem. In: Ribeiro CC, Hansen P, editors. Essays and surveys in metaheuristics. Dordrecht: Kluwer Academic Publishers; 2002. p. 557–588.

[30] Hartmann S. A competitive genetic algorithm for resource-constrained project scheduling. Naval Research Logistics 1998;45:279–302.

[31] J. Coelho and L. Tavares. Comparative analysis of meta–heuricstics for the resource constrained project scheduling problem . Technical report, Department of Civil Engineering, Instituto Superior Tecnico, Portugal, 2003.

[32] A. Schirmer. Case–based reasoning and improved adaptive search for project scheduling . Naval Research Logistics, 47 (2000) 201–222.

[33] R. Kolisch and A. Drexl. Adaptive search for solving hard project scheduling problems . Naval Research Logistics, 43 (1996) 23–40.

[34] Kolisch R. Project scheduling under resource constraints: efficient heuristics for several problem classes. Wurzburg: Physica-Verlag; 1995.

[35] Leon VJ, Ramamoorthy B. Strength and adaptability of problem-space based neighborhoods for resource constrained scheduling. Operations Research Spektrum 1995;17:173–82.

[36] J. Alcaraz and C. Maroto. A robust genetic algorithm for resource allocation in project scheduling . Annals of Operations Research, 102 (2001) 83–109.

Analysis of Cellular Automata and Genetic Algorithm based Test Pattern Generators for Built In Self Test

Balwinder Singh[1], Sukhleen Bindra Narang[2], Arun Khosla[3]

[1]Centre for Development of Advanced Computing (C-DAC), Mohali,

(A scientific Society of Ministry of Communication & Information Technology, Govt. of India)

[2] Electronics Technology Department, Guru Nanak Dev University, Amritsar, India

[3]ECE Department, Dr. B .R. Ambedkar National Institute of Technology, Jalandhar, India

E-mail: balwinder@cdac.in , Sukleen2@yahoo.com, khoslaak@nitj.ac.in

Abstract. In today's semiconductor industry, the increasing growth of sub-micron technology has resulted in the difficulty of VLSI testing. The biology is a rich source of inspiration for designers to solve the problems related to VLSI testing such as high fault coverage, less test time, efficient test pattern generation and to reduce the power consumption during testing. The main goal of this paper is to analyze the bio-inspired test pattern generation mechanisms such as Genetic algorithms and cellular automata for the built in self test. Here we have introduced the concept of cellular automata, and analyzed the parameters (like area and power) obtained from the simulation results of cellular automata and LFSR (type I, II). The experiments are performed for the Genetic algorithm, Random and deterministic cellular automata Test Pattern generation for combinational ISCAS 85 and sequential ISCAS 89 benchmark circuits. Experimental results show that more fault coverage is achieved with less Test Vectors with adequate time.

Keywords: Built in Self Test, Cellular Automata, LFSR, Genetic Algorithms

1 Introduction

Bio inspired algorithms are important part of computational science which are essential to many engineering discipline and applications of these computational methods include Genetic algorithms, Neural network, PSO, Cellular automata etc. The GA is a biological genetic process [21] producing optimal solution by selecting the parents from population which are in the form of binary strings and pro-

J. C. Bansal et al. (eds.), *Proceedings of Seventh International Conference on Bio-Inspired Computing: Theories and Applications (BIC-TA 2012)*, Advances in Intelligent Systems and Computing 201, DOI: 10.1007/978-81-322-1038-2_36, © Springer India 2013

ducing new infants by using cross over and mutation process. The fitness function is written according to the requirement of problem, and best solution is selected from new infants. The cellular automaton is a powerful computing and modeling tools, in which the cell is updated at every clock cycle. The state of the cell is dictated by the immediate neighbors, typically termed as two state three neighborhood cellular automata.

Testing a digital circuit involves applying an appropriate set of input patterns to the circuit and checking for the correct outputs. The conventional approach is to use an external tester to perform the test. However, built-in self-test (BIST) techniques have been developed in which some of the tester functions are incorporated on the chip enabling the chip to test itself. BIST provides a number of well-known advantages. It eliminates the need for expensive testers. It provides fast location of failed units in a system because the chips can test themselves concurrently. The BIST architecture requires a Test Pattern Generator (TPG) to test Circuit Under Test. Traditionally, pseudo-random pattern generators give satisfactory results on combinational CUTs, but generally yield an unacceptably low fault coverage if applied to sequential ones. Deterministic hardware test pattern generators, which generate known output sequences, thus guaranteeing a given test length and fault coverage with considerable area requirements.

This paper concentrates on analysis of bio inspired techniques like genetic algorithm and cellular automata based TPG design. The analysis is made on the basis of power dissipation and area utilization of normal and cellular automata and also the genetic algorithms, random, deterministic techniques are applied to the industry stranded benchmark circuits and analyzed on the basis of fault coverage, time, and test vectors used to test these combinational and sequential circuits.

Section II includes the related work-study done by the researchers during the previous years. In section III , the basics of cellular automata test vector generation is described. Various test patterns generators are described in section IV. Finally the experimental work is described in section V and concluded in the last section.

2. Related work

Many research works have been carried out in study of Cellular Automata. Some of them are discussed here: Yu Yuecheng [2] propose the formal definition of extended cellular automata which provides a new effective approach to describe complicated systems composed of interacting multiple subsystems. Fulvio Corno [3] proposed an algorithm to design a Test Pattern Generator based on Cellular Automata for testing combinational circuits that effectively reduces power consumption while attaining high Fault Coverage. Ding Jianli [4] provides new approach to valid predict flight delay by introducing the multi extended cellular automata into a parallel evolution model to describe the more complicated system. Zhang Chuanwu [5] implements the cellular automata and LFSR CPLD pseudo-random sequence generator demonstrated that the cellular automata have 4 times speed of LFSR I, and 2 times speed of LFSR II. Bei Cao [6] proposes an efficient algorithm to synthesize a built-in TPG from low power deterministic test patterns without inserting any redundancy test vectors. The structure of TPG is based on the non-uniform cellular automata (CA) and is used to test combinational circuits. And the algorithm is based on the nearest neighborhood model, which can find an optimal non-uniform CA topology to generate given low power test patterns.

Cellular Automata (CA) have been deeply studied as effective test pattern generators [19] [20]. Sukanta Das et.al[7] has developed the PRPG for the regular Structure of non-linear Cellular Automata (CA).The quality of the TPG is as good as that designed with the existing schemes, employing maximal of length linear CA incurring O (n3) complexity. Kamki Nakada et al. [8] have proposed the digital VLSI implementation of multiple-value cellular automats for simulating traffic flow. Biplab K. Sikdar[9][16] A special class of cellular automata (CA) referred to as multiple attractors CA (MACA) is employed for the design. Experimental results establish the efficiency of the model with respect to memory overhead, execution speed and percentage of diagnosis. Krishna Kum [10] proposed a method to identify the seed Cellular Automata (CA) that dissipates the minimum energy, during test.

3. Cellular Automata

LFSRs are most commonly used to build TPGs [14] but recently there has been interest in Cellular Automata (CA) for test pattern generation. CA generates test vectors which are more random in nature. A cellular automaton (CA) is a collection of cells arranged in a grid, such that each cell changes state with time according to a defined set of rules that includes the states of neighboring cells [1][13]. Cells are the basic building blocks. Cells are connected with each other in grid where cells connected directly are called neighbor. Some definitions of CA are given as:

1. If same rule is applied to all the cells in a CA, then the CA is said to be Uniform or Regular CA.
2. If different rules are applied to different cells in a CA, then the CA is said to be Hybrid CA.
3. The CA is said to be a Periodic CA if the extreme cells are adjacent to each other.
4. The CA is said to be a Null boundary CA if the extreme cells are connected to logic 0-state.

5. If in CA the neighborhood dependence is on XOR or XNOR only, then the CA is called additive CA, specifically, a linear CA employs XOR rules only.

6. A CA whose transformation is invertible (i.e. all the states in the state transition diagram lie in some cycle) is called a Group CA, otherwise it is called a Non Group CA.

A simple case of cellular automata is shown in figure 1. Here the state of the ith cell is decided by its neighboring cells (i-1 and i+1). Also the rule applied on the both neighboring cells will decide the state.

S_i- State of the ith cell
S_{i-1}- State of the i-1th cell.
S_{i+1}- State of the i+1th cell.
F_i- The cell rule

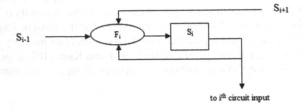

Figure 1: Architecture of ith Stage

CA based generators may provide an alternative to conventional LFSR-based generators. In addition to improved randomness properties these new pseudorandom test pattern generators also have implementation advantages such as they can be designed to require only adjacent neighbor communication and they can be cascaded, i.e., the physical length of the generator can be increased or decreased by simply adding or subtracting cells but it should be noted that the area of each CA cell is somewhat larger than an LFSR cell.

4. Test Pattern Generator

There are basically two methods considered in this paper for test pattern generation (TPG) which are given as
- TPG using LFSR
- TPG using CA

4.1 LFSR

Linear Feedback Shift Register or LFSR is a shift register whose input is the result of XOR of some of its inputs [5]. LFSR are implemented by two different methods first is internal feedback and the second is external feedback. The difference is in the way of applying feedback. Feedback is applied through XOR Gates and the flip flops that feed a XOR gate are known as 'taps'. These taps decide the patterns generated by the LFSR and hence define the characteristic polynomial of an LFSR.

External feedback LFSR

In case of external feedback LFSR, the feedback from the last FF is the input to the first FF of the shift register only as illustrated in figure 2.

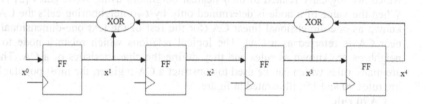

Figure 2: External Feedback LFSR

In case of internal feedback LFSR, the feedback from the last FF is the input to the first FF of the shift register and all the taps are XORed with the feedback to modify the input to the next FF in the shift register as illustrated in figure 3 [11].

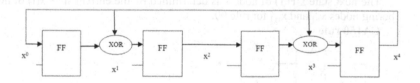

Figure 3: Internal Feedback LFSR

The speed of the internal feedback shift registers is greater than the external feedback shift register due to the number of XOR Gates in the feedback path.

The characteristics polynomial for P(x) for both types of LFSRs shown in figure is:

$$P(x) = \text{``}x^0 + x^1 + x^2 + x^3\text{''} \quad (n=4)$$

n is the degree of polynomial which is defined by the number of bits/nodes of the LFSR. Notice that the terms 'x^0' and 'x^n' are always present and the remaining terms indicate the location of the taps in the circuit. The degree of the polynomial n is equal to the number of bits in an n-bit LFSR pattern. An all zeroes state is invalid for an LFSR as the state would never change if all the bits are '0'. Therefore, the maximum number of unique patterns an n-bit LFSR can generate = $2^n - 1$, where n is the number of bits. The characteristic polynomials of an n-bit LFSR which results in the generation of maximum possible unique are known as primitive polynomials. The primitive polynomials are valid for both types of LFSRs.

4.2 CA-LFSR

Cellular automata consist of collection of cells/nodes formed by flip-flops which are logically related to their nearest neighbors using XOR gates [3] [12]. When the value of a node is determined only by two neighboring cells the CA is known as one-dimensional linear CA (for the rest of the text one-dimensional linear CA is referred as a CA). The logical relations which relate a node to its neighbors are known as rules and they define the characteristics of a CA. There are many rules which can be used to construct a CA register, the most popular being rules 90 and 150 illustrated in figure.

CA90 rule

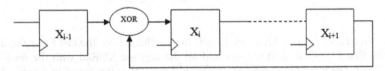

Figure 4: CA rule 90

The next state x(t+1) of node x_i is determined by the current state x(t) of neighboring nodes x_{i-1} and x_{i+1} for rule 90.

CA150 rule

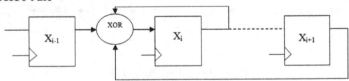

Figure 5: CA rule 150

The next state $x(t+1)$ of node x_i is determined by the current state $x(t)$ of neighboring nodes x_i, x_{i-1} and x_{i+1} for rule 150. All the nodes of a CA register do not have to be implemented with the same rule, different nodes can employ different rules. The first and the last nodes of a CA register have only one neighbor unlike all other nodes which have two; hence normal rules cannot be applied here. One solution is to assume that the missing neighbor is fixed at logic '0' (null boundary condition). The other solution assumes the last and first nodes to be neighboring nodes and is connected using normal rules (cyclic condition). Connection between the end nodes (first and last nodes) introduces a feedback loop in the cyclic boundary condition; this makes null boundary condition a better choice.

5. Experimental Results & Discussion

The conventional LFSR and Cellular Automata based LFSR's discussed in the previous section were coded in the Hardware Descriptive Language (HDL) and implemented on Cadence Design Tool Suite. The synthesis of the code is done in rc tool and the various reports are generated such as power & area as highlighted in Table 1.

Table 1:Power and Area Utilization for LFSR and CA-LFSR

Instances	Cells	Leakage power(nW)	Dynamic power (nW)	Total power (nW)	Cell Area
Internal LFSR	80	1766.066	9324.535	11090.601	623
External LFSR	80	1776.982	11844.083	13621.065	622
CA90	94	2101.158	8221.925	10323.083	701
CA150	125	2662.956	10970.510	13633.466	876

The design of TPG is the most important for Built-In Self-Test (BIST) and must satisfy some main objectives, such as high fault coverage, short test lengths and minimal area overhead. The fault coverage, time and test pattern used to achieve the maximum fault coverage is calculated using BIST analyzer and Turbo tester [17][18].In this, the Cellular Automata based Pseudorandom test patterns are generated and saves in .tst file for each benchmark circuit. Then ATPG is generated for Genetic, deterministic and random algorithms for each benchmark circuit. The results obtained for ISCAS85 combinational and sequential ISCAS89 benchmark circuits are listed in Table 2 and Table 3 respectively. The graphical comparison of each Benchmark circuit is given in figures 6 to 11 It is found that the fault coverage using genetic algorithm is better than deterministic and random.

Table 2: Experimental results for Fault coverage, No. Of Test Vectors used and time taken to test the Combinational Logic Benchmark Circuits ISCAS 89

	Circuit functions	PI	PO	No of Gates	Genetic			Deterministic			Random		
					FC	Vectors	Time	FC	Vectors	Time	FC	Vectors	Time
C17	-	5	2	6	100	6	0.01	100	6	0.01	100	7	0.01
C432	Priority Decoder	36	7	160	93.2	50	0.203	86.20	72	0.015	93.2	76	0.015
C499	ECAT	41	26	202	99.4	85	0.343	99.34	132	0.016	99.34	101	0.406
C880	ALU and control	60	32	383	100	55	0.328	100	77	0.02	99.4	92	0.062
C1355	EACT	41		546	99.50	85	0.484	99.5	126	0.015	99.5	99	0.078
C1908	EACT	33	25	880	99.48	120	0.484	99.36	139	0.015	99.42	184	0.125
C2670	ALU and control	233	140	1193	84.65	62	1.437	95.12	151	.031	84.76	94	.125
C3540	ALU and control	50	22	1669	95.54	155	0.921	94.23	190	.046	95.41	250	.234
C5315	ALU and selector	178	123	2307	98.89	119	1.593	98.41	167	.047	98.89	213	.234
C6288	16 bit multiplier	32	32	2406	99.34	22	2.75	99.34	45	0.046	99.34	47	0.312
C7552	ALU and control	207	108	3512	94.38	181	2.79	94.46	212	0.172	93.59	278	1.359

Table 3: Experimental results for Fault coverage, No. Of Test Vectors used and time taken to test the Sequential Logic Benchmark Circuits ISCAS 89

	PI	PO	FF's	No of Gates	Genetic			Deterministic			Random		
					FC	Vectors	Time	FC	Vectors	Time	FC	Vectors	Time
s298	3	6	14	119	100	32	0.02	99.72	40	0.01	100	53	0.023
s349	9	11	15	161	99.4	21	0.78	99.16	33	0.02	99.18	33	0.016
s382	3	6	21	158	100	35	0.58	100	48	0.04	100	53	0.226
s386	7	7	6	159	100	75	0.21	99.11	81	0.90	100	94	0.33
s420	18	1	16	218	97.1	77	0.387	96.46	73	0.05	84.91	74	0.145
s526	3	6	21	193	100	69	0.38	100	77	0.004	98.2	92	0.282
s641	35	24	19	379	97.5	53	0.53	98.65	63	0.006	95.9	78	0.121
s820	18	19	5	289	99.9	134	56	98.58	156	0.024	99.9	200	0.789
s1238	14	14	18	508	95.13	177	1.00	99.63	180	0.064	94.86	240	2.14
s1494	8	19	6	653	100	136	1.9	99.76	165	1.2	99.64	206	0.112
s5378	35	49	179	2779	98.64	261	5.9	98.89	290	1.6	97.2	319	3.75
s9234	19	22	211	5597	88.6	336	14.5	86.29	410	1.8	85.3	519	17.9
s35932	35	320	1728	16065	88.49	59	22.1	88.49	76	11.3	88.9	77	14.5

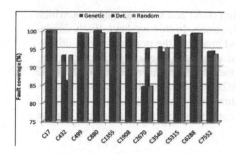

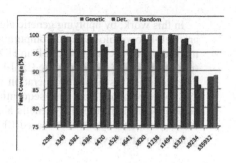

Figure 6: Fault Coverage of Benchmark
ISCAS 85

Figure7:Fault Coverage of Benchmark
ISCAS 89

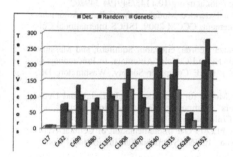

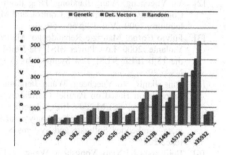

Figure 8 : No. Of Test Vectors used for
Benchmark   ISCAS 85

Figure 9: No. Of Test Vectors used for
Benchmark ISCAS 89

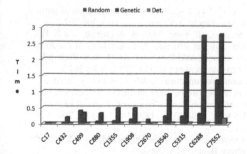

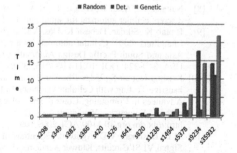

Figure10 : Time taken to test the Benchmark
ISCAS 85

Figure11 : Time taken to test the Benchmark ISCAS 89

6. Conclusion

In this work, by applying genetic algorithm (GA) and Cellular automata, which are well-known bio-inspired computing algorithms. In this we have firstly implemented the LFSR and CA-LFSR in HDL and analyzed the power and area utilization for the BIST. Secondly the CA LFSR is used for Genetic algorithm, random, deterministic Test Pattern generation and then these patterns are applied to the ISCAS85 combinational and sequential ISCAS89 benchmark circuits. It is observed from the results that Genetic algorithm covers more faults with less no. of test vectors and also CA 90 based LFSR consumes less dynamic power for the generation of test vectors.

References

[1]   Palash Sarkar. 2000. A brief history of cellular automata. ACM Comput. Surv. 32, 1 (March 2000), 80-107. DOI=10.1145/349194.349202 http://doi.acm.org/10.1145/349194.349202

[2]   Yu Yuecheng, Wang Jiandong, Ding jianli "An Extended Model of the Cellular Automata" Computing, Communication, Control, and Management, CCCM 2009. ISECS International Colloquium, 2009. Vol. 1 Pp 66 – 69

[3]   Fulvio Corno, Maurizio Rebaudengo, Matteo Sonza Reorda, Giovanni Squillero, and Massimo Violante. 2000. Low Power BIST via Non-Linear Hybrid Cellular Automata. In Proceedings of the 18th IEEE VLSI Test Symposium (VTS '00). IEEE Computer Society, Washington, DC, USA, 29-.

[4]   Ding Jianli, Yu Yuecheng, Wang Jiandong "A Model for Predicting Flight Delay and Delay Propagation Based on Parallel Cellular Automata" Computing, Communication, Control, and Management, CCCM 2009. ISECS International Colloquium 2009.

[5]   Zhang Chuanwu "Performance Analysis of the CPLD/FPGA Implementation of Cellular Automata" Embedded Software and Systems Symposia,. ICESS Symposia '08. International Conference 2008

[6]   Bei Cao, Liyi Xiao, Yongsheng Wang "A Low Power Deterministic Test Pattern Generator for BIST Based on Cellular Automata" IEEE International Symposium on Electronic design, test and applications, 2008. 266 – 269

[7]   Sukanta Das, Debdas Dey, Subhayan Sen, Biplab K Sikdar, and P Pal Chaudhuri. 2004. An efficient design of non-linear CA based PRPG for VLSI circuit testing. In Proceedings of the 2004 Asia and South Pacific Design Automation Conference (ASP-DAC '04). IEEE Press, Piscataway, NJ, USA, 110-112

[8]   Nakada, K., Asai, T., Hirose, T., & Amemiya, Y. (2004). Digital VLSI implementation of ultra-discrete cellular automata for simulating traffic flow. Electrical Engineering, 1(1), 394-397.

[9]   Biplab K. Sikdar, Debesh K. Das, Vamsi Boppana, Cliff Yang, Sobhan Mukherjee, and P. Pal Chaudhuri. 2001. Cellular automata as a built in self test structure. In Proceedings of the 2001 Asia and South Pacific Design Automation Conference (ASP-DAC '01). ACM, New York, NY, USA, 319-324. DOI=10.1145/370155.370367 http://doi.acm.org/10.1145/370155.370367

[10] Krishna Kumar S., Uday Bhaskar P., and Santanu Chattopadhyay. 2009. Low Power Pseudoexhaustive Testing with Cellular Automata. In Proceedings of the 2009 International Conference on Advances in Computing, Control, and Telecommunication Technologies (ACT '09). IEEE Computer Society, Washington, DC, USA, 419-423. DOI=10.1109/ACT.2009.109 http://dx.doi.org/10.1109/ACT.2009.109

[11] M.L. Bushnell, V.D. Agrawal, Essentials of Electronics Testing for Digital, Memory & Mixed Signal VLSI Circuits, Kluwer Academic Publishers, Boston MA, 2000.

[12] P.H. Bardell, W.H. McAnney, J. Savir, Built-in test for VLSI: Pseudorandom Techniques, John Wiley and Sons, New York, 1987.

[13] C. Stroud, A Designer's Guide to Built-In Self-Test, Kluwer Academic Publishers, Bos-ton MA, 2002

[14] S. Zhang, R. Byrne, J.C. Muzio, D.M. Miller, "Why cellular automata are better than LFSRs as built-in self-test generators for sequential-type faults", IEEE International Symposium on Circuits and Systems ISCAS 1994: 69-72, Vol. 1, 1994.

[15] Xiaodeng Zhang ,Roy, K. Peak power reduction in low power BIST. Proceedings. IEEE 2000 First International Symposium on Quality Electronic Design, 2000. ISQED 2000425 – 432

[16] Sikdar, B. K., Ganguly, N., & Chaudhuri, P. P. (2005). Fault diagnosis of VLSI circuits with cellular automata based pattern classifier. IEEE Transactions on Computer-Aided Design of Integrated Circuits and Systems VOL. 24, NO. 7, JULY 2005 pp 1115-1131

[17] Turbo tester Manual , "Turbo Tester Reference Manual", Version 02.10, Tallinn Technical University, Estonia, 2002 http://www.pld.ttu.ee/TT

[18] M.Aarna, E.Ivask, A.Jutman, E.Orasson, J.Raik, R.Ubar, V.Vislogubov, H.D.Wuttke. Turbo Tester - Diagnostic Package for Research and Training. East-West Design & Test Conference - EWDTC'03, Scientific-Technical Journal Radioelectronics and Informatics, No. 3 (24), pp. 69-73, July-Sept. 2000

[19] S. Boubezari, B. Kaminska, "A Deterministic Built-In Self-Test Generator Based on Cellular Automata Structures," IEEE Trans. Computers, V.44, N.6, Jun. 1995, pp. 805-816

[20] P.D. Hortensius, R.D. McLeod, W. Pries, D.M. Miller, H.C. Card, "Cellular Automata-Based Pseudorandom Number Generators for Built-In Self-Test," IEEE Transaction on Computer-Aided Design, V.8, N.8, Aug. 1989, pp. 842-859

[21] K.Paramasivam , "Reordering Algorithm for Minimization of Test power in VLSI Circuits", Engineering Letters, Issues_v14 pp 78-83, 2007

Ant Colony-based System for Retinal Blood Vessels Segmentation

Ahmed.H.Asad[1], Ahmad Taher Azar[2] and Aboul ella Hassaanien[3]

[1] Institute of Statistical Studies and Researches, CS Department, Cairo University
[2] PhD, IEEE Member, Assistant Professor, Faculty of Engineering, Misr University for
Science & Technology (MUST), 6th of October City, Egypt; Scientific Research Group in Egypt
(SRGE)
[3] Faculty of Computer and Information, IT Department, Cairo University

{ah_assad@hotmail.com; ahmad_T_azar@ieee.org; aboitcairo@gmail.com}

Abstract. The segmentation of retinal blood vessels in the eye funds images is crucial stage in diagnosing infection of diabetic retinopathy. Traditionally, the vascular network is mapped by hand in a time-consuming process that requires both training and skill. Automating the process allows consistency, and most importantly, frees up the time that a skilled technician or doctor would normally use for manual screening. Several studies were carried out on the segmentation of blood vessels in general, however only a small number of them were associated to retinal blood vessels. In this paper, an approach for segmenting retinal blood vessels is presented using only ant colony system. It uses eight features; four are based on gray-level and four are based on Hu moment-invariants. The features are directly computed from values of image pixels, so they take about 90 seconds in computation. The performance evaluation of this system is estimated by using classification accuracy. The presented approach accuracy is 90.28% and its sensitivity is 74%.

Keywords: Segmentation; Retinal Blood Vessels; Features Extraction; Ant Colony System; Moment-Invariants; Diabetic Retinopathy (DR).

1 Introduction

For the last two decades, retinal blood vessels segmentation attracts a lot of research in the medical image processing area since it is crucial stage in automated diagnosis of many eye diseases especially the diabetic retinopathy (DR) [1][2]. This disease spreads diabetes on the retina vessels thus they lose blood supply that causes blindness in short time [3]-[5]. Also, retinal blood vessels segmentation is the core stage in automated registration of two retinal blood vessels images of a certain patient to follow and diagnose his disease progress at different periods of time [6]. Screening is vital to preventing visual loss from diabetes because retinopathy is often asymptomatic early in the course of the disease [7]-[14]. If the retinopathy is detected in its early stages, blindness can be prevented in more than 50% of the cases [15] [16]. The automated segmentation faces multiple severe challenges as large number of images varying in modality, quality, noise, disease status, etc. Other challenges inside single image contents are too large number of

J. C. Bansal et al. (eds.), *Proceedings of Seventh International Conference on Bio-Inspired Computing: Theories and Applications (BIC-TA 2012)*, Advances in Intelligent Systems and Computing 201, DOI: 10.1007/978-81-322-1038-2_37, © Springer India 2013

vessels varying in length, with, touristy and branching. It's worth full that segmentation until now is researched. In this paper, an approach for segmenting blood vessels of retina is presented based on ant colony system (ACS) [17]. There is little previous work was done that used also ACS for solving this problem but ACS was used plus the matched filter where their results are combined to construct the final result [18] or standalone but with complex features computed at multiple scales and orientations [19].

ACS standalone based on new features proposed by Marin et al. [20] is used in this work. These features are selected because they are informative; good descriptors of vessels and background, simple; directly computed from pixels values and fast in computation; needn't be computed at multiple scales and orientations. They are mixing from features based on gray-level and others based on Hu moment-invariants [21]. When the same features of Marin et al. work are used, lower performance is obtained than using some different features from the same pool. To determine which features will be left and which others will be entered, features selection technique is performed based on correlation-based feature selection (CFS) approach [22] which is used by Lupacu et al. [23] to reduce features from forty-one to the most significant fourteen features.

The rest of paper is organized as follows: In section 2, an overview of the feature extraction and Ant Colony system is presented. In section 3, the proposed approach is described which segments vessels using ACS. Section 4 reports the results and experimental evaluations of proposed approach. Finally in Section 5, conclusion and directions for future research are presented.

2   Features Extraction and Ant Colony System: an Overview

In this section feature extractions method is described in detail in addition to the Ant Colony System

2.1   Features Extraction

Eight features are extracted from a pool consists of fourteen pixel-wise features; one feature is the gray-level of green channel of RGB retinal image, group of five features are based on gray-level $(f_1, f_2, f_3, f_4, f_5)$ and group of eight features are based on Hu moment-invariants $(Hu_1, Hu_2, Hu_3, Hu_4, Hu_5, Hu_6, Hu_7, Hu_8)$. Most of the vessels segmentation approaches extract and use the green color image of RGB retinal image for further processing since it has the best contrast between vessels and background. So this channel is very significant feature for vessels segmentation. The five gray-level based features group is presented by Marin et al [20] and its features describe the gray-level variation between vessel pixel and its surrounding. They needn't be computed at multiple scales or orientation but they are computed for each pixel in the homogenized background image at the center of window of size 9*9 pixels. So these features need only pre-computation of the homogenized background image which its gray-level is the fifth feature (f_5) in this group. These features are computed according to the following equations [20]:

$$f_1(x,y) = f_5(x,y) - \min_{(s,t)\in S^9_{x,y}} \{ f_5(s,t) \} \tag{1}$$

$$f_2(x,y) = \max_{(s,t)\in S^9_{x,y}} \{ f_5(s,t) \} - f_5(x,y) \tag{2}$$

$$f_3(x,y) = f_5(x,y) - \operatorname*{mean}_{(s,t)\in S^9_{x,y}} \{ f_5(s,t) \} \tag{3}$$

$$f_4(x,y) = \operatorname*{std}_{(s,t)\in S^9_{x,y}} \{ f_5(s,t) \} \tag{4}$$

Where $S^9_{s,t}$ is the sub-image of homogenized background image under window of size 9*9 pixels centered at the pixel of interest and $f_5(x,y)$ is the homogenized background image computed at preprocessing phase. The Hu moment-invariants [21] are best shape descriptors which are invariant to translation, scale and rotation change. So they are used by the second group of eight features to describe vessels have variant widths and angles. These features are computed for each pixel in the vessels-enhanced image at the center of window of size 17*17 pixels. Hu moment-invariants of a sub-region are derived from their central moments of order $(p+q)$ which are computed as follows [20]:

$$\mu_{pq} = \sum_i \sum_j (i-\bar{i})^p (j-\bar{j})^q I_{VE}^{S^{17}_{x,y}}(i,j) \tag{5}$$

Where $I_{VE}^{S^{17}_{x,y}}(i,j)$ is the sub-image of vessels-enhanced image under window of size 17*17 pixels centered at the pixel of interest. The parameters $\bar{i}$ and $\bar{j}$ are the coordinates of center of gravity of the sub-image which are computed as follows [20]:

$$\bar{i} = \frac{m_{10}}{m_{00}}, \qquad \bar{j} = \frac{m_{01}}{m_{00}} \tag{6}$$

As shown they are dependent on 2D moment of order $(p+q)$ which is computed as follows [20]:

$$m_{pq} = \sum_i \sum_j i^p j^q I_{VE}^{S^{17}_{x,y}}(i,j) \quad p,q = 0,1,2,... \tag{7}$$

Here the normalized central moments are used instead of their non-normalized counterpart and they are computed as follows [20]:

$$\bar{\mu}_{pq} = \frac{\mu_{pq}}{(\mu_{00})^\gamma} \quad p,q = 0,1,2, ... \tag{8}$$

Where

$$\gamma = \frac{p+q}{2} + 1 ; \quad (p+q) = 2,3,... \tag{9}$$

The eight Hu moment-invariants are computed and described in detail in Appendix A. The features which are based on these moment-invariants are computed as follows [20]:

$$Hu_i(x,y) = \left| \log(\varphi_i) \right| \quad i = 1,2,3,4,5,6,7,8. \tag{10}$$

Where the logarithm of moment is used instead of its value in order to reduce the range of feature values and using the module of logarithm for preventing complex number that are resulted from computing logarithm of negative moments. Before computing these features for the center pixel of a sub-image, the original sub-image is multiplied by a sub-image of size 17*17 pixels consists of Gaussian values whose mean is 0 and standard deviation is 1.7. The objective of multiplication is for better enhancing the sensitivity of moment's values against central pixels of vessel and background.

2.2 Ant Colony System (ACS)

The ACS as meta-heuristic searching algorithm was first proposed by Dorigo et al. [17] for solving the travelling sales man problem (TSP). The ACS is based on simulating the foraging behavior of real ants in nature. In nature when real ants are searching for foods, multiple ants are going out in random paths. As the ant is moving, it deposits a chemical substance which is called pheromone on its moving path for guiding other subsequent ants to its path. As the time goes, the pheromone is evaporating. So as the path is shorter as its pheromone concentration remains more time and more other ants are attracted to it. Thus the shortest path is the only one which attracts other ants.

In ACS, a predetermined number of artificial ants are initialized randomly at states in solutions space. Each ant constructs a solution by iteratively moving from its current state to one of its neighboring states as long as the termination condition isn't met; i.e., there's at least one unvisited state. Suppose that an ant at state i wants to traverse a new state j from the set of unvisited states Ω, then the random proportional transition rule (RPT) gives the probability distribution of transition from state i to state j; P_{ij} as in Eq. (11). The RPT rule depends on two values of candidate state for transition; its heuristic value η_{ij} (problem-dependant) and its pheromone level τ_{ij}. A uniformly distributed random number $0 \leq q \leq 1$ is generated at each move and compared against a predetermined constant $q\square$ plays as controller which selects between exploitation and biased exploration. If $q \leq q\square$ then the best state is traversed (exploitation), else the traversed state may not be the best one (biased exploration). If there's no at least one state to be traversed then the ant dies.

$$P_{ij} = \begin{cases} \arg \underset{j \in \Omega}{Max}\{\tau_{ij}^{\alpha} \eta_{ij}^{\beta}\} & \text{if } q \leq q_o \\ \dfrac{\tau_{ij}^{\alpha} \eta_{ij}^{\beta}}{\sum\limits_{k \in \Omega} \tau_{ik}^{\alpha} \eta_{ik}^{\beta}} & \text{if } q > q_o \text{ and } k \in \Omega \\ 0 & \text{Otherwise} \end{cases} \tag{11}$$

Where α and β are two predetermined parameters affect on ACS performance. Before transition to new state, the pheromone level at current state is updated by Eq. (12) since the ant deposits pheromone on it; this is called local pheromone update:

$$\tau_{ij} = (1 - \lambda) \tau_{ij} + \lambda \tau_o \tag{12}$$

Where $0 < \lambda < 1$ is a predetermined pheromone evaporation parameter and $\tau\square$ is the predetermined initial pheromone level assigned to all states. The solutions construction process isn't performed by ants only one time but it iterates as the termination condition isn't met, e.g. not reaching the predetermined iterations number. So after each iteration global pheromone update was performed by Eq. (13) only to the states on the best path (solution) traversed in this iteration.

$$\tau_{ij}(t+1) = (1-\rho)\,\tau_{ij}(t) + \frac{\rho}{L_{gbp}} \tag{13}$$

Where $0 < \rho < 1$ is a predetermined pheromone decay parameter and L_{gbp} is the length of the global best path.

3 Vessels Segmentation Using Ant Colony System

The proposed approach consists of three phases; preprocessing, ACS and post-processing. In preprocessing phase, the green channel of RGB retinal image is extracted since it has the best contrast between vessels and background. Its contrast is better enhanced to cover the whole range of intensity [0, 255]. In order to compute the eight features efficiently - the features set is $\{f_2, f_3, f_4, f_5, Hu_1, Hu_3, Hu_4, Hu_5\}$ after features selection process - the following steps are also taken from Marin et al. work [20]; briefly they include removal of central light reflex which runs down the central length of some vessels. After that, the varying illumination in background is corrected by computing the homogenized background image for better discriminating vessels from background and computing gray-level based features. Finally, a vessel enhanced image is computed for better computing Hu moment-invariants based features. The ACS phase classifies each pixel as vessel or background depending on its pheromone level τ and heuristic function η value. Euclidean distance is calculated between the target pixel and both centers of vessels and background clusters in feature space for computing its η value for ACS; Eq. (14):

$$\eta = \frac{\text{Eucld.dist to background cluster center}}{\text{Eucld.dist to vessels cluster center}} \tag{14}$$

The cluster center consists of the eight averages of all eight features over all training pixels; from the final pheromone map image, the vessels are segmented from background. In post-processing phase, linking of disjoint pixels is performed by setting pixel to 1 if it's surrounded at least by four neighboring pixels of 1; otherwise it's sat to 0. All small regions have area less than 20 are filtered out. Finally, a median filter of size 3*3 eliminates all remaining isolated noisy pixels. The algorithm of the presented approach is summarized in Table 1.

4 Experimental Results and Analysis

For computing the parameters values of vessels and background clusters and evaluating the proposed approach, Digital Retinal Images for Vessel Extraction (DRIVE) database is used [24]. It is available for free to scientific research on retinal vasculature. Each image is taken with a Canon CR5 nonmydriatic 3CCD camera with a forty-five field-of-view (FOV). Each image resolution is 584*565 pixels with eight bits per color channel. Since each image colors are encoded into three channels, each image is compressed by LZW TIFF method while it's saved in JPEG format. DRIVE consists of forty retinal images; seven of them are abnormal pathology cases. It's divided into training set of twenty images and testing set of the other half. Inside each set, for each image there is circular FOV mask of diameter that is approximately 540 pixels. Inside training set, for each image one manual segmentation by an ophthalmological expert. Inside testing set, for each image two manual segmentations by two different observers where the first observer segmentation is accepted as the ground-truth for performance evaluation.

Table 1 Algorithm of the presented approach

Step-1: Preprocess the input retinal image
a.   Extract its green channel.
b.   Linear transformation of its intensity to cover the whole intensity range [0, 255].
c.   Remove the central light reflex from it.
d.   Compute its homogenized background image.
e.   Compute its vessels-enhanced image.
Step-2: For each pixel in the image
a.   Compute its gray-level based features from result of step-(1d) according to Eqs. (1), (2), (3) and (4).
b.   Compute its moment-invariants based features from result of step-(1e) according to Eq. (10).
c.   Compute its Euclidean distance to the pre-computed center of vessels' cluster in feature space from the vessels pixels of training images.
d.   Compute its Euclidean distance to the pre-computed center of background cluster in feature space from the background pixels of training images.
e.   Compute its heuristic function value according to Eq. (14).
Step-3: Apply ACS on the image according to Eq.11, 12 and 13.
Step-4: Threshold the resulted ACS pheromone map image to segment vessels from background.
Step-5: Post-process the thresholded image
a.   For each pixel, if it's surrounded at least by four neighboring pixels of 1 then setting it to 1; else setting it to zero.
b.   For each region, if its area is less than 20 then removing it by morphological opening using disk structure element
c.   Apply median filter of size 3*3 to remove all remaining isolated pixels.

Seven features, f_1, f_2, f_3, f_4, f_5, Hu_1 and Hu_2, were used with neural network Marin et al. work [20] and gave the best performance after several experiments. However, when using these features in the proposed approach with ACS instead of NN, the performance was low as shown in Table 3. For this reason, new set of best features were selected from the source pool to discriminate between vessels and background clusters. Correlation-based feature selection (CFS) approach was used which is based on the hypothesis: *Good feature subsets contain features highly correlated with the class, yet uncorrelated with each other* [22]. After extracting the fourteen features and computing the CFS scores for several features sets, best features set were extracted $\{f_2, f_3, f_4, f_5, Hu_1, Hu_3, Hu_7, Hu_8\}$. So that for each pixel there are eight features values; four are based on gray-level and four are based on Hu moment-invariants. Eq. (15) shows how CFS $merit_s$ is computed [22]:

$$Merit_s = \frac{k\,rcf}{\sqrt{k + k\,(k-1)\,rff}} \tag{15}$$

Where "$Merit_s$ is the heuristic merit of a features subset S containing k features, rcf is the average feature-class correlation, and rff is the average feature-feature inter-correlation. The numerator gives an indication of how predictive a group of features are; the denominator of how much redundancy there is among them" [23].

DRIVE was used for computing centers of vessels and background clusters in the feature space. Based on first manual segmentation provided with each training image, the pixels of vessels and background were grouped separately from twenty images. The vessels cluster size was 569415 pixels and the background cluster size was 3971591 pixels. For each cluster, all eight features were computed over all pixels and averaged to form a cluster center. The performance of the presented approach was evaluated by using performance indices such as True Positive Rate *(TPR)*, False Positive Rate *(FPR)* and Accuracy *(Acc)* as the majority of papers in this domain used these measures. For the whole set of testing images, these measures averages were computed. *TPR* is the ratio of well-classified vessel pixels. *FPR* is the ratio of wrong-classified non-vessel pixels. Accuracy is the ratio of well-classified vessel and non-vessel pixels. The main formulations of these indices are as follows:

$$TPR = \frac{TP}{TP + FN} \tag{16}$$

$$FPR = \frac{FP}{TN + FP} \tag{17}$$

$$Acc = \frac{TP + TN}{TP + FN + TN + FP} \tag{18}$$

Where true positive (TP) is the number of pixels classified as vessels pixels and they are vessels pixels in the ground-truth. False negative (FN) is the number of pixels misclassified as background pixels but they are vessels pixels in the ground-truth. True negative (TN) is the number of pixels classified as background pixels and they are background pixels in the ground-truth. False positive (FP) is the number of pixels misclassified as vessels pixels but they are background pixels in the ground-truth. There are a number of parameters affecting ACS performance and their values must be determined at its initialization. These parameters are: number of ants, number of iterations, number of ant moves in single iteration (states visits), initial pheromone level, pheromone decay coefficients in local and global updates, and power coefficients of pheromone and heuristic values in Reference Prediction Tables (RPT). Default values by Dorigo et al. [17] were used except the number of moves in single iteration which was set to the average number of vessel pixels in one image based on training set. Table 2 shows theses parameters values.

Table 2 ACS Parameters Values

Parameter	Symbol	Value
Number of ants	-	10
Number of iterations	-	10
Number of moves in single iteration	-	28490
Initial pheromone level	$\tau\square$	0.01
Pheromone decay coefficient in local and global updates	λ, ρ	0.1
Power coefficient of pheromone value in P_{ij}	A	1.0
Power coefficient of heuristic value in P_{ij}	B	2.0

Table 3 Results of ACS Using Features Set $\{f_1, f_2, f_3, f_4, f_5, Hu_1, Hu_2\}$

Image	TPR	FPR	Acc
1	0.7960	0.0637	0.9179
2	0.7451	0.0552	0.9148
3	0.7098	0.0693	0.8985
4	0.6277	0.0694	0.8902
5	0.6996	0.0770	0.8927
6	0.5691	0.0860	0.8653
7	0.6965	0.0779	0.8923
8	0.6057	0.0876	0.8738
9	0.6248	0.0964	0.8709
10	0.7286	0.0785	0.8985
11	0.6904	0.0804	0.8899
12	0.6840	0.0794	0.8910
13	0.6728	0.0699	0.8936
14	0.7649	0.0830	0.8990
15	0.7490	0.0880	0.8950
16	0.6715	0.0723	0.8941
17	0.6558	0.0983	0.8714
18	0.7200	0.0786	0.8982
19	0.7856	0.0634	0.9184
20	0.7313	0.0880	0.8927
Average	0.6964	0.0784	0.8929

The results of ACS using the seven features of Marin et al. work [20] are shown in Table 3 while the results of ACS using the selected eight features are shown in Table 4. As shown from tables 3 and 4, there is a noticeable improvement in performance especially in average TPR by 4% and average accuracy by 1% which indicates that the selected eight-features set perform better than Marin's seven-features set with ACS. Table 5 shows comparison of the proposed approach accuracy and the accuracies of other two approaches that used ACS not as standalone. The first approach fuses the results of ACS and matcher filter which is convolved with image at multiple orientations. The second approach computes the hessian matrix of image at multiple lengths and the responses of Gabor filters convolved with image at multiple orientations. As shown from Table 5, both approaches outperformed the proposed approach but the difference isn't large and it can be reduced and preceded by more research in future by enhancing the proposed approach performance especially reducing its high false positives rate. Also the proposed approach saves too computation time and complexity with respect to both approaches.

Table 4 Results of ACS Using Features Set $\{f_2, f_3, f_4, f_5, Hu_1, Hu_3, Hu_4, Hu_5\}$

Image	TPR	FPR	Acc
1	0.8148	0.0665	0.9179
2	0.8040	0.0463	0.9312
3	0.7534	0.0600	0.9128
4	0.6961	0.0688	0.8998
5	0.7361	0.0680	0.9054
6	0.6454	0.0715	0.8885
7	0.7323	0.0698	0.9040
8	0.6810	0.0845	0.8859
9	0.6700	0.0855	0.8858
10	0.7595	0.0726	0.9073
11	0.7336	0.0728	0.9021
12	0.7208	0.0780	0.8968
13	0.7215	0.0724	0.8984
14	0.8135	0.0770	0.9101
15	0.7833	0.0806	0.9053
16	0.6990	0.0762	0.8944
17	0.6788	0.0804	0.8899
18	0.7531	0.0831	0.8981
19	0.8168	0.0716	0.9150
20	0.7633	0.0829	0.9007
Average	0.7388	0.0712	0.9028

Table 5 Comparison between Approaches Used ACS and the Proposed Approach

Vessels Segmentation Approach	Accuracy
Matched Filter + ACS [18]	0.9293
Fuzzy ACS [19]	0.9330
Proposed Approach	0.9028

Figure 1 shows the difference in segmentation quality between the two features sets for three pathological testing images in DRIVE database. The selected features set performed well in showing branching and very small capillaries. This is also clear when comparing between the TPRs of three images; numbers 4, 6 and 8 in Table 3 and Table 4. The features computation takes time about one minute and half on PC with Intel Core-i3 at 2.53 GH$_z$ and 3 GB of RAM. Further improvement in the program will decrease the time to about one minute.

5 Conclusion

The eye diseases mainly contribute to blindness and often can't be remedied because the patients are diagnosed too late with the diseases. In this paper, segmentation approach of retinal blood vessels is presented using ant colony system. It uses eight features; four are based on gray-level and four are based on Hu moment-invariants. The results showed that the selected eight-features set perform better than Marin's seven-features set with ACS. Although the results of the proposed approach don't show that it's the best over the all other segmentation approaches, but the proposed approach represents an initial step in using bio-inspired algorithms as ACS for retinal blood vessels segmentation as standalone not in fusion with other approaches or with complex features. The performance of the proposed approach is also considerable as it depends on simple and fast computed features. As a future direction, incorporating other descriptive parameters of vessels and background clusters rather their centers could better improve the performance. In addition, the future investigation will pay much attention to apply this approach on other database systems contain more abnormal retinal images in order to evaluate its performance on pathological images.

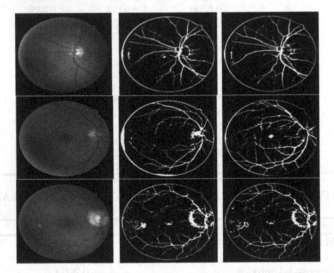

Fig. 1 Segmentation Results of Marin's 7-Features Set and Proposed 8-Features Set with ACS for Testing Images of Numbers 4, 6 and 8 respectively. (Left Column) Input retinal image. (Middle Column) Result of Marin's 7-features set. (Right Column) Result of our proposed 8-features set.

References

[1] Morello CM. Etiology and natural history of diabetic retinopathy: an overview. Am J Health Syst Pharm.; 64 (17 Suppl. 12): S3-7 (2007).

[2] Gardner TW, Antonetti DA, Barber AJ, LaNoue KF, Levison SW. Diabetic retinopathy: more than meets the eye. Surv Ophthalmol.; 47 Suppl 2:S253-62 (2002).

[3] Vijayakumari V, Suriyanarayanan N. Survey on the Detection Methods of Blood Vessel in Retinal Images. Eur. J. Sci. Res. 68, 1, 83-92 (2012).

[4] Serrarbassa PD, Dias AF, Vieira MF. New concepts on diabetic retinopathy: neural versus vascular damage. Arq Bras Oftalmol.; 71(3):459-63 (2008).

[5] Barber AJ.A new view of diabetic retinopathy: a neurodegenerative disease of the eye. Prog Neuropsychopharmacol Biol Psychiatry.; 27(2):283-90 (2003).

[6] Khan MI, Shaikh H, Mohd A. Mansuri, A Review of Retinal Vessel Segmentation Techniques and Algorithms. Int. J. Comp. Tech. Appl. 2(5): 1140-1144 (2011)

[7] Goatman K, Charnley A, Webster L, Nussey S. Assessment of auto-mated disease detection in diabetic retinopathy screening using two-field photography. PLoS One.; 6(12): e27524 (2011).

[8] Verma K, Deep P, Ramakrishnan AG. Detection and classification of diabetic retinopathy using retinal images. Annual IEEE India Conference (INDICON); pp. 1-6. (2011). DOI: 10.1109/INDCON.2011.6139346.

[9] Jones S, Edwards RT. Diabetic retinopathy screening: a systematic review of the economic evidence. Diabet Med.; 27(3):249-56 (2010).

[10] Rodgers M, Hodges R, Hawkins J, Hollingworth W, Duffy S, McKib-bin M, Mansfield M, Harbord R, Sterne J, Glasziou P, Whiting P, Westwood M. Colour vision testing for diabetic retinopathy: a systematic review of diagnostic accuracy and economic evaluation. Health Technol Assess.; 13(60):1-160 (2009).

[11] Farley TF, Mandava N, Prall FR, Carsky C. Accuracy of primary care clinicians in screening for diabetic retinopathy using single-image retinal photography. Ann Fam Med.; 6(5):428-34 (2008).

[12] Bloomgarden ZT. Screening for and managing diabetic retinopathy: current approaches. Am J Health Syst Pharm.; 64 (17 Suppl 12):S8-14 (2007).

[13] Chew EY. Screening options for diabetic retinopathy. Curr Opin Oph-thalmol.; 17(6):519-22 (2006).

[14] Sinclair SH. Diabetic retinopathy: the unmet needs for screening and a review of potential solutions. Expert Rev Med Devices.; 3(3): 301-13 (2006).

[15] Xu J, Hu G, Huang T, Huang H, Chen B. Using multifocal ERG re-sponses to discriminate diabetic retinopathy.Doc Ophthalmol.; 112(3):201-7 (2006).

[16] Jin X, Guangshu H, Tianna H, Houbin H, Bin C. The Multifocal ERG in Early Detection of Diabetic Retinopathy. Conf Proc IEEE Eng Med Biol Soc.; 7:7762-5 (2005).

[17] Dorigo M, Gambardella LM. Ant colony system: a cooperative learning approach to the traveling salesman problem. IEEE Trans. Evol. Comput. 1(1): 53–66 (1997).

[18] Cinsdikici MG, Aydn D. Detection of blood vessels in ophthalmoscope images using MF/ant (matched filter/ant colony) algorithm. Comput Methods Programs Biomed. 96(2): 85-95 (2009).

[19] Hooshyar S, Khayati R. Retina Vessel Detection Using Fuzzy Ant Colony Algorithm. In Proc of Canadian Conference on Computer and Robot Vision (CRV), Ottawa, 239-244 (2010).

[20] Marin D, Aquino A, Gegundez-Arias ME, Bravo JM. A New Supervised Method for Blood Vessel Segmentation in Retinal Images by Using Grey-Level and Moment Invariants-Based Features. IEEE Trans Med Imaging.; 30(1):146-58 (2011).

[21] Hu MK. Visual Pattern Recognition by Moment Invariants. IRE Trans. Inform. Theory. 8(2): 179–187 (1962).

[22] Hall MA. Correlation-based feature selection for discrete and numeric class machine learning. In Proc of 17th International Conference on Machine Learning, San Francisco, CA, 359–366 (2000) ISBN: 1-55860-707-2.

[23] Lupacu CA, Tegolo D, Trucco E. A Comparative Study on Feature Selection for Retinal Vessel Segmentation Using FABC. In Proc of the 13th International Conference on Computer Analysis of Images and Patterns (CAIP 2009), 655–662, Lecture Notes in Computer Science (LNCS) 5702, Springer-Verlag Berlin Heidelberg (2009). DOI: 10.1007/978-3-642-03767-2_80.

[24] Staal JJ, Abramoff MD, Niemeijer M, Viergever MA, van Ginneken B. Ridge based vessel segmentation in color images of the retina. IEEE Transactions on Medical Imaging, 23(4): 501-509 (2004).

Appendix A: Equations of the eight Hu moment-invariants used in the study

$$\varphi_1 = \bar{\mu}_{20} + \bar{\mu}_{02} \tag{A.1}$$

$$\varphi_2 = \left(\bar{\mu}_{20} - \bar{\mu}_{02}\right)^2 + 4\bar{\mu}_{11}^2 \tag{A.2}$$

$$\varphi_3 = \left(\bar{\mu}_{30} - 3\bar{\mu}_{12}\right)^2 + \left(3\bar{\mu}_{21} - \bar{\mu}_{03}\right)^2 \tag{A.3}$$

$$\varphi_4 = \left(\bar{\mu}_{30} + \bar{\mu}_{12}\right)^2 + \left(\bar{\mu}_{21} + \bar{\mu}_{03}\right)^2 \tag{A.4}$$

$$\varphi_5 = \left(\bar{\mu}_{30} + 3\bar{\mu}_{12}\right)\left(\bar{\mu}_{30} + \bar{\mu}_{12}\right)\left[\left(\bar{\mu}_{30} + \bar{\mu}_{12}\right)^2 - 3\left(\bar{\mu}_{21} + \bar{\mu}_{03}\right)^2\right] + \\ \left(3\bar{\mu}_{21} - \bar{\mu}_{03}\right)\left(\bar{\mu}_{21} + \bar{\mu}_{03}\right)\left[3\left(\bar{\mu}_{30} + \bar{\mu}_{12}\right)^2 - \left(\bar{\mu}_{21} + \bar{\mu}_{03}\right)^2\right] \tag{A.5}$$

$$\varphi_6 = \left(\bar{\mu}_{20} - \bar{\mu}_{02}\right)\left[\left(\bar{\mu}_{30} + \bar{\mu}_{12}\right)^2 - \left(\bar{\mu}_{21} + \bar{\mu}_{03}\right)^2\right] \\ + 4\bar{\mu}_{11}\left(\bar{\mu}_{30} + \bar{\mu}_{12}\right)\left(\bar{\mu}_{21} + \bar{\mu}_{03}\right) \tag{A.6}$$

$$\varphi_7 = \left(3\bar{\mu}_{21} - \bar{\mu}_{03}\right)\left(\bar{\mu}_{30} + \bar{\mu}_{12}\right)\left[\left(\bar{\mu}_{30} + \bar{\mu}_{12}\right)^2 - 3\left(\bar{\mu}_{21} + \bar{\mu}_{03}\right)^2\right] - \\ \left(\bar{\mu}_{30} - 3\bar{\mu}_{12}\right)\left(\bar{\mu}_{21} + \bar{\mu}_{03}\right)\left[3\left(\bar{\mu}_{30} + \bar{\mu}_{12}\right)^2 - \left(\bar{\mu}_{21} + \bar{\mu}_{03}\right)^2\right] \tag{A.7}$$

$$\varphi_8 = \bar{\mu}_{11}\left[\left(\bar{\mu}_{30} + \bar{\mu}_{12}\right)^2 - \left(\bar{\mu}_{03} + \bar{\mu}_{21}\right)^2\right] \\ - \left(\bar{\mu}_{20} - \bar{\mu}_{02}\right)\left(\bar{\mu}_{30} + \bar{\mu}_{12}\right)\left(\bar{\mu}_{03} + \bar{\mu}_{21}\right) \tag{A.8}$$

AN EFFICIENT NEURAL NETWORK BASED BACKGROUND SUBTRACTION METHOD

Naveen Kumar Rai[1], Shikha Chourasia[2], Amit Sethi[3]

[1] IIT Guwahati

[2] VIT Tamilnadu

[2] IIT Guwahati

{i.naveen@iitg.ernet.in, shikha.chourasia11@gmail.com, aamitsethi@iitg.ernet.in }

Abstract. The paper presents a neural network based segmentation method which can extract moving objects in video. This proposed neural network architecture is multilayer so as to match the complexity of the frames in a video stream and deal with the problems of segmentation. The neural network combines inputs that exploit spatio-temporal correlation among pixels. Each of these unit themselves produce imperfect results, but the neural network learns to combine their results for better overall segmentation, even though it is trained with noisy results from a simpler method. The proposed algorithm converges from an initial stage where all the pixels are considered to be part of the background to a stage where only the appropriate pixels are classified as background. Results are shown to demonstrate the efficacy of the method compared to a more memory intensive MoG method.

Keywords: Background subtraction, Neural networks, Spatial-temporal correlation, Backpropagation.

1 Introduction

Video Segmentation is a process that involves extraction of moving objects from a video frame. In applications like traffic surveillance, home surveillance background subtraction has been extensively used as an input for object tracking and recognition. And, to perform in such conditions, algorithms need to be real time and robust. Although lots of segmentation techniques have been proposed[1]-[3], still there are some issues that need to be addressed e.g. cluttered backgrounds, aperture problem, noisy video, low resolution video, presence of shadows etc. In

J. C. Bansal et al. (eds.), *Proceedings of Seventh International Conference on Bio-Inspired Computing: Theories and Applications (BIC-TA 2012)*, Advances in Intelligent Systems and Computing 201, DOI: 10.1007/978-81-322-1038-2_38, © Springer India 2013

addition, variation in illumination, non-uniform size variation(which includes the deformability of objects like in the case of human body), slow velocity, foreshortening, closeness of moving objects and the background and objects merge into each other as they recede in the scene also influence the quality of results of various segmentation results. Computational intelligence methodologies such as Artificial Neural Networks (ANNs) have been employed.

In this work, a new background subtraction method is proposed which combines spatio-temporal information given as input to the neural network to extracts the foreground. The aim is to use "right type" of features (information) as inputs to achieve background subtraction in presence of the above mentioned problems with real world videos. This proposed method is tested on traffic surveillance videos, indoor and outdoor people walking videos with naturally cluttered backgrounds. The proposed method makes the following contributions:

1. Reduction of the computation and memory required for storing parameters by using a single NN model that is invariant to pixel location.
2. Utilization of temporal information and spatial consistency.
3. Better results than competing techniques with higher memory requirement.

The paper begins with the description of spatio-temporal features in section two. Neural network architecture is introduced in section three. Finally, in section four and section five, experimental results and a conclusion are represented.

2 Multiple Features

Video is a set of images or frames, wherein each image in the set is correlated through both spatial and temporal domain. The spatial-temporal features or inputs are as follows:

1. Temporal information of each pixel which is defined as the past state of the pixel's neighborhood whether in foreground or background.

2. Color dissimilarity in the neighborhood of the pixel which provides the Edge information.

3. Closeness of a pixel from the corresponding pixel in the reference frame.

We denote pixels by $I(X,t)$ which are RGB vectors. Here, X denotes spatial coordinate and t denotes time. Our aim is to obtain the feature vectors which will acts as input to the neural network. Let us consider the temporal information, to acquire this information, let us define a neighborhood N for a pixel such that N_x is the set containing pixels neighboring pixel at location X within a window of size $w \times w$ for which $x \notin N_x$. Also, $x \in N_y$ commutatively $y \in N_x$. Let $P(X,t) \in \{0,1\}$ represent the background vs. foreground classification for pixel in location X at time t where '0'represents that the pixel is classified as background pixel and '1' represents the pixel is classified as foreground pixel. We define $V_1(X,t)$, $V_2(X,t)$ and $V_3(X,t)$ as three inputs to capture the local spatio-temporal information or features and used as inputs to the neural network. To include local temporal information, we devise input V_1, such that it increases the likelihood of a pixel to belong to the same group as the majority of the pixels in its neighborhood in the previous frame. The potential $V_1(X,t)$ can be written as,

$$V_1(X,t) = \sum_{y \in N_x} P(y, t-1)$$

The input $V_1(X,t)$ ensures the classification of pixels based on the temporal consistency but it will also provide faulty classification result for noisy pixels like thresholding the possibility of the pixel to be foreground or background(neural network only works on equal or more than two inputs). To overcome this, we used spatial information which improves the results. So, let us define $V_2(X,t)$ input which is the closeness of a pixel from the corresponding pixel in the reference frame. Let $G(I(X,t))$ be the median of the pixels calculated which is a function of $I(X,t)$ across the temporal domain. Now, reference frame R is such that $R = \{R \mid \forall G(I(X,t))\}$ contains all the medians and stored as a matrix of size $m \times n$. Now, following reference frames of test videos are shown in Fig.1. Let g_x be the vector such that $g_x = G(I(X,t))$. To acquire this knowledge, let $V_2(X,t)$ be the corresponding potential such that.

$$V_2(X,t) = \| g_x - I(X,t) \|$$

Where, $\| g_x - I(X,t) \|$ is the measure of closeness of the pixel. We consider that static pixels have high degree of closeness than the dynamic pixels. The pixel which is a background pixel will have very less value compared to a pixel which has a changing state from background to foreground and vice-versa (across the temporal domain). Thus this input plays a very important role to support the results of proposed algorithm.

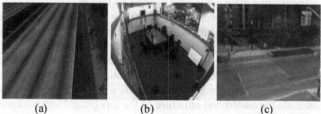

(a)   (b)   (c)

Fig.1. Reference frames of videos of (a) traffic sequence, (b) indoor sequence, (c) outdoor sequence.

With spatial correlation and temporal consistency, the following inputs are sufficient to classify each pixel into foreground or background and it may provide good results but if the video input have similar color pattern of background and moving objects then it will lead to faulty results and oversegment the frame. To tackle this problem, we considered the edge information to be the viable solution as it will extract the boundaries of the objects and improve the classification results greatly. To acquire the edge information, let us define $V_3(X,t)$ potential function such that

$$V_3 = \sum_{y \in N_x} \| I(X,t) - I(Y,t) \|$$

Where, $\| I(X,t) - I(Y,t) \|$ is the Euclidean distance between X and Y pixel. It corresponds to the color dissimilarity in the neighborhood of the pixel which provides the edge information. Finally, it will result into the extraction of boundary information of different meaningful objects in the frame. The spatiotemporal features or information or functions are fed as the inputs into the proposed neural network architecture which will be discussed in the next section.

3 Proposed Architecture

In this work a feed forward single hidden layer neural network architecture is proposed with one hidden layer, one output node and inputs $V_1(X,t)$, $V_2(X,t)$, $V_3(X,t)$ are fed as the three inputs to the neural network. The output node provide Y where $Y \in (0,1)$ and after thresholding new output is label for each pixel describing whether it is foreground or background. An imperfect segmentation using a simpler technique is used as target data for using back propagation algorithm. The neural network architecture is shown in Fig.2. Weights of neural network are randomly initialized and trained using back propagation algorithm. Our NN model reduces the computation and memory required for storing parameters by using a single NN model that is invariant to pixel location. Compared to other models in the literature, we do not form a per-pixel model. We model the background using median pixel color over time of 50 frames.

The first novelty of this proposed algorithm is that it combines spatio-temporal relations as neural network inputs. The algorithm also uses noisy and imperfect segmentation results from a simpler background subtraction scheme also based on NN as training data by R.M.Luque[1] method. The difference between R.M.Luque[1] method and our proposed technique is that we are using spatio-temporal information and we get better results using that training data with these inputs that code the underlying structure of the foreground and background.

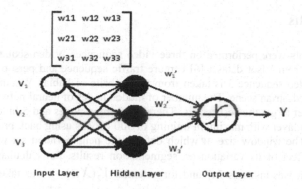

Fig. 2. Neural Network architecture with one hidden layer.

   This requires that the model must be complex enough to capture the details of the desired output. However, it should not be so complex that it overfits that data, resulting in the modeling of the faulty training data. Overfitting is a well-known phenomenon that we have observed for the case of background subtraction. The noisy training data is shown in Fig 3. Another novelty of this algorithm is that the first frame is taken as complete background and as expected it adapts, converges and extracts background and foreground. The noisy training data for the test sequences are shown in Fig. 3.

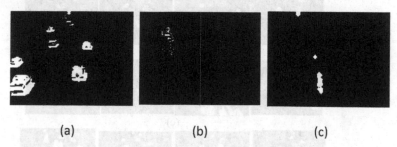

Fig. 3 Noisy training data of (only one frame is shown) (a)Traffic sequence, (b) Indoor sequence, (c) Outdoor sequence.

   To classify pixels either as foreground or background, the NN output is compared to an empirically determined threshold such that thresholding can easily be automated.

4 Results

Experiments were performed on three video sequences. Video sequence 1 and 2 are taken from Visor dataset [6] capture traffic sequence and person walking indoors. Video sequence 3 is taken from the website of Fabian Wauthier [7] is an outdoor pedestrian scene. The number of hidden nodes of neural network was empirically set to get the best results. There are 3 input nodes and 5 hidden nodes in the hidden layer with maximum training of 300 epochs using back propagation algorithm. The window size w which defines the neighborhood set was varied to see the effect of its variation on segmentation results. For calculation of input $V_1(X,t)$ it was taken as 7x7 and for potential $V_2(X,t)$ it was taken as 3x3. In fact, w is very crucial parameter for segmentation. We need large w for $V_1(X,t)$ to capture temporal consistency. And we need a small w for making granular foreground vs. background decision in $V_2(X,t)$. In general, a large w leads to slower execution.

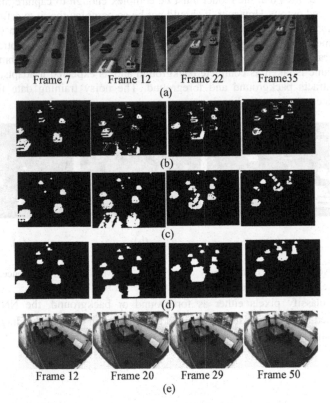

Frame 7   Frame 12   Frame 22   Frame35
(a)

(b)

(c)

(d)

Frame 12   Frame 20   Frame 29   Frame 50
(e)

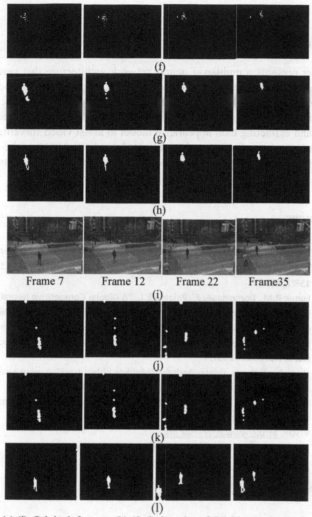

Fig. 4.(a),(e),(i) Original frames, (b),(f),(j) Results of R.M.Luque method[1] used as training data, (c),(g),(k) Results of MoG method, (d),(h),(l) Results of proposed method of traffic, indoor and outdoor sequences respectively (results are downsized for display purpose).

As it can be seen in traffic video that method R.M.Luque[1] fails to segment the foreground objects effectively. Both false positive and false negative pixels can be seen in these results. MoG provides better results than the R.M.Luque method[1] but our method gives the best overall results. This is especially obvious near the horizon in Fig. 4(c). We have applied empirically best manual thresholding and same morphological operations to all three methods.

5 Conclusion

This paper presents a multilayer neural network architecture that performs background subtraction and extracts moving objects. Our proposed method is efficient and gives best overall results compared to widely used and more memory intensive models. In the future, the results of this algorithm can be used for object tracking and extracting their appearance model to assist video surveillance. Tracking and appearance information thus extracted can be fed back to the Neural Network in a feedback mechanism to further improve segmentation results.

References

[1] Luque R.M., Domninguez E., Palomo E.J., Munoz J., "A Neural Network Approach for Video Object Segmentation in Traffic Surveillance", ICIAR 2008, LNCS 5112, 151-158(2008).

[2] Luque R.M., Domninguez E., Palomo E.J., "A Dipolar Competitive Neural Network Video Object Segmentation", IBERAMIA 2008, LNAI 5290, 103-112(2008).

[3] Culbrick Dubravko, Marques Oges,"Neural Network Approach to Background Modeling for Video Object Segmentation", IEEE Transactions on Neural Networks, Vol. 18, No. 6, November 2007.

[4] Humpherys James, Hunter Andrew, "Multiple object tracking using a neural cost function", Image Vision Computing, 417-424, 27(2009).

[5] Owens Jonathan, Hunter Andrew, Fletcher Eric, "A Fast Model Free Morphology Based Object Tracking Algorithm", BMVC 2002.

[6] VISOR, http://www.openvisor.org

[7] Fabian Wauthier, http://www.cs.berkeley.edu/~flw/tracker

Just Think: Smart BCI Applications

Rabie A. Ramadan[1], Ahmed Ezzat AbdElGawad[2], Mohammed Alaa[3]

[1] Assistant Professor at Cairo University

[2,3] Computer Engineering Departement , Masters' Student, Cairo University

{rabie@rabieramadan.org ; ahmed.ezzat.gawad@gmail.com ;
mohamed.eng.alaa@gmail.com }

Abstract. The advances of computer hardware and signal processing made it possible to the usage of the brain signals for communication between human and computers. Extracting electroencephalographic (EEG) signals may help severely disabled individuals with an alternative means of communication and control. The degree of freedom control depends on the quality of the extracted signals. In this paper, we introduce many techniques and algorithms used to extract different EEG signals. To test our developed algorithms, we developed a wheelchair movement simulator as well as virtual keyboard applications. Other contributions in this paper include utilizing of Trie data structure along with T9 and binary search algorithm for efficient virtual keyboard application..

Keywords: Adaptive Filter, Brain Computer Interface (BCI), Machine Learning.

1 Introduction

It is well known that the Brain Computer Interface (BCI) is the interconnection between computer(s) and human brains. It is the most recent development of Human Computer Interface or HCI. Unlike the traditional input devices (keyboard, mouse, pen... etc.), the BCI reads the waves produced from the brain at different locations in the human head, translates these signals into actions, and commands that can control the computer(s). The BCI can have many applications such as: 1) new ways for gamers to play games using their heads, 2) social interactions; enabling social applications to capture feelings and emotions, and 3) helping – partially or fully- disabled people to interact with different computational devices. These applications depend on understanding how the brain works.

Anatomically five basic parts of the brain can be distinguished including Cerebrum, Diencephalon, Cerebellum, Mesencephalon, and Medulla oblongata as shown in Figure 1. The cerebrum which is located directly under the skull surface is the largest part of the brain. Its main functions are the initiation of complex

J. C. Bansal et al. (eds.), *Proceedings of Seventh International Conference on Bio-Inspired Computing: Theories and Applications (BIC-TA 2012)*, Advances in Intelligent Systems and Computing 201, DOI: 10.1007/978-81-322-1038-2_39, © Springer India 2013

movement, speech and language understanding and production, memory, and reasoning. Brain monitoring techniques which make use of sensors placed on the scalp mainly record activities from the outermost part of the cerebrum; the cortex. More inside the cerebrum the basal ganglions can be found which consists of a number of nuclei controlling the direction of slow movements. Also the thalamus is located here which directs sensory information to appropriate parts of the cortex.

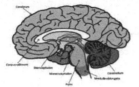   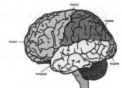

Fig. 1. Brain Anatomy [12]   **Fig. 2.** Hemisphere partitions [12]

The second part of the brain is the Diencephalon. One important function of the diencephalon is the forwarding of sensory information to other brain areas. Besides that, it contains the hypothalamus which controls the body temperature, the water balance and the ingestion to assure the state of homeostasis for the body, i.e. "good working conditions" for all body cells. The coordination among all kinds of movements is , in fact, done in the cerebellum which is the third part. Therefore, it cooperates closely with structures from the cerebrum (e.g. the basal ganglions). Cerebellum and Cerebrum are connected via the Pons. However, the largest part of the reticular system is located in the Mesencephalon where it controls vigilance and the sleep-wake rhythm.

Moreover, the cortex consists of two hemispheres which are connected via a beam called corpus callosum. Each hemisphere is dominant for specific abilities. For right handed persons, the right hemisphere is activated more during the recognition of geometric patterns, spatial orientation, the use nonverbal memory and the recognition of non-verbal noises. More activity in the left hemisphere can be observed during the recognition of letters and words, the use verbal memory and auditory perception of words and language. Each hemisphere is partitioned into five anatomically well-defined regions, the so called lobes as given in Figure 2.

The paper is organized as follows: Different signals that are extracted from the human brains are presented in section 2; our proposed applications and used techniques are elaborated in section 3; finally, the paper concludes in section 4.

2 Extracted Signals

In this section, we introduce the different techniques used to extract different signals from the brain. In order to capture the EEG signals, the Emotiv EPOC head set [9] is used. The set is considered as one of the high resolution, neuro-signal acquisition, and processing wireless neuroheadset. Such signals will be utilized later in the wheelchair control and virtual keyboard applications.

2.1 Gezz/Teeth Pressing Signal

When analyzing the brain wave signals, it is noticed that some signals are produced on nodes FC5 and FC6. These signals are generated when the human with the headset presses on her/his left/right teeth. BCI researchers, in most of the cases, consider such signals as artifacts. Not only that, they develop methods to reduce their effect. Figures 3 (a) and (b) shows the difference between the regular signals and the generated signals on   FC5. We call such signal as "Gezz" signals including the left "Gezz" and right "Gezz". In this paper, we prove that such signals can be beneficial in many applications with careful handling. These signals will be utilized in our applications in the next sections.

One of the challenges faced in extracting the Gezz signals is that they differ from one user to another as given in Figure 4. To make the Gezz signal user independent, we used O1 and O2 stationary nodes as references to FC5 and FC6, respectively.

To classify the signal produced from Gezz actions, first signals coming from the headset are filtered using logical high pass filter with cut off frequency =0.16 Hz in order to remove the DC component (which is highly dependent on the subject) from the signals. Then, the signals coming from FC5, FC6, O1, O2 locations are buffered with a sliding window, to keep the last 20 samples from each of them. The standard deviation of each signal samples is computed, (SD1, SD2, SD3, and SD4) for (FC5, FC6, O1, O2) respectively. Therefore, the final left and right Gezz signals could be extracted using equations 1 and 2.

$$LGezz= SD1/SD3 \qquad\qquad (1)$$
$$RGezz= SD2/SD4 \qquad\qquad (2)$$

Both signals are recognized in 0.2ms.

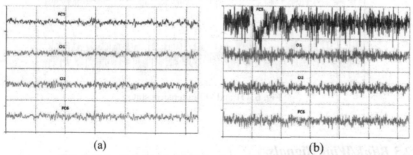

(a)   (b)

Fig. 3. Extracted signals (a) Normal signals, (b) Left Gezz signal

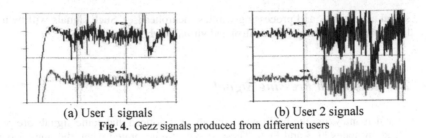

(a) User 1 signals   (b) User 2 signals

Fig. 4. Gezz signals produced from different users

2.2 Tension Signals

Tension signals are other important signals that could be beneficial in many applications. Such signals appear on O1 and O2 nodes; however, they are highly dependent. Therefore, they cannot be used as two separate classes. Figures 5 and 8 show the O1 and O2 signals in normal and tension states. Again high pass filters are used to remove the DC components. The produced signals are then buffered through a sliding window of length 20. Then, the standard deviation is computed for both signals. The signals take 0.2 ms to be recognized; however, it requires some user training.

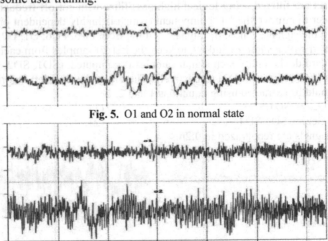

Fig. 5. O1 and O2 in normal state

Fig. 6. O1 and O2 in tension state

2.3 Blink/Wink Signals

When the user closes his/her eyelashes, an instantaneous notch (a peek followed by a drop) appears on the signals produced from both the nodes AF3 and AF4 as shown in Figure 7. Therefore, to extract the blinking signals, the instant increase in the signals is captured. Once more, the logical high pass filter is used to remove the DC components. The produced filtered signals are then buffered into a sliding window, keeping the last 120 samples. The Maximum sliding Difference (MD) is computed twice for each signal based on the following equation:

$$Maximum\ Sliding\ Difference(A[0..n]) = \max_{i,j}(A[i] - A[j])\ where\ 0 \leq i < j \leq n \qquad (3)$$

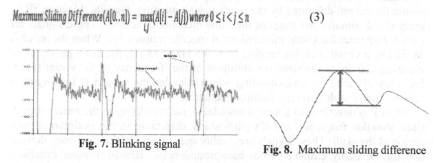

Fig. 7. Blinking signal   **Fig. 8.** Maximum sliding difference

The main idea is to get the two locations, where the difference between the first value minus the second value is a maximum. The condition here is that the smaller value must follow the larger one as shown in Figure 8. That operator helps in detecting a peek followed by a drop. Therefore, computing the following values will extract the blinking signals (D1 and D2):

MD1 = Maximum sliding difference of AF3 [80…120]
MD2 = Maximum sliding difference of AF3 [0…80]
MD3 = Maximum sliding difference of AF4 [80…120]
MD4 = Maximum sliding difference of AF4 [0…80]
D1 = MD1/MD2
D2 = MD3/MD4

When the user closes his/her eyelashes intentionally, an instantaneous notch (a peek followed by a drop) appears on the signals produced from both the nodes AF3 and AF4. This beak is significantly larger than the peak of the eye blinks. In the Wink capturing, we try to capture this instant increase in the signals and try to make this measure a user independent and able to differentiate between it and eye blink. It has the same blink design (reference to blink), but we experimentally modify the blink threshold to able to differentiate between them.

The strong advantage of the Blink signal is it doesn't require from the user any special skills or training. Also this signal is user-independent. The notch can be detected quickly (0.4 ms) but its effect last for about (0.8 ms) (relatively high lag). The Blink signal is highly accurate in detecting true positives, but it can be affected with noise and other actions from the user. So it is recommended to use it in non-critical events.

2.4 SSVEP Signals

The SSVEP (Steady State Visual Evoked Potential) activity is one of the suc-cessfully investigated brain signals. SSVEP [1] is the natural brain response when the retina is excited by flickering visual stimuli. The SSVEP signals are strongly modulated by a selective spatial attention process: these signals are well defined within the extent, delimited by the user's visual attention. Outside this area, flick-ering visual stimuli don't generate the same meaningful activity. These signals are natural responses for visual stimulations at specific frequencies. When the retina is excited by a visual stimulus ranging from 3.5 Hz to 75 Hz, the brain generates an electrical activity at the same (or multiples of the) frequency of the visual stimu-lus. They are used for understanding which stimulus the subject is looking at in case of stimuli with different flashing frequency.

SSVEP is described as a near sinusoidal signal oscillating at the same, or mul-tiple, stimulus frequency and it's particularly detectable in the occipital-parietal region of the skull. These signals are readily quantifiable in the frequency domain and can be easily extracted from background noise. Brain-computer interfaces (BCIs) based on the steady state visual evoked potential (SSVEP) can provide higher bit rates and require shorter training than other BCI [1]. Using two chess-boards: one flickering with 6-Hz and the other with 10-Hz. It then records the data from our head set. The major problems we had were adjusting the exact positions of O1 and O2 electrodes. With our headset, SSVEP was highly dependent on the cap positioning. That represented a major problem in the stability of our results, and so we didn't include an application for this technique in the final delivery. We, however, were able to classify correctly intervals of one second. Figure 9 shows the results of our offline experiments where we can show that 6-Hz spec-trum component magnitude was larger than the magnitude of the 10-Hz fig (1, 2) component when we used data extracted from the standard dataset.

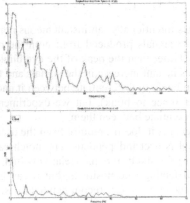

Fig. 9. Results of offline experiments

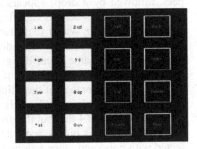

Fig. 10. Binary search approach

2.5 P300 Signal

This signal depends on the appearance of an Event Related Potential (ERP) when a sudden visual or auditory action happens to the subject. The event (a notch in the signal) appears after about 300 ms, and is elicited by rare or significant stimuli [1] when these are interspersed with frequent or routine stimuli. Its amplitude is relatively strongly related to the unpredictability of the stimulus, the more unforeseeable the stimulus is, the higher in the amplitude. Traditionally EEG Signal is averaged to enhance the evoked signal and suppress the background brain activity [7]. Step-wise Discriminant Analysis (SWDA) followed by peak picking and evaluation of the covariance [9]. The discrete wavelet transform (DWT) was also added to the SWDA to localize efficiently the ERP components in both time and frequency [5]. Principal component analysis (PCA) has been employed to assess temporally overlapping EP components [3]. The source signals may be highly correlated, so ICA was first applied.

3 BCI Proposed Applications and Used Techniques

In this section, three applications are presented based on the previous captured signals. In the first application, a virtual keyboard interface is introduced while in the second application, a wheelchair control is described. Finally, a speller application is explained.

3.1 Virtual Keyboard Interface

One of the problems faces disabled persons, is writing documents- especially long ones. Our design to the application doesn't require extra tools other than a computer screen. In the following, we describe different approaches for such application including the naïve approach.

3.1.1 Naïve Approach

The main idea behind this approach is to capture the user Wink signal when letters flash sequentially on a virtual keyboard appeared on the user's monitor. The problem with this approach is the long time that user takes to write a word. Therefore, if there is n letters in the keyboard, a user takes O(n) time to write a letter. Then, for one word with m letters, the required time is O(m*n) times. One might think in ordering the letters according to their frequencies in normal language to reduce the average waiting, but that will make it harder for the user to find the indented letter inside an unordered list of letters.

3.1.2 Binary Search Approach

In this approach, we consider the sequential flashing of letters as a search problem, where the computer is searching for the letter that the user has chosen. The idea behind it is: instead of flashing the letters in sequential fashion (resembles a sequential search that takes $O(n)$), flash the letters in groups in a binary search fashion that takes $O(\log_2(n))$ for one letter as shown in Figure 10. Therefore, for m letters word, it will take $O(m \times \log_2(n))$ time steps instead of $O(m*n)$.

Consider for example that each letter requires 1 second to flash and there are 32 keys (letters + some special functions), then it will take in the worst case only 5 flashes (5 seconds) to write a letter -compared to 32 seconds in the previous approach. Also, for a word length of 6 letters, it will require the user to wait only 30 seconds to write it –compared to 192 second in the previous phase. This approach reduces the waiting time -on average; however, it increases the number of winks to be made by the user to choose a letter. Formally speaking, it will require $O(\log_2(n))$ winks to be made by the user for each letter, (instead of only 1 wink in the previous approach).

3.1.3 Predictive T9 Adaption Approach

In this approach, the predictive T9 algorithm [12][10] is adapted to reduce the number of winks made to write a certain word. The main idea of predictive T9 is to use a smaller set of symbols to do the writing. For example, instead of using a keyboard of 26 letters, one may use a keyboard of only 9 letters as shown in Figure 11. When predictive T9 gets a word with the smaller symbol set, it should return the correct word intended by the user. However, many different words may correspond to the same entry, so predictive T9 is responsible for guessing the most probable word (based on previous knowledge) the user may had meant.

If predictive T9 got the wrong guess, the user informs the application (with a special symbol) to present the next most probable guess and so on until the user gets the indented word. Using predictive T9 reduces the time required for writing a word. Thus, instead of $O(m \times \log_2(n))$ in the previous approach for a word of length m and a set of size n letters, the time required is reduced to $O(m \times \log_2(s) + c)$, where s is the size of the smaller symbol set, and c is the number of wrong predictions.

Using T9 approach, there are two problems that need to be handled. The first problem is collecting and extracting information about word frequencies in languages like English. Up to our knowledge, there is no such information available for free. The second problem is how to store the huge collected amount of information in an efficient way and allowing fast retrieval. The first problem is solved by utilizing the English articles collected from large amount of Wikipedia [14] pages which is available for free. The size of the collected data is almost 1.3 Gigabytes in total. Such data is dealt with offline producing the unigrams (frequencies) for each unique word. Predictive T9 uses these frequencies to decide which words to suggest to the user first.

Our proposal for solving the second problem is to utilize Trie data structure. A Trie is a tree where each link represents a character and each node contains a list of words as shown in Figure 12. A Trie allows searching in linear time, $O(w)$, steps, where w the word's length. That is very efficient in our case, because the words in English are around 8~10 letters except in some extreme cases. Each node has a link represents each symbol. Also a node contains a sorted list of words that are possible with the given state. This list is sorted on frequency, such that more probable words come first.

To summarize the T9 using Trie data structure, the following are the steps:

- The user makes winks trying to choose the correct sequence of symbols.
- When the user chooses a special end symbol, T9 collects the entry and sends it to the Trie.
- The Trie then searches in the tree with each link corresponding to a symbol in the user entry.
- When the Trie reaches the end of the entry, the Trie returns the sorted list of possible words stored in that node.
- Then predictive T9 displays the predictions to the user, ordered on their frequencies.

3.1.4 T9 with Binary Search on Letters and Words Approach

In this Approach, we enhanced the T9 further by applying the binary search on T9. We applied the binary search algorithm once more, aiming at reducing the average number of steps needed to get the correct word from the list of predictions made by T9. So, when T9 suggests a list of predictions for an entry, the application flashes them in halves resembling a binary search. This idea is very useful when the number of suggested words is very large, which happens with small English entries. Using this idea reduces the complexity from $O(m \times \log_2(s) + c)$ time steps in the previous phase, to $O(m \times \log_2(s) + \log_2(c))$ time steps. For example, consider a case where the number of 8 suggestions provided with T9; in the worst case, the previous phase will require 8 time steps, where this phase will require only three, which is an obvious gain. With the T9 with binary search on letters and words, the disabled user can write large documents easily and effectively, reducing the dependence on other people.

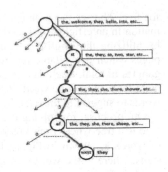

Fig. 11. T9 symbol keyboard   **Fig. 12.** Trie Example in the predictive T9 [12][10]

3.2 Wheel Chair Interface

For fully disabled persons, they usually depend on others to move from one place to another. In this section, we utilized the previously detected signals to help disabled persons to move using a fully controlled wheel chair. However, the application requires the wheel chair to be equipped with two separate motors one for each wheel. This application uses the GEZZ signals (to control the rotation of each wheel of the chair separately, giving rise to four different states as given in Figure 13.

In addition to these four states, the application allows an extra high speed mode using Tension signals as given in Figure 14. In such mode, the two wheels rotates forward with higher speed making the chair move forward in a straight line with higher velocity; that requires each wheel chair motor to be able to operate at two different speeds. This high speed mode can be used when the way is clear to save time reaching the destination. Also the application uses the Blink signals to provide one extra instantaneous state that can be used as a horn or with any other instantaneous alert device.

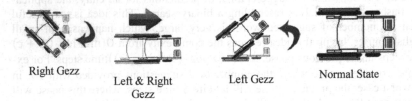

Right Gezz

Left & Right Gezz   Left Gezz   Normal State

Fig. 13. Moving Left - Moving Right – Moving Forward – No Operation

Tension signal

Fig. 14. High speed mode

Fig 15. P300 simple Keyboard

3.3 P300-Based Speller Application

Brain-Computer Interface P300 speller aims at helping patients unable to activate muscles to spell words by means of their brain signal activities. Associated to this BCI paradigm, there is the problem of classifying EEG signals related to responses to some visual stimuli. In the P300 speller, a letter matrix with flashing rows and column is presented on a computer screen as shown in Figure 15. The user focuses on the letter of his/her choosing, and each time it flashes, a P300 response occurs, which is detected by the BCI.

We had trained a LS-SVM on a standard dataset with time domain features in the interval window [d1-k, k+d2] around the Intensification label and the frequency spectrum density feature for this period, and we got high accuracy around 89.9% when the window size [d1+d2] was equal to 600ms. However, the headset in hand doesn't contain the CZ electrodes, so our experiments were based on signals from other nodes collecting data from neighboring electrodes (either using ICA or averaging the values). The application had worked but with undesired accuracy as shown previously. So, we concluded that P300 highly dependable on CZ electrodes.

4. Conclusion

In this paper, we introduced three different applications including virtual keyboard interface, wheel chair Interface, and p300 based speller application. We described and implemented different techniques to efficiently handle such applications. Moreover, we explained how different signals can be extracted.

References

[1] A. Beverina, G. Palmas, S. Silvoni, F. Piccione, S. Giove, "User adaptive BCIs: SSVEP and P300 based interfaces," PsychNology Journal, (2003).

[2] B. Blankertz , G. Dornhege , M. Krauledat , K. M¨uller , V. Kunzmann , F. Losch , and G. Curio , " The Berlin Brain- Computer Interface: EEG-based communication without subject training," IEEE Trans. Neural Systems and Rehabilitation Engineering, (2006).

[3] C. McGillem and J. Aunon, "Measurement of signal components in single visually evoked brain potentials", IEEE Trans. Biomedical. Engineering, (1977).

[4] D. Lange, H. Pratt, and G. Inbar, "Modeling and estimation of single evoked brain potential components, " IEEE Trans. Biomedical Engineering , (1997).

[5] E. Donchin, K. Spencer, and R. Wijesingle, 'The mental prosthesis: assessing the speed of a P300-based brain–computer interface', IEEE Transaction Rehabilitation , (2000).

[6] Emotiv EPOC neuroheadsetuser manual,http://www.emotiv.com/ , accessed in August , 28, 2011.

[7] G. Garcia Molina," Detection of High-Frequency Steady State Visual Evoked Potentials Using Phase Rectified Reconstruction", In 16th European Signal Processing Conference EUSIPCO, 2008.

[8] G. Pfurtscheller, C. Neuper, C. Guger, W. Harkam, H. Ramoser A. Schlögl, B. Obermaier, and M. Pregenzer , "Current trends in graz brain-computer interface (bci) research," IEEE Transactions on Rehabilitation Engineering, (2000).

[9] L. Farwell, and E .Dounchin, "Talking off the top of your heard: toward a mental prosthesis utilizing event-related brain potentials", Electroencephalogr. Clin. Neurophysiol, (1998).

[10] S. Abebe, S. Atnafu, and S. Kassegn; "Ethiopic Keyboard Mapping and Predictive Text Inputting Algorithm in a Wireless Environment."; The International Symposium on ICT Education and Application in Developing Countries, (2004);

[11] S. Roberts and W. Penny, " Real-time brain computer interfacing: a preliminary study using bayesian learning" Medical and Biological Engineering and Computing, (2000).

[12] S.Rajeev, "An Efficient Algorithm for Single Key Stroke Text Entry for Mobile Communication Devices ", International Conference on Wireless Communication Networks, SSN College of Engineering June, (2003).

[13] Scientific Learning, Retrieved from http://www.brainconnection.com/, accessed in August, 28, (2011).

[14] Wikipedia , http://www.wikipedia.org/ (2012)

Interpretability Issues in Evolutionary Multi-Objective Fuzzy Knowledge Base Systems

Praveen Kumar Shukla[1], Surya Prakash Tripathi[2]

[1] Department of Information Technology, Babu Banarasi Das Northern India Institute of Technology, Lucknow, India

[2] Department of Computer Science & Engineering, Institute of Engineering & Technology, Lucknow, India

{praveenshuklaniec@yahoo.co.in, tripathee_sp@yahoo.co.in}

Abstract. Interpretability and accuracy are two conflicting features of any Fuzzy Knowledge Based System during its design and implementation, this conflicting nature leads to Interpretability-Accuracy Trade-Off. Secondly, the assessment of interpretability is another important problem for researchers. Several indexes and methods have been proposed for assessing interpretability but this issue remains an open problem because of its subjective nature. This paper discusses two research issues, interpretability assessment and interpretability-accuracy trade-off in Fuzzy Knowledge Base System design using Evolutionary Multiobjective Optimization by proposing taxonomy for studying and evaluating the interpretability.

Keywords: Interpretability, Accuracy, Interpretability-Accuracy (I-A) Trade-Off, Evolutionary Multiobjective Optimization (EMO), Multi Objective Evolutionary Algorithms (MOEA).

1 Introduction

Interpretability [2, 3, 11, 16, 64] is defined as the capability of the system, showing its working in a comprehensible way. The assessment of interpretability is very complicated issue due to its subjective nature. On the other hand, accuracy is another feature of the system showing quantification of its capability in terms of closeness between real system and modeled system. Alternatively, it is ability of modeled system to faithfully represent the real system.

Interpretability and accuracy [4] are two contradictory features of the system because the enhancement of one can be done at the cost of the other. This contradictory nature leads to Interpretability-Accuracy Trade-Off (I-A Trade-Off) [1, 12, 65, 66] and currently is a burning research issue. For handling this Trade-Off Evolutionary Multiobjective Optimization (EMO) [13, 17-20] approaches are well used. Managing this trade-off enables us to get multiple fuzzy systems with differ-

J. C. Bansal et al. (eds.), *Proceedings of Seventh International Conference on Bio-Inspired Computing: Theories and Applications (BIC-TA 2012)*, Advances in Intelligent Systems and Computing 201, DOI: 10.1007/978-81-322-1038-2_40, © Springer India 2013

ent degrees of accuracy and interpretability and any one of them may be selected depending on user's requirement and needs of application. The integration of EMO in designing and implementing fuzzy systems leads to a new area called, Evolutionary Multi-Objective Fuzzy Systems (EMOFS).

This paper has been divided into 4 sections. Section 2 introduces a new taxonomy of the interpretability. Section 3 introduces evolutionary multi objective optimization. Section 4 introduces interpretability- accuracy trade-off using evolutionary multi objective optimization. Section 5 is the conclusion and provides new research directions.

2   Proposed Taxonomy for Interpretability Assessment of FRBS

Various taxonomies for interpretability assessment are proposed in [5, 6, 7]. In [5], a conceptual framework has been discussed for quantifying the FRBS interpretability, which is based on two issues Description and Explanation. A study of constraints on the design of fuzzy system has been carried out in [6]. Low Level Interpretability and High Level Interpretability based taxonomy has been proposed in [7]. Two axis, "Complexity vs. Semantic Interpretability" and "rule base vs. fuzzy partition" has been given in [8]. For interpretability assessment, two indexes, Average Fired Rules (AFR) and Logical View Index (LVI) has been proposed in [9]. A fuzzy ordering based interpretability index has been proposed in [10].

A taxonomy of interpretability has been proposed (Fig. 1) here under three factors; Knowledge Base Interpretability (KBI), Inference Engine Interpretability (IEI) and User Knowledge Base Interpretability (UKBI). Under KBI, two other subfactors are considered, Data Base Interpretability (DBI) and Rule Base Interpretability (RBI). This taxonomy is fully compatible with the Mamdani Type FRBS model given in [89], (see Fig. 3 in [89]). This taxonomy would be very helpful in studying the interpretability issues of Mamdani Type FRBS. One another feature of this taxonomy is that it includes the importance of user understanding during the quantification of interpretability.

For assessment of interpretability, the parameters from various levels, proposed in the taxonomy are selected for developing the interpretability indexes. To improve the interpretability, various approaches are used, like reduction in number of membership functions by merging, reduction in number of rules by rule selection and by applying the rule learning procedures. According to taxonomy proposed here, interpretability improvement can be done at the level of Data Base, Rule Base, Inference Process and user knowledge base interpretability.

When we try to quantify the interpretability of the system, the level of understanding of the users should be considered because interpretability assessment is directly related to user understanding. For the sake of user interpretability quantification, his/her knowledge should be quantified at the level of RBI, DBI and IEI.

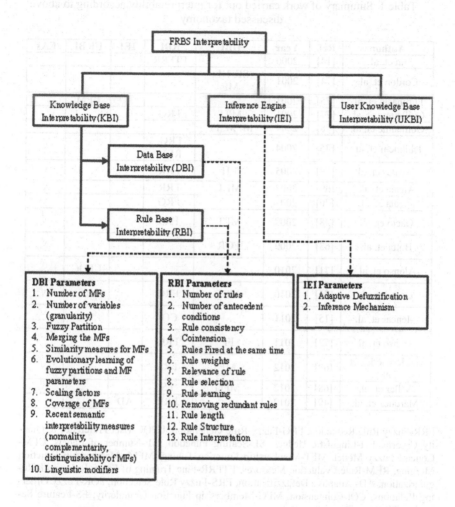

Fig. 1. Taxonomy of Interpretability for FRBS

This user interpretability can be used to estimate the system interpretability with the prospect of user. The interpretability of a system would be different for different users. Also, the system interpretability can be improved by training users for the particular system.

According to above taxonomy, a current state-of-art in the interpretability assessment is discussed in Table 1.

Table 1. Summary of work carried out for interpretability according to above discussed taxonomy

Author	Ref.	Year	DBI	RBI	IEI	UKBI	EAs
Jin et. al.	[14]	2000		FTFRR			
Cordon et. al.	[74]	2001	SF, NL, MF				
Furuhashi et. al.	[75]	2001				CFM	
Guillaume et. al.	[77]	2001	MFT	FRG			
Guillaume et. al.	[78]	2004	MFM, FP				
Ishibuchi et. al.	[45]	2004		FRG, REM			√
Cassilas et. al.	[73]	2005	LH	FRR			√
Alcala et. al.	[69]	2007	MFT	FRR			√
Alcala et. al.	[70]	2007		FRG			√
Gacto et. al.	[38]	2009	MFT	FRS			√
Botta et. al.	[63]	2009	FOR				√
Alonso et. al.	[71]	2010				UPQC	
Gacto et. al.	[76]	2010	MFT	FRG			√
Mencar et. al.	[15]	2011		COI			
Alonso et. al.	[72]	2011	FP	FRG			
G.-Hernandez et. al.	[67]	2012	MF	FRS			√
Villar et. al.	[68]	2012	MFG	FS			√
Marquez et. al.	[49]	2012		FRR	AD		

FRR-Fuzzy Rule Reduction, FRG-Fuzzy Rule Generation, UPQC-User Preference & Quality Criteria, LH-Linguistic Hedges, SF-Scaling Function, NL-Number of Labels, CFM-Concise Fuzzy Model, MFT-Membership Function Tuning, MFM- Membership Function Merging, REM-Rule Evaluation Measures, FTFRR-Fine Training of Fuzzy Rules with Regularization, AD- Adaptive Defuzzification, FRS-Fuzzy Rule Selection, FOR-Fuzzy Ordering Relations, COI-Cointension, MFG-Membership Function Granularity, FS-Feature Selection, FP-Fuzzy Partition.

3 Evolutionary Multiobjective Optimization

Handling I-A Trade-Off during the design of fuzzy systems is identified as a multi-objective optimization problem. Evolutionary Algorithms (EAs) are highly applicable to deal with multiobjective optimization problems because EAs contain the population based procedure to get many solutions in a single run. EAs are also capable to deal with very large uncertain and complex search space. Evolutionary

Multiobjective Optimization includes integration of any of the evolutionary approaches i.e, genetic algorithm [21], evolution strategies [22], genetic programming [23] and evolutionary programming [24] to deal with multi objective problems.

Several EMO algorithms have been developed which are highly applicable in the design of accurate and interpretable fuzzy systems. These algorithms are summarized in Table 2.

Table 2. Overview of Pareto Based Evolutionary Multi-Objective Optimization Algorithms

Generation	MOEA Name	Reference
First Generation MOEA (major features include fitness sharing and niching integrated with Pareto ranking)	NSGA (Non-dominated Sorting Genetic Algorithm)	[82]
	NPGA (Niched Pareto Genetic Algorithm)	[83]
	MOGA (Multi-Objective Genetic Algorithm)	[84]
Second Generation MOEA (integrated with the concept of elitism)	SPEA (Strength Pareto Evolutionary Algorithm)	[85]
	SPEA2 (Strength Pareto Evolutionary Algorithm 2)	[86]
	PAES (Pareto Achieved Evolution Strategies)	[87]
	NSGA-II (Non-dominated Sorting Genetic Algorithm-II)	[88]
	NPGA-II (Niched Pareto Genetic Algorithm II)	[25]
	PESA (Pareto Envelop based Selection Algorithms)	[26]
	Micro Genetic Algorithms	[27, 28]

4. Interpretability-Accuracy Trade-Off

Evolutionary Multi-Objective Optimization Algorithms are highly applicable in the design of fuzzy systems [12, 79, 81] to deal with this situation because they are capable of optimizing multiple conflicting objectives, efficiently. A classification of various issues related to EMO, handling I-A Trade-Off is summarized in Table 3.

Table 3. Handling Interpretability – Accuracy Trade-Off in Fuzzy System Using Evolutionary Multiobjective Optimization

Classification	Major Focus	References
Two objective based approaches for I-A	Feature selection and granularity learning	[29]
	Classification accuracy and number of rules	[30]

Trade-Off	Rule selection and tuning of membership functions	[31-41]
	Root mean squared error and sum of antecedent conditions	[42]
	Fine fuzzy partition, number of antecedent of rules	[43]
Three objective based approaches for I-A Trade-Off	Number of correctly classified training patterns, selected fuzzy rules and antecedents	[44-47]
	Granularity of input and output granularity, membership function parameters and rules	[48-50]
	Number of antecedent of rules, partition integrity and accuracy	[51, 52]
	Semantic quality measures, rule base quality measures and model dimension measure	[53]
	Reduction of input space, granularity of fuzzy system, impact of fuzzification coefficient parameter	[80]
Ensemble classifiers	Three objective based multiobjective formulation of the rule selection	[54]
	Number of classified training patterns, fuzzy rules and antecedent conditions	[55]
User preference	Integration of user preference in EMO, use of satisfaction level functions for user preference	[56,57]
	User preference integrated in rule selection	[58,59]
High dimensional problems	Knowledge extraction from numerical data for high dimensional problem, three objective rule selection	[60]
	Evolutionary training set selection, learning RB and MF parameters for high dimensional and large regression data sets	[61]
Semantic cointension	Interpretable fuzzy rule based classifiers with multiobjective approach	[62]
Context adaptation	Developing context adapted FRBS	[63]

5 Conclusion

In this paper, a new taxonomy of the interpretability has been proposed based on components of Mamdani type Fuzzy Knowledge Base System along with user's role estimation in interpretability assessment. Several major issues about I-A Trade-Off are also discussed.

There are number of open research issues focusing on the above research lines, outlined as;

1. Development of new interpretability indexes using various parameters of DBI, RBI and IEI.

2. Integration of user preferences in the proposed interpretability indexes.
3. The global definition of interpretability is still unclear.
4. Development of new MOEA for handling I-A Trade-Off considering the issues of high dimensional problems, improvement in search ability of MOEA for better finding of Pareto Front and dealing with imbalanced data set.

References

[1] Alcala, R., Alcala-Fdez, J., Cassilas, J., Cordon, O., Herrera, F.: Hybrid learning models to get interpretability-accuracy trade-off in fuzzy modeling. Soft Computing. 10, 717-734 (2006).

[2] Cassilas, J., Cordon, O., Herrera, F., (Eds.): Interpretability Improvements in Linguistic Fuzzy Modeling. Heidelberg, Germany, Springer, (2003)

[3] Alonso, J. M., Magdalena, L.: Special issue on interpretable fuzzy systems. Information Sciences. 181, 4331-4339 (2011)

[4] Cassilas, J., Cordon, O., Herrera, F., Magdalena, L., (Eds.): Accuracy Improvements in Linguistic Fuzzy Modeling. Springer (2003)

[5] Alonso, J. M., Magdalena, L., Rodriguez, G. G.: Looking for a good fuzzy system interpretability index: an experimental approach. International Journal of Approximate Reasoning. 51, 115-134 (2009)

[6] Mencar, C., Fanelli, A. M.: Interpretability constraints for fuzzy information granulation. Information Sciences. 178, 4585-4618 (2008)

[7] Zhou, M., Gan, J. Q., Low level interpretability and high level interpretability: a unified view of data-driven interpretable fuzzy system modeling. Fuzzy Sets and Systems. 159 (23), 3091-3131 (2008)

[8] Gacto, M. J., Alcala, R., Herrera, F.: Interpretability of linguistic fuzzy rule-based systems: an overview of interpretability measures. Information Sciences. 181, 4340-4360 (2011)

[9] Cannone, R., Alonso, J. M., Magdalena, L.: An empirical study on interpretability indexes through Multiobjective evolutionary algorithms. WILF2011, LNAI 6857, 131-138 (2011)

[10] Botta, A., Lazzerini, B., Marcelloni, F., Stefanescu, D. C.: Context adaptation of fuzzy systems through a multi-objective evolutionary approach based on a novel interpretability index. Soft Computing. 13 (5), 437-449 (2008)

[11] Guillaume, S.: Designing fuzzy inference systems from data: an interpretability oriented review. IEEE Transactions of Fuzzy Systems. 9 (3), 426-443 (2001)

[12] Shukla, P.K., Tripathi, S. P., A review on the interpretability-accuracy trade –off in Evolutionary Multi-Objective Fuzzy Systems (EMOFS), Information, 3, 256-277 (2012)

[13] Deb, K., Multi-Objective Optimization Using Evolutionary Algorithms, John Wiley & Sons, Chichester (2001)

[14] Jin, Y.: Fuzzy modeling of high dimensional systems: complexity reduction and interpretability improvement. IEEE Transactions on Fuzzy Systems. 8(2), 212-221 (2000)

[15] Mencar, C., Castiello, C., Cannone, R., Fanelli, A. M.: Interpretability assessment of fuzzy knowledge bases: a cointension based approach. International Journal of Approximate Reasoning. 52, 501-518 (2011).

[16] Mikut, R., Jakel, J., Groll, L.: Interpretability issues in data-based learning of fuzzy systems. Fuzzy Sets and Systems. 150, 179-197 (2005)

[17] Coello, C. A., Van Veldhuizen D. A., Lamont, G. B.: Evolutionary Algorithms for solving multi-objective problems. Kluwer Academic publishers. Newyork (2002)

[18] Coello C.: Recent trends in evolutionary multi-objective optimization, in: A. Abraham, L, Jain, R. Goldberg (Eds.), Evolutionary Multi-objective optimization: Theoretical Advances and Applications (Springer-Verlag, London 2005). 7-32 (2005)

[19] Coello, C., Toscano, G., Mezura, E.: Current and future research trends in evolutionary multi-objective optimization, in: M. Grana, R. Duro, A. d' Anjou, P. P. Wang (Eds.) Information processing and Evolutionary Algorithms: From Industrial Applications to Academic Speculations (Spring-Verlag, London, 2005) 213-231 (2005)

[20] Coello, C.: Evolutionary multiobjective optimization: a historical view of the field. IEEE computational Intelligence Magazine. 1:1, 28-36 (2006)

[21] Goldberg, D.E.: Genetic algorithms in search optimization and machine learning, Addison Wesley Publishing Company Reading Massachusetts (1989)

[22] Schwefel, H.-P.: Evolution and optimization seeking. John Wiley & Sons. Newyork, (1995)

[23] Koza, J. R: Genetic Programming. On the programming of computers by means of natural selection. The MIT Press, Cambridge, Massachusetts (1992)

[24] Fogel, L. J.: Artificial Intelligence through simulated evolution, John Wiley, New York (1966)

[25] Erickson, M., Mayer, A., Horn, J.: The Niched Pareto Genetic Algorithms applied to the design of ground water remediation system. Ist International Conference on Evolutionary Multi Criteria Optimization. 681-695, Springer-Verlag, LNCS No. 1993 (2001)

[26] Corne, D. W., Knowles, J. D., Oates, M. J.: The pareto envelop based selection algorithm for multi-objective optimization. In Proc: VI Conference of Parallel Problem Solving from Nature. Paris, France, Springer LNCS 1917, 839-848 (2000)

[27] Coello, C. A. C., Pulido, G. T.: A microgenetic algorithm for Multiobjective optimization, In: proc. First International Conference on Evolutionary Multi-Criteria Optimization, pp. 126-140, LNCS 1993 (2001)

[28] Coello, C. A. C., Pulido, G. T.: Multi-objective optimization using a micro-genetic algorithm. In Proc.: Genetic and Evolutionary Computation Conference (GECCO' 2001), Morgan Kaufmann Publishers. 274-282 (2001)

[29] Cordon, O., Herrera, F., del Jesus, M. J., Villar, P.: A Multi-objective genetic algorithm for feature selection and granularity learning in fuzzy rule based classification systems. In Proc:. of 9th IFSA World Congress 2001. 1253-1258 (2001)

[30] Ishibuchi, H., Nojima, Y.: Accuracy-complexity trade-off algorithms by Multi-objective rule selection, In Proc:.2005 Workshop on Computational Intelligence in Data Mining, 39-48 (2005)

[31] Alcala. R., Gacto, M. J., Herrera, F.: A Multi-objective genetic algorithm for tuning and Rule selection to obtain accurate and compact linguistic fuzzy rule-based systems.

International Journal of Uncertainty, Fussiness and Knowledge Based Systems. 15(5), 539-557 (2007)

[32] Alcala. R., A-Fdez, J., Gacto, M. J., Herrera, F., A Multi-objective evolutionary algorithm for rule-selection and tuning on Fuzzy rule based systems. In Proc.: of FUZZ-IEEE 2007, 1367-1372 (2007)

[33] Gacto, M. J., Alcala, R., Herrera, F.: An improved multi-objective genetic algorithm for tuning linguistic fuzzy systems. In Proc.: of IPMU'08, 1121-1128 (2008)

[34] Alcala, R., Alcala-Fdez, J., Gacto, M. J,. Herrera, F.: On the usefulness of MOEAs for getting compact FRBSs under parameter tuning and rule selection. Studies in Computational Intelligence (SCI). vol. 98, 91-107 (2008)

[35] Alcala, R., Ducange, P., Herrera, F., Lazzerini, B., Marcelloni, F.: A Multi-objective evolutionary approach to concurrently learn rule and databases of linguistic fuzzy rule based systems. IEEE Transactions on Fuzzy Systems. 17(5), 1106-1121 (2009)

[36] Antonelli, M., Ducange, P., Lazzerini, B., Marcelloni, F.: Learning concurrently partition granularities and rule bases of Mamdani fuzzy systems in a multi-objective evolutionary framework. International Journal of Approximate Reasoning. 50, 1066-1080 (2009)

[37] Gacto, M. J., Alcala, R., Herrera, F.: Multi-objective genetic fuzzy systems: on the necessity of including expert knowledge in the MOEA design process. In Proc.: IPMU 2008, 1446-1453 (2008)

[38] Gacto, M. J., Alcala, R., Herrera, F.: Adaptation and application of multi-objective evolutionary algorithms for rule reduction and parameter tuning of fuzzy rule based systems. Soft Computing. 13, 419-436 (2009)

[39] Di Nuovo, A. G., Catania, V.: Linguistic modifiers to improve the accuracy – interpretability trade-off in multi-objective genetic design of fuzzy rule based classifier systems, 2009 9th International Conference on Intelligent Systems Design and Applications, 128-133 (2009)

[40] Alcala, R., Nojima, Y., Herrera, F., Ishibuchi, H.: Multi-objective genetic fuzzy rule selection of single granularity –based fuzzy classification rules and its interaction with lateral tuning of membership functions. Soft Computing. 15(12), 2303-2318 (2011)

[41] Antonelli, M., Ducange, P., Lazzerini, B., Marcelloni, F.: Multi-objective Evolutionary Generation of Mamdani Fuzzy Rule Based Systems based on rule and condition selection, In Proc:. 5th IEEE International Workshop on Genetic and Evolutionary Fuzzy Systems 2011, 47-53 (2011)

[42] Cococcioni, M., Ducange, P., Lazzerini, B., Marcelloni, F.: A Pareto based multi-objective evolutionary approach to the identification of Mamdani fuzzy systems. Soft Computing. 11, 1013-1031 (2007)

[43] Ishibuchi, H., Nakashima, Y., Nojima, Y.: Effects of fine fuzzy partitions on the generalization ability of evolutionary multi-objective fuzzy rule based classifiers, In Proc:. of FUZZ-IEEE, 1-8 (2010)

[44] Ishibuchi, H., Nakashima, T., Murata, T.: Three-objectives genetics-based machine learning for linguistic rule extraction. Information Sciences. 136, 109-133 (2001)

[45] Ishibuchi, H., Yamamoto, T.: Fuzzy rule selection by multi-objective genetic local search algorithms and rule evaluation measures in data mining. Fuzzy Sets and Systems. 141, 59-88 (2004)

[46] Zhang, Y., Wu, X.-Bei, Xiang, Z.-Yi, Hu, W.-Li: On Generating Interpretable and Precise Fuzzy Systems based on Pareto Multi-objective Cooperating Co-evolutionary Algorithm. Applied Soft Computing. 11, 1289-1294 (2011)

[47] Ishibuchi, H., Nojima, Y.: Analysis of Interpretability- Accuracy Trade –Off of fuzzy systems by Multi-objective fuzzy genetics –based machine learning. International Journal of Approximate Reasoning. 44, 4-31 (2007)

[48] Antonelli, M., Ducange, P., Lazzrini, B., Mareclloni, F.: Multi-objective evolutionary learning of granularity, membership function parameters and rules of Mamdani fuzzy systems. Evolutionary Intelligence. 2, 21-37 (2009)

[49] Marquez, A. A., Marquez, F. A., Peregrin, A.: A mechanism to improve the interpretability of linguistic fuzzy systems with adaptive defuzzification based on the use of a multi-objective evolutionary algorithms. International Journal of Computational Intelligence Systems. 5(2), 297-321 (2012)

[50] Pulkkinen, P., Koivisto, H.: A dynamically constrained multiobjective genetic fuzzy systems for regression problems. IEEE Transactions on Fuzzy Systems, 18 (1), 161-167 (2010)

[51] Antonelli, M., Ducange, P., Lazzerini, B., Marcelloni, F.: Learning knowledge bases of multi-objective evolutionary fuzzy systems by simultaneously optimizing accuracy, complexity and partition integrity. Soft Computing. 15(12), 2335-2354 (2011)

[52] Antonelli, M., Ducange, P., Lazzerini, B.: A three- objective evolutionary approach to gene rate mamdani fuzzy rule based systems, E. Corchado et. al. (eds.): HAIS 2009, LNAI 5572, pp. 613-620 (2009)

[53] Gonzalez, M., Cassilas, J., Morell, C.: Dealing with three uncorrelated criteria by multi-objective genetic fuzzy systems. In Proc.: 5th International Workshop on Genetic and Evolutionary Fuzzy Systems 2011, pp. 39-46 (2011)

[54] Ishibuchi, H., Nojima, Y.: Fuzzy ensemble design thorough multi-objective fuzzy rule selection, Studies in Computational Intelligence (SCI). 16, 507-530 (2006)

[55] Ishibuchi, H., Nojima, Y.: Evolutionary Multi-objective optimization for the design of fuzzy rule based ensemble classifiers, International Journal of Hybrid Intelligent Systems, 3(3), 129-145 (2006)

[56] Branke, J., Deb, K.: Integrating user preferences into Evolutionary Multi-objective Optimization, J. Yaochu (eds.), Knowledge Incorporation in Evolutionary Computation, Springer, vol. 167 (2004)

[57] Nojima, Y., Ishibuchi, H.: Interactive fuzzy modeling by evolutionary Multi-objective optimization with user preferences, In Proc: IFSA-EUSFLAT 2009, pp. 1839-1844 (2009)

[58] Nojima, Y., Ishibuchi, H.: Incorporation of user preference into multi-objective genetic fuzzy rule selection for pattern classification problems. In Proc: 14th International Symposium on Artificial Life and Robotics 2009, pp. 186-189 (2009)

[59] Nojima, Y., Ishibuchi, H.: Interactive genetic fuzzy rule selection through evolutionary multi-objective optimization with user preference. In Proc: 2009 IEEE Symposium on Computational Intelligence in Multicriteria Decision-Making, pp. 141-148 (2009)

[60] Ishibuchi, H., Namba, S.: Evolutionary multi-objective knowledge extraction for high dimensional pattern classification problems, X. Yao et. al. (Eds.), PPSN VIII, LNCS, 3242, pp. 1123-1132 (2004)

[61] Antonelli, M., Ducange, P., Marcelloni, F.: A new approach to handle high dimensional and large data sets in multi-objective evolutionary fuzzy systems. In Proc.: FUZZ- IEEE 2011, 1286-1293 (2011)

[62] Cannone, R., Alonso, J. M., Magdalena, L.: Multi-objective design of highly interpretable fuzzy rule based classifiers with semantic co-intention, In Proc.: 5th International Workshop on Genetic and Evolutionary Fuzzy Systems 2011, 1-8 (2011)

[63] Botta, A., Lazzerini, B., Marcelloni, F., Stefanescu, D.C.: Context adaptation of fuzzy systems through a multi-objective evolutionary approach based on a novel interpretability index. Soft Computing. 13, 437-449 (2009)

[64] Moraga, C.: An essay on interpretability of fuzzy systems. Combining Experimentation and Theory. Studies in Fuzziness and Soft Computing. 271/2012, 61-72 (2012)

[65] Shukla, P. K., Tripathi, S. P.: A Survey on Interpretability-Accuracy (I-A) Trade-Off in Evolutionary Fuzzy Systems. 2011 Fifth International Conference on Genetic and Evolutionary Computation (ICGEC). Xiamen. 97-101 (2011)

[66] Gorzalczany, M. B., Rudziriski, F.: Accuracy vs. interpretability of fuzzy rule based classifiers: an evolutionary approach. Swarm and Evolutionary Computation. LNCS, 7296/2012, 222-230 (2012)

[67] G.-Hernandez, M., S.-Palmero, G. I., F.-Apricio, M. J.: Complexity reduction and interpretability improvement for fuzzy rule systems based on simple interpretability measures and indices by bi-objective evolutionary rule selection. Soft Computing. 16 (3), 451-470 (2012)

[68] Villar, P., Fernandez, A., Carrasco, R. A., Herrera, F., Feature selection and granularity learning in genetic fuzzy rule based classification systems for imbalanced data sets. International Journal of Uncertainty, Fuzziness and Knowledge Based Systems. 20 (3), 369-397 (2012)

[69] Alcalá, R., Alcalá-Fdez, J., Herrera, F.: A proposal for the genetic lateral tuning of linguistic fuzzy systems and its interaction with rule selection. IEEE Transactions on Fuzzy Systems. 15, 616–635 (2007)

[70] Alcalá, R., Alcalá-Fdez, J., Herrera, F., Otero, J., Genetic learning of accurate and compact fuzzy rule based systems based on the 2-tuples linguistic representation. International Journal of Approximate Reasoning. 44, 45–64 (2007)

[71] Alonso, J. M., Magdalena, L., Combining user's preference and quality criteria into a new index for guiding the design of fuzzy systems with a good interpretability-accuracy trade-off, In Proc: IEEE World Congress on Computational Intelligence, 961–968 (2010)

[72] Alonso, J. M., Magdalena, L.: HILK++: an interpretability-guided fuzzy modeling methodology for learning readable and comprehensible fuzzy rule-based classifiers, Soft Computing. 15 (10), 1959-1980 (2011)

[73] Casillas, J., Cordón, O., del Jesus, M. J., Herrera, F.: Genetic tuning of fuzzy rule deep structures preserving interpretability and its interaction with fuzzy rule set reduction. IEEE Transactions on Fuzzy Systems. 13, 13–29 (2005)

[74] Cordón, O., Herrera, F., Magdalena, L., Villar P.: A genetic learning process for the scaling factors granularity and contexts of the fuzzy rule-based system data base. Information Science. 136, 85–107 (2001)

[75] Furuhashi, T., Suzuki, T., On interpretability of fuzzy models based on conciseness measure, In Proc: IEEE International Conference on Fuzzy Systems (FUZZIEEE'01), 284–287 (2001)

[76] Gacto, M. J., Alcalá, R., Herrera, F.: Integration of an index to preserve the semantic interpretability in the multi-objective evolutionary rule selection and tuning of linguistic fuzzy systems. IEEE Transactions on Fuzzy Systems. 18, 515–531 (2010)

[77] Guillaume, S., Designing fuzzy inference systems from data: an interpretability-oriented review. IEEE Transactions on Fuzzy Systems. 9, 426–443 (2001)

[78] Guillaume, S., Charnomordic, B.: Generating an interpretable family of fuzzy partitions from data. IEEE Transactions on Fuzzy Systems. 12, 324–335 (2004)

[79] Fazzolari, M., Alcala, R., Nojima, Y., Ishibuchi, H., Herrera, F.: A review of the application of Multi-Objective Evolutionary Fuzzy systems: current status and further directions. IEEE Transactions on Fuzzy Systems. doi: 10.1109/TFUZZ.2012.2201338

[80] Pedrycz, W., Mingli, S.: A genetic reduction of feature space in the design of fuzzy models. Applied Soft Computing. in press: http://dx.doi.org/10.1016/j.asoc.2012.03.055

[81] Chen, J., Mahfouf, M.: Improving transparency in approximate fuzzy modeling using multi-objective immune inspired optimization. International Journal of Computational Intelligence Systems. 5(2), 322-342 (2012)

[82] Srinivas, N., Deb, K.: Multiobjective optimization using non-dominated sorting in genetic algorithms. Evolutionary Computation. 2(3). 221-248 (1994)

[83] Horn, J., Nafpliotis, N., Goldberg, D.E.: A niched pareto genetic algorithm for multi-objective optimization. In Proc.: Ist IEEE Conference on Evolutionary Computation. IEEE World Congress on Computational Intelligence. 1, 82-87 (1994)

[84] Fonseca, C. M., Fleming, P. J.: Genetic algorithms for multiobjective optimization: formulation, discussion and generalization. In proc.: 5th International Conference on Genetic Algorithms. 416-423 (1993)

[85] Zitzler, E., Thiele, L.: Multiobjective evolutionary algorithms: a comparative case study and the strength pareto approach. IEEE Transactions on Evolutionary Computation. 3 (4), 257-271 (1999)

[86] Zitzler, E., Laumanns, M., Thiele, L.: SPEA2: Improving the strength pareto evolutionary algorithms. Technical Report 103. Computer Engineering & Networks Laboratory (TIK). Swiss Federal Institute of Technology (ETH). Zurich, Switzerland (2001)

[87] Knowles, J. D., Corne, D. W.: Approximating the non-dominated front using the pareto achieved evolution strategy. Evolutionary Computation. 8(2), 149-172 (2000)

[88] Deb, K., Pratap, A., Agarwal, S., Meyarivan, T.: A fast and elitist multiobjective genetic algorithm: NSGA II. IEEE Transactions on Evolutionary Computation. 6 (2), 182-197 (2002)

[89] Herrera, F.: Genetic fuzzy systems: taxonomy, current research trends and prospects. Evolutionary Intelligence. 1, 27-46 (2008)

Hybrid Firefly based Simultaneous Gene Selection and Cancer Classification using Support Vector Machines and Random Forests

Atulji Srivastava[1], Saurabh Chakrabarti[2], Subrata Das[2],Shameek Ghosh[3], V K Jayaraman[3*]

[1]Dr.D.Y.Patil Biotechnology and Bioinformatics Institute, Padmashree Dr. D.Y. Patil University, Pune, Maharashtra, India, Email: atuljisrivastava@gmail.com
[2]Department of Computer Science, University of Pune, Ganeshkhind, Pune – 411007, Maharashtra, India, Email: { saurabhchakrabarti@gmail.com, subrata.bucsd}@gmail.com
[3]Evolutionary Computing and Image Processing Group, Center for Development of Advanced Computing (CDAC), Pune University Campus, Ganeshkhind , Pune – 411007 , Maharashtra, India, Email: {shameekg,jayaramanv}@cdac.in

Abstract. Microarray cancer gene expression datasets are high dimensional and thus complex for efficient computational analysis. In this study, we address the problem of simultaneous gene selection and robust classification of cancerous samples by presenting two hybrid algorithms, namely Discrete firefly based Support Vector Machines(DFA-SVM) and DFA-Random Forests(DFA-RF) with weighted gene ranking as heuristics. The performances of the algorithms are then tested using two cancer gene expression datasets retrieved from the Kent Ridge Biomedical Dataset Repository. Our results show that both DFA-SVM and DFA-RF can help in extracting more informative genes aiding to building high performance prediction models.

Keywords: Cancer Classification, Weighted Gene Ranking, Firefly Algorithm, Support Vector Machines, Random Forests

1 Introduction

Gene expression profiling of cells and tissues has become a major source of data generation for discovering informative patterns towards high impact insights in medicine. Using these microarray datasets, further computational research is carried out to 1) identify diagnostic or prognostic biomarkers, 2) for disease classification, 3) to observe the response to therapy and 4) understand

* Corresponding Author

J. C. Bansal et al. (eds.), *Proceedings of Seventh International Conference on Bio-Inspired Computing: Theories and Applications (BIC-TA 2012)*, Advances in Intelligent Systems and Computing 201, DOI: 10.1007/978-81-322-1038-2_41, © Springer India 2013

the mechanisms involved in related biological processes. But computational analysis of such datasets may turn to be extremely inefficient due to their high dimensionality. These datasets are extremely noisy in general and have a massive number of features. Thus this vast amount of complex data turns out to be a serious impediment towards constructing effective and efficient computational models.

The research, in this context, aims to construct prediction models that can classify cancerous samples with high predictive accuracy based on their gene expression profiles. But the gene (feature) count is much greater than the number of samples (instances) in these datasets. As a result, such composition poses problems to machine learning tasks and makes the model construction problem difficult to solve. This is primarily because, out of thousands of genes, most of the genes do not contribute to the process of classification.

To overcome this problem, one resorts to selecting an informative subset of genes that may be able to build robust classification models consistently. This technique which is known as Gene Selection or Feature Selection helps in getting rid of noisy genes as well as reduces the computational load and increases the overall classification performance.

Gene selection algorithms may thus be categorized as wrappers and filters. Wrappers [1-4] make use of a learning algorithm to estimate the informative power of genes. Methods like Ant Colony Optimization and Genetic Algorithm in combination with a classifier like Support Vector Machines (SVM) come under this category. Filters [5], on the other hand, evaluate the suitability of genes considering the inherent characteristics of the individual genes without making use of a learning algorithm. Methods based on statistical tests come under this category.

In this study, we explore and present a hybrid filter-wrapper based gene selection algorithm and its application to cancer classification. As a filtering process, genes are ranked using a weighted ranking approach. Later a discrete Firefly based wrapper algorithm utilizes the weighted ranking heuristic to traverse the vast gene search space to iteratively obtain more informative gene subsets. For fitness value calculation, we use Support Vector Machines and Random Forests as part of the Firefly based wrapper algorithm.

2 Methodology

2.1 Firefly Algorithm

Firefly algorithm is swarm intelligence based metaheuristic technique inspired by the flashing behaviour of fireflies, which acts as a form of signaling mechanism to attract other fireflies. In 2008, Xin-She Yang [6-7] first proposed the application of this method for continuous optimization problems. It has been observed that the pattern of social communication among various species of fireflies is governed primarily by their brightness which contributes to the attractiveness of a firefly. Accordingly, this attractiveness is the main reason why another firefly moves towards a brighter firefly. At the same time, brightness is relative to the distance between two fireflies. This means that the relative brightness of a firefly will decrease as another firefly draws nearer.

In the context of discrete combinatorial optimization problems, if a candidate solution is represented by a firefly, then one may postulate that fireflies as a whole are dynamically moving towards a brighter fitness landscape, provided we consider brightness as a function of the candidate solution's fitness value. This then is the basis on which the discrete firefly algorithm rests, while solving various combinatorial optimization problems. Earlier work on the application of firefly algorithm has been carried out for travelling salesman problem, knapsack problem, flow shop scheduling, manufacturing process improvements, scheduling task graphs, clustering, and so on[8-13].

For a TSP problem, we normally consider a complete graph G= (V, E) where V= {v_1, v_2,..v_n} and E= {e_1,e_2...e_k}.The objective is to find a circuit in G that contains each vertex exactly once and whose length is minimal. According to the Firefly algorithm for TSP, we first determine the distance between two fireflies (represented as candidate solutions).Conceptually, the distance function(r) should contribute to a numerical difference (relative to the problem's objective function) between two solutions (fireflies in this case)[6,8,10].In case of TSP, this has been earlier considered as the number of different arcs (or edges) while estimating the similarity of two solutions. Later, the distance is used to compute the relative brightness of a firefly as shown in equation (1).

$$\beta = \beta_0 e^{-\gamma r^2} \quad (1)$$

Here γ is the light absorption coefficient, and β_0 is the brightness when r=0. Once the relative brightness of a firefly$_j$ has been computed, firefly$_i$ needs to move towards firefy$_j$. This movement is determined by the following equation (2).

$$x_i = x_i + \beta_0 e^{-\gamma r_{ij}^2}(x_j - x_i) + \alpha(\text{rnd} - \frac{1}{2}) \quad (2)$$

Here x_i is the position of firefly$_i$ and x_j is for firefly$_j$. The third term contributes to a random variable with α being the randomization parameter.

In case of TSP, this movement has been earlier treated as an inverse mutation operation that was dependent on the value of x_i[8]. Thus, depending on a problem, movement is determined based on xi such that a firefly$_i$ (inferior solution) accordingly starts moving towards a firefly$_j$ (superior solution).

The precise formulations relative to gene selection are discussed in subsequent sections.

2.2 Heuristic Information

In large scale data analysis, a system may use some extra knowledge from other data sources that may be provided as heuristic input to guide algorithms for efficient processing and to avoid local minima problems. Likewise, we consider a weighted ranking of genes as a heuristic which is fed to the firefly algorithm for working around such problems. We thus extract the heuristic information for each individual gene by computing the weighted sum of the Information Gain (IG), Chi-Square (CS) and Correlation-based Feature Selection (CFS) scores [14] which are obtained using the WEKA data mining library[15].

IG is an entropy-based measure which assigns a suitability measure to a gene that may have the best capability to separate the samples into individual classes. Genes with higher IG values are thus considered to be "informative". Thus, genes with the top scores may be selected as more relevant genes.

CS is used to estimate the lack of independence between a gene and a class. We scale CS scores in the range of 0-1.

CFS measures the merit of gene subsets by considering the importance of individual genes for predicting the class label along with the level of inter-correlation among them. For best performances, the genes selected by the CFS algorithm are assigned a value of 0.8 and the rest as 0.2.

A high degree of variable , yet informative heuristic ranking is next constructed by assigning weights to the different information components(IG,CS,CFS) as shown in equation (3).

$$Wt_{gene} = wig * IG + wcs * IG + wcfs * CFS \quad (3)$$

Therefore, the weighted rank of a gene is provided by Wt_{gene} and wig, wcs, wcfs are the weights assigned for it's computation. Section 2.5 describes the integration of the Wt_{gene} values with the firefly algorithm.

2.3 Support Vector Machines

Support Vector Machines (SVM) [16-17] fall under a class of learning algorithms based on statistical learning theory that was developed by Vapnik (1995).SVMs are heavily used for purposes of classification and regression.

But the quality of a model generated by an SVM is very much dependent on the features that are provided and the internal parameters taken by SVM.

SVM employs a hyper plane to divide a set of binary-labeled data, maximizing the margin between the nearest data points of each label, in the process. For linearly non-separable data, SVM scales and transforms the input points into a higher dimensional feature space and then finds a suitable linear hyperplane for separating the data points for classification.

For implementation purposes, we have employed the libSVM software library [18].

2.4 Random Forests

Random Forests [19] constitute an ensemble of randomly constructed decision trees. Due to statistical aggregation of information, it is known to report much better predictions. In the process, it brings about a huge amount of variability in the procedure to avoid overfitting of the model.

Random attributes are thus used for node splitting while growing a decision tree. Normally, for every tree, a bootstrap set (with replacement) is drawn from the original training set. This forms the 'in bag' set for a particular tree. About one – third of the samples, on an average, are unused for making the 'in bag' data and are called the 'out of bag' (OOB) data for that particular tree.

The classification tree is constructed using this 'in bag' data based on the CART algorithm [20]. The overall accuracy of the Random Forest is assessed by the OOB data, which is used for cross validation. Thus the k th tree classifies the samples that are OOB for that tree (left out by the k th tree). A majority vote is then used to decide on the class label for each case. Thus, the percentage of times that the voted class label is not equal to the original class of a sample, averaged over all the cases in the training data, is called as the OOB error rate, which gives an estimate of the RF error rate.

We have used the Weka Software suite for RF specific implementation purposes [15].

2.5 Hybrid Firefly based Feature Selection Algorithm

In our model of firefly based feature selection, each firefly represents a candidate gene subset. Accordingly, an initial population of fireflies is thus generated when the model commences operation. A firefly (or a gene subset) is consequently represented by an ordered list of 0/1, based on whether a particular gene is selected or not. For example, with a dataset comprising of 5 genes, one may write firefly$_i$ as 10101 and firefly$_j$ as 11001 for an initial subset size of

3.Each firefly next generates a reduced dataset from their respective input forms, and obtain their corresponding 10 fold cross-validation classification accuracies from SVM or RF. The 10 fold CVA is assigned as the fitness value for each firefly. Later, when an iteration starts, each firefly-i is compared with firefly-j where 'i' is not equal to 'j' and a relative attractiveness is computed. To do this, we first determine the distance(r) between two fireflies based on equation (4).

$$r = \frac{1}{(CVA_j - CVA_i)} \qquad (4)$$

Thus r is given by the inverse of the difference of CVA between two fireflies. Subsequently, the relative brightness of firefly$_j$ with respect to firefly$_i$ may be computed by equation (1)[6-7].To move firefly$_i$ to firefly$_j$ we now make use of the equation (2) to compute x_i, which is the new position of firefly$_i$. The real-valued x_i is thus used to compute $S(x_{ik})$ which is given by equation (5) and 'k' represents a bit in firefly$_i$'s input pattern[10].

$$S(x_{ik}) = \frac{1}{1 + \exp(-x_{ik})} \qquad (5)$$

The value of $S(x_{ik})$ is thus compared against a threshold parameter to decide if the kth bit needs to be flipped. This is repeatedly done till all the bits of firefly$_i$ are processed. This constitutes a single movement of a firefly$_i$. We may thus perform 'm' movements of a firefly$_i$ and store all the 'm' positions for later use. The above process is thus repeated for all the fireflies. Here we implement elitism such that the best (elite_best) solutions are stored, so that they are not corrupted by mutation described next.

To graze over the vast gene search space, we next introduce a mutation process where we integrate the weighted gene ranking as described in section 2.2. A random decision is thus taken to decide whether to take up mutation or not. To provide equal chances for gene selection, we partition the gene search space as informative and non-informative sets based on weighted gene values. As part of mutation, our algorithm either randomly explores newer genes or exploits by selection from the informative gene set. For this, we set a user de-fined exploitation probability as 'q'. Mutation is repeatedly tried on all fire-flies.

Finally, at the end of an iteration, a set of (m*n) +elite_best+n fireflies are ob-tained. To continue with the next iteration, we sort the population and then continue with the 'n' best fireflies again, across several iterations till a termina-tion criteria is reached.

The corresponding hybrid firefly based feature selection algorithm is illustrated in Figure 1.

Initialization
-*Compute the Information Gain (IG), Chi-Square (CS) and CFS scores for each gene (feature) in the original dataset.*
-*Assign wig, wcs and wcfs as weights for weighted importance to IG, CS and CFS for classification.*
-*For each gene, compute the Weighted Index (WI) heuristic function value for a gene i using the weighted sum as shown in equation (3)*
-*Rank features on the basis of their weighted values.*
-*Initialize the Firefly(DFA) parameters.*

Discrete Firefly Algorithm(DFA)
Initialize a population of fireflies x_i (i = 1, 2, ..., n)
Assign CVA as fitness value to x_i fireflies
For each iteration
   For firefly i=1 to n
     For firefly j=1 to n
      Compute r between firefly $_i$ and firefly $_j$
      Compute brightness β
      If i ≠ j then move firefly 'm' times
      Else move firefly randomly

Sort and Store elite_best number of solutions
If rand > mutation probability
   If rand > q
     Perform exploitation using weighted heuristic(use informative gene set)
   Else Perform exploration

Select n brightest fireflies from (m*n) +elite_best+n
Repeat

Figure 1. Discrete Firefly based Feature Selection

3 Results and Discussion

Microarray datasets specify the expression levels of different genes which are available publicly. Two such datasets were obtained from the Kent Ridge Biomedical datasets repository [21] (made available from various other original sources).

The Colon Cancer dataset was retrieved from the Kent Ridge Biomedical dataset repository and consists of 62 instances representing cell lines taken from colon cancer patients. In these, 40 are tumor samples while 22 otherwise [22]. The Leukemia dataset [23]   contains the expression of 7129 genes taken from 72 samples. Among these, 25 belong to the Acute Myeloid Leukemia (AML) class and 47 belong to the Acute Lymphoblastic Leukemia (ALL) class.

The parameters and their corresponding tuned values used in the DFA-SVM and DFA-RF algorithms have been listed in Table I. Extensive simulations had been carried out to maximize algorithmic performance by tuning the relevant algorithm parameters.

Table 1. Firefly Parameters

Algorithm Parameters	Values
Movements(m)	10
Iterations	50
Initial Population Size(n)	20
Gamma	1
β_0	0.2
α	0.5
wig	0.4
wcs	0.3
wcfs	0.3
Mutation probability	0.3
q	0.5

In Table II we list the sizes of gene subsets selected by DFA-SVM and DFA-RF and the 10 – fold cross validation accuracies obtained for the selected gene subsets.

Table 2. DFA-SVM and DFA-RF results

	Colon(Subset Size)	Leukemia(Subset Size)
DFA-SVM	95.16%(15)	97.73%(29)
DFA-RF	91.94 %(13)	90.91%(14)

The results of the DFA-SVM and DFA-RF are encouraging since they compare well against their evolutionary counterparts namely Genetic algorithms, Ant Colony based SVM (ACO-SVM) and RF (ACO-RF) algorithms and Biogeography-based Optimization(BBO-SVM and BBO-RF) methods[2][4]. With reference to literature, DFA-SVM and DFA-RF have fared comparably to the previously best performing algorithms (for the same colon cancer dataset) namely SVMRFE, RG [20], Fisher-RG-SVMRFE [33], ACO-AM (Ant Colony Optimization–Ant Miner) and ACO-RF [2], BBO-SVM and BBO-RF[4] which had demonstrated accuracies of 93.3, 94.7, 95.47, 96.77, 98.39 and 92.34 respectively. Similarly, the best performing algorithms for leukemia cancer classification have shown accuracies in the range of 91–99%, with the best being 99.6% [2], [20], [33], [36], [37]. The DFA-SVM, in this context has fared well by being on the higher side of the accuracy range for both the datasets i.e. 95.16% and 97.73%. Thus our method of using a hybrid filter based DFA wrapper for gene selection in combination with SVM and RF seem to have performed well in comparison to earlier methods.

4 Conclusion

The hybrid DFA-SVM and DFA-RF techniques have shown consistently good results when compared against the highest accuracies for colon cancer and leukemia cancer datasets. The heuristic filters have also been effective in driving the algorithms towards suitable solutions. Both DFA-SVM and DFA-RF are also simple to implement, flexible and robust since we can adapt the DFA to a given problem and domain constraints.

Acknowledgements: VKJ gratefully acknowledges the Department of Science and Technology (DST), New Delhi, India for financial support.

References

[1] Patil, D. , Raj, R., Shingade, P., Kulkarni, B., Jayaraman, V.K.: Feature selection and classification employing hybrid ant colony optimization-random forest methodology. Comb Chem High Throughput Screen, vol. 12, no. 5 . 507–513(2009)

[2] Sharma, S., ,Ghosh, S., Anantharaman, N., Jayaraman, V.K., 2012.: Simultaneous informative gene extraction and cancer classification using aco-antminer and aco-random forests. Advances in Intelligent and Soft Computing. Springer, vol. 132. 755–761) (2012)

[3] Gupta A., Jayaraman V. K., Kulkarni. B. D.: Feature selection for cancer classification using ant colony optimization and support vector machines. Analysis of Biological Data : A Soft Computing Approach. ser. World Scientific, Singapore. 259 –280(2006)

[4] Nikumbh S., Ghosh S., Jayaraman V. K.: Biogeography-Based Informative Gene Selection and Cancer Classification Using SVM and Random Forests. In IEEE World Congress on Computational Intelligence (IEEE WCCI 2012), Australia, In IEEE Press.(2012)

[5] John G. H., Kohavi R., and Pfleger K.: Irrelevant features and the subset selection problem. In Proceedings of the Eleventh International Conference on Machine Learning. 121–129.(1994)

[6] Yang X-S.: Nature-Inspired Metaheuristic Algorithm. Luniver Press(2008).

[7] Yang X-S.: Firefly algorithms for multimodal optimization, in: Stochastic Algorithms: Foundations and Applications, SAGA, *Lecture Notes in Computer Sciences*, 5792, 169-178.(2009)

[8] Jati G. K. and Suyanto S.: Evolutionary discrete firefly algorithm for travelling salesman problem. In ICAIS2011. Lecture Notes in Artificial Intelligence (LNAI 6943). 393-403 (2011)

[9] Palit S., Sinha S., Molla M., Khanra A., Kule M.: A cryptanalytic attack on the knapsack cryptosystem using binary Firefly algorithm. In 2nd Int. Conference on Computer and Communication Technology (ICCCT), 15-17 Sept 2011,India, pp. 428-432 (2011)

[10] Sayadi M. K., Ramezanian R., Ghaffari-Nasab N.: A discrete firefly meta-heuristic with local search for makespan minimization in permutation flow shop scheduling problems. Int. J. of Industrial Engineering Computations 1: 1–10 (2010)

[11] Aungkulanon P., Chai-ead, N., Luangpaiboon P.: Simulated manufacturing process improvement via particle swarm optimisation and firefly algorithms. In Prof. Int. Multiconference of Engineers and Computer Scientists 2: 1123–1128. (2011).

[12] U. Hönig U.: A firefly algorithm-based approach for scheduling task graphs in homogenous systems. Proceeding Informatics. DOI: 10.2316/P.2010.724-033, 724 (2010).

[13] Senthilnath J., Omkar S.N. and Mani V.: Clustering using firefly algorithm: Performance study, Swarm and Evolutionary Computation, June (2011).

[14] Han J., Kamber M., and Pei J., Data Mining: Concepts and Techniques - Information Gain, ser. The Morgan Kaufmann Series in Data Management Systems. Morgan Kaufmann (2011).

[15] Hall M., Frank E., Holmes G., Pfahringer B., Reutemann P., and Witten I. H.: The weka data mining software: An update. SIGKDD Explor, vol. 11. 130–133(2009)

[16] C. N. Shawe-Taylor J.: Support Vector Machines and Other Kernel-based Methods. Cambridge, UK. Cambridge University Press. (2000).

[17] Boser, Bernhard E., Guyon, Isabelle M., and Vapnik, Vladimir N.: Training algorithm for optimal margin classifiers. In 5th Annual ACM Workshop on COLT, 144–152, Pittsburgh, PA, 1992. ACM Press(1992)

[18] Chang, C.-C and Lin, C.-J.: LIBSVM: A library for support vector machines. ACM Transactions on Intelligent Systems and Technology. vol. 2. 27:1–27:27(2011)

[19] Breiman L.: Random forests. Machine Learning. vol. 45. pp. 5–32. (2001)

[20] Breiman L. and Stone F.O.: Classification and regression trees. Chapman and Hall. (1984).

[21] Kent ridge bio-medical dataset. URL: http://datam.i2r.astar.edu.sg/datasets/krbd/.

[22] Alon U., Barkai N., Notterman D.A., Gish K., Ybarra S., Mack D., and Levine A.J. , , .: Broad patterns of gene expression revealed byclustering analysis of tumor and normal colon tissues probed byoligonucleotide arrays. Proceedings of the National Academy of Sciences. vol. 96. no. 12. pp. 6745–6750(1999)

[23] Golub T.R., Slonim D.K., Tamayo P., Huard C., Gaasenbee M., Mesirov J.P., Coller H., Loh M. L., Downing J.R., Caligiuri M.A., Bloomfield C.D., and Lander E.S. : Molecular classification of cancer: Class discovery and class prediction by gene expression monitoring. Science. vol. 286. no. 5439. 531–537.(1999)

[24] Guyon I., Weston J., Barnhill S., and Vapnik V.: Gene selection for cancer classification using support vector machines. Machine Learning. vol. 46. 389–422. (2002)

[25] Mohammad S., Azadeh M. and Mansoor. S.; Identification of disease-causing genes using microarray data mining and gene ontology. BMC Medical Genomics. vol. 4. 4:12 (2011)

[26] Liu Q., Sung A. H., Chen Z., Liu J., Chen L., Qiao M., Wang Z, Huang X. and Deng Y.: Gene selection and classification for cancer microarray data based on machine learning and similarity measures. BMC Genomics. vol. 12. 130–133(2011)

[27] L. Sun, D. Miao, and H. Zhang.: Efficient gene selection with rough sets from gene expression data. In Rough Sets and Knowledge Technology, ser. Lecture Notes in Computer Science. vol. 5009. 164–171(2008)

Recognition of Online Handwritten Gurmukhi Strokes using Support Vector Machine

Mayank Gupta[1], Nainsi Gupta[2] and Rahul Agrawal[1]

[1]M.Tech Student, School of Mathematics and Computer Applications, Thapar Univesity, Patiala
[2]M.Tech Student, Deptt. of Computer Science and Engg., Sriram College of Engineering and Technology and Management, Gwalior
{mayank9854@yahoo.co.in; naincy_negupta@yahoo.com; rahul.thapar021@gmail.com}

Abstract. This paper presents an implementation to recognize Online Handwritten Gurmukhi strokes using Support Vector Machine. This implementation starts with a phase named Preprocessing, which consists of 5 basic algorithms. Prior to these algorithms, a basic step called Stroke Capturing is done, which samples data points along the trajectory of an input device. After preprocessing, recognition of Gurmukhi stroke is done using Support Vector Machine with the help of two cross validation techniques, namely, holdout and k-fold. The recognition is based on the unique IDs identified as the strokes in order to represent a Punjabi *akshar* (word). These strokes are taken from the one hundred Punjabi words written by 3 different writers.

Keywords: Online Handwriting Recognition, Gurmukhi Strokes, Preprocessing, Feature Computation, Support Vector Machines.

1 Introduction

With the development of technology the lifestyle of people is also changing. Problems that were impossible to solve become feasible now. Today, computers have become so much compact that they can easily fit on human hands. Not only their small size, but also their computing power to implement various applications makes them very attractive. This whole thing was started when a company called Palm produced PDAs (Personal Digital Assistant), whose main function was to replace personal organizers. These devices were called Palmtop since they were one handed and fit in one's palm. Though these devices turn out to be common in an increasing number of countries around the world, they are currently almost ignored in major countries of world like in our country India. Even people with full knowledge of operating personal computers do not even know their existence or are not using them due to various reasons. Localizing the applications of the devices could play an important role in making them useful to the Indian society. As it is known that the use of devices like handheld computers and smart phones that have the wireless technology is increasing day by day. When these devices are used, data and commands are needed to be entered while users are moving. The keyboard as well as the mouse does not provide this function of mobility for these devices. Thus, pen-like devices commonly called stylus are used. This is known as Pen Computing and has been very popular today and an important topic in the field of research [1]. Pen Computing includes computers and applications in which electronic pen is the main input device. Pen based input that is used with online handwriting recognition feature allows people to write in a natural way to input data, and provide a pen-paper like interface. These kinds of systems are very useful in terms of mobility, convenience and capability of accessing information at anytime and anywhere.

Online handwriting recognition system is a device that is connected to the computer and is available to the user. The device converts the user's writing motion into a sequence of signals and sends the information to the computer. The most common form of the device is a tablet digitizer. A tablet consists of a plastic or electronic pen and a pressure or electro static sensitive writing surface on which the user forms one's handwriting with the pen. When moving a pen, the digitizer is able to detect information like x and y coordinates of a point, the state of whether the pen touches the surface or not. The information is sent to the connected computer for recognition process as shown in Figure 1.

J. C. Bansal et al. (eds.), *Proceedings of Seventh International Conference on Bio-Inspired Computing: Theories and Applications (BIC-TA 2012)*, Advances in Intelligent Systems and Computing 201, DOI: 10.1007/978-81-322-1038-2_42, © Springer India 2013

A "stroke" in online data is defined as sequence of sampled points from the pen-down state to the pen-up state of the pen, and the completed writing of a character consists of a sequence of one or more strokes.

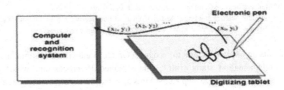

Fig. 1. A tablet digitizer, input sampling and communication to the computer.

The presence of online handwriting recognizer for Devanagiri, Gurmukhi, Bangla and other Asian scripts shall provide a natural way of communication between users and computers and it will increase the usage of personal digital assistant or tablet PCs in Indian languages. The following sections of this paper give a brief overview of Gurmukhi Script, Preprocessing Phase and Feature Computation.

2 Overview of Gurmukhi Script

The word 'Gurmukhi' literally means "from the mouth of guru". Gurmukhi script is used primarily for the Punjabi language, which is the world's 14th most widely spoken language. Gurmukhi characters are shown in Figure 2. Gurmukhi script is written in left to right direction and in top down approach. Most of the Characters have a horizontal line at upper part. The characters of words are connected mostly by this line called head line.

ੳ, ਅ, ੲ, ਸ, ਹ, ਕ, ਖ, ਗ, ਘ, ਙ, ਚ, ਛ, ਜ, ਝ, ਞ, ਟ, ਠ, ਡ, ਢ, ਣ, ਤ, ਥ, ਦ, ਧ, ਨ, ਪ, ਫ, ਬ, ਭ, ਮ, ਯ, ਰ, ਲ, ਵ, ੜ, ਸ਼, ਖ਼, ਗ਼, ਜ਼, ਫ਼, ਲ਼

Fig. 2. Gurmukhi Characters.

A word in Gurmukhi script can be partitioned into three horizontal zones, namely, upper zone, middle zone and lower zone. The upper zone denotes the region above the head line, where some of the vowels and sub-parts of some other vowels reside, while the middle zone represents the area below the head line where the consonants and some subparts of vowels are present. The middle zone is the busiest zone. The lower zone represents the area below middle zone where some vowels and certain half characters lie in the foot of consonants. These zones are shown in Figure 3 [2].

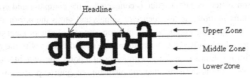

Fig. 3. Three zones and headline in Gurmukhi word.

This paper gives an overview of an online handwriting recognition system for Gurmukhi script using SVM which is rarely used but one of the most important methodologies in data mining [3].

3 Stroke Capturing, Preprocessing and Feature Computation of Gurmukhi strokes

The important phases in an online handwriting recognition system before recognition process are Stroke capturing, preprocessing and feature computation.

3.1 Stroke Capturing

In stroke capturing phase, to collection the data points that represent stroke patterns stylus/pen is used as an input mechanism where as I-BALL digital pen tablet was used to collect data points. These pen traces are sampled at constant rate, therefore these pen traces are evenly distributed in time and not in space. The sequential positions of the stroke are stored dynamically.

The samples of ith stroke $S_i = \{(X_{io}, Y_{io}), (X_{i1}, Y_{i1}), \dots, (X_{io}, Y_{io}), \dots, (X_{in}, Y_{in})\}$ expressed by the time sequence of co-ordinates as shown in Figure 4 represents the generation of the data for a sample of a Gurmukhi characters in which (X_0, Y_0) and (X_n, Y_n) denote the initial and the final point of the generated time sequence of co-ordinates for the particular sample.

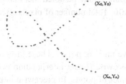

Fig. 4. Stroke of a Gurmukhi character.

3.2 Preprocessing

Preprocessing is the most important phase in an online handwriting recognition system. The main purpose of preprocessing phase in handwriting recognition is to remove noise or distortions present in input text due to hardware and software limitations and convert it into a smooth handwriting. These noise or distortions include different size of text, missing points in the stroke during pen movement, jitter present in stroke, left or right slant in handwriting and uneven distances of points from adjacent positions.

In the preprocessing phase of online handwriting recognition system under consideration, mainly five steps have been given, few of which are sometimes repeated. The steps that are followed in this paper for preprocessing are shown in Figure 5.

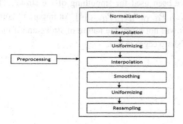

Fig. 5. Steps in preprocessing phase.

In our system the writing area contains a window of size 300×300 pixels. Writer will be able to write inside this window only as shown in Figure. 6.

Fig. 6. Writing Area

3.3 Size Normalization and Centering of Stroke

Size normalization depends on how writer moves the pen on writing pad. Centering is required when the writer moves pen along the boundary of writing pad. Stroke is not generally centered when the pen is moved along the boundary of writing pad. Size normalization and centering of stroke is a necessary process that should be performed in order to recognize a stroke. In this algorithm, it may be noted that number of pixels in the Gurmukhi stroke is n and it is written inside a window of 300×300. Total number of pixels after this algorithm will remain same, *i.e.*, n.

3.4 Interpolation

The stroke written with high speed will have missing points. These missing points can be calculated using various techniques such as Bezier and B-Spline. Piecewise Bezier interpolation is considered in the experiments because it helps to interpolate points among fixed number of points. In piecewise interpolation technique, a set of consecutive four points is considered for obtaining the Bezier curve. The next set of four points gives the next Bezier curve [4].

3.5 Uniformizing of Points

Uniformizing of points in a stroke is required to remove excessive points that are aggregated at a particular point in a stroke while writing. The main reason for this problem is the variation in the speed of writing for different writers. Whenever writer writes slowly, points tend to aggregate in the stroke and when writer writes fast, some points are missed in the stroke. In this uniformizing only those points are taken into consideration that has distance between them greater than a particular threshold value.

3.6 Smoothing

Smoothing of input handwriting is required to remove jitter in handwriting. Flickers exist in handwriting because of individual handwriting style and the hardware used. Smoothing usually averages a point with its neighbors. Figure 7. shows how 2-neighbors from each side can be considered for this purpose. In this figure five points of the list, generated in the previous step, have been used for smoothing of the stroke. The point P_i has been modified or smoothes with the help of points $P_{(i-2)}, P_{(i-1)}, P_{(i+1)}, P_{(i+2)}$ [4]. An important point that should be noted here is that if three points are considered then it will not affect the nature of stroke and if more than five points are considered then the nature of stroke in terms of sharp edges is lost.

Fig. 7. Formation of angle α at point P_i

3.7 Resampling of Stroke

When the writer is writing slowly, more number of points will be present in the stroke than when the writer is writing quickly. To get rid of such variations resampling is applied. Resampling is the process of resampling the data such that the distance between adjacent points is approximately equal. Besides the removal of the variation, this step essentially reduces the number of points in a stroke, to a desired value. After resampling, the data is significantly reduced and the irregularly placed data points that create jitter on the trajectory of the stroke are removed. This makes the resampling step very useful in noise elimination as well as data reduction. In this phase new data points are calculated on the basis of the original points of list. After this phase, only 64 equidistant points will be present in the stroke.

3.8 Feature Computation

In the process of handwriting recognition, it is important to identify correct features [5]. Computational complexity of a classification problem can also be reduced if suitable features are selected. Features vary from one script to another script and the method that gives better result for a particular script cannot be applied for other scripts [6].

In current system, points generated after preprocessing phase are used as a feature for recognition. These points are always fixed for each stroke and are equidistant as far as possible. Number of points in a stroke is fixed to 64. The next section of this paper gives recognition method used for the current system.

4   Gurmukhi Stroke Recognition

Different strokes of Gurmukhi characters from three different writers are collected using the I-Ball electromagnetic digital pen Tablet which is having 1024 levels of pressure sensitivity, resolution 400 lpi, 200 rps report rate, and have ±0.01 inch accuracy.

Each of the three writers has written same 100 words of Gurmukhi script. The words are stored as a collection of strokes in an xml format and the snapshot of an xml file is shown below in Figure 8. There is a single xml file for each word and the strokes of each word are stored as a collection of points. It should be noted that only the strokes of each word are considered for recognition.

```
- <wordSetDef>
  - <word>
    <wordNo>1</wordNo>
    <totalStrokes>12</totalStrokes>
    - <wordDesc>
      ਅੲਿਕਆਸੀ
      - <stroke>
        <strokeNo>1</strokeNo>
        <strokeId>144</strokeId>
        - <point>
          <X>9</X>
          <Y>36</Y>
          </point>
        - <point>
          <X>30</X>
          <Y>48</Y>
          </point>
        - <point>
          <X>45</X>
          <Y>61</Y>
          </point>
        - <point>
          <X>60</X>
          <Y>79</Y>
          </point>
        - <point>
```

Fig. 8. An xml file for a word written in Gurmukhi.

There are total 2428 strokes of the 100 words written by three different writers and each unique stroke out of these, is assigned a unique stroke ID.

4.1 Support Vector Machine

A Support Vector Machine (SVM) [7] is a concept in statistics and computer science for a set of related supervised learning methods that analyze data and recognize patterns, used for classification and regression analysis.

The standard SVM takes a set of input data and predicts, for each given input, which of two possible classes forms the input. SVM in its basic form implement two class classifications. The advantage of SVM, is that it takes into account both experimental data and structural behavior for better generalization capability based on the principle of Structural Risk Minimization (SRM). The basic SVM formulation is for linearly separable datasets. It can be used for non-linear datasets by indirectly mapping the nonlinear inputs into to linear feature space where the maximum margin decision function is approximated. The mapping is done by using a kernel function. A Support Vector Machine includes Support Vector Classifier (SVC) and Support Vector Regressor (SVR). Commonly used kernels in SVM include:

1. **Linear Kernel:** $K(x, y) = x.y$
2. **Radial Basis Function (Gaussian) Kernel:** $K(x, y) = \exp\left(-\|x - y\|^2/2\sigma^2\right)$
3. **Polynomial Kernel:** $K(x, y) = (x.y + 1)^d$

4.2 Recognition using MATLAB

MATLAB is a high performance language for technical computing, created by The MathWorks. It features a family of add-on application-specific solutions called toolboxes *(i.e.,* comprehensive collections of functions) that extend the MATLAB environment to solve particular classes of problems. In this research work MATLAB version 7.10.0.499(R2010a) with bioinformatics toolbox is used for recognition of Gurmukhi stroke. Bioinformatics toolbox offers an integrated software environment for genome and proteome analysis. The key feature of the basic categories in the bioinformatics toolbox is a general-purpose technical computing language and development environment that is widely used in scientific and engineering applications. In this work two methods holdout and k-fold cross validation has been used and are discussed in next sections.

4.2.1 Hold out cross validation for training and testing

The holdout method is the simplest kind of cross validation. The data set is separated into two sets, called the training set and the testing set. The errors it makes are accumulated as before to give the mean absolute test set error, which is used to evaluate the model. The advantage of this method is that it is usually preferable to the residual method and takes no longer to compute. However, its evaluation can have a high variance. The evaluation may depend heavily on which data points end up in the training set and which end up in the test set, and thus the evaluation may be significantly different depending on how the division is made. In MATLAB crossvalind function is used. This function is used to generate cross validation indices. Prototype of this cross validation function is crossvalind ('HoldOut', N, P). It returns logical index vectors for cross-validation of N observations by randomly selecting P*N (approximately) observations to hold out for the evaluation set. P must be a scalar between 0 and 1. P defaults to 0.5 when omitted, corresponding to holding 50% out.

4.2.2 k-Fold cross-validation for Training and Testing

K-fold cross validation is one way to improve over the holdout method. The data set is divided into k subsets, and the holdout method is repeated k times. Each time, one of the k subsets is used as the test set and the other $k - 1$ subsets are put together to form a training set. Then the average error across all k trials is computed as shown in Figure 9. The advantage of this method is that it matters less how the data gets divided. Every data point gets to be in a test set exactly once, and gets to be in a training set $k - 1$ times. The variance of the resulting estimate is reduced as k is increased.

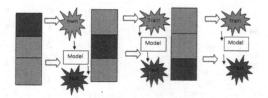

Fig. 9. Functioning of 3-fold cross validation

In the next section the result after applying two partitioning strategies of cross validation in SVM is applied on the data collected from three different writers consisting of 2428 strokes is given.

5 Results and Discussions

This section contains the results of the experiments carried out for recognition of Gurmukhi strokes written by three writers. Each writer has written 100 words in Gurmukhi script. These words are stored in an xml format as described in previous section. Table 1, given below, shows the frequency of occurrences of each stroke in these 100 words written by three writers. Each stroke is identified by a unique ID in this Table.

Table 1. Frequency of strokes for three writers

Stroke ID	Writer 1	Writer 2	Writer 3	Stroke ID	Writer 1	Writer 2	Writer 3
101	40	40	40	165	10	35	1
102	19	20	20	170	2	-	2
104	-	-	1	171	-	2	-
105	2	-	-	172	1	-	1
121	91	26	96	173	1	1	-
122	15	15	15	174	11	12	12
124	2	2	2	179	5	-	6
126	10	10	10	181	1	6	-
127	12	11	12	182	1	9	-
128	12	12	12	183	3	-	3
129	5	6	10	184	-	3	-
141	-	-	23	185	4	4	-
142	20	23	-	186	-	-	4
144	53	55	-	189	11	5	11
145	2	-	56	191	25	25	1
146	32	11	34	192	-	-	28
147	14	-	14	193	14	-	12
148	9	23	9	194	2	16	-
149	10	23	9	197	12	12	12
151	37	3	40	199	45	6	58
152	5	-	-	202	4	4	4
154	17	94	-	204	2	-	2
155	20	-	20	205	2	2	-
156	-	20	-	212	3	1	1
157	23	1	26	213	1	1	1
158	4	26	1	215	27	19	28
159	-	-	2	216	27	27	26
161	18	23	17	226	4	43	6
162	85	48	144	229	-	9	-
163	9	9	9	**Total**	810	743	875
164	26	-	34				

For the recognition process, SVM classifier has been used and this is implemented in MATLAB. Recognition is done for each writer separately as well as for the combined data of all the writers. Writer 1 has used 810 strokes, writer 2 has used 743 strokes and writer 3 has used 875 strokes for writing 100 words in Gurmukhi scripts as shown in Table 1. Recognition has also been done using 1000, 2000 and 3000 strokes of preprocessed data for each writer. Recognition has been done for only those strokes that are written by writer 1 only and are present in the strokes written by writer 2 and writer 3 also.

For preprocessed data, three types of partitions (70% training, 60% training and 50% training) have been used for recognition. These partitions are used for recognition of stroke IDs, for individual writer

Graph 1. Recognition accuracy of three writers for preprocessed strokes

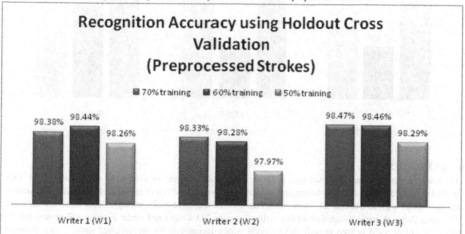

From the above Graph, it has been observed that writer 2 is the most inconsistent writer amongst all writers, writer 1 is slightly better than writer 2, and writer 3 is the best one. So here it can be determined that there is minimum variation in handwriting of writer 3.

The recognition process is then applied to specific number of stroke for each writer. In this study 1000, 2000 and 3000 preprocessed strokes are considered from each writer. These preprocessed strokes were taken randomly from the stroke database. The results are calculated only for those stroke IDs present in 1000 strokes. T_{s1000}, T_{s2000} and T_{s3000} represents 1000, 2000 and 3000 strokes respectively.

For each writer, recognition using 3000 preprocessed strokes gives maximum accuracy as compared to 1000 and 2000 preprocessed strokes. Out of the three writers, writer 3 gives highest accuracy for 1000, 2000 and 3000 preprocessed strokes.

Graph 2. Recognition accuracy of three writers for 1000, 2000 and 3000

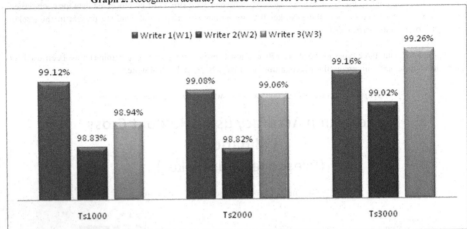

6 CONCLUSION

This research work focuses on developing a stroke recognition system for Gurmukhi script using SVM. One hundred words from three different writers are considered in this study. Each of these words is divided into strokes and each stroke is given a unique ID. These strokes are divided into three zones, namely, Upper Zone, Middle Zone and Lower Zone. Writer 1 writes 810 strokes, writer 2 writes 743 strokes and writer 3 writes 875 strokes for the same 100 words. SVM is used to perform Gurmukhi stroke recognition for preprocessed strokes. Two different partitioning techniques viz. holdout and 10-fold cross validation have been used for generating training and testing sets. Recognition is done for 70%, 60% and 50% training sets in holdout cross validation preprocessed data.

For 70%, 60% and 50% training set of preprocessed stroke in holdout cross validation method, writer 3 has better accuracy 98.47%, 98.46% and 98.29% respectively. For 10-fold cross validation writer 3 attains 98.53% for preprocessed strokes. 1000, 2000 and 3000 preprocessed strokes are considered and recognition accuracy for each writer is examined. For writer 1, writer 2 and writer 3 recognition accuracies 99.16%, 99.02% and 99.26% are achieved with 3000 preprocessed strokes which are higher as compared to 1000 and 2000 preprocessed strokes. Writer 3 attains better recognition accuracy 99.26% amongst all the writers for 1000, 2000 and 3000 preprocessed strokes. From this, it can be concluded that writer 3 has achieved maximum accuracy for Gurmukhi strokes in each situation. So it can be concluded that writer 3's handwriting is better writer as compared to others.

References

1. Shimeles, A.: Online Handwriting Recognition for Ethiopic characters, Master of Science Thesis, Dept. of Computer Science, Ababa University (2005).
2. Sharma, A., Sharma, R. and Sharma, R. K.: Online Handwritten Gurmukhi Character Recognition Using Elastic Matching, Congress on Image and Signal Processing, pp. 391-396 (2008).
3. Ward, J.R. and Kuklinski, T.: A model for variability effects in handwriting with implications for design of handwriting character recognition systems, IEEE Transactions on systems, Man, and Cybernetics, vol. 18, no. 3, pp. 438-451 (1988).
4. Sharma A.: Online handwritten Gurmukhi character recognition, Ph.D. thesis, School of Mathematics and Computer Applications, Thapar University, Patiala (2009).
5. Gader, P. D. and Khabou, A. M.: Automatic feature generation for handwritten digit recognition. *IEEE Transactions on Pattern Analysis and Machine Intelligence*, 18(12):1256– 1261 (1996).
6. Jaeger, S., Manke, S., Reichert, J., and Waibel, A.: Online handwriting recognition: the npen++ recognizer. *IJDAR*, 3:169–180 (2001).
7. Al-Emami, S. and Usher, M.: Online recognition of handwritten Arabic characters. IEEE Transactions on Pattern Analysis and Machine Intelligence, vol. 12, no. 7, pp. 704-710 (1990).
8. Aparna, K.H., Subramanian, V., Kasirajan, M., Prakash, G. V., Chakravarthy, V.S., and Madhvanath, S.: Online handwriting recognition for Tamil. Ninth International Workshop on Frontiers in Handwriting Recognition, pp. 438-443 (2004).
9. Beigi, H. S. M., Nathan K., Clary G. J. and Subrahmo, J.: Size normalization in online unconstrained handwriting, IEEE International Conference on Image Processing, vol. 1, pp. 169-173 (1994).
10. Bharath, A. and Sriganesh M.: On the significance of stroke size and position for online handwritten Devanagari word recognition: An empirical study, 20th International Conference on Pattern Recognition (ICPR), pp. 2033-2036 (2010).
11. Higgins, C. A. and Ford, D. M.,: On-Line Recognition of Connected and writing by Segmentation and Template Matching, IEEE International Conference on Pattern Recognition, vol. 2, pp. 200-203 (1992).
12. Hosny, I., Abdou, S. and Fahmy, A.: Using advanced hidden markov models for online Arabic handwriting recognition, First Asian Conference on Pattern Recognition, pp.565-569 (2011).
13. Huang, B. Q., Zhang, Y. B. and Kechadi, M. T.: Preprocessing techniques for online handwriting recognition, Seventh International Conference on Intelligent Systems Design and Applications, pp. 793-800 (2007).
14. Yura, K., Hayama, T., Hidai, Y., Minamikawa, T., Tanaka, A. and Masuda, S.: Online recognition of freely handwritten Japanese characters using directional feature densities, 11th IEEE International Conference on Pattern Recognition, vol. 2, pp. 183-186 (1992).
15. Khan, K. U. and Haider I.: Online Recognition of Multi-Stroke Handwritten Urdu Characters, IEEE International Conference on Image Analysis and Signal Processing, pp. 284-290 (2010).
16. Kim, M., Park, M. and Kwon, Y. B.: A cursive on-line Hangul recognition system based on the combination of line segments, Second International Conference on Document Analysis and Recognition, pp. 200-203 (1993).
17. Tappert, C.C. and Suen, C.Y., Wakahara, T.: The state of the art in online handwriting recognition, IEEE Transactions on Pattern Analysis and Machine Intelligence, vol. 12, no. 8, pp. 787-808 (1990).
18. Wei, W. and Guanglai, G.: Online handwriting Mongolia words recognition with Recurrent Neural Networks, Fourth International Conference on Computer Sciences and Convergence Information Technology, pp. 165-167 (2009).
19. Artieres, T. and Gallinari, P.: Stroke level HMMs for online handwriting recognition. Proceedings of IWFHR, pp. 227-232 (2002).
20. Basu, M., Bunke, H. and Bimbo, A. D.: Guest editors' introduction to the special section on syntactic and structural pattern recognition, IEEE Transactions on Pattern Analysis and Machine Intelligence, vol. 27, no. 7, pp. 1009-1012 (2005).
21. Bontempi, B. and Marcelli, A.: A genetic learning system for on-line character recognition, Proceedings of 12th IAPR International Conference on Pattern Recognition, vol. 2, pp. 83-87 (1995).
22. Brakensiek, A., Kosmala, A., and Rigoll G.: Evaluation of Confidence Measures for On-Line Handwriting Recognition, Springer-Berlin Heidelberg, pp. 507–514 (2002).
23. Brown M. K. and Ganapathy S.: Preprocessing techniques for cursive script word recognition, Pattern Recognition, vol. 16, pp. 447-458 (1983).
24. Chan, K. F. and Yeung, D. Y.: Recognizing on-line handwritten alphanumeric characters through flexible structural matching, Pattern Recognition, vol. 32, no. 7, pp. 1099-1114 (1999).
25. Cho, S.: Neural-network classifiers for recognizing totally unconstrained handwritten numerals, IEEE Transactions on Neural Networks, vol. 8, no. 1, pp. 43-53 (1997).
26. Connell, S. D. and Jain A. K.: Template-based online character recognition, Pattern Recognition, vol. 34, no. 1, pp. 1-14 (2001).
27. Connell, S. D. and Jain, A. K.: Writer adaptation for online handwriting recognition, IEEE Transactions on Pattern Analysis and Machine Intelligence, vol. 24, no. 3, pp. 329-346 (2002).
28. Duneau, L. and Dorizzi, B.: Online cursive script recognition: a system that adapts to an unknown user, Proceedings of International Conference on Pattern Recognition, vol. 2, pp. 24-28 (1994).

29. Shimodaira, H., Sudo, T., Nakai, M., Sagayama, S.: Online overlaid-handwriting recognition based on substroke HMMs, Proceedings of International Conference on Document Analysis and Recognition, pp. 1043-1047 (2003).
30. Shintani, H., Akutagawa, M., Nagashino, H., Kinouchi, Y.: Recognition mechanism of a neural network for character recognition, Proceedings of Engineering in Medicine and Biology 27th Annual Conference, pp. 6540-6543 (2005).
31. Shu, H.: On-Line Handwriting Recognition Using Hidden Markov Models, M.E. Thesis, Electrical Engineering and computer Science Department, MIT, Massachusetts (1996).
32. Spitz, A.L.: Shape-based word recognition, International Journal of Document Analysis and Recognition, vol. 1, pp. 178-190 (1999).
33. Suen, C. Y., Berthod, M., and Mori, S.: Automatic Recognition of Hand printed Character-the State of the Art, Proceedings of IEEE, vol. 68, no. 4, pp. 469-487 (1980).
34. Veltman, S. R. and Prasad, R.: Hidden Markov Models applied to online handwritten isolated character recognition, IEEE Transactions on Image Processing, vol. 3, no. 3, pp. 314-318 (1994).
35. Wakahara, T. and Odaka, K.: Online cursive kanji character recognition using stroke-based affine transformation. IEEE Transactions on Pattern Analysis and Machine Intelligence, vol. 19, no. 12, pp. 1381-1385 (1997).
36. Ward, J.R. and Kuklinski, T.: A model for variability effects in handwriting with implications for design of handwriting character recognition systems, IEEE Transactions on systems, Man, and Cybernetics, vol. 18, no. 3, pp. 438-451 (1988).
37. Zhou, J., Gan, Q., Krzyzak, K., Suen, C.Y.: Recognition of handwritten numerals by quantum neural network with fuzzy features, International Journal of Document Analysis and Recognition, vol. 2, pp. 30-36 (1999).

An Optimal Fuzzy Logic Controller tuned with Artificial Immune System

S. N.Omkar[1a], Nikhil Ramaswamy[1b],R.Ananda[2],Venkatesh N.G.[3] and J.Senthilnath[1]

[1] Department of Aerospace Engineering, Indian Institute of Sciene, Bangalore-560012

[2] Department of Aerospace Engineering, Indian Institute of Technology, Kharagpur-721302

[3]Department of Electronics and Communication Engineering, National Institute of Technology Karnataka Surathkal, Mangalore -575025

{[a]omkar@aero.iisc.ernet.in , [b]nrghatta@gmail.com}

Abstract. In this paper, a method for the tuning the membership functions of a Mamdani type Fuzzy Logic Controller (FLC) using the Clonal Selection Algorithm(CSA) a model of the Artificial Immune System(AIS) paradigm is examined. FLC's are designed for two problems, firstly the linear cart centering problem and secondly the highly nonlinear inverted pendulum problem. The FLC tuned by AIS is compared with FLC tuned by GA. In order to check the robustness of the designed FLC's white noise was added to the system, further, the masses of the cart and the length and mass of the pendulum are changed. The FLC's were also tested in the presence of faulty rules. Finally, Kruskal Wallis test was performed to compare the performance of the GA and AIS. An insight into the algorithms are also given by studying the effect of the important parameters of GA and AIS.

Keywords: Fuzzy logic Controller,Artificial Immune System, Genetic Algorithms

1 Introduction

Fuzzy Logic Controllers (FLC) have gained prominence in recent years because of its ability to control devices which tend to imitate human decision making. This allows for a control design which is simple and accurate. A FLC is also efficient as it captures the approximate and qualitative boundary conditions of a system's variables by fuzzy sets with a membership function. The FLC controls a system by a set of linguistic IF-THEN rules and has been shown to be robust and straightforward to implement [1,2]. Due to these advantages FLC's have been successfully applied to various industrial applications [3-6].

However, the main challenge of designing the FLC lies in choosing optimal fuzzy parameters for its membership functions (MF). Previously, the task of generation of MF's were done by trial and error techniques, unfortunately this is a te-

J. C. Bansal et al. (eds.), *Proceedings of Seventh International Conference on Bio-Inspired Computing: Theories and Applications (BIC-TA 2012)*, Advances in Intelligent Systems and Computing 201, DOI: 10.1007/978-81-322-1038-2_43, © Springer India 2013

dious approach. Since this task is a natural candidate for stochastic approaches like Genetic Algorithms (GA). Various researchers have extensively used GA's for tuning the MF's of a FLC[7-9]. However , despite their advantages of evolving optimal solutions for wide variety of problems, as reported in [10], GA has the disadvantage of getting stuck in the local minima and may take large time to obtain a near optimal solution. This has motivated researchers to study other evolutionary algorithms to overcome the disadvantages of GA.

In this paper a relatively new evolutionary computation paradigm—Artificial immune system (AIS)[11] is used to tune the Mamdani type fuzzy controller's anecedent and consequent parameters. In this work the fuzzy inference system(i.e Mamdani) tuned by AIS, is applied to a FLC to show the efficacy of the proposed method. In our study we use the Clonal selection algorithm (CSA) which is one of the most extensively used AIS model [12]. This method of CSA has been successfully used for solving various real world and optimization problems [13].To show the efficacy of AIS in tuning of the FLC, two problems are considered. The first is the cart centering problem (CCP), in which the objective of the controller is to bring a cart with initial displacement and velocity to rest [14]. The second problem is the classic benchmark problem in control i.e the inverted pendulum (IP) problem [15] in which the objective is to bring a pendulum on a cart to rest from an initial angular displacement and angular velocity. On designing the FLC's using GA and AIS we compare the performance of the FLC tuned by the two algorithms based on statistical comparison, robustness, stability and the ability of the FLC to work in the presence of faulty fuzzy rules. Further, a performance comparison between the two algorithms is presented and an insight into determining the optimal parameters for each algorithm is analyzed.

2 Fuzzy Logic Controller

FLC is based on the concept of fuzziness in which, rather than allowing a system to have a value of 0 or 1, fuzzy allows the system to have degrees of membership functions over the range [0, 1]. The basis for design of a FLC is the linguistic IF-THEN rules for eg: *IF I_1 is B_{i1} AND I_2 is B_{i2} THEN O is D_i* , where I_1, I_2 are the inputs and O is the output of the FLC. B_{i1}, B_{i2} and D_i, $i= \{1,...,n\}$ are linguistic values presented in the fuzzy subsets of the universe of discourse.

In this paper, for the CCP, the values of x(displacement)(m) and v(velocity)(m/s) are scaled to the interval of [-5,5] and F(Force)(N) from [-75,75]. For the IP, the values of θ(angular displacement)(rad) and $\dot{\theta}$(angular velocity)(rad/s)are scaled to the interval of [-0.5,0.5] and F from [-5,5]. The inputs and outputs of the FLC are composed of seven linguistic terms NB(Negative Big), NM(Negative Medium), NS(Negative Small), Z(Zero), PS(Positive Small), PM(Positive Medium) and PB(Positive Big). This set of linguistic terms forms a fuzzy partition of input and output spaces. In our study we use the Gaussian symmetrical membership functions(GMF).The fuzzy IF-THEN rules for the two problems examined here i.e. CCP and IP are the same as shown in Table1. The defuzzification method used is the centroid of area.

Table 1. Rule Base

I_1 / I_2	NB	NM	NS	Z	PS	PM	PB
PB	Z	NS	NM	NB	NB	NB	NB
PM	PS	Z	NS	NM	NM	NM	NB
PS	PM	PS	Z	NS	NM	NM	NB
Z	PB	PM	PS	Z	NS	NM	NB
NS	PB	PM	PM	PS	Z	NS	NM
NB	PB	PB	PB	PM	PS	Z	NS
NM	PB	PB	PM	PB	PS	PM	Z

3 Problem Formulation

In this section we discuss the two problems for which FLC's are designed.

3.1 Cart Centering Problem

In this problem a cart of mass M moves on a one dimensional track .The input variables for this problem are the cart's location on the track, x, and the cart's velocity, v. The objective of the controller is to apply a suitable force F which will bring the cart to rest, i.e x=0 and v=0 from an initial displacement and velocity. The equations of motion for the cart are

$$x(t+\tau)=x(t)+ \tau v(t) \qquad (1)$$
$$v(t+\tau)=v(t)+\tau F(t)/M \qquad (2)$$

where τ is the time step. In this paper τ is 0.02 s and M is 1 kg and the initial conditions for x and v is 1m and 1m/s respectively.

3.2 Inverted Pendulum Problem

The IP is a classic benchmark problem used in control literature due to its high nonlinear dynamics. Moreover this system has fewer control inputs than the degrees of freedom, making its control challenging. The IP is also a representative of a class of altitude control problems whose goal is to maintain the desired vertically oriented position at all times [2].

The input variables for this problem are the angular displacement of the pole, θ and the angular velocity, $\dot{\theta}$. The objective of the controller is to apply a suitable force F on the cart which will bring the pendulum to rest at vertical position i.e $\theta = 0$ and, $\dot{\theta} = 0$ from an initial θ and, $\dot{\theta}$. Here we are concerned with the control of the angular position of the pendulum and not the position or velocity of the cart. The cart travels in one direction along a frictionless track. The equations describing the angle dynamics of the IP are:

$$\dot{\theta} = \frac{d\theta}{dt} \tag{3}$$

$$\ddot{\theta} = \frac{gsin(\theta) - (cos(\theta)/ m_p + m_c)\,(m_p l\dot{\theta}^2\,sin(\theta) + F(t)}{\frac{4l}{3} - (m_p l\,(cos(\theta))^2/\,m_p + m_c)} \tag{4}$$

where g is 9.8 m/s2, mass of pendulum(m_p) is 0.1 kg , mass of cart(m_c) is 0.9 kg and length of pendulum is 0.5m. the initial condition for θ and, $\dot{\theta}$ are 0.1 rad and 0.1rad/s respectively.

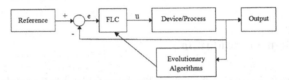

Fig.1. Block structure of the FLC optimizing process using evolutionary algorithms

Fig. 1. shows the block structure of the FLC optimizing process using evolutionary algorithms here, GA or AIS. 'e' represent the error between the output and reference and 'u' represents the control input. Device/Process refers to either CCP or IP . Since there are 21 MF's and 49 rules , there are a total of 42 decision variables to be optimized . In the next two sections we discuss the applicability of GA and AIS to tune the FLC's.

5 Genetic Algorithm

The GA are based on the Darwinian principle of survival of fittest and the natural process of evolution through reproduction. In GA the solution to a given problem is encoded in the form of strings called chromosomes. Each chromosome consists of a set of genes that contain a set of values for the optimization variables. GA works by composing a random initial population P consisting of a set of chromosomes each representing a possible solution to the problem. The fitness of each chromosome is then computed by evaluating it against the objective function. The present population then evolves towards better solutions through genetic operators namely reproduction, cross over and mutation. For tuning the FLC using GA, all the antecedent and consequent parameters in the d-dimensional space is searched. The solution representation in GA is as shown below:

$P_a = [\sigma_{i1}{}^k \sigma_{i2}{}^k \sigma_o{}^k c_{i1}{}^k c_{i2}{}^k c_o{}^k]$, where $k \in [1, ...7]$

Where σ_{i1}, σ_{i2}, c_{i1} and c_{i2} represents the deviation and centre for the two inputs of the GMF's and σ_o and c_o represents the deviation and centre for the outputs of the GMF's. The most important step in applying evolutionary algorithms is to establish a measure of its fitness, to accurately evaluate the performance of the controller. However the task of defining the fitness function is application specific. In this paper we use the Root Mean Square Error(RMSE) given by Eq.(5).

$$\text{RMSE} = \sqrt{\frac{\sum_i^N e^2}{N}} \tag{5}$$

Where e is the error in trajectory of displacement for CCP and e is error in trajectory of angular displacement for IP and N is the number of points. The pseudocode for optimization of FLC using GA is as follows:

1. The initial population is initialized with random antecedent and consequent parameters of the FLC in the d-dimensional space
2. The fitness of each chromosome is calculated according to the objective function using Eq(5)
3. The population is arranged in order of their fitness values and the chromosome with best fitness is reproduced.
4. Crossover and mutation operators are carried out on consecutive chromosomes. This is done ensuring that no chromosome is chosen twice.
5. The fitness value of the new population is calculated and it is ensured that only the fittest chromosomes enter the next generation.
6. Steps 2-5 are carried out till maximum number of generations is reached.

6   Artificial Immune System

AIS is composed of intelligent methodologies inspired by the natural immune system which are used to solve the real world problems [16]. The clonal selection algorithm is part of AIS based on clonal expansion and affinity maturation [11]. The clonal selection theory describes that when an antigen (Ag) is detected, antibodies (Ab) that best recognize this Ag will proliferate by cloning. This immune response is specific to each Ag. The immune cells will reproduce in response to a replicating Ag until it is successful in recognizing and fighting against this Ag. Some of the new cloned cells will be differentiated into plasma cells and memory cells. The plasma cells produce Ab and promotes genetic variation when it undergo hypermutation. The memory cells are responsible for the immunologic response for future Ag invasion. Subsequently, the selection mechanism will keep the cells with the best affinity to the Ag in the next generation.

Based on the clonal selection principle, an algorithm is developed in which various immune system aspects are taken into account such as: maintenance of the memory cells, selection and cloning of the most stimulated cells, death of non-stimulated cells and re-selection of the clones with higher affinity and generation and maintenance of diversity.

Since our problem is in continuous domain space, the antibody chains are binary coded. During simulations, each value in the chain is encoded by 23 bits. The fitness of each chain is calculated after it was decoded to real values. The initial population is cloned and the cloned chains undergo the process of hypermutation and re-selection.

Hypermutation is an important parameter similar to the mutation operator in GA. Assuming n chains are selected for cloning and n_c clones are created per chain then $n \times n_c$ chains are subjected to hypermutation operation according to the probability p_m. The probability p_m is adaptive in nature and varies for certain number of iterations after being set to its original value.

In reselection process the n best clones are selected from $n \times n_c$ clones and placed in the current population. The replacement can be either greedy or non greedy. If we use a greedy method we replace the worst n in the current population P with the n best clones. If we use a non greedy strategy then we replace the worst in particular intervals.

Fig.1. shows the block structure of the FLC optimizing process using evolutionary algorithms, in this case AIS. Similar to GA optimization of FLC, there are a total of 42 decision variables to be optimized. AIS searches for all the antecedent and consequent parameters in the d-dimensional space.

The pseudo-code for optimization of FLC using CSA model of AIS is as follows:

1. Initialize the population with antibody chains with each chain being a solution of the FLC.
2. Evaluate the fitness of each of the antibody chains using Eq(5).
3. Select antibodies for cloning. Each antibody would have the same clone size, not privileging anyone for their affinity.
4. The antigen affinity(corresponding to fitness value) is used to determine the p_m which is adaptive in nature.
5. For re-selection process, the best clones are selected to replace the current population and move on to the next generation.
6. Steps 2-5 are repeated till maximum number of generations is reached

7 Results and Discussion

In this section the FLC's tuned by GA and AIS are analyzed

7.1 Cart Centering Problem

For optimization of FLC for CCP using GA, a random population of chromosomes are created. The parameter values for GA are: Number of chromosomes per generation = 10, Genetic operators: Cross over rate 60%, Mutation rate 20%, reproduction rate 20% and Number of generations = 1000. The best RMSE value obtained was 0.1407 using GA.

Similar to GA, an initial random population of antibodies are created for AIS, for optimization of FLC for CCP. The parameter values for AIS are: Number of antibodies = 22, Number of antigens = 22, Number of clones per candidate = 10, p_m = 0.1 and Number of generations = 1000. The best RMSE value obtained was 0.0911 using AIS.

To compare the efficacy of AIS over GA a statistical test is done, in which each algorithm is run twenty times and the maximum, minimum, average and standard deviation of the RMSE values are noted, as shown in Table 2. We observe that AIS succeeds to a large extent in tuning a FLC over GA. It is also observed that AIS has a lower average RMSE value indicating that it is also consistent for the CCP. A low standard deviation obtained for AIS also indicates that there is very little divergence in the RMSE values obtained for the twenty trials. Fig2. shows the plot of cart position for the best design of GA and AIS. It can be observed that the FLC tuned by AIS brings the cart to zero position in under 0.5 seconds unlike GA which takes over 1.5 seconds. Fig. 3. Shows the plot of RMSE values for the best design of GA and AIS for CCP.

Once the FLC are tuned by GA and AIS, a study was made to determine their performance in the presence of noise. For this test we consider the best design of GA and AIS and add noise of power 10^{-6} and 10^{0}. For FLC tuned by GA we obtain RMSE values of 0.1408 and 0.1910 for noise power of 10^{-6} and 10^{0} respectively. And for FLC tuned by AIS we obtain RMSE values of 0.0912 and 0.0946 for noise power of 10^{-6} and 10^{0} respectively. Thus indicating the superiority of AIS in obtaining a FLC which works well even in the presence of noise. The robustness of the FLC tuned by GA and AIS was tested for robustness by changing the mass of the cart .The RMSE values obtained for GA are 0.270607, 0.3834 and 0.579 and the RMSE values obtained for AIS are 0.105, 0.2277 and 0.3441 for mass of cart 5kg, 10kg and 20kg respectively. This indicates that the FLC tuned by AIS is also robust. It was also necessary to know , how the FLC's performed in the presence of faulty rules. For this test three rules were chosen and purposely made faulty. Rule at location(row, coloumn) (1,1), (2,2) and (4,4) with respect to Table 1 are selected. The divergence errors(DE), which is the difference between the actual RMSE obtained with correct rule and RMSE obtained for faulty rule is noted. This is shown in Table 4 and Table 5 for GA and AIS respectively. It is noted that by changing rules on the edges (1,1) will result in large DE while rules well placed within the rule base do not change the RMSE values. This can be attributed to the fact that rule (1,1) do not have the luxury of having rules completely surrounding them and able to compensate for their mistakes.

The comparison of the performance of GA and AIS was also carried out using the Kruskal Wallis (KW) rank test. In which the ranks for the respective algorithms are assigned. The lower the rank the better the algorithm. For CCP the ranks obtained for AIS and GA are 10.55 and 30.45 respectively and the Chi- sq is 28.976 and p=0 (p<0.05). Thus through KW test it is observed that AIS is better than GA for CCP.

A study was also made to understand the effect of the most important parameters of the algorithms. For GA, cross over rate and mutation rate play an important role and for AIS, it is the p_m and the number of clones. For this study the random initial population was kept same for both GA and AIS so that comparison on a common platform is possible and the effect of individual parameters can be studied. For GA, by changing the cross over rates to 20%, 40%, 60% and 80% the RMSE values obtained is 0.240318, 0.167296, 0.126362 and 0.139249 respectively. And by changing the number of genes undergoing mutation per chromosome to 1, 2, 3 and 4 RMSE values of 0.614012, 0.305291, 0.365119 and 0.216281 were obtained. For AIS, the p_m is changed to 0.1, 0.15, 0.2, 0.25, 0.3, 0.35 and 0.4 and the RMSE values obtained are 0.132462, 0.148254, 0.15662, 0.173147, 0.17584, 0.225132and 0.206485 respectively. The number of clones are changed to 3, 5, 7 and 10 and the RMSE values obtained are 0.122387, 0.109719, 0.107266 and 0.101245 respectively. It is observed that the optimal crossover rate for GA is 60% and fitness value linearly becomes better with increase in mutation. It is also observed that for AIS the optimal p_m is around 0.1 and fitness value linearly becomes better with increase in number of clones.

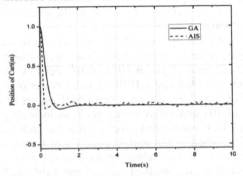

Fig. 2. Plot of Cart Position for the best design of GA and AIS.

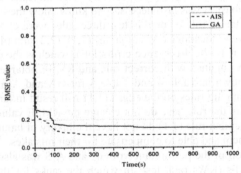

Fig. 3. Plot of RMSE values for the best design of GA and AIS for CCP

Table 2. Statistical Tests of fitness values for AIS and GA

Algorithms	Maximum	Minimum	Average	Standard Deviation
AIS	0.1461	0.0911	0.109	0.0162
GA	0.5116	0.1407	0.223	0.0836

Table 3. Presence of Faulty rules for FLC tuned by GA for CCP

(1,1)	4-1	4-2	4-3	4-5	4-6	4-7
DE	0.7432	0.0045	0.1391	1.0043	0.3087	0.7432
(2,2)	4-1	4-2	4-3	4-5	4-6	4-7
DE	0.0044	0	0.008	0.0053	0.002	0.0044
(4,4)	4-1	4-2	4-3	4-5	4-6	4-7
DE	0	0	0	0	0	0

Where 1,2,3,4,5,6,7 are NB, NM, NS, Z, PS, PM and PB respectively.

Table 4. Presence of Faulty rules for FLC tuned by AIS for CCP

(1,1)	4-1	4-2	4-3	4-5	4-6	4-7
DE	0.0004	0.0896	0.0896	6.8238	0.0896	6.8782
(2,2)	4-1	4-2	4-3	4-5	4-6	4-7
DE	0	0	0	0	0	0
(4,4)	4-1	4-2	4-3	4-5	4-6	4-7
DE	0	0	0	0	0	0

7.2 Inverted Pendulum

For optimization of FLC for IP using GA and AIS the parameters are the same as chosen for CCP. The total number of generations chosen is 100. To compare the efficacy of AIS over GA for the IP a statistical test is done similar to CCP as shown in Table 5. We observe that here too AIS succeeds to a large extent in tuning an optimal FLC over GA as the maximum, minimum and average RMSE for AIS being lower than GA. Further the Standard deviation of AIS is also lesser than GA. Fig.4. shows the plot of pendulum position for the best design of GA and AIS. It can be observed that the FLC tuned by AIS brings the pendulum to zero position in under 0.5 seconds, while GA stabilizes the pendulum to a value slightly less than 0 in over 0.75 seconds.

A study to determine the performance of FLC tuned by GA and AIS in the presence of noise is made. For this test we consider the best design of GA and AIS and add noise of power 10^{-6} and 10^{-9}. For FLC tuned by GA we obtain RMSE values of 0.019677 and0.01737 for noise power of 10^{-6} and 10^{-9} respectively. And for FLC tuned by AIS we obtain RMSE values of 0.018067 and 0.016725 for noise power of 10^{-6} and 10^{-9} respectively. Thus indicating the superiority of AIS in obtaining a FLC which works well even in the presence of noise. The robustness of the FLC tuned by GA and AIS was tested for robustness by changing the mass of the pendulum and length of the pendulum to three times its initial value .The RMSE values obtained for GA are 0.020708 and 0.0207 and the RMSE values obtained for AIS are 0.015168 and 0.0189 for mass of pendulum 0.3kg and length of pendulum 1.5 m respectively. This indicates that the FLC tuned by AIS is also robust. A test of the FLC's performance in the presence of faulty rules was also conducted. For this test the same three rules were chosen as in the case of CCP. This is shown in Table 6 and 7 for GA and AIS respectively. It is noted that by changing rules on the edges (1,1) will result in large DE while rules well placed within the rule base do not change the RMSE values. For IP we obtained the ranks for AIS and GA to be 7.30 and 13.70 respectively and the chi-sq is 5.8157, p=0.016(p<0.05). Thus for IP, AIS is better than GA due to its lower rank.

A study was also made to understand the effect of the most important parameters of the algorithms similar to the CCP. For GA, by changing the cross over rates to 20%, 40%, 60% and 80% the RMSE values obtained is 0.01735, 0.022517, 0.023607 and 0.026612 respectively. And by changing the number of genes undergoing mutation per chromosome to 1, 2, 3 and 4 RMSE values of 0.017977, 0.021193, 0.022882 and 0.019445 were obtained. For AIS, the p_m is changed to 0.1, 0.15, 0.2, 0.25, 0.3, 0.35 and 0.4 and the RMSE values obtained are 0.020537, 0.018752, 0.017945, 0.021381, 0.018301, 0.019726 and 0.023414 respectively. The numbers of clones are changed to 3, 5, 7 and 10 and the RMSE values obtained are 0.01843, 0.014206, 0.014151 and 0.01335 respectively. It is observed that the optimal crossover rate for GA is 20% and optimal mutations per chromosome is 1. It is also observed that for AIS the optimal p_m is around 0.2 and fitness value linearly becomes better with increase in number of clones as was observed for CCP. With these tests, it is clear that the optimal parameters of either GA or AIS are problem dependent and no definitive pattern can be obtained, except for the number of clones in AIS. This is logical as increase in number if clones increases the local search around an optimal point which could result in a better solution.

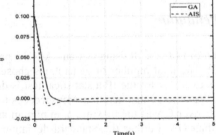

Fig.4 Trajectory Plot of Cart Position for the best design of GA and AIS

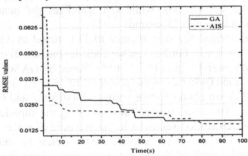

Fig.5 Plot of RMSE values for the best design of GA and AIS for IP

Table 5. Statistical Tests of fitness values for AIS and GA

Algorithms	Maximum	Minimum	Average	Standard Deviation
AIS	0.0151	0.0256	0.0196	0.0027
GA	0.0182	0.02510	0.0229	0.0018

Table 6. Presence of Faulty rules GA for IP

(1,1)	4-1	4-2	4-3	4-5	4-6	4-7
DE	0.1067	0.8994	0.0162	0.0012	0.0076	0
(2,2)	4-1	4-2	4-3	4-5	4-6	4-7
DE	0.0061	0.007	0.0002	0	0.0094	0.0005
(4,4)	4-1	4-2	4-3	4-5	4-6	4-7
DE	0	0	0	0	0	0

Table 7. Presence of Faulty rules AIS for IP

(1,1)	4-1	4-2	4-3	4-5	4-6	4-7
DE	0.4023	0.4015	0.0024	0	0.2571	0.0011
(2,2)	4-1	4-2	4-3	4-5	4-6	4-7
DE	0.0002	0	0	0	0	0.0001
(4,4)	4-1	4-2	4-3	4-5	4-6	4-7
DE	0	0	0	0	0	0

7 Conclusions

In this paper, the Clonal selection Algorithm(CSA) a model of the Artificial Immune System(AIS) paradigm for tuning the antecedent and consequent parameters of the membership functions of a Mamdani type fuzzy Logic controller is examined. Two problems were considered, firstly the cart centering problem and secondly the classic benchmark in control, the inverted pendulum problem. The FLC tuned by AIS was compared with a FLC tuned by GA for both the problems considered. The FLC once tuned by the two algorithms were then checked for robustness by adding white noise of varying power densities to the system. Robustness was also checked by varying the mass of the cart and the mass and the length of the pendulum. The FLC's were also tested in the presence of faulty rules. It was observed that the FLC tuned by AIS outperforms the FLC tuned by GA for all cases. The Kruskall Wallis test performed, has also proved that AIS is a better algorithm for optimization compared to GA. An insight into the effect of individual parameters of GA and AIS was also conducted.

References

[1] Zafer Bingul, Oguzhan Karahan: A Fuzzy Logic Controller tuned with PSO for 2 DOF robot trajectory control. Eng. Appl. Artif. Intel. 38 (1), 1017–1031(2011)

[2] Lal Bahadur Prasad, Hari Om Gupta, Barjeev Tyagi: Intelligent control of nonlinear inverted pendulum dynamical system with disturbance input using fuzzy logic systems. International Conference on Recent Advancements in Electrical, Electronics and Control Engineering (2011)

[3] Arup Kumar Nandi, J. Paulo Davim: A study of drilling performances with minimum quantity of lubricant using fuzzy logic rules. Mechatronics.19,218–232 (2009)

[4] Olga Diamante, Giovanna Fargione, Antonino Risitano, Domenico Tringali: A soft computing approach to fuzzy sky-hook control of semiactive suspension. IEEE T. Contr. Syst. T. 11 (6), (2003)

[5] Un-Chul Moon; Kwang Y; Lee: Hybrid algorithm with fuzzy system and conventional PI control for the temperature control of TV glass furnace. IEEE T. Contr. Syst. T. 11(4), (2003)

[6] Timothy J. Ross: Fuzzy logic with engineering applications. John Wiley (2004)

[7] Belarbi, K., Titel F., Bourebi W., Benmahammed, K.: Design of mamdani fuzzy logic controllers with rule base minimisation using genetic algorithm, Eng. Appl. Artif. Intel. 18(7),875–880 (2005)

[8] Lam, H.K., Leung, F.H.F. Tam, P.K.S.: Design and stability analysis of fuzzy model based nonlinear controller for nonlinear systems using genetic algorithm. Proceedings of the IEEE International Conference on Fuzzy Systems 1,232 - 237 (2002)

[9] Leung, F.H.F. , Lam, H.K. , Ling, S.H. , Tam, P.K.S. :Optimal and stable fuzzy controllers for nonlinear systems based on an improved genetic algorithm. IEEE T. Ind Electron. 51(1), 182 – 192(2004)

[10] Emad Elbeltagi, Tarek Hegazy, Donald Grierson,: Comparison among five evolutionary based optimization algorithms. Adv. Eng. Info. 19,43-53(2005)

[11] Omkar, S.N., Rahul Khandelwal, Santosh Yathindra,Narayana Naik,G., Gopalakrishnan :Artificial immune system for multi-objective design optimization of composite structures. Eng. Appl. Artif. Intel. 21 (8), 1416–1429(2008)

[12] Pengfei Liu, Ping Xu, Jinyang Zheng:Artificial Immune System for optimal design of composite hydrogen vessel. Comput. Mater. Sci. 47 (1), 261-267(2009)

[13] Zhuhong Zhang, Shuqu Qian: Artificial immune system in dynamic environments solving time-varying non linear constrained multi-objective problems. Soft Comput. 15, 1333–1349(2011)

[14] Taur,J.S, Tzuen Wuu Hsieh, Tsai,C.L.: design of a Fuzzy Controller with Fuzzy Swing Up and Parallel distributed Pole Assignment Schemes for an Inverted Pendulum and Cart System. IEEE T. Contr. Syst. T. 16 (6), 1277-1288(2008)

[15] El-Hawwary, M.I: Adaptive Fuzzy Control of the Inverted Pendulum Problem. IEEE T. Contr. Syst. T. 14(6), 1135-1144(2006)

[16] Henry Lau, Y.K., Vicky Wong, W.K.: An immunity approach to strategic behavioral control. Eng. Appl. Artif. Intel.20 (3), 289–306(2007)

Comparisons of Different Feature Sets for Predicting Carbohydrate-Binding Proteins from Amino Acid Sequences Using Support Vector Machine

Suchandra Payal[1], Piyali Chatterjee[2], Subhadip Basu[1], Mahantapas Kundu[1], Mita Nasipuri[1]

[1]Department of Computer Science and Engineering, Jadavpur University, Kolkata, WB, India

[2]Netaji Subhas Engineering College, Kolkata, WB, India

{suchan.payal@gmail.com; chatterjee_piyali@yahoo.com; subhadip8@yahoo.com; mahantapas@gmail.com; mitanasipuri@yahoo.com}

Abstract. Proteins which can interact with sugar chains but do not modify them are called as Carbohydrate-binding proteins. These proteins have several biological importance. To predict them computationally with SVM classifier, we have developed different feature sets – based on secondary structures and selective physicochemical properties of the constituent amino acids. The feature set formed with combination of both the secondary structures and physicochemical properties gives better prediction accuracy (up to 89.19%). We have also prepared an up-to-date dataset of carbohydrate-binding proteins and non-carbohydrate-binding proteins in this work.

Keywords: Carbohydrate-binding proteins, Feature sets, Secondary structures, Selective physicochemical properties.

1 Introduction

Carbohydrate-binding proteins have importance in developmental processes, ligand recognition, cell to cell signaling and adhesion process which have an important role in cancer growth and progression. Therefore in cancer therapeutics, Carbohydrate ligands play important roles [1]. There are active researches on glycosyltransferases [2] to understand the roles of carbohydrates on cancerous cell.

J. C. Bansal et al. (eds.), *Proceedings of Seventh International Conference on Bio-Inspired Computing: Theories and Applications (BIC-TA 2012)*, Advances in Intelligent Systems and Computing 201, DOI: 10.1007/978-81-322-1038-2_44, © Springer India 2013

There are several experimental works to identify carbohydrate-binding proteins but these are very costly and time consuming. Here, lies the importance of comparatively fast and less costly computational methods to identify them.

Literature survey shows that there are several works to identify the Carbohydrate-binding sites in proteins [3-4] (already annotated as carbohydrate-binding proteins), but very few works could be found on prediction of carbohydrate-binding protein [5-6].

Someya et al [5] first provide (in 2010) a clear and efficient method of defining and detecting carbohydrate-binding proteins from Amino Acid sequences. They first construct the datasets of Carbohydrate-binding proteins of various kinds including enzymes and also the proteins, which are not specifically annotated as Carbohydrate-binding proteins in biological databases for which they also specified a set of search conditions. Finally, they have obtained 345 Carbohydrate-binding protein sequences or positive datasets and 7827 Non-Carbohydrate-binding protein sequences or negative dataset from UniProtKB [8]. Then they have set two types of encoding methods - direct encoding (AA-20) and group encoding (Levitt-6 [11] and Someya-7) [5] and have trained SVM with these encoded data samples and tested on the same dataset by applying leave one out method. In these prediction methods the used positive datasets are too small to compare the negative datasets.

Most recently there is another work published in 2012 [6], in which they have used a large set of data from three up-to-date databases of CAZy, CGF and Swiss-Prot. This data set CBPDS consists of 2380 positive and negative sequences with sequence identity 25% by removing sequence redundancy. Then they have proposed a SVM-based prediction obtained by using an inheritable bi-objective genetic algorithm. They have actually obtained 17 physicochemical properties by this method. Using GA, they have designed a methodical the selection of informative physiochemical properties for the carbohydrate-binding proteins prediction and obtained 17 such properties. Prediction accuracy of this approach is 79.45%.

In this work, at the beginning Carbohydrate-binding protein sequences are collected with SRS queries from the SRS query server of EMBL-EBI [7]. The non-Carbohydrate-binding protein sequences are collected from UniProtKB [8] randomly with some conditions. Then sequence redundancy of these collected sequences has been removed with the help of two bioinformatics tool BLASTclust [9] and ClustalW2 [10]. In this approach the redundant sequences are discarded to form a redundancy free good training samples. Then from the remaining samples the test samples are constructed. After constructing the database, the encoding of the sequences for feature extraction of SVM is done in different approaches including those used by Someya et al in their work [5]. Here we also propose three different approaches for feature extraction: 1) SS (Secondary Structure) based, 2) PP (Selective physicochemical properties) based and 3) SS+PP (combination of secondary structure and selective physicochemical properties) based We have compared all these feature sets including Someya's on a uniform balanced and standard training as well as test samples and it is found that our feature sets give better test accuracy with a SVM classifier.

2 Materials

2.1 Dataset Preparation

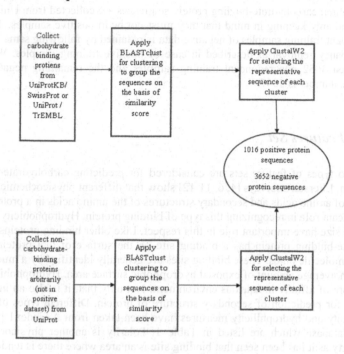

Fig.1. Steps followed in Training Dataset Preparation

Someya *et al* first provide a clear definition of carbohydrate-binding proteins irrespective of whether it is annotated as carbohydrate-binding proteins in the available biological databases. We have followed their carbohydrate-binding protein sequence collection method which seems to be very convenient. First, we submit SRS queries in SRS server of European Bioinformatics Institute at Hinxton, UK [7]. Keywords of the queries are more or less same as Someya's with little server-based modifications. Then we have used BLASTclust tool of Max-Plank Institute for Developmental Biology [9] to group the sequences with some similarity score matching. This similarity matching score is obtained by setting two parameters: 1) Sequence length to be covered is 40% and 2) percentage of identity threshold is 35% at least. Then from each group only one single representative sequence is selected for which the sum of its pairwise aligned matching scores with all other sequence of that group or cluster is maximum with the help of ClustalW2 tool of EBI [10]. Finally, 1016 total number of positive instances is selected as training samples. An overview of the various steps involved in the dataset preparation is shown in Fig. 1.

The positive instances of test samples are obtained from the sequences which are discarded as having some kind of redundancy. So the number of positive test sample is 6231.

The non carbohydrate-binding protein sequences are collected from UniProtKB [8] randomly keeping in mind that they must not be in positive samples. The non redundant training samples of negative data is obtained by following same steps of redundancy removal as described in case of positive training samples. We have obtained 3652 non redundant training samples and the remaining negative test sample number is 2047.

2.2 Features Set

Two types of feature sets are considered for predicting carbohydrate-binding protein. Literature surveys [4, 6, 11-12] show that different physicochemical properties of amino acids and secondary structures of the amino acids in a protein have significant role in recognizing this type of binding protein. Hydrophobicity, polarity and size have important role in this respect. Like other binding proteins, carbohydrate-binding protein has a binding site on the surface of that protein where sugar molecule binds. These binding sites are generally identified by a much higher than average amount of exposed hydrophobic surface area. Hydrophobicity is a measure of affinity in aqueous environment. Not only that, it is also an important factor for prediction of secondary structure of protein. Different types of hydrophobicity and hydrophilicity measures have been taken from AAindex1 [13] feature database which are listed in Table 1. Polarity is another physicochemical property as it has been seen that binding site is an area where there is tendency for non-polar groups to be clustered on the protein surface. Polarity values from AAindex1 [13] listed in Table 1 have been also used in this work. The complete list of physicochemical properties (27 in number) collected from AAindex1 together with their accession numbers are shown in Table 1.

It has been seen that the secondary structures of carbohydrate-binding protein also have similar patterns. They can have variety of secondary structures consisting of β-folds, the β-trefoil and the β-barrel (β-strands interrupted by loop regions, i.e., jelly roll) [5] [14-17].

So, inspecting hydrophobic, polar or non-polar regions as well as searching for similar patterns in secondary structure elements is relevant in prediction of carbohydrate-binding protein.

The secondary structures of the collected sequences are required for the feature extraction based on secondary structures. So we collect secondary structures of the already prepared data samples from PSIPRED server of Bioinformatics Group, UCL Department of Computer Science [18]. We managed to collect only 820 positive and 849 negative training samples' and 35 positive and 39 negative test samples' secondary structures.

Table 1. Selected physicochemical properties of AAindex1

Accession number(H)	Data description(D)
ARGP820101	Hydrophobicity index
CHAM820101	Polarizability parameter
CHAM820102	Free energy of solution in water, kcal/mole
EISD860101	Solvation free energy
EISD860102	Atom-based hydrophobic moment
EISD860103	Direction of hydrophobic moment
ENGD860101	Hydrophobicity index
FASG890101	Hydrophobicity index
GOLD730101	Hydrophobicity factor
GOLD730102	Residue volume
GRAR740102	Polarity
GRAR740103	Volume
HOPT810101	Hydrophilicity value
JANJ780101	Average accessible surface area
JOND750101	Hydrophobicity
NOZY710101	Transfer energy, organic solvent/water
PONP800101	Transfer energy, organic solvent/water
PONP800102	Surrounding hydrophobicity in folded form
PONP800103	Average gain in surrounding hydrophobicity
PONP800104	Surrounding hydrophobicity in alpha-helix
PONP800105	Surrounding hydrophobicity in beta-sheet
PONP800106	Surrounding hydrophobicity in turn
PRAM900101	Hydrophobicity
RADA880108	Mean polarity
ROSM880102	Side chain hydropathy, corrected for solvation
ZIMJ680101	Hydrophobicity
ZIMJ680103	Polarity

3 Methods

Preparing positive and negative data from UniProtKB [8] database using SRS queries [5] and thereafter applying ClustalW2 tool [10] for grouping them is significant part of our work. In order to improve better prediction accuracy, selection of features is also important. In our work, we have used two types of features set utilizing relevant physicochemical properties of amino acids as well as secondary structure elements. Finally, we have combined these two different features to get better prediction accuracy. We have also implemented the different approaches used by Someya *et al* [5] for encoding the amino acid sequences for feature extraction.

In AA-20 [5], simply the occurrence frequency of each of the amino acid triplet patterns is calculated as features. As these triplet patterns can be 20^3 so the feature vector or input vector size here becomes 8000 which makes the prediction system too slow so another two group encoding methods i.e. Levitt-6 and someya-7 are also introduced by Someya *et al* [5] on the basis of polarity and secondary structure of amino acid mainly. So in Levitt-6 feature vector size become 6^3 i.e. 216 and in Someya-7 it is 7^3 i.e. 343. Though the training accuracy with our training samples of these three approaches are higher but testing accuracy not good enough with our test samples. So we go for other approaches.

For feature extraction on the basis of secondary structures only, the secondary structure information of a given amino acid sequence is collected from PSIPRED server [18]. Then we represent the input amino acid sequences with these secondary structures. The PSIPRED predicts the secondary structures of the amino acid residues of the protein sequence as either helix (H) or strands (E) or coil (C). So a triplet of three consecutive secondary structures in a sequence can be of 3^3 or 27 types. The occurrence frequency of each of these triplets in a protein sequence with respect to the length of that particular sequence is considered as an element of the feature vector for this prediction system. Here the total number of features is 27.

In case of PP based approach, we find 27 properties are found useful for prediction of Carbohydrate-binding protein prediction based on literature surveys [5] [11-13]. For computing the feature values, a histogram representing the frequencies of occurrences of 20 different amino acids in a given protein (amino acid sequence) is prepared first. In doing this, for each of the 20 amino acids, the occurrence of a particular amino acid in an amino acid sequence is divided by its length. In AAindex [13], for each type of physicochemical properties, there is a table of numeric values representing that property values for different amino acids. Using the above mentioned amino acid occurrence frequency histogram for an amino acid sequence, and a given AAindex1 [13] table, we have computed an average value of that physicochemical property for that sequence by multiplying the property value of each of the twenty amino acids with the frequency values for that amino acid in the corresponding histogram and finally summing up these twenty values. This results in one feature value. Thus, there are 27 feature values in the feature vector.

Support vector machine is a well-established binary classifier for pattern classification problems which maps feature vectors in higher dimension and classifies the two classes with a maximum margin generating hyperplane. In this work, the huge samples as well as their features require higher dimensional feature spaces to be distinguished between the positive instances and negative instances. For this purpose support vector machine is most suitable. Here, freely available LIBSVM [19] package is used.

3.1 SVM Training and Prediction

Then with these different types of feature sets the SVM is trained and tested. For training we choose the nu-svm and kernel-type as RBF and vary the gamma and nu values (which lies between 0 to1).

3.2 Performance Measurement Metrics

To evaluate the performances of the predictors, different performance metrics have been used as mentioned below:

True Positive (TP): The number of positive samples predicted as positive.
True Negative (TN): The number of negative samples predicted as negative.
False Positive (FP): The number of negative samples predicted as positive.
False Negative (FN): The number of positive samples predicted as negative.

Sensitivity: True Positive Rate is a performance measure of binary classification test. True positive rate or recall rate or sensitivity is the probability of correct prediction of an example of that class. TPR is defined in the following equation. Higher value of this measure represents perfect prediction of positive class.

$$\text{Sensitivity} = \frac{TP}{TP + FN} \times 100$$

Specificity: It measures the percentage of actual positives out of predicted positive samples False Positive Rate (also known as False Alarm Rate) for a class is the probability that an example which does not belong to the class is classified as belonging to the class.

$$\text{Specificity} = \frac{FP}{FP + TP} \times 100$$

Matthews Correlation Coefficient (MCC): The Matthews Correlation Coefficient (MCC) is used in machine learning as a measure of quality of binary (two class) classifications. It takes into account true and false positive and negatives. MCC is generally regarded as a balanced measure which can be used even if the classes are of different sizes. The MCC is a correlation coefficient between the observed and predicted binary classifications. It returns a value between (-1) and (+1). A coefficient of (+1) represents a perfect prediction, (0) an average random prediction and (-1) an inverse prediction.

$$\text{MCC} = \frac{TP \times TN - FP \times FN}{\sqrt{(TP + FN) + (TP + FP) + (TN + FP) + (TN + FN)}}$$

Accuracy: The accuracy is the overall probability that prediction is correct

$$Accuracy = \frac{TP + TN}{TP + TN + FP + FN}$$

4 Results and Discussions

The effectiveness of the different feature sets, considered here, in predicting Carbohydrate-binding proteins are compared on a common data set. This data set contains 1669 (820 positive samples and 849 negative samples) training samples and 74 test samples (35 positive samples and 39 negative samples). All of the mentioned approaches for feature extraction are applied to these data samples. With these extracted features, different SVM classifiers are designed for predicting Carbohydrate-binding proteins. We have chosen the nu-svm and kernel-type as RBF and varied the gamma and nu values (which lies between 0 to1). The prediction performances of these classifiers are shown in Table 2.

AA-20 feature set is of so huge volume (8000 and most of the cases containing 0 values) that the prediction system becomes too slow. So it is very difficult to handle this prediction system and also gives comparatively low accuracy (78.37%). The classifier with Levitt-6 (216 features) dataset is comparatively faster and gives reasonable accuracy (81.08%).Same classifier with Someya-7 dataset (343 features) gives test accuracy in between AA-20 and Levitt-6 (79.73%).Secondary structure(27 features) based prediction system gives better accuracy (82.43%) and also it is faster. Selective Physicochemical properties(27 features) based system gives better accuracy (85.14%) than all the previous techniques and as faster as secondary structure based (see Fig 2).The best predictive accuracy (89.19%)is obtained by combining Secondary structure and physicochemical properties based prediction system(54 features) which is also reasonably fast as well. Various prediction performance metrics (discussed in section 3.2) are evaluated for different feature sets. These values are shown in Table 2 and also represented graphically in Fig 3. From fig 3, it is clear that the proposed feature sets achieve better prediction performances than AA-20, Levitt-6 and Someya-7.

Table 2. Prediction Performance metrics for different feature sets

Feature set	Feature size	TP	FN	FP	TN	Accuracy	Sensitivity	Specificity	MCC
AA-20	8000	30	5	11	28	78.37	0.86	0.72	0.58
Levitt-6	216	31	4	10	29	81.08	0.89	0.74	0.63
Someya-7	343	30	5	10	29	79.73	0.86	0.74	0.6
SS based	27	29	6	7	32	82.43	0.83	0.82	0.65
PP based	27	29	6	5	34	85.14	0.83	0.87	0.7
SS + PP based	54	32	3	5	34	89.19	0.91	0.87	0.78

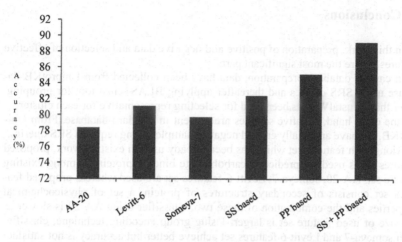

Fig. 2. Bar charts showing prediction accuracy of using existing feature sets and proposed feature sets

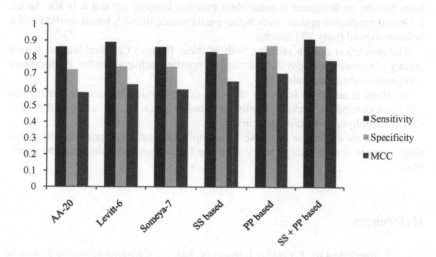

Fig.3. Performance measures of SVM classifiers of using existing feature sets and proposed feature sets

5 Conclusions

In this work, preparation of positive and negative data and selection of effective features set are the most significant part.

In case of database preparation, data have been collected from UniprotKB database using SRS queries and thereafter applying BLASTclust tool for grouping. After that, ClustalW2 has been used for selecting representative for each cluster. On the other hand, negative samples are present in standard database. From UniprotKB, we have artificially created negative samples using required SRS queries.

Along with features set which has been already used in existing work, proposed features set is used for predicting carbohydrate binding proteins. Among existing feature set, AA-20, someya-7, Levitt-6 features set are used whereas proposed features set consists of secondary structures of protein, a set of physicochemical properties and the combination of these two. Classifier using AA-20 is slower as the size of used feature set is larger. Using group encoding technique, classifier with someya-7 and Levitt-6 features set achieve better but accuracy is not satisfactory in prediction of test samples.

In SS based prediction method, though the feature vector has become reduced here, but the performance is better than existing features set and it is also faster. PP based prediction system gives better performance than SS based method and it is faster as well (only 27 features).

The prediction system utilizing both of these features i.e. combination of secondary structures and physicochemical properties achieves better performance compared to these methods.

So, there is an ample scope for studying other physicochemical properties like solvent accessibility, packing density, protrusion index etc. for successful prediction of carbohydrate-binding proteins.

In this work all feature sets have been applied on subset of prepared data samples, so there is also a scope to try all these feature sets with the huge data samples.

References

[1] Nangia-Makker, P. Conklin, J. Hogan, V. Raz, A.: Carbohydrate-binding proteins in cancer, and their ligands as therapeutic agents. Trends in Molecular Medicine. pp. 187-192. Cell Press, USA (2002). doi: 10.1016/S1471-4914(02)02295-5

[2] Tumor   Research   centre,   Tag   Archive   for   "glycotransferase". http://www.tumorres.com/tag/glycotransferase

[3] Shionyu-Mitsuyama, C., Shirai, T., Ishida, H. and Yamane, T.: An empirical approach for structure-based prediction of carbohydrate-binding sites on proteins. Protein Engineering, vol. 16.(7)(2003), 467-478.

[4] Malik, A. and Shandar, A. : Sequence and structural features of carbohydrate binding in proteins and assessment of predictability using a neural network. BMC Structural Biology(2007), 1-14.

[5] Someya, S. Kakuta, M. Morita, M. Sumikoshi, K. Cao W.: Prediction of Carbohydrate-Binding Proteins from Sequences Using Support Vector Machines. Advances in Bioinformatics. Hindawi Publishing Corporation. Japan (2010). Article ID 289301, 9 pages .doi:10.1155/2010/289301.

[6] Huang, H. Lee, H. Liou,Y. Li, M. Ho, S. : Designing predictors of carbohydrate-binding proteins using informative physicochemical properties. International Conference on Bioinformatics and Biomedical Technology. IACSIT Press. Singapore (2012). www.ipcbee.com/vol29/28-ICBBT2012-H047.pdf

[7] SRS@EBI. European Bioinformatics Institute .Hinxton.UK. http://srs.ebi.ac.uk/

[8] UniprotKB. http://www.uniprot.org/

[9] BLASTclust. Bioinformatics Toolkit, Max-Planck Institute for Developmental Biology (2008). http://toolkit.tuebingen.mpg.de/blastclust

[10] Align Sequences ClustalW2 | EBI. European Bioinformatics Institute .Hinxton.UK (2012). http://www.ebi.ac.uk/Tools/msa/clustalw2/

[11] Levitt, M.: Conformational preferences of amino acids in globular proteins. Biochemistry. pp. 4277-4285 (1978).

[12] Ringe, D.: Protien Sturcture and Function. New Science Press Ltd. London (2004).

[13] Kawashima, S. Kanehisa, M.:AAindex: amino acid index database. Nucleic Acids Research. V. 28(1): 374. PMC. Oxford Journals (2000). http://www.genome.jp/aaindex/ ; ftp://ftp.genome.ad.jp/pub/db/community/aaindex/aaindex1

[14] Sharon, N. Lis, H.: Lectins. Springer.,Kluwer Academic Publishers. Dordrecht. The Netherlands. USA (2003).

[15] Chandra, N.R. Prabu, M.M. Suguna, K. Vijayan, M.:Structural similarity and functional diversity in proteins containing the legume lectin fold. Protein Engineering. pp. 857–866, Oxford Jurnals (2001).

[16] Wright, L.M. Van Damme, E. J. M. Barre, A. et al.: Isolation, characterization, molecular cloning and molecular modelling of two lectins of different specificities from bluebell (Scilla campanulata) bulbs. Biochemical Journal. pp. 299–308 (1999).

[17] Hamelryck, T. W. Loris, R. Bouckaert, J. Wyns, L.:Structural features of the legume lectins. Trends in Glycoscience and Glycotechnology. pp. 349–360 Belgium (1998).

[18] psipred : index.UCL(1999-2008) http://bioinf.cs.ucl.ac.uk/psipred/

[19] Chung, C. C. and Jen, L. C. : LIBSVM: a library for support vector machines. ACM Transactions on Intelligent Systems and Technology, vol. 2(27) 2011.

A PSO based Smart Unit Commitment strategy for Power Systems Including Solar Energy

Ravikanth Reddy Gaddam[1], Amit Jain[2], Lingamurthy Belede[3]

[1] Research student with Power Systems Research Center, IIIT Hyderabad, India, 500032

[2] Lead consultant in utilities at Infotech Enterprises Ltd., Hyderabad, India, 500081

[3] Research student with Power Systems Research Center, IIIT Hyderabad, India, 500032

{ravikanth.gaddam@research.iiit.ac.in; amit.jain@infotech-enterprises.com; lingamurthy.b@research.iiit.ac.in}

Abstract. Today's large scale utilization, faster depletion of fossil fuels and energy security encouraged the world towards renewable sources and smart power systems. The problem of Unit Commitment (UC) has in itself become increasingly complex with understandably growing system sizes and involvement of imperative reliability measures. Given the scenario, integration of non-conventional energy sources into the system exhibits challenges in both economic and secure system operations. The work here presents a technique for the secure commitment of thermal generating units in a power system integrated with a solar powered plant. A solar thermal plant (STP) is modeled employing concentrated solar power (CSP) technology of parabolic trough type collectors. Hourly output expected from STP is calculated based on forecasted solar insolation levels. The UC solution of conventional thermal units accounts for the intermittent nature of power output from STP present in the same grid. The problem is solved in two stages with the primary one attending to UC and the sub problem catering to the optimal dispatch of load among the units committed. The evolutionary technique of Particle Swarm Optimization (PSO) is used in solving the main UC problem. To account for the optimal and secure system operation system, the subroutine of optimal power flow (OPF) is run every interval (hour) of the UC planning period. The optimal dispatch cost obtained from this execution is submitted as a fitness value or criterion to the PSO routine. The effectiveness of the proposed technique is demonstrated on the standard IEEE 30 bus system.

Keywords: Optimal Power Flow, Parabolic Trough Collector, Particle Swarm Optimization, Solar Thermal Plant, Unit Commitment.

J. C. Bansal et al. (eds.), *Proceedings of Seventh International Conference on Bio-Inspired Computing: Theories and Applications (BIC-TA 2012)*, Advances in Intelligent Systems and Computing 201, DOI: 10.1007/978-81-322-1038-2_45, © Springer India 2013

1 Introduction

Economic operation of power system is very important to return profit on the capital invested and to subside apart of investment itself through proper planning. Power system is subjected to independently varying loads and in turn forces the operating utilities to plan for generation-demand balance optimally. Thus, utilities have to plan for the units that are to be run and how much each should generate in catering the load demand. Problem with committing enough number of units and leave them online to meet the varying demand is one of the economics [1]. In view of this, a UC problem decides which among the available units are to be committed and the problem of Economic Dispatch (ED) determines optimal sharing of load among the committed units. Utilities are more interested in the problem of UC because in a typical power system there are a variety of units available for power generation, and each has its own performance characteristics. The obvious choice is to first turn on the more efficient and cheaper units as demand increases and to turn off the least efficient and costly units as demand decreases. Owing to many system operational constraints the UC problem takes the form of an optimization problem with non-linear as well as discontinuous constraints like minimum generator up-down times. The sub problem to be solved for optimal dispatch from each unit at minimum production cost should also account for the system operational constraints.

The fast depletion of fossil fuels and increased awareness towards environment in recent times however has led to the employment of renewable energy like solar and wind to a significant level in modern intelligent grids. Since the time of installation of solar energy generating systems in California [2], solar thermal plants (STP) using parabolic trough collectors have proven their potential in harnessing solar power in large scale. But the intermittent nature of solar power availability is always a challenge to be accounted for. This made integration of renewable power sources in to conventional gird a tedious task.

In literature the UC problem is addressed mainly by two techniques called the deterministic and stochastic evolutionary techniques. Deterministic are the numerical optimization techniques such as priority list [3], dynamic programming [4], integer programming [5], branch and bound [6], Lagrangian relaxation [7]. The drawbacks with these techniques are computational burden with increase in system size and difficulties in numerical convergence.

The other class called stochastic evolutionary techniques is known to handle complicated constraints assuring good quality of solution. They are Artificial Neural Networks [8], Genetic Algorithms (GA) [9], Simulated Annealing [10], Ant Colony optimization [11] and Particle swarm optimization (PSO) [12]. All these are nature and biologically inspired techniques and share many similarities in behavior. The difference among them lies in problem representation, solution updation and adjustments they adopt in reaching optimal solutions. PSO stands apart from the other evolutionary techniques in the way it traverses the potential solutions through hyperspace, accelerating towards "better" solution, while others operate directly on potential solutions which are represented as locations in hyperspace [13]. It is simple to implement besides being robust and efficient. A PSO technique that uses binary variables was proposed in [14]. The technique is more suitable in this context due to the binary representation of ON/OFF status of generators and has thus been adopted here.

Renewable energy sources are being integrated more earnestly into modern power system today due to their positive effect on environment besides being economical in the long run and several methodologies have been proposed towards the same. More significantly, solar and wind generation have played their part. Reference [15] proposed a distributed stochastic scheme for smart grids with intermittent renewable energy sources. A solar thermal plant is always a more suitable alternative in geographic locations bound to have adequate solar insolation levels and hours of sun shine. Reference [16] proposes a method to optimize the daily operation of a small autonomous system fed by a mix of diesel generators, wind turbine generators and photovoltaic panels. A GA operated PSO technique is presented in [17] for commitment of thermal units with solar and wind energy systems. GA operators are applied within the higher potential solutions to generate new population for PSO.

The technique proposed here aims at finding a secure and optimal UC schedule of conventional thermal units by integrating an STP of appreciable capacity into the grid. An OPF which considers all system operational violations is implemented for every hour commitment as sub-problem in the place of usually employed economic dispatch problem. The OPF here also accounts for the hourly intermittent output of STP.

2 Unit Commitment Formulation

Unit Commitment has been a standard problem investigated given the uneven pattern of fuel consumption by various generators. It is a cost minimization problem that determines the ON/OFF status of all units during every hour of the planning period T. In generation of power, fuel cost accounts for the major part of capital requirements. On the other hand there is a cost incurred in starting up and shut down of units called the transitional cost which is small compared to fuel cost. The objective function of UC problem is as shown:

$$\text{Total Cost} = \sum_{t=1}^{T} \sum_{i=1}^{NG} \left[U_i^t F(Pg_i) + U_i^t \left(1 - U_i^{t-1}\right) STC_i + U_i^{t-1} \left(1 - U_i^t\right) STD_i \right] \tag{1}$$

Where NG is number of units in the system, U_i^t is the ON/OFF (1-On, 0-Off) status of ith unit during t^{th} hour. $F(Pg_i)$ is the fuel cost of unit expressed as a second order polynomial in terms of unit's power generation and STC_i is the startup cost as given in (2) and (3) respectively. STD_i is the shunt down cost.

$$F(Pg_i) = a_i Pg_i^2 + b_i Pg_i + c_i \tag{2}$$

$$STC_i = \begin{cases} \text{Hot start cost} & \text{if } T_i^{down} \leq T_i^{off} \leq \left(T_i^{cold} + T_i^{down}\right) \\ \text{Cold start cost} & \text{if } T_i^{off} > \left(T_i^{cold} + T_i^{down}\right) \end{cases} \tag{3}$$

Where a_i, b_i, c_i are cost coefficients, T_i^{up}, T_i^{down} are minimum up and down times, T_i^{cold} is cold start hour, T_i^{on}, T_i^{off} are continuously on and off timings in hours, Pg_i is the active power generation of ith unit.

In determination of UC schedule, generator technical constraints and system power balance set certain limitations. The technical constraints that govern generator status and behavior in UC are the unit minimum start-up and shut down timings, maximum and minimum generation limits, the possible instantaneous change in power level called the ramping rates.

Therefore, above given UC objective function should be optimized, while satisfying the unit and system operational constraints that are listed below during every hour of the UC time horizon

Power balance
$$\sum_{i=1}^{ng} Pg_i^t = P_d^t \tag{4}$$

Minimum capacity committed
$$\sum_{i=1}^{ng} Pmax_i \geq P_d^t + P_r^t \tag{5}$$

Minimum up-down time
$$T_i^{on} \geq T_i^{up}$$
$$T_i^{off} \geq T_i^{down} \tag{6}$$

Generation limits
$$Pmin_i \leq Pg_i^t \leq Pmax_i \tag{7}$$

Ramp rate limits
$$Pg_i^{t-1} \leq Pg_i^t \leq Pg_i^{t-1} + Rup_i$$
$$Pg_i^{t-1} \geq Pg_i^t \geq Pg_i^{t-1} - Rdn_i \tag{8}$$

Where ng is the number of committed units, P_d^t, P_r^t represents active power demand and reserve during hour t respectively. $Pmin_i$, $Pmax_i$ are the ith unit minimum and maximum active power generation limits, Rup_i, Rdn_i represents i^{th} unit generation ramp up and ramp down constants.

In the proposed methodology, UC is solved at two levels. At the outer level, PSO is used to search the minimum production cost among feasible commitment schedules which satisfy the constraints listed above. At the second level, a complete optimal power flow is run for every hour's commitment to determine the optimum dispatch and provide production cost to the outer level.

3 Solar Thermal Plant (STP)

In a typical solar thermal plant the solar energy is transferred to a thermal fluid at an outlet temperature that is high enough to feed a heat engine or a turbine which produces electricity. In CSP there exist technologies like parabolic trough collectors, linear Fresnel reflector systems, power towers or central receiver systems, and parabolic dish/engine systems. Among these the parabolic trough collector technology is mature and had demonstrated its potential in conversion of solar thermal energy into electricity [2]. A parabolic trough collector consists of a reflector which focuses the incident solar beam radiation on to a receiver tube located exactly at the focal line. The receiver tube carries working fluid which collects the thermal energy. Area of receiver is very small compared to that of reflector resulting in high concentration ratio and ultimately heating up the working fluid to high temperatures. Once the working fluid is heated up, the extraction of heat energy from it and converting to electrical output is a standard procedure.

The study models a solar thermal plant using parabolic trough collector technology. The data of a single collector module considered is as follows. i) Aperture of the collector 5.7m, ii) Length of the collector 99m, iii) Area of the collector 545m^2, iv) Optical efficiency 0.8. A suitable figure of efficiency is chosen for Rankine cycle conversion, which extracts heat from working fluid and converts water to steam.

The power output from each module purely depends on the level of solar insolation. For a suitable capacity STP the number of collectors and field area required are calculated based on maximum solar insolation availability and metrological conditions of the selected geographic region. The below expressions (9) and (10) gives the idea about the amount of heat energy collected q^t and electrical power generated P_s^t during hour t.

$$q^t = I^t A \eta_{opt} \eta_{th} \tag{9}$$

$$P_s^t = q^t \eta_R \eta_{TG} \tag{10}$$

Where I^t is the solar insolation during hour t, A is the collector field area, η_{opt} is the optical efficiency, η_{th} is solar field thermal efficiency and η_R, η_{TG} are rankine cycle and turbine generator set efficiencies respectively.

In order to make the model more practical and accurate, the solar insolation levels are taken from [18] for the city of Hyderabad in India. The average value of solar to electrical efficiency of large solar thermal systems with parabolic trough technology is about 17% [19].

4 Particle Swarm Optimization

Particle Swarm Optimization is proposed by Kennedy and Eberhart in 1995 [13]. It is a nature inspired evolutionary search technique where in the psychological behavior of birds flock is simulated towards application for solving complex optimization problems.

PSO uses the fact of information sharing between birds to attain better position in a flock while they travel. Each individual in the swarm has the idea of its position and that of its neighbor which assists in collision free movement towards better positions. This behavior of birds' movement in a flock is imitated in PSO algorithm. It is an iterative technique initialized with a set of solutions called initial population in the search space. Each member of the population called as *particle* is a valid solution for the problem intended. The method assigns a velocity for every particle in population with which it is free to move in the multidimensional search space. Each particle is influenced by its own behavior and that of its neighbor's. This is possible because over the course of travel a particle remembers its own best position called '*pbest*' and the best among all individual best positions in the population called global best as '*gbest*'. The further movement of a particle in the search space is guided by its velocity which is dependent on the deviation of current position from its individual best and the overall global best. The below given expression clearly illustrates the movement of particle x in every iterative step using the derived velocity v. Where k is the particle, j is the iteration, w is a weighting factor, c_1 and c_2 are constants, $rand$ () is a random function which assumes uniformly distributed values between 0 and 1.

$$v_k^{j+1} = w*v_k^j + c_1*\text{rand}(\ )*\left(pbest - x_k^j\right) + c_2*\text{rand}(\ )\left(gbest - x_k^j\right) \tag{11}$$

$$x_k^{j+1} = x_k^j + v_k^{j+1}$$

Velocity term is the factor that is responsible for the movement of particle towards optimal position. Particle in the search space is moved to a new position by adding the derived velocity term to the current position as shown in (11). Over the iterative process the deviations between particle's current position from *pbest* and *gbest* decreases and gives raise to some minimum velocity. This implies the convergence of PSO and the *gbest* particle corresponds to an optimal solution.

A suitable strategy for solving UC problem that pose discreteness is proposed by Kennedy and Eberhart [14]. At first the velocities are found and a function called '*Sigmoid*' is introduced to map these real valued velocities to [0, 1]. Equation (13) shows the updation of PSO particle according to the sigmoid function evaluation.

$$Sig = \frac{1}{1 + e^{\left(-v_k^{j+1}\right)}} \tag{12}$$

$$x_k^{j+1} = \begin{cases} 0, & Sig \le rand() \\ 1, & Sig > rand() \end{cases} \tag{13}$$

5 Implementation

ON and OFF status of a unit is represented with binary 1 and a binary 0 respectively [12]. PSO solution starts with an initial population of desired size. Particles in the population are valid solutions making up various unit commitment schedules for the time horizon of T hours. Thus, particle is a matrix of dimension (T X NG) with 0's and 1's as its entries with each row making up the commitment of NG units for an hour. If K represents population size then the whole population is represented as a (K X T X NG) matrix. For the updation of particle iteratively, a velocity matrix v of the same size is initialized randomly within feasible range with each (T X NG) matrix pertaining to one particle. Fitness value of each particle is obtained by the cost evaluated using (1). The particles corresponding to *pbest* and *gbest* are determined based on these costs.

The proposed methodology is illustrated on IEEE 30 bus system. In order to ensure that the main layout of standard test is not disturbed with the addition of STP, the work done here models one of the conventional thermal units in the system as a solar thermal unit for the sake of simulation. The methodology would work equally well with STP added as an extra unit or any existing conventional generator in the system being assigned as the solar unit irrespective of size. However, the main concern with the option of assigning a conventional unit as solar thermal unit would remain the ability of remaining generators to supply the load demand throughout the time horizon. This is owing to the intermittent output from the STP over the considered time frame. During the absence or low output periods of solar generation, load demand has to be catered by the other units without violating any operational constraints. Thus, the choice of generator bus on which the STP is modeled is required to be judicious. Without any loss of generality, the lowest capacity unit is chosen for the same. In the standard IEEE 30 bus system, unit on bus 11 is the lowest capacity of 30MW and has been assigned as STP. The capacity of STP modeled is same as that of the conventional thermal unit.

The UC problem with the integration of STP is approached in four stages. At the first stage the required amount of generation to be committed is found with respect the possible STP output, losses in the system and reserve level requirements. The second stage proceeds with the creation of initial population for PSO implementation with consideration of unit dynamic behavior like ramp rates. The third stage finds fitness value of each particle with OPF executions. Fourth stage is the updation of particles according to PSO procedure until convergence.

STAGE 1: Forecasted load demand given is not the exact amount of power requirement that needs to be committed. Losses in the system, spinning reserve requirements and STP output should also be accounted for generation requirement.

STP output: Depending upon the solar insolation level, the hourly output expected from STP model is calculated. It has been observed that the output from STP is in usable quantities from 09.00 to 17.00 hours. During early hours and evenings, the STP output is too low to run the unit and is taken as zero.

Losses: In order to commit the units to cater primarily to system demand and in addition, the system losses, a rough loss estimate is determined by performing ED employing B coefficients [20]. Since ED is run only to assure that a major aspect like system loss is never ignored, it is run assuming all the generators are available for dispatch every hour of the commitment period. This is seemingly better than assuming losses to be an arbitrary percentage of system's load.

Spinning reserve: For secure operation of system under abnormal conditions a reserve is considered to be 10% of system's load.

Since one of the generators is treated as a STP, its hourly fixed output power is subtracted from the total MW demand. In this regard the modified or new net demand to be committed is

$$P_{d(new)}^t = P_d^t + P_L^t + P_r^t - P_S^t \tag{14}$$

P_L^t is the estimated rough loss during hour t. The UC schedule is determined every hour only for the remaining (NG-1) units to cater to this new net demand.

STAGE 2: An initial population of desired size, K, with each particle of dimension $T X (NG$-1) is generated. Each of the $T X (NG$-1) elements of the particle needs a velocity to update the status of a unit, thus requiring a randomly initiated velocity matrix of the same size. This initial population needs to at least satisfy the primary constraints (6), (8) and (14). Ramp limits are one of the most practical unit constraints which are imposed to keep the change in generation between consecutive hours within feasible limits. This constraint is modeled by modifying the feasible minimum and maximum limits of units every hour accordingly as given in (16) [21]. All such feasible particles generated, each of which corresponds to a T-

hr schedule, are submitted for optimization.

$$\sum_{i=1}^{ng} Pmax_i \geq P_{d(new)}^t \tag{15}$$

$$Pmin_i = \max\left\{ Pmin_i, Pg_i^{t-1} - Rdn_i \right\} \tag{16}$$

$$Pmax_i = \min\left\{ Pmax_i, Pg_i^{t-1} + Rup_i \right\}$$

STAGE 3: Fitness value of a particle is obtained when hourly transition costs are added to the costs of generation over the entire commitment period T. OPF is run as a sub problem to determine the optimum generation cost for every hour's commitment. The STP is modeled as a generator bus in the OPF process. However, it is not considered to be voltage controlled for the evident reason that the capacity of the STP varies with every hour and thus the reactive power required to support the constant voltage may not be sufficient. The OPF model implemented is given below explaining various constraints imposed and state variables involved. System losses also get updated to actual values in OPF execution from rough estimates made in stage 1.

The active power injection P must be matched at every bus while the reactive power injection Q is matched only for load buses because of the unavailability of information regarding the reactive generation Qg. However, Qg is computed for the generator buses and checked for limit violations. The power flow Pl on each line is also constrained to be within the specified limit.

Equality Constraints: Matching of Bus active power injection $Pinj_i$ at all N buses and reactive power injection $Qinj_i$ at all Nb load buses.

$$Pg_i - Pload_i - Pinj_i = 0 \qquad i = 1 \text{ to } N \tag{17}$$

$$Qg_i - Qload_i - Qinj_i = 0 \qquad i = 1 \text{ to } Nb$$

Inequality Constraints: Reactive power generation limits on committed generators ng, flow limits on all Nl transmission lines.

$$Qmin_i \leq Qg_i \leq Qmax_i \qquad\qquad i = 1 \text{ to } ng \tag{18}$$

$$Plmin_i \leq Pl_i \leq Plmax_i \qquad\qquad i = 1 \text{ to } Nl$$

State variable vector 'X' and its bounds: The active power generation is a variable for all generator buses except the STP. Voltage magnitudes V are variable for all load buses and the STP, load angle is a variable for all buses except slack. Tap and • are state variables for those lines having transformers with tap and phase shifter. Each state variable is constrained to be within its upper and lower bounds. The augmented state vector is given below.

$$X = \left[\left[x_{Pg} \right]^T \left[x_V \right]^T \left[x_\delta \right]^T \left[x_{tap} \right]^T \left[x_\phi \right]^T \right]^T \tag{19}$$

$$X_{min} \leq X \leq X_{max}$$

STAGE 4: Each particle's commitment with least fitness value over the iterations is remembered as its *pbest*. The best among the personal best of all particles is stored as *gbest*. The entire population is updated by using (11), (12) and (13). Velocity and inertia weight are important parametric values in PSO which govern the particle movement. Instead of allowing the particle to fly in a large hyperspace, the maximum and minimum values of velocity and inertia weight are fixed between -0.5 and 0.5, 0.8 and 1.0 respectively.

The convergence criterion usually followed is a limit on maximum iterations. A more logical criterion is adopted here. The iterations continue till no further improvement in the *gbest* is evident over a given number of iterations. This will be a good trade-off between speed and optimal requirements. Stages 3 and 4 are repeated till the convergence criterion is met.

6 Simulation Studies

Table 1. Unit Data

Bus No.	Pmax (MW)	Pmin (MW)	Cost Coefficients			Min. ON (Hr)	Min. OFF (Hr)	Cold start (Hr)	Start up costs ($)	
			a	b	c				Hot	Cold
1	200	50	.00375	2.00	0	1	1	2	70	176
2	80	20	.01750	1.75	0	2	2	1	74	187
5	50	15	.06250	1.00	0	1	1	1	50	113
8	35	10	.00834	3.25	0	1	2	1	110	267
11	30	10	.02500	3.00	0	2	1	1	72	180
13	40	12	.02500	3.00	0	1	1	1	40	113

Table 1 shows the unit data of IEEE 30 bus system [22] [23]. Unit at bus number 11 in the system is considered to be STP given that its capacity is lowest. Hourly output of STP is determined beforehand from the model using forecasted solar insolation and is shown in Table 2. Solar insolation is considered for a given summer day [18]. The cost coefficients viz. a, b, c for STP unit are made zero.

Table 2. UC Schedules with 24 hr load and STP output (Base 100MVA)

Hour	Load MW	UC without STP						UC with STP				11 (STP)		13
		1	2	5	8	11	13	1	2	5	8	Status	MW Output (p.u.)	
1	1.66	1	1	0	0	0	1	1	0	0	1	0	0	0
2	1.96	1	1	1	0	0	0	1	0	1	1	0	0	0
3	2.29	1	1	0	0	0	0	1	1	1	0	0	0	1
4	2.67	1	1	0	1	0	1	1	1	0	0	0	0	1
5	2.834	1	1	1	1	0	0	1	1	0	1	0	0	1

6	2.72	1	1	1	1	1	1	1	1	1	0	0	0	0
7	2.46	1	1	0	1	1	0	1	0	1	0	0	0	1
8	2.13	1	1	0	1	1	1	1	0	0	1	0	0	1
9	1.92	1	1	0	0	0	1	1	1	0	1	1	0.1385	1
10	1.62	1	0	1	0	1	1	1	1	1	1	1	0.1779	0
11	1.74	1	0	0	1	1	1	1	1	1	0	1	0.2163	1
12	1.60	1	0	0	1	1	0	1	0	0	0	1	0.2393	1
13	1.70	1	1	1	0	1	1	1	0	1	1	1	0.2534	1
14	1.85	1	1	0	0	0	1	1	0	0	0	1	0.2595	0
15	2.08	1	0	1	1	1	0	1	1	0	0	1	0.2632	0
16	2.32	1	0	1	0	1	1	1	1	1	1	1	0.2327	1
17	2.46	1	1	1	0	1	0	1	0	1	1	1	0.1000	1
18	2.41	1	1	1	1	1	1	1	0	0	1	0	0	1
19	2.36	1	0	1	1	1	1	1	1	1	1	0	0	0
20	2.25	1	0	1	0	0	1	1	1	1	0	0	0	0
21	2.04	1	1	1	0	0	0	1	0	1	0	0	0	0
22	1.82	1	1	1	0	0	1	1	0	1	0	0	0	1
23	1.61	1	0	1	0	0	1	1	0	0	1	0	0	1
24	1.31	1	0	1	0	1	1	1	1	0	0	0	0	1
Total Cost ($)		17860						15899						

Simulations are carried out with a PSO population of size 50. Thus the size of the whole population without STP is (50 X 24 X 5) and that with STP is (50 X 24 X 4).

Table 2 shows UC schedules for the system without and with STP along with costs incurred. The maximum and minimum limits on voltage magnitudes at non regulated buses are assumed to be 0.95 and 1.1 per unit respectively. The commitment schedule obtained is checked for secure operation within limits using a full-fledged OPF. The STP unit at bus 11 is committed between 09.00 and 17.00 hours during which solar insolation is high and off during remaining hours. The remaining units at buses 1, 2, 5, 8 and 13 are subjected to commitment procedure accordingly. The PSO iterative process of updation is continued until there is no further improvement in *gbest* for 5 consecutive iterations. This convergence criterion is more logical as compared to a limit on maximum iterations and enables the swarm to traverse the search space well and reduces the chances of settling to a local minimum.

5 Conclusion

The UC method proposed is suitable to modern day power systems. It is an attempt towards integration of renewable energy sources in finding UC schedules which may assists future smart systems. To have the practical effect of renewable source on conventional system a STP is modeled with a CSP technology and integrated into the grid. The UC schedule obtained is operationally secure and economical for solar integrated power systems with executions of full-fledged OPF for every hour unlike conventional ED.

References

[1]   A. J. Wood, B. F. Wollenberg: Power Generation, Operation and Control, second edition, New York: Wiley, 1996

[2]   Solar Electric Generating Stations (SEGS): IEEE, Power Engineering Review, vol.9, no.8, pp.4-8, Aug. 1989.doi: 10.1109/MPER.1989.4310850

[3]   T. Senjyu, K. Shimabukuro, K. Uezato and T. Funabashi.: A fast technique for unit commitment problem by extended priority list, IEEE/PES ,T & D Conference and Exhibition 2002, vol.1, pp. 244- 249, 6-10 Oct. 2002.doi: 10.1109/TDC.2002.1178301

[4]   Lowery.P.G.: Generating Unit Commitment by Dynamic Programming, IEEE Transactions on Power Apparatus and Systems, vol.PAS-85, no.5, pp.422-426, May 1966. doi: 10.1109/TPAS.1966.291679

[5]   T. S. Dillon, K. W. Edwin, H.D. Kochs, R.J. Taud.: Integer programming approach to the problem of optimal unit commitment with probabilistic reserve determination, IEEE Transactions on Power Apparatus and Systems, vol. 6, pp. 2154-2166, Nov 1978. doi: 10.1109/TPAS.1978.354719

[6]   A. I. Cohen, M. Yoshimura.: A branch-and-bound algorithm for unit commitment, IEEE Transactions on Power Apparatus and Systems, vol. PAS-102, no. 2, pp. 444–451, Feb 1983. doi: 10.1109/TPAS.1983.317714

[7]   F. Zhuang, F. D. Galiana.: Towards a more rigorous and practical unit commitment by Lagrangian relaxation, IEEE Transactions on Power Systems, vol. 3, no. 2, pp. 763–773, May 1988. doi: 10.1109/59.192933

[8]   H. Sasaki, M. Watanabe, J. Kubokawa, N. Yorina, R.Yokoyama.: A solution method of unit commitment by artificial neural networks, IEEE Transactions on Power Systems, vol.7, no.3, pp.971-981, Aug 1992. doi: 10.1109/59.207310

[9]   D. Dasgupta, D. R. McGregor.: Thermal unit commitment using genetic algorithms, IEEE Proceedings on Generation, Transmission and Distribution, vol. 141, pp. 459–465, Sept. 1994. doi: 10.1049/ip-gtd:19941221

[10] A. H. Mantawy, Y. L. Abdel-Magid, S. Z. Selim.: A simulated annealing algorithm for unit commitment, IEEE Transactions on Power Systems, vol.13, no.1, pp.197-204, Feb 1998. doi: 10.1109/59.651636

[11] N. S. Sisworahardjo, A. A. El-Keib.: Unit commitment using the ant colony search algorithm, Large Engineering Systems Conference on Power Engineering, LESCOPE 02, pp. 2- 6, 2002. doi: 10.1109/LESCPE.2002.1020658

[12] Zwe-Lee Gaing.: Discrete particle swarm optimization algorithm for unit commitment, IEEE Power Engineering Society General Meeting, vol.1, 13-17 July 2003. doi: 10.1109/PES.2003.1267212

[13] J. Kennedy, R. Eberhart.: Particle swarm optimization, IEEE International Conf. on Neural Networks, vol.4, pp.1942-1948 Dec 1995. doi: 10.1109/ICNN.1995.488968

[14] J. Kennedy, R. C. Eberhart.: A discrete binary version of the particle swarm algorithm, IEEE International Conference on Computational Cybernetics and Simulation, vol.5, pp.4104-4108, 12-15 Oct 1997. doi: 10.1109/ICSMC.1997.637339

[15] Shengrong Bu, F. Richard Yu and Peter X. Liu.: Stochastic Unit Commitment in Smart Grid Communications, IEEE INFOCOM 2011.

[16] A. G. Bakirtzis, P. S. Dokopoulos.: Short term generation scheduling in a small autonomous system with unconventional energy sources, IEEE Transactions on Power Systems, vol.3, no.3, pp.1230-1236, Aug 1988. doi: 10.1109/59.14586

[17] T. Senjyu, S. Chakraborty, A. Y. Saber, H. Toyama, A. Yona, T. Funabashi.: Thermal unit commitment strategy with solar and wind energy systems using genetic algorithm operated particle swarm optimization, IEEE International Power and Energy Conference, 2008, pp.866-871, 1-3 Dec. 2008. doi: 10.1109/PECON.2008.4762597

[18] Ajit P. Tyagi.: Solar Radiant Energy over India, Indian Meteorological Department, Ministry of Earth Sciences, New Delhi, 2009.

[19] Frank Kreith, D. Yogi Goswami.: Hand book of Energy Efficiency and Renewable Energy, New York: CRC press, 2007.

[20] D. P. Kothari, J .S. Dhillon.: Power System Optimization, Eastern Economy Edition, New Delhi: Prentice Hall of India, 2004.

[21] D. Simopoulos, S. Kavatza.: Consideration of ramp rate constraints in unit commitment using simulated annealing, IEEE Power Tech, Russia, pp.1-7, 27-30 June 2005. doi: 10.1109/PTC.2005.4524359

[22] Raglend. I.J, Padhy N.P: Solutions to practical unit commitment problems with operational, power flow and environmental constraints, IEEE Power Engineering Society General Meeting, 2006. doi: 10.1109/PES.2006.1708996

[23] Lee K.Y, Park Y.M, Ortiz J.L: A United Approach to Optimal Real and Reactive Power Dispatch, IEEE Transactions on Power Apparatus and Systems, vol.PAS-104, no.5, pp.1147-1153, May 1985. doi: 10.1109/TPAS.1985.323466

A User-oriented Content Based Recommender System Based on Reclusive Methods and Interactive Genetic Algorithm

Vibhor Kant and Kamal K. Bharadwaj

School of Computer and Systems Sciences, Jawaharlal Nehru University (J.N.U.), New Delhi, India

{vibhor.kant;kbharadwaj}@gmail.com

Abstract. Due to the unprecedented proliferation of textual contents such as blog articles, news, research papers and other things like movies, books and restaurants etc. on the web in recent years, the development of content-based recommender systems (CB-RSs) has become an important research area which aims to provide personalized suggestions about items to users while interacting with the large spaces on the web using items' contents and users' preferences. However, item representation and responds to changing user preferences are still major concerns. In our work, we have employed Reclusive Methods (RMs) to deal with the uncertainty associated with item representation and Interactive Genetic Algorithm (IGA) is used to adapt the system with changing user preferences. First of all, a fuzzy theoretic approach to content based recommender system (FCB-RS) is designed to generate initial population for IGA using reclusive methods. Second, K-means algorithm is employed for clustering the items in order to handle the time complexity of IGA algorithm. Thereafter, a user-oriented content based recommender system (UCB-RS) is developed by incorporating IGA into FCB-RS through user evaluation. Experimental results show that the proposed system (UCB-RS) outperforms both the classical CB-RS and the FCB-RS.

Keywords: Recommender Systems; Content-based Filtering; Reclusive Methods; Fuzzy Theoretic Approach; Interactive Genetic Algorithm

1 Introduction

During the last decade, the rapid advances of Internet and Web technologies have produced a large number of items in various areas such as entertainment, education, medicine and e-commerce etc. As a result, it is getting more difficult for searching suitable items or products to users related to their preferences among enormous amount of items available in online environments [9]. Recommender

J. C. Bansal et al. (eds.), *Proceedings of Seventh International Conference on Bio-Inspired Computing: Theories and Applications (BIC-TA 2012)*, Advances in Intelligent Systems and Computing 201, DOI: 10.1007/978-81-322-1038-2_46, © Springer India 2013

System (RS) has been established as an important information exploration paradigm that retrieves relevant items to users based on their preferences expressed implicitly or explicitly while interacting with the large spaces on online environments to cope with the information and product overload [1, 8]. Two different types of filtering techniques, i.e. collaborative and content based filtering (CBF), are usually employed in the area of RSs. In collaborative filtering (CF), items are recommended to target users based on only similar users while content based recommender system (CB-RS) recommends items to target users based on items' content and users' past behaviors [2]. Three main steps involved in CB-RS are as follows:

- *Feature Extraction:* First of all, it extracts sets of features from items which are assessed by the users.
- *Similarity Computation:* After the first step, the CB-RS computes the similarity between items in users' profile and the other remaining items.
- *Recommendation Generation:* Finally, it recommends the most similar items to the target user

The CB-RSs focus mainly on item representation and modeling of user preferences. But features of items and user preferences are highly subjective in nature [16]. Therefore representation of items and user modeling are major concerns in CB-RSs. Furthermore, CB-RS is limited by only recommending those items which are similar to the specific items that the user rated highly in the past and it is not prompt to immediate changes in user preferences. In order to compute effective similarity between items, Yager [15] proposed reclusive methods for CB-RSs in order to handle the uncertainty associated with item features and user preferences by providing a framework for the representation of item features and user preferences through fuzzy logic.

Most of CB-RSs are based on the similarity computation step to find out a specific item or a group of items that are similar to target user and similarity computation is fully dependent on the better representation of item features and user preferences. We consider that the incorporation of fuzzy logic approaches into CB-RS system can provide the better representation of items by handling the uncertainty associated with items. In this paper, we propose a user oriented CB-RS that uses not only interactive genetic algorithm (IGA) for providing the ability of responding to immediate changes in user preferences to infer which item in the item repository would be of the most interest to the target user according to the current situation but also employs reclusive approach for content based filtering in order to handle the uncertainties associated with item features.

Interactive genetic algorithm (IGA) is an optimization technique [14] that adopts genetic algorithm based on subjective human evaluation to trace the user's uncertain and time varying preferences. It is simply a genetic algorithm (GA) whose fitness function is replaced by a user. IGA has been successfully employed in several real world problems in various diverse areas such as pattern recognition, machine learning, RS [2, 10] and image retrieval [11] etc under steady state. However, it may run a long while to find out a very near optimal solution under the given condition. Time complexity is not a major concern of IGA as most real world applications of IGA operate in steady state. As soon as more interactions happen, it would gradually converge to the best optimal solution [10]. Consequently, IGA seems to be more suitable to the design of content based recommender system for providing the ability of responding to user's time varying preferences. The main contribution of our proposed work is three fold:

- Design of fuzzy theoretic approach to CB-RS (FCB-RS) to deal with the uncertainty associated with item representation and to provide initial population for Interactive Genetic Algorithm (IGA)
- Clustering of items using k-means algorithm in order to achieve better computational efficiency of our proposed system
- Development of user-oriented CB-RS (UCB-RS) by incorporating IGA into FCB-RS through user subjective evaluation

The rest of the paper is structured as follows: Section 2 provides the background related to our work. In Section 3, the proposed scheme is presented. Computational experiments and results are given in Section 4. Finally, in the last Section we conclude our work with some future research directions.

2 Background

In this section, we describe some background on RSs in various domains using different recommendation techniques, reclusive methods, K-means and interactive genetic algorithm.

2.1 Recommender System

With the rapid advances of information and Web technologies in recent years, it is very difficult task for online users to find suitable items or information while interacting with large corpus on online environments. Because of this fast advances in web technologies, the elaboration of intelligent automatic filtering system becomes a key issue. RS is one of the promising tools to assist the users in suggesting appropriate items dealing with the information and product overload in web base services. RS recommends everything from CDs, news, websites and movies to more complex items such as digital cameras, e-governance and finance services [1].The major concern of an RS is how to provide items tailored with user preferences.

The two basic entities which appear in any RSs are the user and the item. On the basis of these two entities, two information filtering algorithms are widely used one is collaborative filtering and another one is content based filtering [13]. In CF, the user will be recommended items people with similar preferences in the past while CBF recommends items similar to the ones liked in the past [2].

2.2 Reclusive Methods

Yager, 2003[15] provided a fuzzy logic based methodology for designing recommender systems called reclusive methods. These approaches utilize an individual's preferences for generating recommendations as opposed to CF based method that uses preferences of other similar users. In this methodology, he offered a way of representing items, user preferences and subsequent construction of justification as well as recommendation rules. However, he did not conduct an empirical study to support or refute the effectiveness of using fuzzy modeling. Consequently, Zenebe and Norcio [16] provided an empirical assessment of reclusive methods in the domain of movies.

2.3 K-means Algorithm

Clustering is a widely implemented methodology in various application fields such as location selection, routing, scheduling, image processing, network partitioning, , data mining etc. It is used to group all objects into various mutually exclusive clusters in such a way, objects within same cluster are more similar to each other than objects belonging to different clusters. The simplest and most popular clustering algorithm is the K-means clustering algorithm [5].

In our proposed system, we apply data clustering by employing K-means algorithm for improving the computational efficiency in terms of accuracy and quality of recommendations. It starts by choosing k initial cluster centers, and then iteratively refines the given data set.

2.4 Interactive Genetic Algorithm

In the field of evolutionary computation, genetic algorithms (GAs) are robust, computational and stochastic search approaches inspired from the mechanics of natural evolution and genetic inheritance which start with an initial population of candidate solutions generated randomly or based on problem specific knowledge [12]. These candidate solutions of the search spaces are encoded as binary or real number strings called chromosomes [11, 14]. A new population is constructed from the preceding one through evolution, selection and mating procedures in each generation. Each candidate solution has a fitness value representing its closeness to an optimal solution. The solutions having high fitness value are selected, survive to the next generation and generate better offspring through genetic operators. The methods can discover, preserve, and propagate promising sub-solutions [6, 12]

Interactive genetic algorithm (IGA) [14], a branch of evolutionary computation differs from GA in the construction of fitness function. There is no role of predefined fitness function likewise GA; here fitness value of each possible solution is directly evaluated by users according to their own preferences. A user can interactively determine which members of the population will survive and automatically generate the next generation. Through repeated rounds, IGA enables unique evolved item that suits the user's preferences [10]. Our proposed system employs IGA to recognize user preferences and to respond in immediate changes in time varying user preferences.

3 Proposed User Oriented Content based Recommender System

Our proposed system is based on the interactive genetic algorithm (IGA) which aims to recognize user preferences. Our proposed system presents an effective way of representation of item features and user preferences. We start by presenting an innovative fuzzy theoretic approach to CB-RS [9] based on Reclusive methods to deal with the uncertainties associated with item features.

3.1 Fuzzy Theoretic Approach to CB-RS Based on Reclusive Methods (FCB-RS)

Content based approach provides recommendations by comparing representations of content contained in an item. But features of items are highly subjective, imprecise and vague in nature. In relation to items, the uncertainty is associated to the extent in which the items have some features [16]. For instance, to what extent does a movie have romantic content or is it highly romantic? To deal with such type of uncertainty, fuzzy theoretic approach is developed for designing CB-RS.

3.1.1 Item Representation

An item can be represented in the terms of their features. In MovieLens dataset, movies are represented in the terms of genres. If a genre is present in a movie then it is represented by 1 otherwise 0. But it is not a realistic situation as a movie does not contain equal amounts of different genres. Let $M = \{m_1, m_2, ... m_n\}$ be the set of items and f is a feature of an item that can take multiple values from a set $A = \{f_1, f_2, ... f_s\}$. The Gaussian membership function [16] of an item m_l to value f_j is described as follows:

$$\mu_{f_j}(m_l) = \frac{r_k}{2\sqrt{\rho * \gamma(r_k - 1)}} \tag{1}$$

where γ the number of values of A that is associated to an item m_l, r_k is the rank position of f_j and $\rho > 1$.

3.1.2 Similarity Computation between Items

Let items m_l and m_p are defined as $\left\{\left(f_j, \mu_{f_j}(m_l)\right) \mid j = 1,2, ... s\right\}$ and $\left\{\left(f_j, \mu_{f_j}(m_p)\right) \mid j = 1,2, ... s\right\}$
respectively. The similarity between these items as computed as

$$Sim(m_l, m_p) = \left(1 - \frac{\sum_{j=1}^{s}\left|\mu_{f_j}(m_l) - \mu_{f_j}(m_p)\right|}{s}\right) \tag{2}$$

3.1.3 Recommendation Score

In order to compute the recommendation score of an unseen item m to an active user a $(R_{a,m})$ through FCB-RS approach, the following formula is used:

$$R_{a,m} = \frac{\sum_{t \in L} Sim(m,t) * r_{a,t}}{\sum_{t \in L} |Sim(m,t)|} \tag{3}$$

where L is the set of liked items by user a. These items are computed on the basis of average rating of user a and $r_{a,t}$ is the rating of an item t given by user a.

3.2 Item Clustering

Once the item representation is completed, our proposed system generates a set of clusters for the given items by employing K-means algorithm to improve the computational efficiency of our scheme. This technique aims to partition the given items into k different clusters. It starts by choosing k initial cluster centers, and then iteratively refines the given data set.

Based on extracted features of an individual item, the system stores each cluster's information regarding a list of items within the same cluster. In addition, our system uses the stored cluster information to find similar items when making a recommendation through IGA search process.

3.3 Proposed Recommendation Framework

We design a user-oriented CB-RS based on IGA as shown in Fig. 1. Our system operates in four phases:

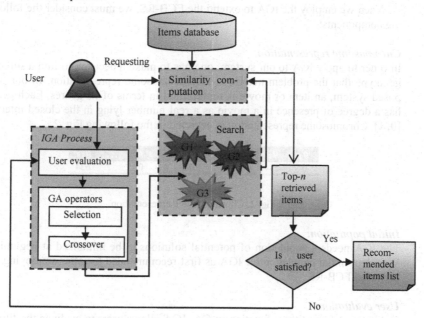

Fig. 1. System Flowchart of the Proposed Approach

- **Requesting:** The target user provides his past experiences on the items in the forms of ratings known as user preferences and requests for the recommendations.
- **Similarity Computation:** The system computes the similarity between the user's liked items and items in the databases through fuzzy similarity measures.
- **Recommendation:** the system presents a sequence of items ranked in the decreasing order of recommendation scores. As a result, user can find easily relevant items by getting the top ranked items first.

- *Incremental Search:* After providing some relevant items to the target user, the system provides an interactive mechanism via IGA in which user evaluates the top recommended items as more or less relevant to his preference and then the system updates its recommendation results by including other relevant items through genetic operators in the next generation of IGA. The search process is repeated until the user is satisfied with the result or predefined number of generations.

When we employ the IGA to extend the FCB-RS, we must consider the following components:

Chromosome representation:

In order to apply IGA to our system, one has to make a decision to find a suitable genotype that the problem needs i.e. the chromosome representation. In our proposed system, an item or movie is represented in terms of its genres. Each genre has a degree of presence in a movie as a real number lying in the closed interval [0, 1]. Chromosome representation is depicted in the following Fig. 2.

g_1	g_2	g_3	g_4	g_5	g_6	g_7	...	g_{18}
0	.234	0	0	.675	0	0	...	.634

Fig. 2. Chromosome Representation

Initial population:

The IGA needs a population of potential solutions to be initialized at beginning. We adopt initial population for IGA as first recommended list after employing the proposed FCB-RS.

User evaluation:

Unlike predefined fitness function in GA, IGA allows users to evaluate the fitness value for each possible solution in each generation. Each user assigns his rating score to each solution presented in the current or subsequent generations. Our system evolves a new population based on the user evaluation data through genetic operators.

Genetic operators:

The selection operator determines which chromosomes are chosen for crossover or mutation to produce a number of offspring. Here we adopt the truncated selection operator. We do not consider the mutation because it can destroy the common pattern of candidate solutions generated by IGA or fuzzy theoretic approach (initially). We consider BLX-α crossover [7] method as the real numbers are included in the representation of chromosomes. After applying genetic operators, the system computes similarity between items resulted from operators and all the remaining items in the item database using equation (2). When the system conducts the search process, the system computes similarity with specific items based on cluster information.

4 Experiments and Results Analysis

In order to demonstrate the effectiveness of our proposed scheme for recommendations, we have carried out several experiments on MovieLens dataset.

4.1 Design of Experiments

Based on MovieLens dataset, we considered only those users who have rated at least 50 movies. Out of 943 users, only 568 users satisfied this condition and contributed their ratings out of 100,000. In this dataset, the rating scale is as follows 1-bad, 2- average, 3-good, 4-very good and 5-excellent. For our experiments, we select three subsets form this dataset containing 50, 100 and 150 users called ML50, ML100 and ML150 respectively. Each selected dataset is randomly separated into 60% training and 40% testing. From the testing dataset, we have randomly taken as top T % movies as an initial population for IGA after employing fuzzy theoretic approach. All experiments are run 10 times to eliminate the effect of any biasness in the data.

4.2 Performance Evaluation

In order to test the performance of our recommendation strategy, we measure system's accuracy using two evaluation metrics namely, Precision and Recall.

- *Precision:* Precision, measuring correctness of recommendation, is defined as the ratio of the number of selected items to the number of recommended items:

$$Precision = \frac{Number\ of\ relevant\ items\ recommended}{Total\ number\ of\ items\ recommended} \tag{4}$$

- *Recall:* Recall can be used as a measure of the ability of our system to all relevant resources.

$$Recall = \frac{Number\ of\ relevant\ items\ recommended}{Total\ number\ of\ relevant\ items} \tag{5}$$

4.3 Experiments

To demonstrate the feasibility and effectiveness of proposed scheme UCB-RS with CB-RS and FCB-RS schemes, we compared these schemes via precision, recall using equation (4) and (5) in all experiments. The results are presented in Table 1. The precision and recall are computed based on the average over 10 runs of the experiments over different datasets. Higher values of precision and recall imply the better performance of the proposed scheme.

Table 1. Precision and Recall Comparison of Proposed Scheme UCB-RS with FCB-RS and CB-RS Schemes

Datasets		Schemes		
	Measures	CB-RS	FCB-RS	UCB-RS
ML50	Precision	0.3824	0.5957	0.7153
	Recall	0.4568	0.6502	0.7652
ML100	Precision	0.2903	0.5418	0.6195
	Recall	0.4862	0.7169	0.8206
ML150	Precision	0.4258	0.5696	0.6583
	Recall	0.4685	0.6822	0.7696

Results presented in Table 1 show the relative performances of CB-RS, FCB-RS schemes with our proposed scheme UCB-RS. It is clear from Table 1, the proposed scheme UCB-RS considerably performed better than any of the other techniques in terms of both the precision and recall. The precision and recall for the different runs of the experiment for ML100 are depicted in Fig. 3 and Fig. 4 respectively. For all the runs, the proposed scheme UCB-RS outperforms the other schemes in terms of precision as well as recall metrics.

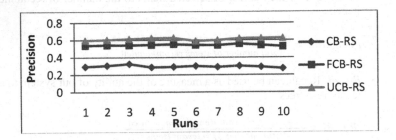

Fig.3. Precision for ML100 over 10 Runs

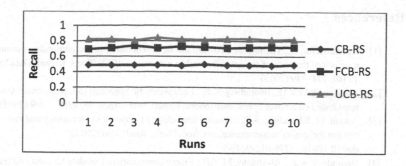

Fig.4. Recall for ML100 over 10 Runs

5 Conclusions

Our work in this paper is an attempt towards introducing a user-oriented content based recommender system (UCB-RS) for recommending items to users by employing Interactive Genetic Algorithm (IGA) that aims to bridge the gap between item features and user preferences. Incorporation of IGA into our proposed system has resulted in providing quality recommendations by adaptively responding to the changing user preferences. To deal with the uncertainties associated with item features, we have also incorporated fuzzy theoretic approach for designing fuzzy content based recommender system (FCB-RS) based on reclusive methods. To evaluate the effectiveness of our proposed scheme UCB-RS, we conducted an experimental study comparing the proposed approach with the traditional CB-RS and the FCB-RS. Our results indicate that proposed scheme consistently outperforms both the CB-RS and FCB-RS.

In our future work, we plan to extend the current approach by including other features of items e.g. directors, actors etc. for further enhancing the accuracy of our system. Our approach is specific to movie RS and it would be interesting to explore the feasibility of extending the current approach to other domains e.g. jokes, books, music etc. Also our proposed system is specific to only content based filtering and therefore it would be investigate the possible hybridization of both the UCB-RS and collaborative filtering by exploiting the notions of trust, reputation and risk [3, 4].

References

[1] Adomavicius, G., Tuzhilin, A.: Toward the next generation of recommender systems: a survey of the state-of-the-art and possible extensions. IEEE Trans. Knowl. Data Eng. 17 (6), 734–749 (2005)

[2] Al-Shamri, M.Y.H., Bharadwaj, K.K.: Fuzzy-genetic approach to recommender systems based on a novel hybrid user model. Expert Syst. Appl. 35, 1386–1399 (2008)

[3] Anand, D., Bharadwaj, K.K.: Pruning trust-distrust network via reliability and risk estimates for quality recommendations. Soc. Netw. Anal. Min.(2012). doi:10.1007/s13278-012-0049-9

[4] Bharadwaj, K.K., Al-Shamri, M.Y.H.: Fuzzy computational models for trust and reputation Systems. Elect. Comm. Research Appl. 8(1), 37–47 (2009)

[5] David, J.C.M.: An Example Inference Task: Clustering Information Theory, Inference and Learning Algorithm. Cambridge University Press (2003)

[6] Goldberg, D.E., Holland, J.H.: Genetic algorithms and machine learning. Mach. Learn. 3(2-3), 95–99 (1988)

[7] Herrera, F., Lozano, M., Verdegay, J.L.: Tackling real-coded genetic algorithms: operators and tools for behavioural analysis. Artif. Intell. Review. 12, 265–319 (1998)

[8] Kant, V., Bharadwaj, K.K.: Incorporating fuzzy trust in collaborative filtering based recommender systems. In: Panigrahi, B.K. et al. (Eds.) SEMCCO, Part I, LNCS 7076, pp. 433–440. Springer-Verlag, Berlin, Heidelberg (2011)

[9] Kant, V. and Bharadwaj, K.K.: Enhancing recommendation quality of content-based filtering through collaborative predictions and fuzzy similarity measures, Accepted for the publication in Journal Procedia Engineering , Elsevier , 2012 (To appear).

[10] Kim, H.T., Kim, E., Lee, J.H., Ahn, C.W.: A recommender system based on genetic algorithm for music data. *In Proceedings of the 2nd International Conference on Computer Engineering and Technology (ICCET)* vol.6, pp. 414-417 (2010)

[11] Lai, C.C., Chen, Y.C.: A user-oriented image retrieval system based on interactive genetic algorithm. IEEE trans. Instrument. Measure. 60 (10), 3318-3325 (2011)

[12] Mitchell, M.: An Introduction to Genetic Algorithm, MIT Press (1998)

[13] Sarwar, B., Karypis, G., Konstan, J., Riedl, J.: Analysis of recommendation algorithms for ecommerce. *In Proceedings of the 2nd ACM conference on Electronic commerce*, pp.158-167. Minneapolis, Minnesota, United States (2000)

[14] Takagi, H.: Interactive evolutionary computation: fusion of the capabilities of EC optimization and human evaluation. Proc. the IEEE 89(9), 1275–1296 (2001)

[15] Yager, R.R.: Fuzzy logic methods in recommender systems. Fuzzy Sets Syst. 136, 133-149 (2003)

[16] Zenebe, A., Norcio, A. F.: Representation, similarity measures and aggregation methods using fuzzy sets for content-based recommender systems. Fuzzy Sets Syst.160, 76 – 94 (2009)

Author Index

J. C. Bansal et al. (eds.), *Proceedings of Seventh International Conference on Bio-Inspired Computing: Theories and Applications (BIC-TA 2012)*, Advances in Intelligent Systems and Computing 201, DOI: 10.1007/978-81-322-1038-2, © Springer India 2013